Classics in Mathematics

Max Karoubi

K-Theory

Max Karoubi received his PhD in mathematics (Doctorat d'Etat) from Paris University in 1967, while working in the CNRS (Centre National de la Recherche Scientifique), under the supervision of Henri Cartan and Alexander Grothendieck. After his PhD, he took a position of "Maître de Conférences" at the University of Strasbourg until 1972. He was then nominated full Professor at the University of Paris 7-Denis Diderot until 2007. He is now an Emeritus Professor there.

Max Karoubi

K-Theory

An Introduction

Reprint of the 1978 Edition

With a New Postface by the Author and a List of Errata

With 26 Figures

Springer

Max Karoubi
Université Paris 7 – Denis Diderot
UFR de Mathématiques
2 place Jussieu
75251 Paris CEDEX 05
France
max.karoubi@gmail.com

Originally published as Vol. 226 of the series *Grundlehren der mathematischen Wissenschaften*

ISBN 978-3-540-79889-7 ISBN 978-3-540-79890-3 (eBook)

DOI 10.1007/978-3-540-79890-3

Classics in Mathematics ISSN 1431-0821

Library of Congress Control Number: 2008931976

Mathematics Subject Classification (1970):
Primary 55B15, 18F25, 55F45, 55F10, 55D20
Secondary: 10C05, 16A54, 46H25, 55E20, 55E50, 55E10, 55J25, 55G25

Cover design: WMXDesign GmbH, Heidelberg

Printed on acid-free paper

9 8 7 6 5 4 3 2 1

springer.com

Grundlehren der mathematischen Wissenschaften 226

A Series of Comprehensive Studies in Mathematics

Max Karoubi

K-Theory

An Introduction

With 26 Figures

Springer-Verlag
Berlin Heidelberg New York 1978

Max Karoubi

Université Paris VII, U.E.R. de Mathématiques, Tour 45–55,
5e Etage, 2 Place Jussieu
F-75221 Paris Cedex 05

AMS Subject Classification (1970)

Primary: 55B15, 18F25, 55F45, 55F10, 55D20
Secondary: 10C05, 16A54, 46H25, 55E20, 55E50, 55E10, 55J25, 55G25

ISBN 978-3-540-79889-7 Springer-Verlag Berlin Heidelberg New York

Library of Congress Cataloging in Publication Data. Karoubi, Max. K-theory. (Grundlehren der mathematischen Wissenschaften; 226). Bibliography: p. Includes index. 1. K-theory. I. Title. II. Series: Die Grundlehren der mathematischen Wissenschaften in Einzeldarstellungen; 226. QA612.33.K373. 514'.23. 77–23162.

A Pierre et Thomas

Foreword

K-theory was introduced by A. Grothendieck in his formulation of the Riemann–Roch theorem (cf. Borel and Serre [2]). For each projective algebraic variety, Grothendieck constructed a group from the category of coherent algebraic sheaves, and showed that it had many nice properties. Atiyah and Hirzebruch [3] considered a topological analog defined for any compact space X, a group $K(X)$ constructed from the category of vector bundles on X. It is this "topological K-theory" that this book will study.

Topological K-theory has become an important tool in topology. Using K-theory, Adams and Atiyah were able to give a simple proof that the only spheres which can be provided with H-space structures are S^1, S^3 and S^7. Moreover, it is possible to derive a substantial part of stable homotopy theory from K-theory (cf. J. F. Adams [2]). Further applications to analysis and algebra are found in the work of Atiyah–Singer [2], Bass [1], Quillen [1], and others. A key factor in these applications is Bott periodicity (Bott [2]).

The purpose of this book is to provide advanced students and mathematicians in other fields with the fundamental material in this subject. In addition, several applications of the type described above are included. In general we have tried to make this book self-contained, beginning with elementary concepts wherever possible; however, we assume that the reader is familiar with the basic definitions of homotopy theory: homotopy classes of maps and homotopy groups (cf. collection of spaces including projective spaces, flag bundles, and Grassmannians. Hilton [1] or Hu [1] for instance). Ordinary cohomology theory is used, but not until the end of Chapter V. Thus this book might be regarded as a fairly self-contained introduction to a "generalized cohomology theory".

The first two chapters ("Vector bundles" and "First notions in K-theory") are chiefly expository; for the reader who is familiar with this material, a brief glance will serve to acquaint him with the notation and approach used. Chapter III is devoted to proving the Bott periodicity theorems. We employ various techniques following the proofs given by Atiyah and Bott [1], Wood [1] and the author [2], using a combination of functional analysis and "algebraic K-theory".

Chapter IV deals with the computation of particular K-groups of a large The version of the "Thom isomorphism" in Section IV.5 is mainly due to Atiyah, Bott and Shapiro [1] (in fact they were responsible for the introduction of Clifford algebras in K-theory, one of the techniques which we employ in Chapter III).

Chapter V describes some applications of K-theory to the question of H-space structures on the sphere and the Hopf invariant (Adams and Atiyah [1]), and to the solution of the vector field problem (Adams [1]). We also present a sketch of the theory of characteristic classes, which we apply in the proof of the Atiyah–Hirzebruch integrality theorems [1]. In the last section we use K-theory to make some computations on the stable homotopy groups of spheres, via the groups $J(X)$ (cf. Adams [2], Atiyah [1], and Kervaire–Milnor [1]).

In spite of its relative length, this book is certainly not exhaustive in its coverage of K-theory. We have omitted some important topics, particularly those which are presented in detail in the literature. For instance, the Atiyah–Singer index theorem is proved in Cartan–Schwartz [1], Palais [1], and Atiyah–Singer [2] (see also appendix 3 in Hirzebruch [2] for the concepts involved). The relationship between other cohomology theories and K-theory is only sketched in Sections V.3 and V.4. A more complete treatment can be found in Conner–Floyd [1] and Hilton [2] (Atiyah–Hirzebruch spectral sequence). Finally algebraic K-theory is a field which is also growing very quickly at present. Some of the standard references at this time are Bass's book [1] and the Springer Lecture Notes in Mathematics, Vol. 341, 342, and 343.

I would like to close this foreword with sincere thanks to Maria Klawe, who greatly helped me in the translation of the original manuscript from French to English.

Paris, Summer 1977

Max Karoubi

Table of Contents

Remarks on Notation and Terminology

The following notation is used throughout the book: $\mathbf{Z}$ integers, $\mathbb{Q}$ rational numbers, $\mathbb{R}$ real numbers, $\mathbb{C}$ complex numbers, $\mathbb{H}$ quaternions; $\mathrm{GL}_n(A)$ denotes the group of invertible $n \times n$ matrices with coefficients in the ring A. The notation $*\cdots*$ signifies an assertion in the text which is not a direct consequence of the theorems proved in this book, but which may be found in the literature; these assertions are not referred to again, except occasionally in exercises.

If $\mathscr{C}$ is a category, and if E and F are objects of $\mathscr{C}$, then the symbol $\mathscr{C}(E, F)$ or $\mathrm{Hom}_{\mathscr{C}}(E, F)$ means the set of morphisms from E to F.

More specific notation is listed at the end of the book.

A reference to another part of the book is usually given by two numbers (e.g. 5.21) if it is in the same chapter, or by three numbers (e.g. IV.6.7) if it is in a different chapter.

Interdependence of Chapters and Sections

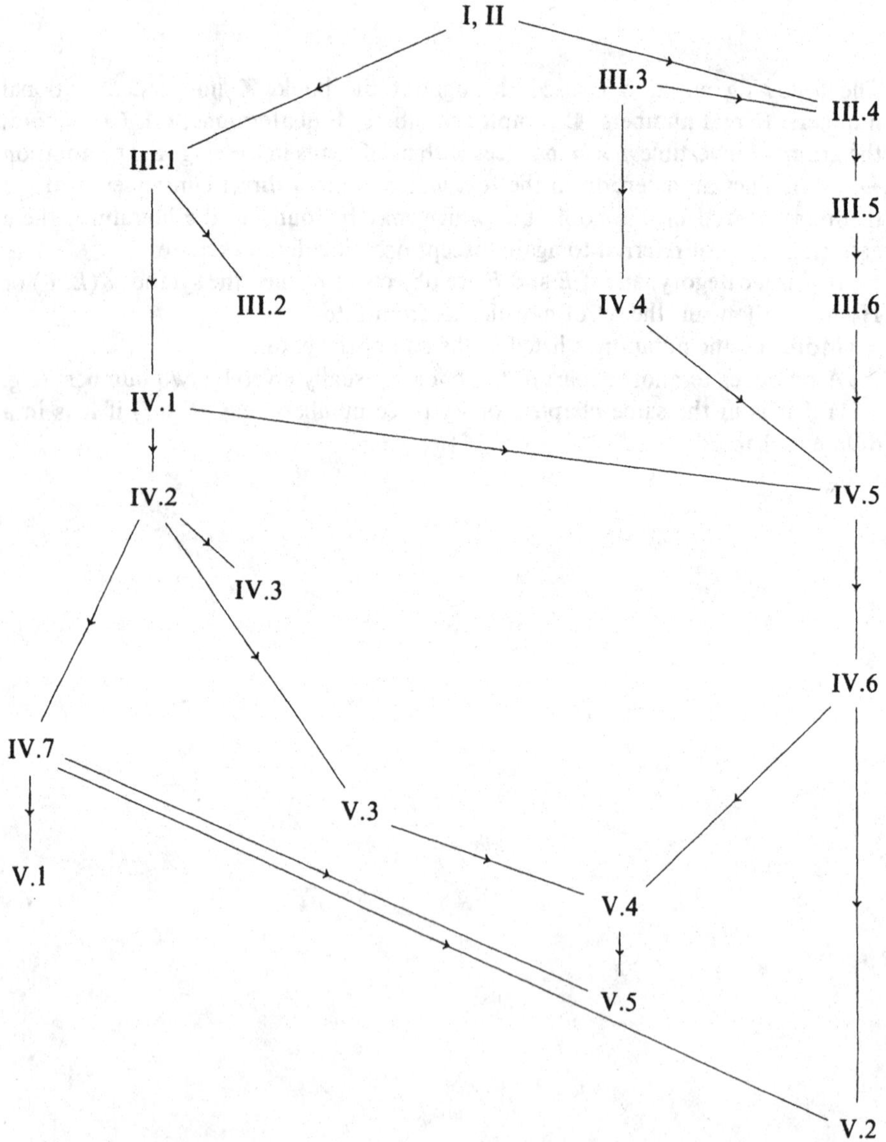

Summary of the Book by Sections

Chapter I. Vector Bundles

1. *Quasi-vector bundles.* This section covers the general concepts and definitions necessary to introduce Section 2. Theorem 1.12 is particularly important in the sequel.

2. *Vector bundles.* The "vector bundles" considered here are locally trivial vector bundles whose fibers are finite dimensional vector spaces over $\mathbb{R}$ or $\mathbb{C}$. To be mentioned: Proposition 2.7 and Examples 2.3 and 2.4 will be referred to in the sequel.

3. *Clutching theorems.* This technical section is necessary to provide a bridge between the theory of vector bundles and the theory of "coordinate bundles" of N. Steenrod [1]. The clutching theorems are useful in the construction of the tangent bundle of a differentiable manifold (3.18) and in the description of vector bundles over spheres (3.9; see also I.7.6).

4. *Operations on vector bundles.* Certain "continuous" operations on finite dimensional vector spaces: direct sum, tensor product, duality, exterior powers, etc. . . . can be also defined on the category of vector bundles.

5. *Sections of vector bundles.* Only continuous sections are considered here. The major topic concerns the solution of problems involving extensions of sections over paracompact spaces.

6. *Algebraic properties of the category of vector bundles.* In this section we prove that the category $\mathscr{E}(X)$ of vector bundles over a compact space X, is a "pseudo-abelian additive" category. Essentially this means that one has direct sums of vector bundles (the "Whitney sum"), and that every projection operator has an image. From this categorical description (6.13), we deduce the theorem of Serre and Swan (6.18): The category $\mathscr{E}(X)$ is equivalent to the category $\mathscr{P}(A)$, where A is the ring of continuous functions on X, and $\mathscr{P}(A)$ is the category of finitely generated projective modules over A.

7. *Homotopy and representability theorems.* This section is essential for the following chapters. We prove that the problem of classification of vector bundles with compact base X depends only on the homotopy type of X (7.2). We also prove that $\Phi_n^k(X)$ (the set of isomorphism classes of k-vector bundles, over X of rank n for $k=\mathbb{R}$ or $\mathbb{C}$), considered as a functor of X, is a direct limit of representable functors. This takes the concrete form of Theorems 7.10 and 7.14.

8. *Metrics and forms on vector bundles.* It is sometimes important to have some additional structure on vector bundles, such as bilinear forms, Hermitian forms, etc. With the exception of Theorem 8.7, this section is not used in the following chapters (except in the exercises).

Chapter II. First Notions of K-Theory

1. *The Grothendieck group of an additive category. The group* K(X). Starting with the simple notion of symmetrization of an abelian monoid, we define the group $K(\mathscr{C})$ of an additive category using the monoid of isomorphism classes of objects of $\mathscr{C}$. Considering the case where $\mathscr{C}$ is $\mathscr{E}(X)$ and X is compact, we obtain the group $K(X)$ (actually $K_{\mathbb{R}}(X)$ or $K_{\mathbb{C}}(X)$ according to which theory of vector bundles is considered). We prove that $K_{\mathbb{R}}(X)\approx[X,\mathbb{Z}\times BO]$ and $K_{\mathbb{C}}(X)\approx[X,\mathbb{Z}\times BU]$ (1.33).

2. *The Grothendieck group of an additive functor. The group $K(X, Y)$.* In order to obtain a "reasonable" definition of the Grothendieck group $K(\varphi)$ for an additive functor $\varphi:\mathscr{C}\to\mathscr{C}'$, which generalizes the definition of $K(\mathscr{C})$ when $\mathscr{C}'=0$, we assume some topological conditions on the categories $\mathscr{C}$ and $\mathscr{C}'$ and on the functor φ (2.6). Since these conditions are satisfied by the "restriction" functor $\mathscr{E}(X)\to\mathscr{E}(Y)$ where Y is closed in X, we then define the "relative group" $K(X, Y)$ to be the K-group of this functor. In fact, $K(X, Y)\approx\tilde{K}(X/Y)$ (2.35). This isomorphism shows that essentially we do not obtain a new group; however, the groups $K(\varphi)$ and $K(X, Y)$ will be important technical tools later on.

3. *The group K^{-1} of a Banach category. The group $K^{-1}(X)$.* This section represents the first step towards the construction of a cohomology theory h^* where the term h^0 is the group $K(X, Y)$ (also denoted by $K^0(X, Y)$) considered in II.2. The group $K^{-1}(\mathscr{C})$, where $\mathscr{C}$ is a Banach category, is obtained from the automorphisms of objects of $\mathscr{C}$. Again, if we consider the case where $\mathscr{C}$ is $\mathscr{E}(X)$, we obtain the group called $K^{-1}(X)$. We prove that if Y is a closed subspace of X then the sequence

$$K^{-1}(X)\to K^{-1}(Y)\to K(X, Y)\to K(X)\to K(Y) \text{ is exact.}$$

We also prove that $K_{\mathbb{R}}^{-1}(X)\approx[X, 0]$ and $K_{\mathbb{C}}^{-1}(X)\approx[X, \mathrm{U}]$ (3.19).

4. *The groups $K^{-n}(X)$ and $K^{-n}(X, Y)$.* The aim of this section is to define the groups $K^{-n}(X, Y)$ for $n\geqslant 2$ and to establish the exact sequence

$$K^{-n-1}(X)\to K^{-n-1}(Y)\to K^{-n}(X, Y)\to K^{-n}(X)\to K^{-n}(Y), \quad \text{for } n\geqslant 1$$

One possible definition is $K^{-n}(X, Y) = \tilde{K}(S^n(X/Y))$ (4.12). We prove some "Mayer–Vietoris exact sequences" (4.18 and 4.19) which will be very useful later on.

5. *Multiplicative structures.* The tensor product of vector bundles provides the group $K(X)$ with a ring structure. It is more difficult to define a "cup-product"

$$K(X, Y) \times K(X', Y') \to K(X \times X', X \times Y' \cup Y \times X')$$

or more generally

$$K^{-n}(X, Y) \times K^{-n'}(X', Y') \to K^{-n-n'}(X \times X', X \times Y' \cup Y \times X')$$

when Y and Y' are non-empty. This is accomplished in a theoretical sense in proposition 5.6; however, in applications it is often useful to have more explicit formulas. For this we introduce another definition of the group $K(X, Y)$ by putting metrics on the vector bundles involved (5.16). This will not be used before Chapter IV. The existence of such cup-products shows that there is a direct splitting $K(X) \approx H^0(X; \mathbb{Z}) \oplus K'(X)$ where $K'(X)$ is a nil ideal (cf. 5.9; note that $K'(X) \approx \tilde{K}(X)$ if X is connected).

Chapter III. Bott Periodicity

1. *Periodicity in complex K-theory.* In this section we define an isomorphism $K_{\mathbb{C}}^{-n}(X, Y) \approx K_{\mathbb{C}}^{-n-2}(X, Y)$. The method (due to Atiyah, Bott, and Wood) is to reduce this isomorphism for general n, to a theorem on Banach algebras (1.11): If A is a complex Banach algebra, the group $K(A)$ (defined as $K(\mathscr{P}(A))$ is naturally isomorphic to $\pi_1(\mathrm{GL}(A))$ where $\mathrm{GL}(A) = \text{inj lim} \mathrm{GL}_n(A)$. This theorem is proved using the Fourier series of a continuous function with values in a complex Banach space, and some classical results in Algebraic K-theory on Laurent polynomials. The original theorem follows when we let A be the ring of complex continuous functions on a compact space.

2. *First applications of Bott periodicity theorem in the complex case.* As a first application we obtain the classical theorem of Bott: for $n > i/2$, we have $\pi_i(\mathrm{U}(n)) \approx \mathbb{Z}$ if i is odd and $\pi_i(\mathrm{U}(n)) = 0$ for i even. We also prove that real K-theory is periodic of period 4 mod. 2-torsion: $K_{\mathbb{R}}^{-n}(X, Y) \otimes_{\mathbb{Z}} \mathbb{Z}' \approx K_{\mathbb{R}}^{-n-4}(X, Y) \otimes_{\mathbb{Z}} \mathbb{Z}'$, where $\mathbb{Z}' = \mathbb{Z}[\frac{1}{2}]$. This theorem will be strengthened in III.5.

3. *Clifford algebras.* These algebras play an important role in real K-theory and will be used in Chapter IV in both real and complex K-theory. This section is purely algebraic. The essential result is Theorem 3.21, which establishes a kind of periodicity for Clifford algebras. This "algebraic" periodicity will be effectively used in III.5 to prove the "topological" periodicity of real K-theory and at the same time give another proof of the periodicity of complex K-theory.

4. *The functors $K^{p,q}(\mathscr{C})$ and $K^{p,q}(X)$.* The idea of this section is to use the Clifford algebras $C^{p,q}$ to algebraicly define new functors $K^n(X)=K^{p,q}(X)$ for $n=p-q\in\mathbb{Z}$. We prove that these functors are *by definition* periodic, of period 8 in the real case, and of period 2 in the complex case, and that $K^0(X)$ and $K^{-1}(X)$ are indeed the functors defined in Chapter II. Bott periodicity will then be proved if we show that the two definitions of $K^n(X)$ agree for negative values of n. This is done in the next two sections.

5. *The functors $K^{p,q}(X, Y)$ and the isomorphism t. Periodicity in real K-theory.* After some preliminaries introducing the relative groups $K^{p,q}(X, Y)$ we present the fundamental theorem of this chapter: The groups $K^{p,q+1}(X, Y)$ and $K^{p,q}(X\times B^1, X\times S^0\cup Y\times B^1)$ are isomorphic. Assuming this theorem (the proof follows in Section III.6), we prove that $K_{\mathbb{R}}^{-n}(X, Y)\approx K_{\mathbb{R}}^{-n-8}(X, Y)$ with the definitions of Chapter II. At the same time we prove the periodicity in complex K-theory (5.17) once more. Moreover, using Propositions 4.29 and 4.30 we prove the existence of weak homotopy equivalences between the iterated loop spaces $\Omega^r(0)$ and certain homogeneous spaces (5.22). We also compute the homotopy groups $\pi_i(0(n))$ for $n>i+1$ (5.19) with the help of Clifford algebras.

6. *Proof of the fundamental theorem.* The pattern of this section is analogous to that of Section III.1, since the main theorem is likewise a consequence of a general theorem on Banach algebras (6.12). Moreover the proof of this general theorem uses the same ideas as the proof of Theorem 1.11.

Chapter IV. Computations of Some K-Groups

1. *The Thom isomorphism in complex K-theory for complex vector bundles.* The purpose of this section is to compute the complex K-theory of the Thom space of a complex vector bundle (1.9). In this computation a key role is played by bundles of exterior algebras. Theorem 1.3. is particularly important in the sequel.

2. *Complex K-theory of complex projective spaces and complex projective bundles.* In this section (classical in style), we construct a method which may also be used for ordinary cohomology (see V.3). Using the technical Proposition 2.4 we are able to compute the K-theory of $P_n=P(\mathbb{C}^{n+1})$ and more generally of $P(V)$ where V is a complex vector bundle (2.13). The "splitting principle" (2.15) is used frequently later on. With this principle we are able to make the multiplicative structure of $K^*(P(V))$ explicit (2.16).

3. *Complex K-theory of flag bundles and Grassmann bundles. K-theory of a product.* This section is also classical in style, but is not essential to the sequel. We explicitly compute $K^*(F(V))$ where $F(V)$ is the flag bundle of a complex vector bundle V. We also compute $K^*(G_p(V))$ where $G_p(V)$ is the fiber bundle of p-subspaces in V (3.12). These results are used to compute $\mathcal{K}(BU(n))$ = proj lim $K(G_p(\mathbb{C}^n)$ (3.22), and the K-theory of a product (3.27).

4. *Complements in Clifford algebras.* The concept of "spinors" was not introduced in Section III.3, since it is not essential in proving Bott periodicity. However we now need this concept to prove the analog of Thom's theorem in K-theory (for real or complex vector bundles). After some algebraic preliminaries we study the possibilities of lifting the structural group of a real vector bundle to the spinorial group Spin(n) or Spin^c(n). Theorem 4.22 is particularly important for our purpose.

5. *The Thom isomorphism in real and complex K-theory for real vector bundles.* As in IV.1, the purpose of this section is to compute the K-theory of the Thom space of a vector bundle, but now the vector bundle is real, and the K-theory used is real or complex. With an additional spinorial hypothesis, we prove that $K(V) \approx K^{-n}(X)$ if n is the rank of V. If the base is compact and n is a multiple of 8 (of 2 in complex K-theory), we prove that $K(V)$ is a $K(X)$-module of rank one generated by the "Thom class" T_V. Finally, if $f: X \to Y$ is a proper continuous map between differentiable manifolds and if $\mathrm{Dim}(Y) - \mathrm{Dim}(X) = 0 \bmod 8$ (mod 2 in the complex case), we define, with an additional spinorial hypothesis, a "Gysin homomorphism" $f_*: K(X) \to K(Y)$ which is analogous to the Gysin homomorphism in ordinary cohomology. This homomorphism is only used in V.4.

6. *Real and complex K-theory of real projective spaces and real projective bundles.* This section is much more technical than the others (the results are only used in V.2). After some easy but tedious lemmas making systematic use of Clifford algebras, we are able to compute (up to extension) the real and complex K-theory of a real projective bundle (6.40 and 6.42). In the case of real projective spaces, the K-theory is completely determined (6.46 and 6.47).

7. *Operations in K-theory.* One of the charms of K-theory is that we are able to define some very nice operations. For example, there are the exterior power operations λ^k (due to Grothendieck). By a method due to Atiyah we determine all the operations in complex K-theory. With this method we show that the "Adams operations" ψ^k are the only ring operations in complex K-theory (7.13). They will be very useful in applications.

The operations λ^k and ψ^k may also be defined in real K-theory. However, their properties are more difficult to prove. We must refer to Adams [3] or Exercise 8.5 for a complete proof. From the operations ψ^k, we obtain the operations ρ^k, which will be very useful in V.2 and V.5.

Chapter V. Some Applications of *K*-Theory

1. *H-space structures on spheres and the Hopf invariant.* Using the Adams operations in complex K-theory, we prove that the only spheres which admit an H-space structure are S^1, S^3, and S^7. In fact, we prove more: if $f\colon S^{2n-1} \to S^n$ is a map of odd Hopf invariant, then n must be 2, 4 or 8.

2. *The solution of the vector field problem on the sphere.* Let us write every integer t in the form $(2\alpha-1)\cdot 2^\beta$, for $\beta=\gamma+4\delta$ with $0 \leqslant \gamma \leqslant 3$, and define $\rho(t)=2^\gamma+8\delta$. Then the maximum number of independent vector fields on the sphere S^{t-1} is exactly $\rho(t)-1$ (2.10). The proof of this classical theorem is "elementary" (in the context of this book) and uses essentially the operations ρ^k in the real K-theory of real projective spaces.

3. *Characteristic classes and the Chern character.* For each complex vector bundle V, we define "Chern classes" $c_i(V) \in H^{2i}(X; \mathbb{Z})$ in an axiomatic way (3.15). The construction of these classes is analogous to the construction of classes done in Section IV.3. By means of these classes, we construct a fundamental homomorphism, the "Chern character", from $K_{\mathbb{C}}(X)$ to $H^{\text{even}}(X; \mathbb{Q})$. The Chern character induces an isomorphism between $K_{\mathbb{C}}(X) \otimes_{\mathbb{Z}} \mathbb{Q}$ and $H^{\text{even}}(X; \mathbb{Q})$ for every compact X.

4. *The Riemann–Roch theorem and integrality theorems.* To each complex stable vector bundle (resp. ^{c}spinorial real stable bundle) we associate an important characteristic class $\tau(V)$, called the Todd class (resp. $A(V)$, called the Atiyah–Hirzebruch class). These classes play an important role in the "differentiable Riemann–Roch theorem": For each suitably continuous map $f\colon X \to Y$ and for each element x of $K_{\mathbb{C}}(X)$, we have the formula $ch(f_*^K(x)) = f_*^H(A(v_f)\cdot ch(x))$ where $A(v_f)$ denotes the Atiyah–Hirzebruch class of the stable bundle $f^*(TY)-TX$ (assuming that $\mathrm{Dim}(Y)=\mathrm{Dim}(X)$ mod 2 and that there is a stable ^{c}spinorial structure on v_f). From this theorem we obtain integral theorems for characteristic classes (4.21) and the homotopy invariance of certain characteristic classes (4.24).

5. *Applications of K-theory to stable homotopy.* In this section we explain how K-theory may be applied to obtain some interesting information about the stable homotopy groups of spheres. We only include those partial results which can be obtained from the material in this book. More complete results are found in the series of J. F. Adams on the groups $J(X)$ [2], and in Husemoller's book [1].

Chapter I

Vector Bundles

1. Quasi-Vector Bundles

1.1. Let k be the field of real numbers or complex numbers[1], and let X be a topological space.

1.2. Definition. A *quasi-vector bundle with base X* is given by

1) a finite dimensional k-vector space E_x for every point x of X,

2) a topology on the disjoint union $E = \bigsqcup E_x$ which induces the natural topology on each E_x, such that the obvious projection $\pi: E \to X$ is continuous.

1.3. Example. Let X be the sphere $S^n = \{x \in \mathbb{R}^{n+1} \mid \|x\| = 1\}$. For every point x of S^n we choose E_x to be the vector space orthogonal to x. Then $E = \bigsqcup E_x$ is naturally a subspace of $S^n \times \mathbb{R}^{n+1}$ and may be provided with the induced topology.

1.4. Example. Starting from the preceding example, let us arbitrarily choose a vector space $F_x \subset E_x$ for each $x \in S^n$; then if F is given the induced topology again we have a quasi-vector bundle on X.

More examples are given in the following sections.

1.5. A quasi-vector bundle is denoted by $\xi = (E, \pi, X)$ or simply by E if there is no risk of confusion. The space E is the *total space* of ξ and E_x is the *fiber* of ξ at the point x.

1.6. Let $\xi = (E, \pi, X)$ and $\xi' = (E', \pi', X')$ be quasi-vector bundles. A *general morphism* from ξ to ξ' is given by a pair (f, g) of continuous maps $f: X \to X'$ and $g: E \to E'$ such that

1) the diagram

$$\begin{array}{ccc} E & \xrightarrow{g} & E' \\ {\scriptstyle \pi}\downarrow & & \downarrow{\scriptstyle \pi'} \\ X & \xrightarrow{f} & X' \end{array}$$

is commutative.

---

[1] In general, these are the most interesting cases; however, sometimes we will use the field of quaternions $\mathbf{H}$.

2) The map $g_x: E_x \to E'_{f(x)}$ induced by g is k-linear.

General morphisms can be composed in an obvious way. In this way we construct a category whose objects are quasi-vector bundles and whose arrows are general morphisms.

1.7. If ξ and ξ' have the same base $X=X'$, a *morphism* between ξ and ξ' is a general morphism (f, g) such that $f=\mathrm{Id}_X$. Such a morphism will be simply called g in the sequel. The quasi-vector bundles with the same base X are the objects of a subcategory, whose arrows are the morphisms we have just defined.

1.8. Example. Let us return to Example 1.3, and let $n=1$. Let $\xi'=(E', \pi', X')$ where $X=X'=S^1$, and $E'=S^1 \times \mathbb{R}$ with the product topology. If we identify $\mathbb{R}^2$ with the complex numbers as usual, we can define a continuous map $g: E \to E'$ by the formula $g(x, z)=(x, iz/x)$ (this is well defined because x is orthogonal to z in $\mathbb{R}^2=\mathbb{C}$). In fact g is an isomorphism between E and E' in the category described in 1.7.

1.9. Example. Let E'' be the quotient of $E'=S^1 \times \mathbb{R}$ by the equivalence relation $(x, t) \sim (y, u)$ if $y=\varepsilon x$ and $u=\varepsilon t$ with $\varepsilon=\pm 1$. Then E'' is the total space of a quasi-vector bundle over $P_1(\mathbb{R})$ called the *infinite Moebius band.* By identifying $P_1(\mathbb{R})$ with S^1 by the map $z \mapsto z^2$, we see easily that E'' is also the quotient of $I \times \mathbb{R}$ by the equivalence relation which identifies $(0, u)$ with $(1, -u)$. If we restrict u to have norm less than 1, we obtain the classical Moebius band.

We claim that the bundles E' and E'' over S^1 are not isomorphic. Suppose there exists an isomorphism $g: E' \to E''$; then we must have $E'-X'$ homeomorphic to $E''-X''$ where X' (and X'') denote the set of points of the form $(x, 0)$ with $x \in S^1$ (note that $X'' \approx X'$). But $E''-X''$ is connected and $E'-X'$ is not.

1.10. Let V be a finite dimensional vector space (as always over k). The preceding examples show the importance of quasi-vector bundles of the form $E=X \times V$, as models. To be more precise, $E_x=V$ and the total space may be identified with $X \times V$ provided with the product topology. Such bundles are called trivial quasi-vector bundles or simply *trivial vector bundles.*

1.11. Let $E=X \times V$ and $E'=X \times V'$ be trivial vector bundles with base X. We want to explicitly describe the morphisms from E to E' (again in the category defined in 1.7). Since the diagram

$$\begin{array}{ccc} X \times V & \longrightarrow & X \times V' \\ & \searrow \quad \swarrow & \\ & X & \end{array}$$

is commutative, for each point x of X, g induces a linear map $g_x: V \to V'$. Let $\breve{g}: X \to \mathscr{L}(V, V')$ be the map defined by $\breve{g}(x)=g_x$.

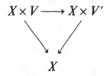

1.12. Theorem. *The map* $\check{g}: X \to \mathscr{L}(V, V')$ *is continuous relative to the natural topology of* $\mathscr{L}(V, V')$. *Conversely, let* $h: X \to \mathscr{L}(V, V')$ *be a continuous map, and let* $\hat{h}: E \to E'$ *be the map which induces* $h(x)$ *on each fiber. Then* $\hat{h}$ *is a morphism of quasi-vector bundles.*

Proof. To prove this theorem we choose a basis $e_1, \ldots, e_n$ of V and a basis $\varepsilon_1, \ldots, \varepsilon_p$ of V'. With respect to this basis, g_x may be regarded as the matrix $(\alpha_{ij}(x))$ where $\alpha_{ij}(x)$ is the ith coordinate of the vector $g_x(e_j)$. Hence the function $x \mapsto \alpha_{ij}(x)$ is obtained from the composition of the following continuous maps.

$$X \xrightarrow{\beta_j} X \times V \xrightarrow{g} X \times V' \xrightarrow{\gamma} V' \xrightarrow{p_i} k,$$

where $\beta_j(x) = (x, e_j)$, $\gamma(x, v') = v'$, and p_i is the ith projection of $V' \supseteq k^p$ on k. Since the functions $\alpha_{ij}(x)$ are continuous, the map $\check{g}$ which they induce is also continuous according to the definition of the topology of $\mathscr{L}(V, V')$.

Conversely, let $h: X \to \mathscr{L}(V, V')$ be a continuous map. Then $\hat{h}$ is obtained from the composition of the continuous maps

$$X \times V \xrightarrow{\delta} X \times \mathscr{L}(V, V') \times V \xrightarrow{\varepsilon} X \times V',$$

where $\delta(x, v) = (x, h(x), v)$ and $\varepsilon(x, u, v) = (x, u(v))$. Hence $\hat{h}$ is continuous and defines a morphism of quasi-vector bundles. □

1.13. *Remark.* Clearly we have the identities $\hat{\check{g}} = g$ and $\check{\hat{h}} = h$.
* The reader may also note that the second part of the theorem can be generalized to Banach bundles (see Lang [2]), but not the first part.*

1.14. *Remark.* As we have seen in Example 1.9, it is not obvious whether or not a given quasi-vector bundle is isomorphic to a trivial bundle. Let TS^n denote the quasi-vector bundle considered in 1.3 (this is the "tangent bundle" of the sphere). Then it is only at the end of this book that we are able to show that TS^n is *not* isomorphic to a trivial bundle unless $n = 1$, 3, or 7 (cf. Section V.2).

1.15. Let $\xi = (E, \pi, X)$ be a quasi-vector bundle, and let X' be a subspace of X. The triple $(\pi^{-1}(X'), \pi|_{\pi^{-1}(X')}, X')$ defines a quasi-vector bundle ξ' which is called the *restriction* of ξ to X'. We denote it by $\xi|_{X'}$, $E|_{X'}$, or even simply $E_{X'}$. The fibers of ξ' are just the fibers of ξ over the subspace X'. If $X'' \subset X' \subset X$, we have $(\xi|_{X'})|_{X''} = \xi|_{X''}$.

1.16. More generally, let $f: X' \to X$ be any continuous map (X' is not necessarily a subspace of X). For every point x' of X', let $E'_{x'} = E_{f(x')}$. Then the set $E' = \bigsqcup_{x' \in X'} E'_{x'}$ may be identified with the *fiber product* $X' \times_X E$, i.e. with the subset of $X' \times E$ formed by the pairs (x', e) such that $f(x') = \pi(e)$. If $\pi': E' \to X'$ is defined by $\pi'(x', e) = x'$, it is clear that the triple (E', π', X') defines a quasi-vector bundle over X', when we provide E' with the topology induced by $X' \times E$. We write ξ' as

$f^*(\xi)$ or $f^*(E)$: this is the *inverse image* of ξ by f. We have $f^*(\xi)=\xi$ for $f=\mathrm{Id}_X$, and also $(f\cdot f')^*(\xi)=f'^*(f^*(\xi))$ if $f': X''\to X'$ is another continuous map. If $X'\subset X$ and f is the inclusion map, then $f^*(\xi)=\xi|_{X'}$.

1.17. Let $(f, g): E_1'\to E$ be a general morphism of quasi-vector bundles with $f: X'\to X$ (1.6). This general morphism induces a morphism $h_1: E_1'\to E'=f^*(E)$ as shown in the diagram

$$\begin{array}{ccccc} E_1' & \xrightarrow{h_1} & E'\approx X'\times_X E & \xrightarrow{h} & E \\ {\scriptstyle \pi_1'}\downarrow & & {\scriptstyle \pi'}\downarrow & & \downarrow{\scriptstyle \pi} \\ X' & \xrightarrow{\mathrm{Id}_{X'}} & X' & \xrightarrow{\quad f \quad} & X, \end{array}$$

where h is induced by the projection of $X'\times E$ on its second factor. The general morphism (f, g) is called *strict* if h_1 is an isomorphism.

1.18. Let us now consider two quasi-vector bundles over X and a morphism $\alpha: E\to F$. If we let $E'=f^*(E)$ as in 1.16 and $F'=f^*(F)$, we can also define a morphism $\alpha'=f^*(\alpha)$ from E' to F' by the formula $\alpha'_{x'}=\alpha_{f(x')}$. If we identify E' with $X'\times_X E$ and F' with $X'\times_X F$, then α' is identified with $\mathrm{Id}_{X'}\times_X \alpha$, which proves the continuity of the map α'.

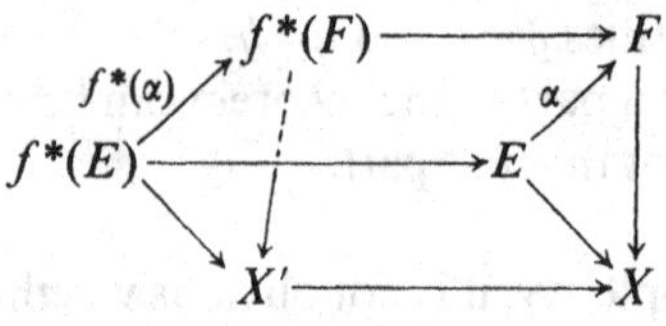

In particular, if $X'\subset X$ and if f is the inclusion map, then $f^*(\alpha)$ is the restriction of α. We denote it by $\alpha|_{X'}$ or simply $\alpha_{X'}$. The proof of the next proposition is easy and is left as an exercise for the reader:

1.19. Proposition. *Let $f: X'\to X$ be a continuous map. Then the correspondence $E\mapsto f^*(E)$ and $\alpha\mapsto f^*(\alpha)$ induces a functor between the category of quasi-vector bundles over X and the category of quasi-vector bundles over X'.*

Exercises (Section I.9) 1–4 and 6.

2. Vector Bundles

A vector bundle is a quasi-vector bundle which is locally isomorphic to a trivial vector bundle. The next definition will make this idea more precise.

2.1. Definition. Let $\xi=(E,\pi,X)$ be a quasi-vector bundle. Then ξ is said to be "*locally trivial*" or a "*vector bundle*" if for every point x in X, there exists a neighbourhood U of x such that $\xi|_U$ is isomorphic to a trivial bundle.

2.2. The last condition may be expressed in the following way: there exists a finite dimensional vector space V and a *homeomorphism* $\varphi: U\times V\to\pi^{-1}(U)$ such that the diagram

$$\begin{array}{ccc} U\times V & \xrightarrow{\ \varphi\ } & \pi^{-1}(U) \\ & {\scriptstyle \mathrm{pr}_1}\searrow \quad \swarrow{\scriptstyle \pi|_{\pi^{-1}(U)}} & \\ & U & \end{array}$$

commutes, and such that for every point y in U, the map $\varphi_y: V\to E_y$ is k-linear. We call U a *trivialization domain* of the vector bundle ξ. A cover (U_i) of X is called a *trivialization cover* if each U_i is a trivialization domain.

Of course, there exist quasi-vector bundles which are not locally trivial (1.4).

2.3. Example. Let us prove that Example 1.3, where $E=TS^n$, is in fact a vector bundle. Let $x\in S^n$ and let U be the neighbourhood of x defined by $U=\{y\in S^n\,|\,\langle y,x\rangle\neq 0$

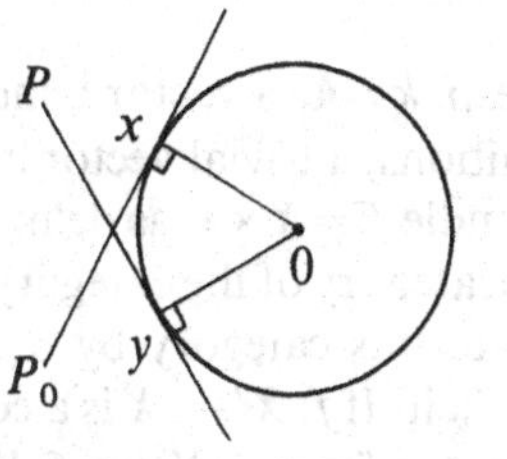

Fig. 1

where $\langle\ ,\ \rangle$ denotes the usual scalar product in $\mathbb{R}^{n+1}$. Let P_0 be the subspace of $\mathbb{R}^{n+1}$ which is orthogonal to x, and let $\varphi: TS^n|_U\to U\times P_0$ be the map taking the pair (y,v) to the pair (y,w), where w is the orthogonal projection of v on P_0. Explicitly $w=v-\langle x,v\rangle x$. Conversely, v may be obtained from w by the formula $v=w-\dfrac{\langle y,w\rangle}{\langle x,y\rangle}x$, showing that φ is a homeomorphism, and hence that TS^n is locally trivial.

2.4. Example. Let V be a finite dimensional vector space over k, and let $P(V)$ be the associated projective space (provided with the quotient topology). The subspace E of $P(V)\times V$ which consists of pairs (D,e) where $D\in P(V)$ and $e\in D$, is fibered over $P(V)$ by the first projection. More precisely, the fiber E_D, where $D\in P(V)$, is the one-dimensional vector space whose elements are the vectors e such that $e\in D$. We prove now that E is actually a vector bundle. If we provide V with a positive Hermitian form when $k=\mathbb{C}$, or a positive quadratic form when

$k=\mathbb{R}$, for each line D we can consider the neighbourhood U_D which consists of the lines Δ which are not orthogonal to D. Now a trivialization of $E|_{U_D}$ is given by the map $\varphi: E|_{U_D} \to U_D \times D$ defined by $\varphi(\Delta, e) = (\Delta, e')$, where e' is the orthogonal projection of e on D. By exhibiting explicit formulas for these projections as in 2.3, one shows that φ is a homeomorphism. This bundle E is called the *canonical line bundle* over $P(V)$.

2.5. There are other ways to deal with Example 2.4. For the real case it is well known that $P(V) \sim S^n/\mathbb{Z}_2$, where the dimension of $V=n+1$ (explicitly $P(V)$ is the quotient of S^n by the equivalence relation $x \sim \pm x$). Let F be the quotient of $S^n \times \mathbb{R}$ by the equivalence relation $(x, t) \sim (x', t') \Leftrightarrow (x', t') = (\varepsilon x, \varepsilon t)$ where $\varepsilon = \pm 1$. Then F is a quasi-vector bundle over $P(V)$, and thus we can define a morphism $f: F \to E$ by the formula $f(x, t) = (\pi(x), tx)$ where $\pi: S^n \to P(V)$ is the natural projection, and $tx \in \pi(x)$. One can also define a morphism $g: E \to F$ by the formula $g(D, v) = (x, t)$ where $x \in D \cap S^n$ and t is the scalar such that $tx = v$. (Of course in these formulas (x, t) represents the class of the pair (x, t) in $S^n \times \mathbb{R}/\sim$.) Then f and g are isomorphisms, with $f = g^{-1}$.

In the complex case, $P(V) \approx S^{2n+1}/U$ where the dimension of $V=n+1$, and where U is the group of complex numbers of norm 1 (explicitly $P(V)$ is the quotient of S^{2n+1} by the equivalence relation $x \sim \lambda x$ if $|\lambda| = 1$). The vector bundle E may be identified in a similar fashion with the quotient of $S^{2n+1} \times \mathbb{C}$ by the equivalence relation $(x, t) \sim (x', t') \Leftrightarrow (x', t') = (\varepsilon x, \bar{\varepsilon} t)$ for $\varepsilon \in U$.

2.6. Now for some terminology. When $k=\mathbb{R}$ (resp. $k=\mathbb{C}$) a vector bundle will be called *real* (resp. *complex*). By abuse of our definitions, a trivial vector bundle will mean a vector bundle which is *isomorphic* to a bundle $E = X \times V$ as defined in 1.10. Vector bundles are in fact the objects of a full subcategory of the category of quasi-vector bundles considered in 1.7. We will denote this category by $\mathscr{E}(X)$, or by $\mathscr{E}_k(X)$ when we want to make the basic field k explicit. If $f: X' \to X$ is a continuous map, the functor f^* defined in 1.19 induces a functor from $\mathscr{E}(X)$ to $\mathscr{E}(X')$. To see this it suffices to show that $f^*(\xi)$ is locally trivial whenever ξ is locally trivial over X. Let $x' \in X'$ and let U be a neighbourhood of $f(x')$ such that $\eta = \xi|_U$ is trivial. Then $\xi'|_{U'} = g^*(\eta)$ where $U' = f^{-1}(U)$ and $g: U' \to U$ is the map induced by f. Hence we have $\eta \approx U \times V$ and $g^*(\eta) \approx U' \times_U (U \times V) \approx U' \times V$ which is trivial over U'. In particular, if X' is a subspace of X, then $\xi|_{X'}$ is a vector bundle.

2.7. Proposition. *Let E and F be two vector bundles over X and let $g: E \to F$ be a morphism of vector bundles such that $g_x: E_x \to F_x$ is bijective for each point x in X. Then g is an isomorphism in the category $\mathscr{E}(X)$.*

Proof. Let $h: F \to E$ be the map defined by $h(v) = g_x^{-1}(v)$ for $v \in F_x$. It suffices to prove that h is continuous. Consider a neighbourhood U of x and isomorphisms $\beta: E_U \to U \times M$ and $\gamma: F_U \to U \times N$. If we let $g_1 = \gamma \cdot g_U \cdot \beta^{-1}$ we have $h_U = \beta^{-1} \cdot h_1 \cdot \gamma$ where h_1 is defined by $\check{h}_1(x) = (\check{g}_1(x))^{-1}$ (cf. 1.12). Since the map from $\operatorname{Iso}(M, N)$ to $\operatorname{Iso}(N, M)$ defined by $\alpha \mapsto \alpha^{-1}$ is continuous, h_1 is continuous. Thus

h is continuous on a neighbourhood of each point of F; hence h is continuous on all of F. □

2.8. Let $\xi=(E, \pi, X)$ be a vector bundle. We define two maps (where $E\times_X E$ is the fiber product)

$$s: E\times_X E\to E \quad \text{and} \quad p: k\times E\to E$$

by the formulas $s(e, e')=e+e'$ and $p(\lambda, e)=\lambda e$, where e and e' are vectors of the same fiber. These maps are continuous. To see this, it is enough to consider the case where $E=X\times V$, since continuity is a local condition as before. In this case, $E\times_X E\approx X\times V\times V$ and under this isomorphism s becomes the map from $X\times V\times V$ to $X\times V$ defined by $(x, v, v')\mapsto(x, v+v')$ which is clearly continuous. The continuity of p is proved in the same way.

2.9. We define the *rank* of a vector bundle $\xi=(E, \pi, X)$ to be the locally constant function $r: X\to\mathbb{N}$ given by $r(x)=\mathrm{Dim}(E_x)$. The rank of ξ is equal to an integer n if $r(x)=n$ for each point x of X. When the base is connected the rank is constant.

Exercise (I.9.5).

3. Clutching Theorems

In the preceding section we defined vector bundles as locally trivial quasi-vector bundles. Now we would like to construct vector bundles using their restrictions to suitable subsets.

3.1. Theorem ("clutching of morphisms"). *Let $\xi=(E, \pi, X)$ and $\xi'=(E', \pi', X)$ be two vector bundles on the same base X. Let us consider also*

a) *A cover of X consisting of open subsets U_i (resp. a locally finite cover of X of closed subsets U_i).*

b) *A collection of morphisms $\alpha_i: \xi|_{U_i}\to\xi'|_{U_i}$ such that $\alpha_i|_{U_i\cap U_j}=\alpha_j|_{U_i\cap U_j}$.*

Then there exists a unique morphism $\alpha: \xi\to\xi'$ such that $\alpha_{U_i}=\alpha_i$.

Proof. The proof naturally breaks into two parts:

(i) *Uniqueness.* Let e be a point of E. Since (U_i) is a cover of X, the point e belongs to some E_{U_i}. Hence $\alpha(e)=r_i'(\alpha_i(e))$ where $r_i': E'_{U_i}\to E'$ is the inclusion map.

(ii) *Existence.* To simplify the notation, let us identify E_{U_i} and E'_{U_i} with subsets of E and E' respectively. For $e\in E$, let $\alpha(e)=\alpha_i(e)$ for $e\in E_{U_i}$. It follows from b) that this definition is independent of the choice of i. The subsets $E_{U_i}=\pi^{-1}(U_i)$ form an open cover (resp. a locally finite cover of closed subsets) of the space E; hence α is continuous. Since $\alpha_x: E_x\to E'_x$ is linear, the map α defines a morphism of vector bundles. □

3.2. Theorem ("clutching of bundles"). *Let (U_i) be an open cover of a space X (resp. a locally finite closed cover of a paracompact space X). Let $\xi_i=(E_i, \pi_i, U_i)$ be a vector bundle over each U_i, and let $g_{ji}:\xi_i|_{U_i\cap U_j} \to \xi_j|_{U_i\cap U_j}$ be isomorphisms which satisfy the compatibility condition $g_{ki}|_{U_i\cap U_j\cap U_k}=g'_{kj}\cdot g'_{ji}$, where $g'_{kj}=g_{kj}|_{U_i\cap U_j\cap U_k}$ and $g'_{ji}=g_{ji}|_{U_i\cap U_j\cap U_k}$. Then there exists a vector bundle ξ over X and isomorphisms $g_i:\xi_i\to \xi|_{U_i}$ such that the diagram*

$$\begin{array}{ccc} \xi_i|_{U_i\cap U_j} & \xrightarrow{g_{ji}} & \xi_j|_{U_i\cap U_j} \\ & {\scriptstyle g_i|_{U_i\cap U_j}}\searrow \quad \swarrow{\scriptstyle g_j|_{U_i\cap U_j}} & \\ & \xi|_{U_i\cap U_j} & \end{array} \qquad \textit{(Diagram 1)}$$

is commutative.

Proof. For simplicity we use the same letter to denote a morphism and its restriction to a subspace. In the topologically disjoint union $\bigsqcup E_i$, consider the equivalence relation $e_i\sim e_j \Leftrightarrow g_{ji}(e_i)=e_j$, and let $E=\bigsqcup E_i/\sim$ be given the quotient topology. The continuous map $\bigsqcup E_i\to X$ induced by the π_i defines a continuous map $\pi: E\to X$. For $x\in U_i$, the structure of the vector space $E_x=\pi^{-1}(\{x\})$, which is induced by the isomorphism $E_x\approx E_i|_{\{x\}}$, does not depend on the choice of i since g_{ji} is linear on each fiber. Let $g_i: E_i\to \pi^{-1}(U_i)$ be the map defined by $g_i(e)=\bar{e}$, where $\bar{e}$ is the class of e in E. Then g_i is continuous, bijective, open, and induces a linear isomorphism on each fiber. Therefore g_i defines an isomorphism between the quasi-vector bundles (E_i, π_i, U_i) and $(E_{U_i}, \pi|_{E_{U_i}}, U_i)$, where $E_{U_i}=\pi^{-1}(U_i)$.

Suppose that (U_i) is an *open* cover of X. Let x be a point of U_i, and let V be a neighbourhood of x contained in U_i such that $\xi_i|_V$ is trivial. If ξ is the quasi-vector bundle (E, π, X) as defined above, we have $\xi_{U_i}\approx\xi_i$. Hence $\xi|_V\approx\xi_i|_V$ is trivial, which proves that ξ is locally trivial.

Let us assume now that X is paracompact and that (U_i) is a closed cover which is locally finite. Let x_0 be a point of X. Since the cover (U_i) is locally finite, there exists a closed neighbourhood V of x_0 which meets only a finite number of subsets $U_{i_1},\dots, U_{i_p}$, and such that the bundles $\xi_j|_{V_j}$ are trivial, where $V_j=U_{i_j}\cap V$ for $j=1,\dots,p$. Without loss of generality we may assume that $x_0\in V_1$ and that $\xi_j|_{V_j}\approx V_j\times k^n$. Starting with an arbitrary isomorphism $\alpha_1:\xi|_{V_1}\approx V_1\times k^n$ we are going to define by induction on r, a morphism α_r between $\xi|_{V_1\cup\cdots\cup V_r}$ and the trivial bundle $(V_1\cup\cdots\cup V_r)\times k^n$. Since $\xi|_{V_r}$ is trivial, this is equivalent to defining a continuous map $\beta_r: V_r\to \mathscr{L}(k^n, k^n)$ which extends $\check{\gamma}_r$ with $\gamma_r=\alpha_{r-1}|_{V_r\cap(V_1\cup\cdots\cup V_{r-1})}$. This extension is possible due to the Tietze extension theorem (Kelley [1], Bourbaki [1]). Let $\alpha:\xi|_V\to V\times k^n$ be the morphism thus obtained. Since $\mathrm{Iso}(k^n, k^n)$ is open in $\mathscr{L}(k^n, k^n)$, Theorem 1.12 shows that the set of points x of V_s, such that α_x is an isomorphism, is an open subset of V_s. Since the sets V_s are finite in number, the set of points x of V such that α_x is an isomorphism is a neighbourhood W of x_0. The map $\alpha_W: E|_W\to W\times k^n$ induces a homeomorphism $E|_{V_s\cap W}\to (V_s\cap W)\times k^n$ for each s. Hence α_W is a homeomorphism itself. Since this holds for every point x_0 in X, we see that ξ is locally trivial in this case also. □

3.3. *Remark.* Moreover one may say that the bundle ξ which we just constructed is "unique" in the following sense. Let ξ' be another vector bundle, and let $g_i': \xi_i \to \xi'|_{U_i}$ be isomorphisms which make the diagram

$$\begin{array}{ccc} \xi_i|_{U_i\cap U_j} & \xrightarrow{g_{ji}} & \xi_j|_{U_i\cap U_j} \\ & {\scriptstyle g_i'|_{U_i\cap U_j}} \searrow \quad \swarrow {\scriptstyle g_j'|_{U_i\cap U_j}} & \\ & \xi'|_{U_i\cap U_j} & \end{array} \qquad \text{(Diagram 2)}$$

commutative. Then there exists a unique isomorphism $\alpha: \xi \to \xi'$ which makes the following diagram commutative.

$$\begin{array}{ccc} & \xi_i & \\ & {\scriptstyle g_i} \swarrow \quad \searrow {\scriptstyle g_i'} & \\ \xi|_{U_i} & \xrightarrow{\alpha|_{U_i}} & \xi'|_{U_i} \end{array}$$

In fact, one may construct α in the following way. The morphism $\alpha_i = g_i' \cdot g_i^{-1}$ is an isomorphism from ξ_{U_i} to ξ'_{U_i}, and over $U_i \cap U_j$, we have the identity $g_{ji} = g_j^{-1} \cdot g_i = g_j'^{-1} \cdot g_i'$ according to diagrams (1) and (2). Therefore, over $U_i \cap U_j$ we have $\alpha_i = g_i' \cdot g_i^{-1} = g_j' \cdot g_j^{-1} = \alpha_j$. The existence of α is then guaranteed by Theorem 3.1. Its uniqueness is obvious.

3.4. Example. Let S^n be the sphere of $\mathbb{R}^{n+1}$, i.e. the set of points $x = (x_1, \ldots, x_{n+1})$ such that $\|x\|^2 = \sum_{i=1}^{n+1} (x_i)^2 = 1$. Let S_+^n (resp. S_-^n) be the subset of S^n whose points x satisfy $x_{n+1} \geqslant 0$ (resp. $x_{n+1} \leqslant 0$). Then S^n is compact, S_+^n and S_-^n are closed subsets, and $S_+^n \cap S_-^n = S^{n-1}$.

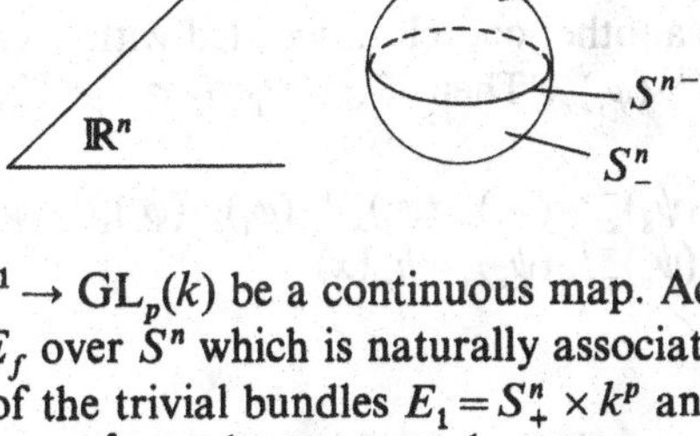

Fig. 2

Let $f: S^{n-1} \to \mathrm{GL}_p(k)$ be a continuous map. According to Theorem 3.2, there is a bundle E_f over S^n which is naturally associated with f. It is obtained from the clutching of the trivial bundles $E_1 = S_+^n \times k^p$ and $E_2 = S_-^n \times k^p$ by the "transition function" $g_{21} = \hat{f}: S^{n-1} \times k^p \to S^{n-1} \times k^p$ (g_{11} and g_{22} are the identity map). We see later (7.6) that all bundles over S^n are isomorphic to bundles of this type.

3.5. Theorem 3.2 is related to the problem of classification of "*G-principal bundles*", where G is the topological group $\mathrm{GL}_n(k)$. To be more precise, let us consider an arbitrary topological group G and a topological space X. A *G-cocycle* on X is given by an open cover (U_i) of X, and continuous maps $g_{ji}: U_i \cap U_j \to G$ such that $g_{kj}(x) \cdot g_{ji}(x) = g_{ki}(x)$ for $x \in U_i \cap U_j \cap U_k$.

Two cocycles (U_i, g_{ji}) and (V_r, h_{sr}) are equivalent if there exist continuous maps $g_i^r: U_i \cap V_r \to G$ such that $g_j^s(x) \cdot g_{ji}(x) \cdot g_i^r(x)^{-1} = h_{sr}(x)$ for $X \in U_i \cap U_j \cap V_r \cap V_s$. Let us check that this relation is an equivalence relation. The symmetry and reflexivity are obvious (note that $g_{ii} = \mathrm{Id}$ and $g_{ij} = g_{ji}^{-1}$). If (W_u, l_{vu}) is a cocycle equivalent to (V_r, h_{sr}) we can find continuous maps $h_r^u: V_r \cap W_u \to G$ such that $h_s^v(x) \cdot h_{sr}(x) \cdot h_r^u(x)^{-1} = l_{vu}(x)$ for $x \in V_r \cap V_s \cap W_u \cap W_v$. For $i=j$ the first identity gives the relation $g_i^s(x) \cdot g_i^r(x)^{-1} = h_{sr}(x)$ for $x \in U_i \cap V_r \cap V_s$. For $u=v$ the second identity gives the relation $h_s^u(x)^{-1} \cdot h_r^u(x) = h_{sr}(x)$ for $x \in V_r \cap V_s \cap W_u$. From this it follows that $h_r^u(x) \cdot g_i^r(x) = h_s^u(x) \cdot g_i^s(x)$ for $x \in U_i \cap V_r \cap V_s \cap W_u$. By glueing together the continuous functions $h_r^u(x) \cdot g_i^r(x)$ as r varies, this defines another continuous map $l_i^u: U_i \cap W_u \to G$. Moreover $l_{vu}(x) = h_s^v(x) \cdot h_{sr}(x) \cdot h_r^u(x)^{-1} = l_j^v(x) \cdot g_{ji}(x) \cdot l_i^u(x)^{-1}$ for $x \in U_i \cap U_j \cap V_r \cap V_s \cap W_u \cap W_v$. Since this relation is true for every pair (r, s), it also holds for $x \in U_i \cap U_j \cap W_u \cap W_v$. Hence the equivalence relation in the set of G-cocycles is well-defined. The quotient set will be denoted by $H^1(X; G)$ (see Hirzebruch [2] and Greenberg [1] for the justification of this terminology). This set depends contravariantly on X and covariantly on G.

3.6. Theorem. *Let $\Phi_n^k(X)$ be the set of isomorphism classes of k-vector bundles of rank n over the topological space X. Then $\Phi_n^k(X)$ is naturally isomorphic to the set $H^1(X; G)$, where $G = \mathrm{GL}_n(k)$.*

Proof. We define two maps

$$h: \Phi_n^k(X) \longrightarrow H^1(X; G) \quad \text{and} \quad h': H^1(X; G) \longrightarrow \Phi_n^k(X)$$

such that $h' = h^{-1}$.

Let $\xi = (E, \pi, X)$ be a vector bundle, and let (U_i) be a trivialization cover of X. Choose isomorphisms $\varphi_i: U_i \times k^n \to E_{U_i}$, and let g_{ji} be the map from $U_i \cap U_j$ to $G = \mathrm{GL}_n(k)$ defined by $g_{ji}(x) = (\varphi_j)_x^{-1} \cdot (\varphi_i)_x$. In this way we obtain a G-cocycle on X. Its class in the set $H^1(X; G)$ is independent of the choice of the trivialization cover and of the φ_i. In fact, if (V_r, h_{sr}) is another cocycle associated with trivializations $\psi_r: V_r \times k^n \to E_{V_r}$, let $g_i^r(x) = (\psi_r)_x^{-1} \cdot (\varphi_i)_x$. Then if $x \in U_i \cap U_j \cap V_r \cap V_s$, we have

$$\begin{aligned} g_j^s(x) \cdot g_{ji}(x) \cdot (g_i^r(x))^{-1} &= (\psi_s)_x^{-1} \cdot (\varphi_j)_x \cdot (\varphi_j)_x^{-1} \cdot (\varphi_i)_x \cdot (\varphi_i)_x^{-1} \cdot (\psi_r)_x \\ &= (\psi_s)_x^{-1} \cdot (\psi_r)_x = h_{sr}(x). \end{aligned}$$

This shows that h is well-defined.

Conversely, let (U_i, g_{ji}) be a G-cocycle, and let E be the vector bundle over X obtained by clutching the trivial bundles $E_i = U_i \times k^n$ with the "*transition functions*" g_{ji} (Theorem 3.2). Then the class of E in $\Phi_n^k(X)$ depends only on the class of the cocycle in $H^1(X; G)$. In fact, consider a cocycle (V_r, h_{sr}) equivalent to (U_i, g_{ji}), and let F be the vector bundle obtained from this cocycle by clutching the trivial bundles $F_r = V_r \times k^n$. Let $\alpha: E \to E'$ be the unique morphism which makes the following diagram commutative for each pair (i, r):

$$
\begin{array}{ccc}
E_i|_{U_i \cap V_r} & \xrightarrow{\hat{g}_i^r} & F_r|_{U_i \cap V_r} \\
\downarrow{\scriptstyle g_i|_{U_i \cap V_r}} & & \downarrow{\scriptstyle h_r|_{U_i \cap V_r}} \\
E|_{U_i \cap V_r} & \xrightarrow{\alpha|_{U_i \cap V_r}} & F|_{U_i \cap V_r}
\end{array}
$$

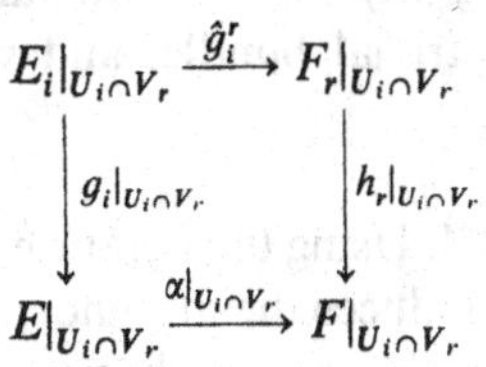

(in this diagram the morphisms g_i and h_r are the morphisms defined by clutching as in Theorem 3.2). To see that α is well-defined, we note the following identities for $x \in U_i \cap U_j \cap V_r \cap V_s$:

$$h_{sr}(x) = g_j^s(x) \cdot g_{ji}(x) \cdot (g_i^r(x))^{-1},$$

$$h_{sr}(x) \cdot g_i^r(x) \cdot g_{ij}(x) = g_j^s(x),$$

$$(h_s(x))^{-1} \cdot h^r(x) \cdot g_i^r(x) \cdot g_i(x))^{-1} \cdot g_j(x) = g_j^s(x),$$

and finally

$$h_r(x) \cdot g_i^r(x) \cdot (g_i(x))^{-1} = h_s(x) \cdot g_j^s(x) \cdot (g_j(x))^{-1}.$$

This shows that h' is also well-defined. The fact that $h \cdot h'$ and $h' \cdot h$ are the identities of $H^1(X; G)$ and $\Phi_n^k(X)$ respectively, follows directly from their definitions. $\square$

3.7. Remark. When X is paracompact, one could equally well work with locally finite covers of closed subsets, and thus obtain another set $H_f^1(X; G)$ analogous to $H^1(X; G)$. The above argument shows that $H_f^1(X; G)$ is also naturally isomorphic to $\Phi_n^k(X)$ if $G = \mathrm{GL}_n(k)$.

3.8. Theorem. *Let (U_i, g_{ji}) and (U_i, h_{ji}) be two cocycles relative to the same open cover of a space X (resp. locally finite closed cover of a paracompact space X). Then The associated vector bundles E and F are isomorphic if and only if there exist continuous functions $\lambda_i: U_i \to G = \mathrm{GL}_n(k)$, such that $h_{ji}(x) = \lambda_j(x) \cdot g_{ji}(x) \cdot (\lambda_i(x))^{-1}$ for $x \in U_i \cap U_j$. In particular, the vector bundle E is trivial if and only if $g_{ji}(x) = \lambda_j(x)^{-1} \lambda_i(x)$ for suitable choices of the λ_i.*

Proof. Let $\alpha: E \to F$ be an isomorphism. Then we have the commutative diagram

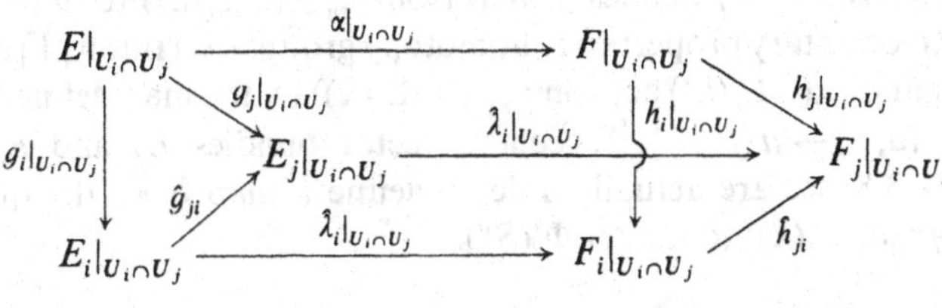

where $E_i = F_i = U_i \times k^n$ and $\hat{\lambda}_i = h_i \cdot \alpha|_{U_i} \cdot (g_i)^{-1}$ (using the notation of 1.12). From this diagram, we obtain the relation $h_{ji}(x) = \lambda_j(x) \cdot g_{ji}(x) \cdot (\lambda_i(x))^{-1}$. In particular, if we choose $h_{ji} = 1$, then F is isomorphic to the trivial bundle, and we obtain $g_{hi}(x) = \lambda_j(x)^{-1} \cdot \lambda_i(x)$ as desired. □

3.9. Let us apply the above theorem to Example 3.4. Using the notation of 3.4, we let λ be an automorphism of $E_1 = S^n_+ \times k^p$ which induces the automorphism μ of $E_1|_{S^{n-1}} = S^{n-1} \times k^p$. Then using the notation of 1.12, we see that the vector bundles E_f and $E_{f\mu}$ are isomorphic. To see this, we apply the above theorem, set $\lambda_1 = \lambda$ and $\lambda_2 = 1$, and visualize the situation with the diagram

$$\begin{array}{ccc} E_1 & \xrightarrow{\lambda^{-1}} & E_1 \\ {\scriptstyle f}\vdots & & \vdots{\scriptstyle f\mu} \\ E_2 & \xrightarrow{\text{Id}} & E_2 \end{array}$$

(the dotted lines denote morphisms defined only over S^{n-1}). In the same way, we can prove that E_f is isomorphic to E_{vf}, when v is an automorphism of $E_2|_{S^{n-1}}$ which can be extended to an automorphism of E_2.

Now consider two continuous maps f_0 and f_1 from S^{n-1} to $\mathrm{GL}_p(k)$, which are homotopic. Let $\alpha: S^{n-1} \to \mathrm{GL}_p(k)$ be the map defined by $\alpha(x) = (f_1(x))^{-1} \cdot f_0(x)$. Then α is homotopic to 1; more precisely there exists a continuous map $\beta: S^{n-1} \times I \to \mathrm{GL}_p(k)$ such that $\beta(x, 0) = \alpha(x)$ and $\beta(x, 1) = 1$. We parametrize the upper half hemisphere S^n_+ of S^n by writing each element w of S^n_+ in the form $v \cos\theta + e_{n+1} \sin\theta$, where $v \in S^{n-1}$.

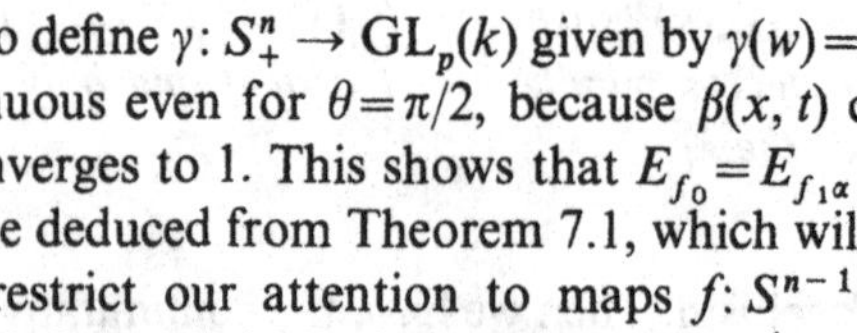

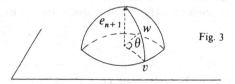

Fig. 3

We use β to define $\gamma: S^n_+ \to \mathrm{GL}_p(k)$ given by $\gamma(w) = \beta(v, \sin\theta)$. This is well-defined and continuous even for $\theta = \pi/2$, because $\beta(x, t)$ converges to 1 uniformly in x when t converges to 1. This shows that $E_{f_0} = E_{f_1\alpha}$ is isomorphic to E_{f_1} (this fact may also be deduced from Theorem 7.1, which will be proved independently).

If we restrict our attention to maps $f: S^{n-1} \to \mathrm{GL}_p(k)$ such that $f(e) = 1$, where $e = (1, 0, \ldots, 0)$ is the base point of S^{n-1}, the above discussion shows that the correspondence $f \mapsto E_f$ defines a map from $\pi_{n-1}(\mathrm{GL}_p(k))$ to $\Phi_p^k(S^n)$ (for the definition and elementary properties of homotopy groups see Hilton [1] or Hu [1]). On the other hand $\pi_0(\mathrm{GL}_p(k))$ acts on $\pi_{n-1}(\mathrm{GL}_p(k))$ by the map defined on representatives by $(a, f) \to a \cdot f \cdot a^{-1}$. Since the vector bundles E_f and $E_{a \cdot f \cdot a^{-1}}$ are isomorphic by 3.8, we are actually able to define a map from the quotient *set* $\pi_{n-1}(\mathrm{GL}_p(k))/\pi_0(\mathrm{GL}_p(k))$ to the set $\Phi_p^k(S^n)$.

3.10. Theorem. *The map*

$$\pi_{n-1}(\mathrm{GL}_p(k))/\pi_0(\mathrm{GL}_p(k)) \to \Phi_p^k(S^n)$$

is injective.

Proof. Let f and g be continuous maps from S^{n-1} to $\mathrm{GL}_p(k)$ with $f(e)=g(e)=1$ (where e is the base point of S^{n-1}) such that the vector bundles E_f and E_g are isomorphic. We have a commutative diagram

$$\begin{array}{ccc} E_1 & \xrightarrow{\alpha_1} & E_1 \\ {\scriptstyle \hat{f}}\,\vdots & & \vdots\,{\scriptstyle \hat{g}} \\ E_2 & \xrightarrow{\alpha_2} & E_2 \end{array}$$

where the dotted arrows denote morphisms which are defined only over S^{n-1}. The maps $\beta_i=\check{\alpha}_i|_{S^{n-1}}$ are maps from S^{n-1} to $\mathrm{GL}_p(k)$ which are homotopic to constant ones, and such that $g(x)=\beta_2(x)\cdot f(x)\cdot\beta_1(x)^{-1}$. Moreover, since $f(e)=g(e)=1$, β_1 and β_2 are homotopic to the *same* constant map a (since they are restrictions of maps defined on a contractible subset i.e. $S^n-\{p\}$ where $p\notin S^{n-1}$). This implies that g and $a\cdot f\cdot a^{-1}$ are homotopic. If $h\colon S^{n-1}\times I\to \mathrm{GL}_p(k)$ is this homotopy, then the homotopy $l\colon S^{n-1}\times I\to \mathrm{GL}_p(k)$ defined by $l(x,t)=h(x,t)h(e,t)^{-1}$, shows that g and $a\cdot f\cdot a^{-1}$ have the same class in the homotopy group $\pi_{n-1}(\mathrm{GL}_p(k))$, since $l(e,t)=e$. □

3.11. *Remark.* It will be shown later (7.6) that the above map is also surjective.

3.12. *Remark.* More generally, it can be shown by the same method that $H^1(S^n;G)\approx\pi_{n-1}(G)/\pi_0(G)$ for any topological group G (Steenrod [1]).

***3.13.** Some comments on Theorem 3.10. If $k=\mathbb{C}$, then the group $\mathrm{GL}_p(k)=\mathrm{GL}_p(\mathbb{C})$ may be regarded as the topological product of $\mathrm{U}(p)$ by $\mathbb{R}^q$, where $q=p^2$ (see Chevalley [1]). Since $\mathrm{U}(p)$ is arcwise connected, $\pi_0(\mathrm{U}(p))=\pi_0(\mathrm{GL}_p(\mathbb{C})=0$. Hence $\Phi_p^{\mathbb{C}}(S^n)\approx\pi_{n-1}(\mathrm{U}(p))$. Now we have the locally trivial fibration

$$\mathrm{U}(p)\to\mathrm{U}(p+1)\to S^{2p+1},$$

hence the exact sequence of homotopy groups

$$\pi_{i+1}(S^{2p+1})\to\pi_i(\mathrm{U}(p))\to\pi_i(\mathrm{U}(p+1))\to\pi_i(S^{2p+1}).$$

Since $\pi_j(S^r)=0$ for $j>r$, it follows that for $p>i/2$, we have $\pi_i(\mathrm{U}(p))\approx\pi_i(\mathrm{U}(p+1))$ and $\pi_i(\mathrm{U}(p))\approx\operatorname{inj\,lim}\pi_i(\mathrm{U}(m))$. We shall prove later (III.2) that $\operatorname{inj\,lim}\pi_i(\mathrm{U}(m))=0$ for i even and $\operatorname{inj\,lim}\pi_i(\mathrm{U}(m))=\mathbb{Z}$ for i odd. It follows from this theorem that the problem of classification of complex vector bundles of rank p over the sphere S^r is completely solved when $p>(r-1)/2$. When $p\leqslant(r-1)/2$ the problem is still open in general.

If $k=\mathbb{R}$, the group $\mathrm{GL}_p(k)=\mathrm{GL}_p(\mathbb{R})$ may be regarded as the topological product $\mathrm{O}(p)\times\mathbb{R}^q$, where $q=p(p+1)/2$ (see Chevalley [1]). Hence $\pi_i(\mathrm{GL}_p(\mathbb{R}))\approx\pi_i(\mathrm{O}(p))$ and $\pi_0(\mathrm{GL}_p(\mathbb{R}))\approx\mathbb{Z}/2$. The homotopy exact sequence associated with the locally trivial fibration

$$\mathrm{O}(p)\to\mathrm{O}(p+1)\to S^p,$$

i.e.

$$\pi_{i+1}(S^p)\longrightarrow\pi_i(\mathrm{O}(p))\longrightarrow\pi_i(\mathrm{O}(p+1))\longrightarrow\pi_i(S^p),$$

shows that $\pi_i(\mathrm{O}(p))\approx\pi_i(\mathrm{O}(p+1))$ and that $\pi_i(\mathrm{O}(p))\approx\operatorname{inj\,lim}\pi_i(\mathrm{O}(m))$, when $p>i+1$. We prove later (III.5) that the groups $\pi_i=\operatorname{inj\,lim}\pi_i(\mathrm{O}(m))$ are respectively isomorphic to $\mathbb{Z}/2$, $\mathbb{Z}/2$, 0, $\mathbb{Z}$, 0, 0, 0 when $i\equiv 0, 1, 2, 3, 4, 5, 7 \bmod 8$. Hence the homotopy groups $\pi_i(\mathrm{O}(p))\approx\pi_i(\mathrm{GL}_p(\mathbb{R}))$ are completely known for $p>i+1$. Moreover, in this case, the action of $\pi_0(\mathrm{GL}_p(\mathbb{R}))=\mathbb{Z}/2$ is trivial. When p is odd, we have nothing to prove since $\mathrm{Det}(-1)=-1$. When p is even, the isomorphism between $\pi_i(\mathrm{O}(p))$ and $\pi_i(\mathrm{O}(p+1))$ is compatible with the action. Hence $\pi_i(\mathrm{O}(p))/(\mathbb{Z}/2)\approx\pi_i(\mathrm{O}(p+1))/(\mathbb{Z}/2)\approx\pi_i(\mathrm{O}(p+1))\approx\pi_i(\mathrm{O}(p))$. From this discussion we see that the problem of classification of real vector bundles of rank p over S^n is completely solved for $p>n$. As in the complex situation, the case where $p\leqslant n$ is still open in general.*

3.14. The case of vector bundles over the sphere may be generalized in the following way. If X is a paracompact space one considers the double cone over X: it is defined as the quotient of $X\times[-1,1]$ by the equivalence relation which identifies $X\times\{1\}$ with a single point, and $X\times\{-1\}$ with another single point.

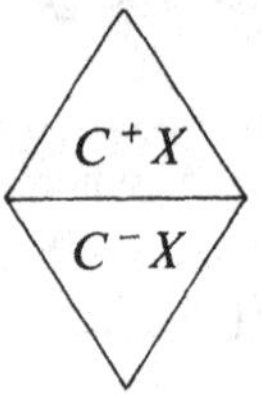

Fig. 4

We write $C^+(X)$ (resp. $C^-(X)$) for the image of $X\times[0,1]$ (resp. $X\times[-1,0]$) in the quotient and let $S'(X)=C^+(X)\cup C^-(X)$. The space $S'(X)$ (which is paracompact) is also called the *suspension* of X. The argument used to parametrize S^n_+ in 3.9 shows that $S'(S^{n-1})$ is homeomorphic to S^n.

Let $f\colon X\to\mathrm{GL}_p(k)$ be a continuous map. Then one may define a bundle E_f over $S'(X)$ by clutching the trivial bundles $E_1=C^+(X)\times k^p$ and $E_2=C^-(X)\times k^p$ with the transition function $g_{21}=f$. As in 3.9, one can prove that E_{f_0} is isomorphic to E_{f_1}, when f_0 is homotopic to f_1. If we choose a base point e in X, let $[X,\mathrm{GL}_p(k)]'$ denote the set of homotopy classes of maps f such that $f(e)=1$. Then as in 3.10 $\Phi_p^k(S'(X))$ contains the quotient of $[X,\mathrm{GL}_p(k))]'$ by the action of $\pi_0(\mathrm{GL}_p(k))$. This result will be strengthened later (7.6).

We are now going to use the clutching theorems to define the tangent bundle of a manifold. First we briefly review some basic ideas and definitions in order to acquaint the reader with the notation (cf. Lang [2]).

1.15. Let A be an open set in $\mathbb{R}^n$ and let $TA = A \times \mathbb{R}^n$, regarded as the trivial bundle over A. If B is another open set in $\mathbb{R}^p$, and if $f\colon A \to B$ is a differentiable map of class C^m, for $m \geqslant 1$, we can associate f with a general morphism of vector bundles (in the sense of 1.6)

$$\begin{array}{ccc} TA & \xrightarrow{Tf} & TB \\ \downarrow & & \downarrow \\ A & \xrightarrow{f} & B \end{array}$$

by the formula $(Tf)_x(v) = f'(x)(v)$. In this formula, $f'\colon A \to \mathscr{L}(\mathbb{R}^n, R^p)$ is the differential of f and v is a vector of $\mathbb{R}^n$.

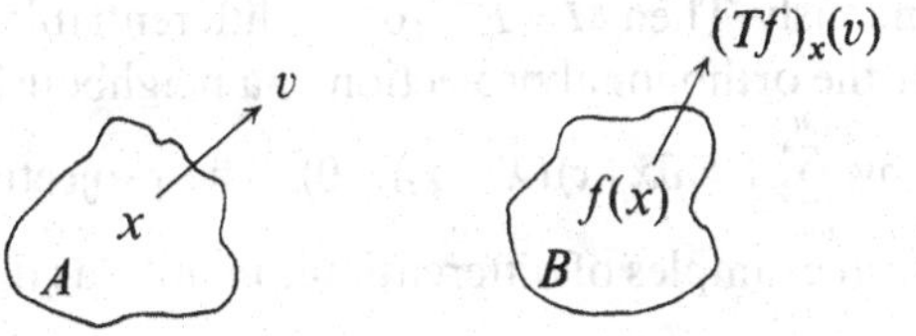

Fig. 5

If $g\colon B \to C$ is another differentiable map of class C^m between B and an open set C in $\mathbb{R}^q$, one has $T(g \cdot f) = T(g) \cdot T(f)$ (composition of general morphisms) following from the theorem which gives the derivative of the composition of two differentiable maps. Hence the reader can easily verify that the correspondence $A \mapsto TA$ is a "functor" from the category of differentiable maps to the category of trivial bundles.

1.16. Definition of a manifold of class C^m. Let M be a topological space and let (U_i) for $i \in I$, be an open cover of M. For each $i \in I$ let φ_i be a homeomorphism of U_i with an open set A_i of $\mathbb{R}^{n_i}$. We write $U_{ij} = U_i \cap U_j$ and $A_{ji} = \varphi_i(U_i \cap U_j)$.

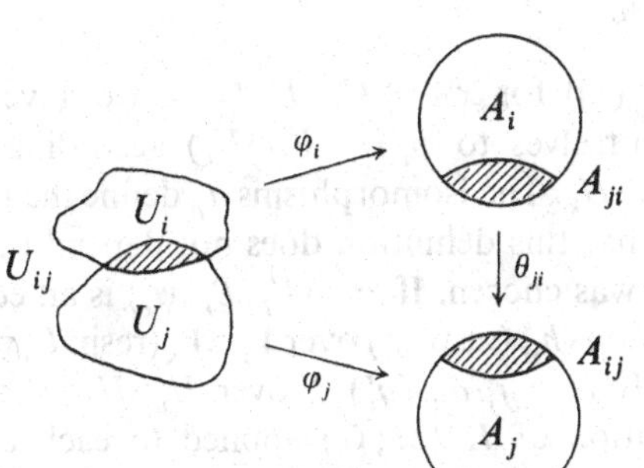

Fig. 6

We say that (U_i, A_i, φ_i) is an atlas of class C^m for $m \geqslant 1$, if $\theta_{ji} = \varphi_j|_{U_{ij}} \cdot \varphi_i^{-1}|_{A_{ji}}$ is a diffeomorphism from A_{ji} to A_{ij} of class C^m. Two atlases are equivalent if their union in the obvious sense is again an atlas. A *differentiable structure of class C^m* on M is given by a "class" of atlases. To avoid logical difficulties [the covers of a space are not the elements of a set!], we shall always consider that the index set I is contained in some fixed set, for example $\bigsqcup_{\substack{U \in S \\ n \in \mathbb{N}}} F(U, \mathbb{R}^n)$, where S is the set of open subsets of M and where $F(U, \mathbb{R}^n)$ is the set of all continuous maps from U to $\mathbb{R}^n$. There is a canonical way to associate a differentiable structure (U_i, A_i, φ_i) on M with an atlas (this atlas is called *maximal*). It is defined as the set of all triples (V, B, φ), where V is open in M, B is open in some $\mathbb{R}^q$, and $\varphi: V \to B$ is a homeomorphism, which satisfy the following condition. For each index i, let us put $V_i = V \cap U_i$ and $B_i = \varphi(V_i)$; then $\varphi_i|_{V_i} \cdot \varphi^{-1}|_{B_i}$ must be a diffeomorphism of class C^m from B_i to $\varphi_i(B_i)$. It is easy to verify that two atlases are equivalent if and only if the maximal atlases associated with them are equal. A *chart* of a differentiable manifold M is simply an element $\varphi: V \to B$ of the maximal atlas.

3.17. Examples. Let $F: \mathbb{R}^n \to \mathbb{R}$ be a differentiable function such that $\partial F/\partial x_1, \ldots, \partial F/\partial x_n$ are not all zero simultaneously. Then $M = F^{-1}(0)$ is a differentiable manifold (at each point x of M, consider the orthogonal projection of a neighbourhood of x onto the hyperplane defined by $\sum_{i=1}^{n} \partial F/\partial x_i(x)(X - x_i) = 0$). The projective spaces $P_n(\mathbb{C})$ and $P_n(\mathbb{R})$ are also classical examples of differentiable manifolds (Godbillon [2], Spivak [1]).

3.18. The tangent bundle of a differentiable manifold. Let $\mathscr{A} = (U_i, A_i, \varphi_i)$ be an atlas on M. We want to define a bundle TM on M by clutching the trivial bundles $TU_i = U_i \times \mathbb{R}^{n_i}$ over the covering (U_i). Let $\bar{\varphi}_i: U_i \times \mathbb{R}^{n_i} \to A_i \times \mathbb{R}^{n_i}$ be the isomorphism from TU_i to TA_i defined by $\bar{\varphi}_i(x, v) = (\varphi_i(x), v)$. Let $g_{ji}: TU_i|_{U_i \cap U_j} \to TU_j|_{U_i \cap U_j}$ be the isomorphism which makes the diagram

$$\begin{array}{ccc} TU_i|_{U_i \cap U_j} & \xrightarrow{\bar{\varphi}_i|_{U_i \cap U_j}} & TA_i|_{A_{ji}} \\ \Big\downarrow g_{ji} & & \Big\downarrow T(\theta_{ji}) \\ TU_j|_{U_i \cap U_j} & \xrightarrow{\bar{\varphi}_j|_{U_i \cap U_j}} & TA_j|_{A_{ij}} \end{array}$$

commutative. Since $\theta_{ki}(x) = \theta_{kj}(\theta_{ji}(x))$ for $x \in \varphi_i(U_i \cap U_j \cap U_k)$, we have $T(\theta_{ki}) = T(\theta_{kj}) \cdot T(\theta_{ji})$ when we restrict ourselves to $\varphi_i(U_i \cap U_j \cap U_k)$ according to 3.15. Therefore $g_{ki} = g_{kj} \cdot g_{ji}$ over $U_i \cap U_j \cap U_k$. The isomorphisms g_{ji} define the transition functions for TM. Let us prove that this definition does not depend (up to isomorphism) on the atlas $\mathscr{A}$ which was chosen. If $\mathscr{B} = (V_r, B_r, \psi_r)$ is an equivalent atlas, we also have transition functions h_{sr} (resp. g_i^r) over $V_r \cap V_s$ (resp. $U_i \cap V_r$) such that $g_j^s \cdot g_{ji} = h_{sr} \cdot g_i^r$, or equivalently $h_{sr} = g_j^s \cdot g_{ji} \cdot (g_i^r)^{-1}$ over $U_i \cap U_j \cap V_r \cap V_s$. The assertion follows from the definition of $H^1(M; G)$ applied to each connected component of M, and from Theorem 3.6. We call TM the *tangent bundle* of M.

3.19. The Functor T. In order to make the above construction functorial, one must choose, once and for all, an atlas for each manifold M (for example the maximal atlas). Now let M and N be two differentiable manifolds of class C^m, (U_i, A_i, φ_i) the atlas of M, and (V_r, B_r, ψ_r) the atlas of N. Let $f\colon M \to N$ be a continuous map. We say that f is of class C^α, $1 \leqslant \alpha \leqslant p$, if $\forall i, \forall x \in U_i$ and for $f(x) \in V_r$, the map $f_i^r = \psi_r \cdot f \cdot \varphi_i^{-1}$ is differentiable of class C^α on a neighbourhood of $\varphi_i(x)$ which is small enough to have the composition of the three functions, $\psi_r \cdot f \cdot \varphi_i^{-1}$ make sense. We want to define a general morphism $Tf\colon TM \to TN$ over $f\colon M \to N$, which generalizes the general morphism defined in 3.15 for open sets in Euclidean spaces. On TU_i we define Tf over a neighbourhood of x, for $x \in U_i$, so that the following diagram commutes

$$\begin{array}{ccc} TU_i & \xrightarrow{\bar{\varphi}_i} & TA_i \\ {\scriptstyle Tf}\downarrow & & \downarrow{\scriptstyle T(f_i^r)} \\ TV_r & \xrightarrow{\bar{\psi}_r} & TB_r, \end{array}$$

where $T(f_i^r)$ is defined as in 3.15. We must show that this definition is compatible with the transition functions g_{ji} and h_{sr} defined on M and N, respectively. We have the identities $f_j^s = (\psi_s \cdot \psi_r^{-1}) \cdot (\psi_r \cdot f \cdot \varphi_i^{-1}) \cdot (\varphi_i \cdot \varphi_j^{-1}) = \theta_{sr}^N \cdot f_i^r \cdot \theta_{ij}^N$, where $\theta_{sr}^N\colon B_{sr} \xrightarrow{\approx} B_{rs}$ and $\theta_{ij}^M\colon A_{ij} \xrightarrow{\approx} A_{ji}$ (all morphisms are defined on suitable neighbourhoods). From these identities we see that $T(f_j^s) = T(\theta_{sr}^N) \cdot T(f_i^r) \cdot T(\theta_{ij}^M)$, and hence the commutativity of the diagram

$$\begin{array}{ccc} TA_i & \xrightarrow{g_{ji}^N} & TA_j \\ {\scriptstyle T(f_i^r)}\downarrow & & \downarrow{\scriptstyle T(f_j^s)} \\ TB_r & \xrightarrow{g_{sr}^N} & TB_s \end{array}$$

since $T(\theta_{sr}^N) = g_{sr}^N$ and $T(\theta_{ij}^M) = g_{ij}^M$ (where g_{ij}^M and g_{sr}^N are the transitions functions of TM and TN respectively). According to Theorem 3.1, this shows that Tf is well-defined. Moreover, if $g\colon N \to P$ is another differentiable map, it can be verified that $T(g \cdot f) = T(g) \cdot T(f)$.

3.20. Example. Let $f\colon M \to \mathbb{R}^p$ be an imbedding (Lang [2]). Then M is given *locally* at a neighbourhood of every point $a \in M$ by equations

$$\begin{aligned} &f_1(x_1, \ldots, x_p) = 0 \\ &f_2(x_1, \ldots, x_p) = 0 \\ &\quad\cdot \quad \cdot \quad \cdot \quad \cdot \quad \cdot \\ &f_n(x_1, \ldots, x_p) = 0, \quad n < p, \end{aligned}$$

where the matrix

$$\begin{pmatrix} \partial f_1/\partial x_1 & \cdots & \partial f_n/\partial x_1 \\ \cdot & \cdot & \cdot \\ \partial f_1/\partial x_p & \cdots & \partial f_n/\partial x_p \end{pmatrix}$$

is of rank n on a neighbourhood of the point $a=(a_1,\ldots,a_p)$. Moreover $T_a(M)$ is the subspace of $T_a(\mathbb{R}^p)$ of dimension $n-p$ which is orthogonal to the "gradient vectors" $V_1,\ldots,V_n$ where $V_i=(\partial f_i/\partial x_1(a),\ldots,\partial f_i/\partial x_p(a))$.

Exercises (Section I.9) 7, 8, 28.

4. Operations on Vector Bundles

As is the common practice in this book we use $\mathscr{E}$ to denote the category of finite dimensional vector spaces, and $\mathscr{E}(X)$ to denote the category of vector bundles over X. When we want to make the basic field $k=\mathbb{R}$ or $\mathbb{C}$ precise, we write $\mathscr{E}_k$ or $\mathscr{E}_k(X)$.

4.1. Definition. A functor $\varphi:\mathscr{E}\to\mathscr{E}$ is called *continuous* if for each pair (M,N) of objects in $\mathscr{E}$, the natural map $\varphi_{M,N}:\mathscr{E}(M,N)\to\mathscr{E}(\varphi(M),\varphi(N))$ is continuous (with respect to the usual topology on finite dimensional vector spaces).

4.2. Examples. There are many well known examples of such functors: a) $\varphi(M)=\underbrace{M\oplus\cdots\oplus M}_{i}$, b) $\varphi(M)=\underbrace{M\otimes\ldots\otimes M}_{i}$, c) $\varphi(M)=\lambda^i(M)$ (i^{th} exterior power), d) $\varphi(M)=S^i(M)$ (i^{th} symmetric power). To see that all these functors are continuous, we choose basis on M and N, and notice that $\varphi_{M,N}(\alpha)$ is given by a matrix which depends continuously on the matrix of α. As an example of a functor which is *not* continuous, one may take a discontinuous automorphism of the complex numbers $\mathbb{C}$. This induces a functor $\mathscr{E}_{\mathbb{C}}\to\mathscr{E}_{\mathbb{C}}$ which is not continuous.

4.3. The purpose of this section is to associate any such functor φ with a functor $\varphi'=\varphi(X):\mathscr{E}(X)\to\mathscr{E}(X)$, which coincides with φ when X is reduced to a point. If $\xi=(E,\pi,X)$ is a vector bundle over X, we first define the *set* $E'=\varphi'(E)$ to be the disjoint union $\bigsqcup_{x\in X}\varphi(E_x)$, provided with the obvious projection $\pi':\varphi'(E)\to X$. In order to supply $\varphi'(E)$ with a topology so that it becomes a vector bundle, we need the following lemma.

4.4. Lemma. *Let U and V be open subsets of X and let $\beta:E_U\to U\times M$ and $\gamma:E_V\to V\times N$ be trivializations of E over U and V respectively. Let $\beta':E'_U\to U\times\varphi(M)$ and $\gamma':E'_V\to V\times\varphi(N)$ be the bijections induced by functoriality on each fiber. If we* *give E'_U and E'_V the topologies induced by these bijections, these two topologies agree on $E'_U \cap E'_V = E'_{U \cap V}$, and $E'_{U \cap V}$ is open in both E'_U and E'_V.*

Proof. We have the commutative diagram

$$\begin{array}{ccc} E'_{U\cap V} & \xrightarrow{\beta'|_{U\cap V}} & (U\cap V)\times\varphi(M) \\ \| & & \downarrow \delta \\ E'_{U\cap V} & \xrightarrow{\gamma'|_{U\cap V}} & (U\cap V)\times\varphi(N), \end{array}$$

where δ is the composition of the continuous maps

$$U\cap V \xrightarrow{\hat{s}} \mathscr{E}(M,N) \xrightarrow{\varphi_{M,N}} \mathscr{E}(\varphi(M),\varphi(N))$$

with $\hat{s} = \gamma|_{U\cap V}\cdot\beta^{-1}|_{U\cap V}$. Since δ is continuous, $\hat{\delta}$ also is continuous (1.12). For the same reason δ^{-1} is continuous, and this shows that the two topologies on $E'_{U\cap V}$ agree. Moreover, the projection $\pi'_U: E'_U \to U$ is continuous, with respect to the topology induced by β'. Hence $E'_{U\cap V} = \pi'^{-1}_U(U\cap V)$ is open in E'_U, and similarly $E'_{U\cap V}$ is open in E'_V. □

4.5. We are now able to define the topology on $E' = \varphi'(E)$. Let (U_i) be an open cover of X, and let $\beta_i: E_{U_i} \to U_i \times M_i$ be a trivialization of E over U_i for each i. By functoriality, the isomorphisms β_i induce a bijection $E'_{U_i} \to U_i \times \varphi(M_i)$, and in this way E'_{U_i} may be provided with a topology. Now we provide E' with the largest topology making the inclusions $E'_U \to E'$ continuous. This is possible because according to the previous lemma for each pair (i, j) the topologies on E'_{U_i} and E'_{U_j} agree on $E'_{U_i\cap U_j}$, making it an open subset of E'_{U_i} and E'_{U_j}.

This topology does not depend on the choice of covering, nor on the choice of trivializations. In fact, if (V_r) is another covering and if $\psi_r: E_{V_r} \to V_r \times N_r$ are other trivializations, the same argument as before shows that the two possible topologies on $E'_{U_i\cap V_r}$ coincide and that $E'_{U_i\cap V_r}$ is open in both E'_{U_i} and E'_{V_r}. Hence the two topologies on E' coincide. Finally E' is locally trivial since $E'_{U_i} \approx U_i \times \varphi(M_i)$ is a trivial bundle, for each i.

4.6. In order to completely define the functor φ', we have to define what $f' = \varphi'(f): \varphi'(E) \to \varphi'(F)$ must be when $f: E \to F$ is a morphism of vector bundles. We simply define f' on each fiber by $f'_x = \varphi(f_x): \varphi(E_x) \to \varphi(F_x)$ which is linear. To prove that f' is continuous, we look at the commutative diagrams

$$\begin{array}{ccc} E_U & \xrightarrow[\approx]{\beta} & U\times M \\ {\scriptstyle f|_U}\downarrow & & \downarrow{\scriptstyle g} \\ F_U & \xrightarrow[\approx]{\delta} & U\times N \end{array} \quad \text{and} \quad \begin{array}{ccc} E'_U & \xrightarrow{\beta'} & U\times\varphi(M) \\ {\scriptstyle f'|_U}\downarrow & & \downarrow{\scriptstyle g'} \\ F'_U & \xrightarrow{\delta'} & U\times\varphi(N), \end{array}$$

where β', δ', and g' are again induced from β, δ, and g respectively by functoriality on each fiber. The map $\check{g}'$, induced from g', is the composition of the continuous maps

$$U \xrightarrow{\check{g}} \mathscr{E}(M, N) \longrightarrow \mathscr{E}(\varphi(M), \varphi(N)).$$

According to Theorem 1.12 the map g' is continuous, and hence f' is continuous.

4.7. Generalization. Let $\mathscr{C}$ be the category

$$\underbrace{\mathscr{E}^0_{\mathbb{R}} \times \cdots \times \mathscr{E}^0_{\mathbb{R}}}_{p_1} \times \underbrace{\mathscr{E}^0_{\mathbb{C}} \times \cdots \times \mathscr{E}^0_{\mathbb{C}}}_{p_2} \times \underbrace{\mathscr{E}_{\mathbb{R}} \times \cdots \times \mathscr{E}_{\mathbb{R}}}_{q_1} \times \underbrace{\mathscr{E}_{\mathbb{C}} \times \cdots \times \mathscr{E}_{\mathbb{C}}}_{q_2}$$

and let $\mathscr{C}'$ be the category

$$\underbrace{\mathscr{E}^0_{\mathbb{R}} \times \cdots \times \mathscr{E}^0_{\mathbb{R}}}_{p'_1} \times \underbrace{\mathscr{E}^0_{\mathbb{C}} \times \cdots \times \mathscr{E}^0_{\mathbb{C}}}_{p'_2} \times \underbrace{\mathscr{E}_{\mathbb{R}} \times \cdots \times \mathscr{E}_{\mathbb{R}}}_{q'_1} \times \underbrace{\mathscr{E}_{\mathbb{C}} \times \cdots \times \mathscr{E}_{\mathbb{C}}}_{q'_2}$$

where the notation ^{0} means the opposite category (same objects but arrows reversed). A functor $\varphi: \mathscr{C} \to \mathscr{C}'$ is called continuous if for each pair (R, S) of objects of $\mathscr{C}$, the map $\mathscr{C}(R, S) \to \mathscr{C}'(\varphi(R), \varphi(S))$ is continuous. Then the same method as before shows how to define a functor $\varphi' = \varphi(X): \mathscr{C}(X) \to \mathscr{C}'(X)$ where

$$\mathscr{C}(X) = \underbrace{\mathscr{E}^0_{\mathbb{R}}(X) \times \cdots \times \mathscr{E}^0_{\mathbb{R}}(X)}_{p_1} \times \underbrace{\mathscr{E}^0_{\mathbb{C}}(X) \times \cdots \times \mathscr{E}^0_{\mathbb{C}}(X)}_{p_2}$$
$$\times \underbrace{\mathscr{E}_{\mathbb{R}}(X) \times \cdots \times \mathscr{E}_{\mathbb{R}}(X)}_{q_1} \times \underbrace{\mathscr{E}_{\mathbb{C}}(X) \times \cdots \times \mathscr{E}_{\mathbb{C}}(X)}_{q_2}$$

and

$$\mathscr{C}'(X) = \underbrace{\mathscr{E}^0_{\mathbb{R}}(X) \times \cdots \times \mathscr{E}^0_{\mathbb{R}}(X)}_{p'_1} \times \underbrace{\mathscr{E}^0_{\mathbb{C}}(X) \times \cdots \times \mathscr{E}^0_{\mathbb{C}}(X)}_{p'_2}$$
$$\times \underbrace{\mathscr{E}_{\mathbb{R}}(X) \times \cdots \times \mathscr{E}_{\mathbb{R}}(X)}_{q'_1} \times \underbrace{\mathscr{E}_{\mathbb{C}}(X) \times \cdots \times \mathscr{E}_{\mathbb{C}}(X)}_{q'_2}$$

If the composition $\varphi_2 \cdot \varphi_1$ is well defined, we have $(\varphi_2 \cdot \varphi_1)(X) \approx \varphi_2(X) \cdot \varphi_1(X)$. Finally, if φ_1 and φ_2 are isomorphic functors, then $\varphi_1(X)$ and $\varphi_2(X)$ are also isomorphic.

4.8. Examples. a) The functor $\varphi: \mathscr{E} \times \mathscr{E} \to \mathscr{E}$ given by $\varphi(M, N) = M \oplus N$ induces $\varphi(X): \mathscr{E}(X) \times \mathscr{E}(X) \to \mathscr{E}(X)$. If E and F are vector bundles on X, then the bundle $\varphi(X)(E, F)$ is denoted by $E \oplus F$, and is called the *Whitney sum* of the bundles

E and F. It is easy to see that $E \oplus F$ is isomorphic to the fiber product $E \times_X F$. Moreover, the classical identities for vector spaces imply canonical isomorphisms $(E \oplus F) \oplus G \approx E \oplus (F \oplus G)$ and $E \oplus F \approx F \oplus E$ from 4.7.

b) Let $\varphi: \mathscr{E}_k \times \mathscr{E}_k \to \mathscr{E}_k$ be the functor defined by $\varphi(M, N) = M \otimes_k N$. Then $\varphi(X)(E, F) = E \otimes F$ is the *tensor product* of E and F. Again we have canonical isomorphisms $(E \otimes F) \otimes G \approx E \otimes (F \otimes G)$ and $E \otimes F \approx F \otimes E$.

c) If $\varphi: \mathscr{E}^0 \times \mathscr{E} \to \mathscr{E}$ is the functor $(M, N) \mapsto \mathscr{E}(M, N)$, then the object $\varphi(X)(E, F) = \mathrm{HOM}(E, F)$ is called the *vector bundle of homomorphisms* between E and F (the fiber over a point x of X is $\mathscr{E}(E_x, F_x) = \mathrm{Hom}_{\mathscr{E}}(E_x, F_x)$).

d) The "duality functor" $M \mapsto M^*$ from $\mathscr{E}_k^0$ to $\mathscr{E}_k$ induces another *duality functor* $E \mapsto E^*$ from $\mathscr{E}_k^0(X)$ to $\mathscr{E}_k(X)$. Of course, we have the identity $\mathrm{HOM}(E, F) \approx E^* \otimes_k F$.

e) We also have the *conjugate* functor t from $\mathscr{E}_{\mathbb{C}}(X)$ to $\mathscr{E}_{\mathbb{C}}(X)$, induced by the functor $M \mapsto \overline{M}$ which associates each complex vector space with its conjugate. Let $c: \mathscr{E}_{\mathbb{R}}(X) \to \mathscr{E}_{\mathbb{C}}(X)$ be the "*complexification functor*", induced by the functor $\mathscr{E}_{\mathbb{R}} \to \mathscr{E}_{\mathbb{C}}$ defined by $M \mapsto M \otimes_{\mathbb{R}} \mathbb{C}$. Let $r: \mathscr{E}_{\mathbb{C}}(X) \to \mathscr{E}_{\mathbb{R}}(X)$ be the "*realification functor*", which is induced by the functor $\mathscr{E}_{\mathbb{C}} \to \mathscr{E}_{\mathbb{R}}$ which associates each complex vector space with its underlying real vector space. Then $(rc)(E)$ is naturally isomorphic to $E \oplus E$ and $(cr)(E)$ is naturally isomorphic to $E \oplus \overline{E}$.

f) Examples c) and d) in 4.2 enable us to define operations λ^i and S^i in the category of vector bundles. These operations are very important in the application of K-theory described in the last chapter of this book.

4.9. Let E and F be vector bundles with bases X and Y respectively. We define the "*external Whitney sum*" of E and F as the vector bundle $E \boxplus F$ on $X \times Y$, where $E \boxplus F = \pi_1^*(E) \oplus \pi_2^*(F)$ with $\pi_1: X \times Y \to X$ and $\pi_2: X \times Y \to Y$. Obviously we have $E \boxplus F = E \times F$ and $(E \boxplus F)_{(x,y)} = E_x \oplus F_y$. In the same way, the "*external tensor product*" of E and F is the vector bundle $E \boxtimes F$ on $X \times Y$, where $E \boxtimes F = \pi_1^*(E) \otimes \pi_2^*(F)$. Similarly $(E \boxtimes F)_{(x,y)} = E_x \otimes F_y$.

Exercises (Section I.9) 9–12, 30.

5. Sections of Vector Bundles

5.1. Definition. Let $\xi = (E, \pi, X)$ be a vector bundle. Then a *section* of ξ is a map $s: X \to E$ such that $\pi \cdot s = \mathrm{Id}_X$. A section s is called *continuous* if s is a continuous map (very often sections will refer to continuous sections since we do not consider other types of sections in this book).

5.2. Example. Let $s: X \to E$ be the map which associates each point x of X with the vector 0 of the vector space E_x. If $\beta: E_U \to U \times M$ is a trivialization of E over an open set U, we have $(\beta \cdot s)(x) = (x, 0)$ for $x \in U$. It follows that s is a continuous section of E. This section is called the *zero section* of the vector bundle.

5.3. Example. Let us suppose that E is the trivial bundle $X \times M$. Then a continuous section of E may be written as $x \mapsto (x, s_1(x))$, where $s_1 : X \to M$ is a continuous map. Conversely, any such continuous map induces a continuous section of E. In this way, we see that the notion of a section of a vector bundle is, in some sense, a generalization of the notion of a continuous map with values in a vector space.

5.4. Let $s_1, \ldots, s_n$ be n (continuous) sections of the vector bundle E. Let $\alpha : X \times k^n \to E$ be the morphism defined by $\alpha(x, \lambda_1, \ldots, \lambda_n) = \sum_{i=1}^{n} \lambda_i s_i(x)$ (where the sum is taken in the vector space E_x). These sections are called linearly independent if $s_1(x), \ldots, s_n(x)$ are linearly independent for each point x. If the rank of E is equal to n, then α induces an isomorphism on each fiber and hence is an isomorphism by 2.7.

5.5. Example. Let us consider once more the vector bundle $E = TS^{n-1}$ of 2.3 (with a slight change of notation). In V.2 we prove that TS^{n-1} admits exactly $\rho(n)-1$ linearly independent sections and not $\rho(n)$, where $\rho(n)$ is the arithmetic function of n defined by $\rho(n) = 2^\gamma + 8\delta$, where $n = k(2^{\gamma+4\delta})$, for k odd and $0 \leqslant \gamma < 3$. In particular, for $n-1$ even, there is no section $\neq 0$ everywhere, and TS^{n-1} is a trivial bundle if and only if $n-1 = 1$, 3, or 7

$n-1$	1	2	3	4	5	6	7	8	9	10	11
$\rho(n)-1$	1	0	3	0	1	0	7	0	1	0	3

On the other hand, if θ_1 is the trivial bundle of rank one on S^{n-1}, then $TS^{n-1} \oplus \theta_1$ is isomorphic to the trivial bundle θ_n of rank n over S^{n-1}. The isomorphism $TS^{n-1} \oplus \theta_1 \to \theta_n$ is given by $((x, v), \lambda) \mapsto (x, v + \lambda x)$, where $(x, v) \in TS^{n-1}$ and $TS^{n-1} \oplus \theta_1$ is identified with $TS^{n-1} \times \mathbb{R}$. Hence $TS^{n-1} \oplus \theta_1$ admits n continuous linearly independent sections.

5.6. Let E be a vector bundle with base X. We denote the set of continuous sections of E by $\Gamma(X, E)$. It is obviously a vector space under the operations $(s+t)(x) = s(x) + t(x)$ and $(\lambda s)(x) = \lambda s(x)$. Now let i: $Y \to X$ be a continuous map. By composition with i, a section s of E induces a section of the induced bundle $i^*(E) = Y \times_X E$ by the formula $t(y) = (y, (si)(y))$. The map $\Gamma(X, E) \to \Gamma(Y, i^*(E))$ is obviously k-linear. In particular, if Y is a subspace of X, the section t is the "restriction" of s to Y, and we denote it by $s|_Y$ or simply s_Y.

5.7. Theorem. *If X is paracompact and Y is closed in X, the restriction homomorphism $\Gamma(X, E) \to \Gamma(Y, E_Y)$ is surjective.*

Proof. The proof of this theorem splits into three parts:

a) *Suppose $E = X \times M$.* From Example 5.3, we see that the space of sections $\Gamma(X, E)$ may be identified with the vector space $F(X, M)$ of continuous maps

$s_1: X \to M$. From this point of view, the restriction homomorphism $\Gamma(X, E) \to \Gamma(Y, E_Y)$ may be interpreted as the restriction homomorphism of functions

$$F(X, M) \longrightarrow F(Y, M),$$

which is surjective due to the Tietze extension theorem (Bourbaki [1], Kelley [1]).

b) *Suppose E is isomorphic to $X \times M$.* Let us consider an isomorphism $E \approx F = X \times M$. Then we have the commutative diagram

$$\begin{array}{ccc} \Gamma(X, E) & \xrightarrow{\approx} & \Gamma(X, F) \\ \downarrow & & \downarrow \\ \Gamma(Y, E_Y) & \xrightarrow{\approx} & \Gamma(Y, F_Y) \end{array}$$

and the assertion follows from a).

c) *General case.* Let (U_i), $i \in I$, be a locally finite open cover of X such that E_{U_i} is trivial and let (V_i) be an open cover of X such that $\overline{V}_i \subset U_i$. Let $T_i = \overline{V}_i \cap Y$. If t is a section of E_Y, we let $t_i = t|_{T_i}$; then by *b*) we can choose a section s_i of $E|_{\overline{V}_i}$ such that $s_i|_{T_i} = t_i$. If (α_i) is a partition of unity associated with the cover (V_i), we set $s_i'(x) = \alpha_i(x) s_i(x)$ for $x \in \overline{V}_i$ and $s_i'(x) = 0$ otherwise. Then s_i' is a continuous section which is zero over all but a finite number of the V_j. Therefore, the sum $\sum_{i \in I} s_i'(x)$ (taken in each fiber) is actually a finite sum on a neighbourhood of each point x, and defines a continuous section s of E. For $x \in Y$, we have

$$s(x) = \sum_{i \in I} \alpha_i(x) s_i'(x) = \sum_{i \in I} \alpha_i(x) t(x) = \Big(\sum_{i \in I} \alpha_i(x)\Big) t(x) = t(x).$$

Hence $s|_Y = t$. □

5.8. If E and F are vector bundles over X, let us write $\mathrm{Hom}(E, F)$ for the set of morphisms from E to F. This set is obviously a vector space under the operations $(f+g)_x = f_x + g_x$ and $(\lambda f)_x = \lambda f_x$. The correspondence $(E, F) \mapsto \mathrm{Hom}(E, F)$ induces a functor from $\mathscr{E}_k(X)^0 \times \mathscr{E}_k(X)$ to the category of k-vector spaces (of arbitrary dimension). On the other hand, the correspondence $(E, F) \mapsto \Gamma(X, \mathrm{HOM}(E, F))$ (cf. 4.8.c) also induces a functor from $\mathscr{E}_k(X)^0 \times \mathscr{E}_k(X)$ to the same category.

5.9. Theorem. *The functors $(E, F) \mapsto \mathrm{Hom}(E, F)$ and $(E, F) \mapsto \Gamma(X, \mathrm{HOM}(E, F))$ are isomorphic.*

Proof. Let $\alpha: E \to F$ be a morphism. Then the map $x \mapsto \alpha_x$ defines a section s of the vector bundle $\mathrm{HOM}(E, F)$. To see that s is continuous, let us consider trivializations $\beta: E_U \xrightarrow{\approx} U \times M$ and $\gamma: F_U \xrightarrow{\approx} U \times N$, where U is an open set in X.

Then we have the commutative diagram

$$\begin{array}{ccc} U\times M & \xrightarrow{\hat{g}} & U\times N \\ \uparrow{\scriptstyle\beta} & & \uparrow{\scriptstyle\gamma} \\ E_U & \xrightarrow{\alpha|_U} & F_U \end{array}$$

which defines a continuous map g by 1.12. Carrying the structure, we see that $\mathrm{HOM}(E,F)|_U \approx \mathrm{HOM}(E_U, F_U) \approx U\times \mathscr{E}(M,N)$. Thus $s|_U$ may be identified with the section of the trivial bundle $U\times\mathscr{E}(M,N)$ defined by $x\mapsto(x,g(x))$. Hence s is continuous.

Conversely, if s is a continuous section of $\mathrm{HOM}(E,F)$, then s defines a map $\alpha: E\to F$ by the formula $\alpha_x = s(x)$. The same method as before shows that α is continuous. □

5.10. Theorem. *Let E and F be vector bundles over a paracompact space X, let Y be a closed subset of X, and let $\alpha: E_Y\to F_Y$ be a morphism of vector bundles. Then there exists a morphism $\tilde{\alpha}: E\to F$ such that $\tilde{\alpha}|_Y=\alpha$* ($\tilde{\alpha}$ is called an "*extension*" of α to X).

Proof. This theorem follows from Theorem 5.7 applied to the vector bundle $\mathrm{HOM}(E,F)$, due to the commutativity of the diagram

$$\begin{array}{ccc} \mathrm{Hom}(E,F) & \xrightarrow{\approx} & \Gamma(X, \mathrm{HOM}(E,F)) \\ \downarrow & & \downarrow \\ \mathrm{Hom}(E_Y,F_Y) & \xrightarrow{\approx} & \Gamma(Y, \mathrm{HOM}(E_Y,F_Y)) \end{array}$$

which follows from 5.9. □

5.11. Corollary. *With the notation of Theorem 5.10, let us suppose that α is an isomorphism. Then there exists a neighbourhood V of Y and an isomorphism $\alpha': E_V\to F_V$ such that $\alpha'_Y=\alpha$.*

Proof. Let $\tilde{\alpha}: E\to F$ be an extension of α and let V be the set of points x of X such that $\tilde{\alpha}_x$ is an isomorphism. Let us prove first that V is an open neighbourhood of Y. In fact, if $x\in V$, we can find an open neighbourhood W_x of x such that $E|_{W_x}\approx F|_{W_x}\approx W_x\times k^n$. Inducing the vector bundle structure, we see that $\beta_x=\tilde{\alpha}|_{W_x}$ is a morphism from $W_x\times k^n$ to $W_x\times k^n$ which is represented by a continuous map $\check{\beta}_x: W_x\to\mathscr{E}(k^n,k^n)$. Hence $V\cap W_x$ may be identified with the set of points v of W_x such that $\check{\beta}_x(v)\in\mathrm{Iso}(k^n,k^n)$. Since $\mathrm{Iso}(k^n,k^n)$ is an open subset of $\mathscr{E}(k^n,k^n)$, we see that $V\cap W_x$ is open in W_x, and hence in X. It follows that $V=\bigcup_{x\in V} V\cap W_x$ is an open subset of X which contains Y. Finally $\alpha'=\tilde{\alpha}|_V$ is an isomorphism by 2.7. □

5.12. For an application, one may read in advance, 7.1 to 7.6.

5.13. Theorem. *Let X be a paracompact space, let E and F be vector bundles over X, and let $\alpha: E \to F$ be a morphism such that $\alpha_x: E_x \to F_x$ is surjective for each point x of X. Then there exists a morphism $\beta: F \to E$ such that $\alpha\beta = \mathrm{Id}_F$.*

Proof. Let x be a point of X, and let U be a neighbourhood of x such that E_U and F_U are trivial. Then we may identify E_U and F_U with $U \times M$ and $U \times N$ respectively. Under this identification, the morphism $\alpha_U: U \times M \to U \times N$ may be written as $\hat{\theta}$, where $\theta: U \to \mathscr{E}(M, N)$ is a continuous map. If we write M in the form $N \oplus \mathrm{Ker}(\theta(x))$, then the map $\theta(y): N \oplus \mathrm{Ker}(\theta(x)) \to N$ may be represented by the matrix

$$(\theta_1(y), \theta_2(y)),$$

where θ_1 and θ_2 are continuous functions of y such that $\theta_1(x) = 1$ and $\theta_2(x) = 0$. Since $\mathrm{Aut}(N)$ is open in $\mathrm{End}(N)$, there exists a neighbourhood V_x of x such that $\theta_1(y)$ is an isomorphism for $y \in U_x$. Let us now consider the map $\theta'_x: V_x \to \mathscr{E}(N, M)$, which is represented by the matrix

$$\theta'_x(y) = \begin{pmatrix} \theta_1(y)^{-1} \\ 0 \end{pmatrix}.$$

Then θ'_x induces a morphism $\hat{\theta}'_x: F|_{V_x} \to E|_{V_x}$ such that $\alpha_{V_x} \cdot \hat{\theta}'_x = \mathrm{Id}$. Varying the point x, we construct a locally finite open cover (V_i) of X and morphisms $\beta_i: F_{V_i} \to E_{V_i}$ such that $\alpha_{V_i} \cdot \beta_i = \mathrm{Id}\, F_{V_i}$. Let (η_i) be a partition of unity associated with the cover (V_i), and let $\beta: F \to E$ be the map defined by $\beta(e) = \sum_{i \in I} \eta_i(x)\beta_i(e)$ for $e \in E_x$. In this formula we use the convention that $\eta_i(x)\beta_i(e) = 0$ for $x \notin V_i$. Therefore, in a suitable neighbourhood of x, we have $\eta_i(x)\beta_i(e) = 0$ except for a finite number of indices i, and thus β is continuous. Finally,

$$(\alpha \cdot \beta)(e) = \sum_{i \in I} \eta_i(x)(\alpha \cdot \beta_i)(e) = \Big(\sum_{i \in I} \eta_i(x)\Big)(\alpha \cdot \beta_i(e)) = e. \square$$

5.14. Theorem. *Let X be a paracompact space, let E and F be two vector bundles over X, and let $\alpha: E \to F$ be a morphism such that $\alpha_x: E_x \to F_x$ is injective for each point x. Then there exists a morphism $\beta: F \to E$ such that $\beta \cdot \alpha = \mathrm{Id}_E$.*

Proof. The proof of this theorem is completely analogous to the proof above and is left as an exercise for the reader. $\square$

Exercises (Section I.9) 13–15, 27.

6. Algebraic Properties of the Category of Vector Bundles

6.1. Theorem. *The category $\mathscr{E}(X)$ of vector bundles with base X is an additive category.*

Proof. If E and F are vector bundles, we already know that $\mathrm{Hom}(E, F) = \mathscr{E}(X)(E, F)$ is a vector space, and a fortiori, an abelian group (5.8). It is clear that the map

$$\mathrm{Hom}(M, N) \times \mathrm{Hom}(N, P) \longrightarrow \mathrm{Hom}(M, P)$$

given by the composition of morphisms, is bilinear, and that there exists a zero object which is the trivial bundle of rank 0 $\xi = (X, \mathrm{Id}_X, X)$. The main point of the proof is to show that the Whitney sum $E_1 \oplus E_2$ of E_1 and E_2 is actually the "sum" of E_1 and E_2 in the category $\mathscr{E}(X)$. For this, we need the canonical morphisms $i_\alpha: E_\alpha \to E_1 \oplus E_2$ for $\alpha = 1, 2$, given by the obvious homomorphisms $(E_\alpha)_x \to (E_1)_x \oplus (E_2)_x$. Let $f_\alpha: E_\alpha \to F$ be arbitrary morphisms in $\mathscr{E}(X)$. We must prove that there exists a unique morphism $f: E_1 \oplus E_2 \to F$ which makes the diagram

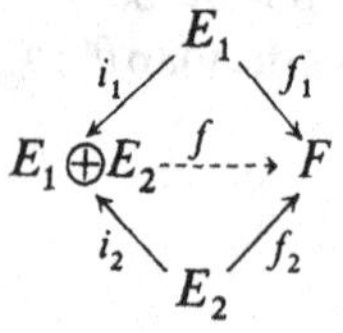

commutative.

a) *Uniqueness of f.* Over each point x of X, we must have $(f_\alpha)_x = f_x \cdot (i_\alpha)_x$. This implies $f_x(e_1, e_2) = (f_1)_x(e_1) + (f_2)_x(e_2)$ for each $(e_1, e_2) \in (E_1)_x \oplus (E_2)_x = (E_1)_x \times (E_2)_x$.

b) *Existence of f.* Let us define a map $f: E_1 \oplus E_2 \to F$ by the formula above. To verify that f is continuous, we consider an open set U such that $E_\alpha|_U \approx U \times M_\alpha$ and $F|_U \approx U \times N$. Inducing the vector bundle structure, $f_\alpha|_U$ becomes $\hat{g}_\alpha$ where $g_\alpha: U \to \mathscr{E}(M, N)$ is a continuous map (1.12 again). In the same way, $f|_U$ becomes the map from $U \times (M_1 \oplus M_2)$ to $U \times N$ defined by $(x, (m_1, m_2)) \mapsto (x, g_1(m_1) + g_2(m_2))$, which is clearly continuous. Hence f is continuous. □

6.2. *Remarks and application.* Since $\mathscr{E}(X)$ is an additive category, $E_1 \oplus E_2$ is also the product of E_1 and E_2 in $\mathscr{E}(X)$ (but not in the category of topological spaces unless X is a point or $\varnothing$). If $E_1, \ldots, E_n$ and $F_1, \ldots, F_p$ are vector bundles over X, a morphism from $\bigoplus_{i=1}^{n} E_i$ to $\bigoplus_{j=1}^{p} F_j$ may be represented by a matrix (α_{ji}) where $\alpha_{ji} \in \mathrm{Hom}(E_i, F_j)$.

The following theorem will be very useful in this book.

6.3. Theorem. *Let E be a vector bundle over X, and let p be a projector of E (i.e. an endomorphism of E such that $p^2 = p$). Then the quasi-vector bundle* $\mathrm{Ker}\, p = \bigsqcup_{x \in X} \mathrm{Ker}\, p_x$ *(provided with the topology induced by the topology of E) is locally trivial.*

Proof. Since the question is "local", we may suppose that E is the trivial bundle $X \times M$. Let x_0 be a point of X and let $f: X \to \mathscr{E}(M, M)$ be the map defined by $f(x) = 1 - p_x - p_{x_0} + 2p_x p_{x_0}$. Since $p_{x_0} \cdot f(x) = f(x) \cdot p_x$, we have the commutative diagram

$$\begin{array}{ccccccc}
0 & \longrightarrow & \operatorname{Ker} p & \longrightarrow & X \times M & \xrightarrow{\ p\ } & X \times M \\
 & & \vdots & & \Big\downarrow \hat{f} & & \Big\downarrow \hat{f} \\
0 & \longrightarrow & X \times \operatorname{Ker} p_{x_0} & \longrightarrow & X \times M & \xrightarrow{\ p_0\ } & X \times M
\end{array}$$

where $p_0 = \mathrm{Id}_X \times p_{x_0}$. Since $f(x_0) = 1$, there exists a neighbourhood $V(x_0)$ of x_0 such that $f(x)$ is an automorphism for $x \in V(x_0)$ (because $\mathrm{Aut}(M)$ is open in $\mathrm{End}(M)$). Over $V(x_0)$, $\hat{f}$ induces a homeomorphism between $\operatorname{Ker} p$ and $V(x_0) \times \operatorname{Ker} p_{x_0}$, since one can define a continuous map $(\hat{f}_{V(x_0)})^{-1} = (\widehat{f^{-1}_{V(x_0)}})$ by 1.12. □

6.4. *Remarks.* It is easy to verify that the vector bundle $\operatorname{Ker} p$, which is defined in this way, is the kernel of p in the categorical sense. If $p^2 \neq p$, the kernel of p does not exist in general.

* The more advanced reader will notice that the proof of Theorem 6.3 also holds for Banach bundles (Lang [2]).*

6.5. Theorem. *Let E be a vector bundle with compact base X. Then there exists a vector bundle E' such that $E \oplus E'$ is trivial.*

Proof. Let (U_i) for $i = 1, \ldots, r$ be a finite open cover of X such that $E_{U_i} \approx U_i \times k^{n_i}$, and let (η_i) be a partition of unity associated with the cover (U_i). According to 5.4, there exist n_i linear independent sections $s_i^1, s_i^2 \ldots, s_i^{n_i}$ of E_{U_i}. The sections $\eta_i s_i^1, \eta_i s_i^2, \ldots, \eta_i s_i^{n_i}$, extended by 0 outside U_i, are also n_i linearly independent sections of $E|_{V_i}$, where $V_i = \eta_i^{-1}((0, 1])$. Let σ_i^j be the sections $\eta_i s_i^j$, for $1 \leqslant j \leqslant n_i$. Then the vectors $\sigma_i^j(x)$ generate E_x as a vector space, and by 5.4, we have a morphism

$$\alpha: T = X \times k^n \longrightarrow E$$

for $n = \sum_{i=1}^{r} n_i$, such that $\alpha_x: T_x \to E_x$ is surjective for each point x of X. According to Theorem 5.13, there exists a morphism $\beta: E \to T$ such that $\alpha \cdot \beta = \mathrm{Id}_E$. Let E' be the kernel of the projector $p = \beta \cdot \alpha$. Now we may conclude the proof in two slightly different ways:

(i) Since $\mathscr{E}(X)$ is an additive category and since $E \approx \operatorname{Ker}(1-p)$, we have $E \oplus E' \approx T$.

(ii) We define a morphism from $E \oplus E'$ to T by the sum of the inclusion of E' in T and the map β. This morphism induces an isomorphism on each fiber because $E'_x \approx \operatorname{Ker} p_x$ and $E_x \approx \operatorname{Ker}(1 - p_x)$. Hence the morphism is an isomorphism by 2.7. □

*6.6. *Remark*. According to Hurewicz and Wallmann [1], a topological space X is of dimension $\leqslant p$, if for every finite open cover $\mathcal{U}$ of X, there exists a finer cover (V_i), such that each point of X belongs to at most $(p+1)$ sets of the cover (V_i). A space is of dimension p if it is of dimension $\leqslant p$ and not of dimension $\leqslant p-1$. A differential manifold modelled on $\mathbb{R}^p$ is of dimension p. A CW-complex with cells contained in $\mathbb{R}^p$ is also of dimension $\leqslant p$. If E is a vector bundle over a compact connected space of dimension p, there exists a finite cover (V_i), $i=1,\ldots,p$, such that E_{V_i} is trivial. Hence E is a direct factor of a trivial bundle of rank $\leqslant p \times \operatorname{rank}(E)$.*

6.7. Definition. Let $\mathscr{C}$ be an additive category. Then $\mathscr{C}$ is called *pseudo-abelian*, if for every object E of $\mathscr{C}$ and for every morphism $p: E \to E$ such that $p^2 = p$, the kernel of p exists.

6.8. Examples. As we have just shown, the category $\mathscr{E}(X)$ is pseudo-abelian. From another point of view, if A denotes an arbitrary ring with unit, the category $\mathscr{P}(A)$ of finitely generated projective modules over A[2)] is pseudo-abelian since a direct factor of a projective module is again projective. The category $\mathscr{L}(A)$ of finitely generated *free* modules is not pseudo-abelian in general (*ex*: $A = k \times k$, $M_n(k), \ldots$).

6.9. Proposition. *Let $\mathscr{C}$ be a pseudo-abelian category, let E be an object of $\mathscr{C}$ and let $p: E \to E$ be such that $p^2 = p$. Then the object E splits into the direct sum $E = \operatorname{Ker}(p) \oplus \operatorname{Ker}(1-p)$. Relative to this decomposition, the endomorphism p takes the matrix form*

$$p = \begin{pmatrix} 0 & 0 \\ 0 & 1 \end{pmatrix}$$

(i.e. $p = 0_{\operatorname{Ker}(p)} \oplus \operatorname{Id}_{\operatorname{Ker}(1-p)}$).

Proof. Let $i_1: \operatorname{Ker}(p) \to E$ and $i_2: \operatorname{Ker}(1-p) \to E$ be the canonical inclusions. By a well known lemma in the theory of additive categories (cf. Mitchell [1]), we must show the existence of $j_1: E \to \operatorname{Ker}(p)$ and $j_2: E \to \operatorname{Ker}(1-p)$ such that $j_1 \cdot i_1 = \operatorname{Id}_{\operatorname{Ker}(p)}$, $j_2 \cdot i_2 = \operatorname{Id}_{\operatorname{Ker}(1-p)}$, $j_1 \cdot i_2 = 0$, $j_2 \cdot i_1 = 0$ and finally $i_1 \cdot j_1 + i_2 \cdot j_2 = \operatorname{Id}_E$. We define j_1 and j_2 as the unique morphisms which make the diagrams

$$\begin{array}{ccccc} \operatorname{Ker}(p) & \xrightarrow{i_1} & E & \xrightarrow{p} & E \\ \uparrow{\scriptstyle \operatorname{Id}} & \nwarrow{\scriptstyle j_1} & & & \uparrow{\scriptstyle 1-p} \\ \operatorname{Ker}(p) & & \overset{i_1}{\dashrightarrow} & & E \end{array} \qquad \text{and} \qquad \begin{array}{ccccc} \operatorname{Ker}(1-p) & \xrightarrow{i_2} & E & \xrightarrow{1-p} & E \\ \uparrow{\scriptstyle \operatorname{Id}} & \nwarrow{\scriptstyle j_2} & & & \uparrow{\scriptstyle p} \\ \operatorname{Ker}(1-p) & & \overset{i_2}{\dashrightarrow} & & E \end{array}$$

---

2) Here we are considering right A-modules.

commutative. By the universal property of kernels, $i_1 j_1 i_1 = i_1$ and $i_2 j_2 i_2 = i_2$; hence, $j_1 \cdot i_1 = \mathrm{Id}_{\mathrm{Ker}(p)}$ and $j_2 \cdot i_2 = \mathrm{Id}_{\mathrm{Ker}(1-p)}$. Moreover, the universal property of kernels also shows that the following diagrams are commutative.

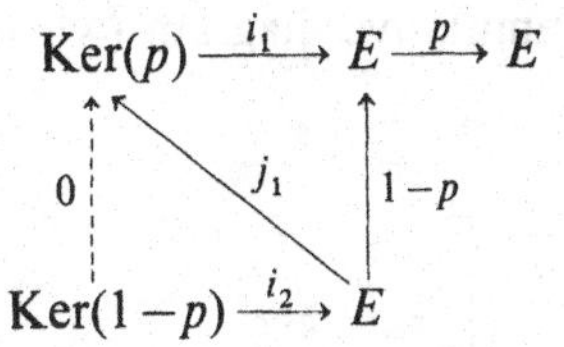

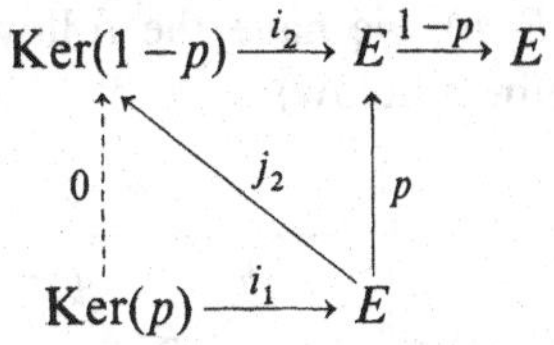

Hence $j_1 \cdot i_2 = 0$ and $j_2 \cdot i_1 = 0$. Finally $i_1 \cdot j_1 + i_2 \cdot j_2 = p + (1-p) = 1$. Since $p = i_2 \cdot j_2$, the matrix of p is necessarily of the form stated. $\square$

Now we are going to describe a universal procedure of imbedding an additive category in a pseudo-abelian category. Its main application is Theorem 6.18.

6.10. Theorem. *Let $\mathscr{C}$ be an additive category. Then there exists a pseudo-abelian category $\tilde{\mathscr{C}}$, and an additive functor $\varphi: \mathscr{C} \to \tilde{\mathscr{C}}$ which is fully faithful and which satisfies the following universal property: For every pseudo-abelian category $\mathscr{D}$ and for every additive functor $\psi: \mathscr{C} \to \mathscr{D}$, there exists an additive functor $\psi': \tilde{\mathscr{C}} \to \mathscr{D}$, which is unique up to isomorphism and which makes the following diagram*

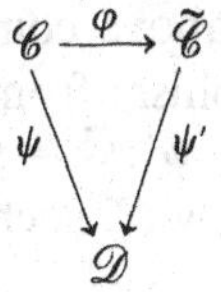

commutative up to isomorphism. Moreover, the pair $(\varphi, \tilde{\mathscr{C}})$ is unique up to equivalence of categories.

Proof. It is clear that the pair $(\varphi, \tilde{\mathscr{C}})$, as the solution of a "universal problem", is unique up to equivalence of categories (exercise left to the reader). Thus it suffices to explicitly construct $\tilde{\mathscr{C}}$ and φ.

The objects of $\tilde{\mathscr{C}}$ are the pairs (E, p) where $E \in \mathrm{Ob}(\mathscr{C})$ and p is a projector in E (when $\mathscr{C}$ is a category of modules for instance, one may think of this pair as the "image" of p). The morphisms from source (E, p) with target (F, q) are the $\mathscr{C}$-morphisms $f: E \to F$ such that $f \cdot p = q \cdot f = f$ (to understand "why" we make this definition, again one may think of (E, p) as the image of p so that in the decompositions $E = \mathrm{Im}\,p \oplus \mathrm{Im}(1-p)$, $F = \mathrm{Im}\,q \oplus \mathrm{Im}(1-q)$, the map f appears as a matrix

$$\begin{pmatrix} f_1 & 0 \\ 0 & 0 \end{pmatrix},$$

and hence actually defines a morphism f_1 from $\mathrm{Im}\,p$ to $\mathrm{Im}\,q$).

The composition of morphisms in $\tilde{\mathscr{C}}$ is induced by the composition of morphisms in $\mathscr{C}$, the identity morphisms of (E, p) is p, and the sum of two objects (E, p) and (F, q) is the object $(E \oplus F, p \oplus q)$. With these definitions it is clear that $\tilde{\mathscr{C}}$ is an additive category. Let us show that $\tilde{\mathscr{C}}$ is pseudo-abelian. If f is a projector of the object (E, p), we have the following commutative diagram (where F, q, g, and h are defined below).

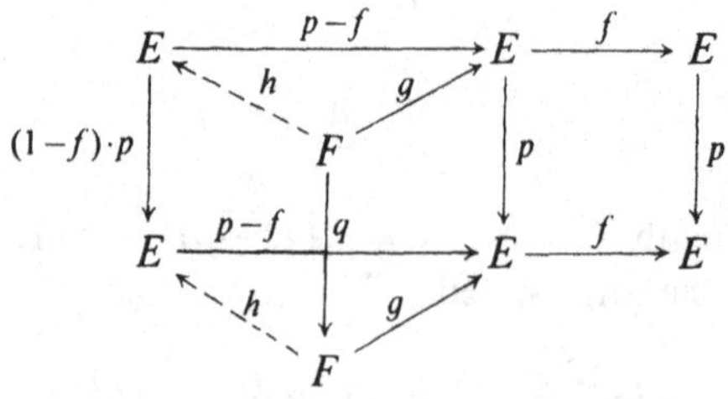

The pair $(E, (1-f)\cdot p)$ is an object of $\tilde{\mathscr{C}}$, and $p-f$ defines a morphism from this object to the object (E, p). In fact, the pair formed by the object $(E, (1-f)\cdot p)$ and the morphism $p-f$ is a kernel of f in the categorical sense. To see this, let us consider a third object (F, q), and a morphism $g\colon (F, q) \to (E, p)$ such that $f\cdot g = 0$. If $h\colon (F, q) \to (E, (1-f)\cdot p)$ is a morphism which makes the diagram commutative, we must have $h = (1-f)\cdot ph = p(1-f)h = pg = g$, which shows the uniqueness of h. Conversely, if we let $h = g$, then the diagram is obviously commutative.

Let us now define the functor $\varphi\colon \mathscr{C} \to \tilde{\mathscr{C}}$ by the formulas $\varphi(E) = (E, \mathrm{Id}_E)$ and $\varphi(f) = f$, on the objects and morphisms respectively. The above computation shows that (E, p) is the kernel of $1-p$, interpreted as a morphism from $\varphi(E)$ to $\varphi(E)$. Hence $\varphi(E) \approx (E, p) \oplus (E, 1-p)$. If $\psi\colon \mathscr{C} \to \mathscr{D}$ (resp. $\psi'\colon \tilde{\mathscr{C}} \to \mathscr{D}$) is an additive functor from $\mathscr{C}$ (resp. $\tilde{\mathscr{C}}$) to a pseudo-abelian category $\mathscr{D}$, such that the diagram

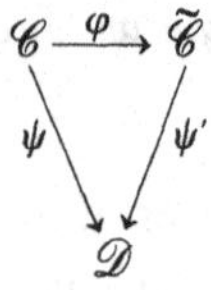

is commutative up to isomorphism, we have $\psi'(\operatorname{Ker} f) \approx \operatorname{Ker}(\psi'(f))$ for every projector f. Hence $\psi'(E, p) = \operatorname{Ker} \psi(1-p)\colon \psi(E) \to \psi(E)$ and $\psi'(f) = \psi(f)_{\operatorname{Ker}\psi(1-p)}$ on the objects and morphisms respectively. Conversely, these formulas define ψ' (up to isomorphism). $\square$

6.11. *Notation.* We call $\tilde{\mathscr{C}}$ the *pseudo-abelian category* associated with $\mathscr{C}$.

6.12. Theorem. *Let $\mathscr{C}$ be an additive category, $\mathscr{D}$ a pseudo-abelian category, and $\psi\colon \mathscr{C} \to \mathscr{D}$ an additive functor which is fully faithful such that every object of $\mathscr{D}$ is a direct factor of an object in the image of ψ. Then the functor ψ' defined in 6.9 is an equivalence between the categories $\tilde{\mathscr{C}}$ and $\mathscr{D}$.*

Proof. Let us first show that ψ' is essentially surjective. If G is an object of $\mathscr{D}$, there exists by hypothesis an object E of $\mathscr{C}$ and a projector $q: \psi(E) \to \psi(E)$, such that $G \approx \operatorname{Ker}(q)$. Since ψ is fully faithful, we may write q as $\psi(p)$, where p is a projector of E. Then G is isomorphic to $\psi'(E, 1-p)$ according to the formula for kernels given in 6.10 (substitute $1-p$ for p).

To prove that ψ' is fully faithful, let us consider two objects H and H' of $\tilde{\mathscr{C}}$ which are direct factors of $\varphi(E)$ and $\varphi(E')$ respectively. Then the diagram

$$\begin{array}{ccc} \mathscr{C}(\varphi(E), \varphi(E')) \approx \mathscr{C}(E, E') & \rightleftarrows & \mathscr{C}(H, H') \\ \searrow \psi'_{\varphi(E), \varphi(E')} \quad \downarrow \psi_{E, E'} & & \downarrow \psi'_{H, H'} \\ \mathscr{D}(\psi(E), \psi(E')) & \rightleftarrows & \mathscr{D}(\psi'(H), \psi'(H')) \end{array}$$

where the horizontal arrows are induced by the direct sum decompositions $\varphi(E) = H \oplus H_1$ and $\varphi(E') = H' \oplus H'_1$, shows that $\psi'_{H,H'}$ is an isomorphism, since $\psi_{E,E'}$ is an isomorphism by the hypothesis. Hence ψ' is a category equivalence. □

6.13. Theorem. *Let $\mathscr{C} = \mathscr{E}_T(X)$ be the full subcategory of $\mathscr{E}(X)$ whose objects are the trivial bundles. Then if X is compact, the pseudo-abelian category $\tilde{\mathscr{C}}$ associated with $\mathscr{C}$ is equivalent to the category $\mathscr{E}(X)$ of all vector bundles over X.*

Proof. Let $\mathscr{D}$ be the category $\mathscr{E}(X)$, and let $\psi: \mathscr{C} \to \mathscr{D}$ be the inclusion functor. According to Theorem 6.5, every object of $\mathscr{D}$ is a direct factor of some $\psi(G)$. Hence, by Theorem 6.12, the functor ψ' induces an equivalence $\widetilde{\mathscr{E}_T(X)} \sim \mathscr{E}(X)$. □

6.14. *Remark.* Theorem 6.12 gives us a purely algebraic way to describe vector bundles over a compact space X as images of projection operators. It is possible to prove all the main theorems of K-theory in Chapters II and III from this point of view. However, the vector bundles that we are generally interested in (for example the tangent bundle of a differentiable manifold) are not defined in this way. Nevertheless, this point of view will sometimes be convenient for theoretical purposes.

***6.15.** *Remark.* It follows from Remark 6.6, that Theorem 6.13 can be refined for X connected of finite dimension.*

6.16. Theorem. *Let A be an arbitrary ring with unit, and let $\mathscr{C} = \mathscr{L}(A)$ be the category considered in 6.8. Then $\tilde{\mathscr{C}}$ is equivalent to the category $\mathscr{P}(A)$ of finitely generated projective A-modules.*

Proof. This theorem is another consequence of Theorem 6.12 applied to the categories $\mathscr{D} = \mathscr{P}(A)$, since every object of $\mathscr{P}(A)$ is a direct factor of some A^n. □

6.17. Let $A = C_k(X)$ be the ring of continuous functions on a compact space X with values in k. If E is a k-vector bundle with base X, the set $\Gamma(X, E)$ of continuous sections of E is an A-module under the operation $(s \cdot \lambda)(x) = s(x)\lambda(x)$ where $\lambda \in A$

and $s \in \Gamma(X, E)$. If E is the trivial bundle $X \times k^n$, then $\Gamma(X, E)$ may be identified with A^n as an A-module. If E is an arbitrary vector bundle, and if $E \oplus E' \approx X \times k^n$, then $\Gamma(X, E) \oplus \Gamma(X, E') \approx \Gamma(X, E \oplus E') \approx A^n$ as A-modules. Hence $\Gamma(X, E)$ is an object of $\mathscr{P}(A)$, and the correspondence $E \mapsto \Gamma(X, E)$ induces an additive functor denoted by Γ from $\mathscr{E}(X)$ to $\mathscr{P}(A)$.

6.18. Theorem (Serre [2]–Swan [1]). *Let $A = C_k(X)$ be the ring of continuous functions on a compact space X with values in k. Then the section functor Γ induces an equivalence of categories $\mathscr{E}(X) \sim \mathscr{P}(A)$.*

Proof. The functor Γ induces a functor $\Gamma_T : \mathscr{E}_T(X) \to \mathscr{L}(A)$, where $\mathscr{E}_T(X)$ is defined in 6.13. Since $A^n \approx \Gamma_T(E)$ with $E = X \times k^n$, Γ_T is essentially surjective. If $F = X \times k^p$ and $f: E \to F$ is a morphism, then $\Gamma_T(f)$ is represented by the matrix $M(x) = (a_{ji}(x))$, for $i = 1, \ldots, n$ and $j = 1, \ldots, p$, and the map $x \mapsto M(x)$ coincides with $\check{f}: X \to \mathscr{E}_k(k^n, k^p)$ using the notation of 1.12. Moreover, Theorem 1.12 shows that Γ_T is fully faithful, and hence a category equivalence.

Let us put $\mathscr{C} = \mathscr{E}_T(X)$, $\mathscr{D} = \mathscr{P}(A)$, and let $\psi : \mathscr{C} \to \mathscr{D}$ be the composition of Γ_T and the inclusion factor of $\mathscr{L}(A)$ in $\mathscr{P}(A)$. Since the diagram

$$\begin{array}{ccc} \mathscr{E}_T(X) = \mathscr{C} & \longrightarrow & \tilde{\mathscr{C}} \sim \mathscr{E}(X) \\ \Big\downarrow \psi & \swarrow \Gamma & \\ \mathscr{P}(A) = \mathscr{D} & & \end{array}$$

is commutative, Γ may be identified with ψ' of Theorem 6.10. From Theorem 6.12, it now follows that Γ is a category equivalence. $\square$

∗ **6.19.** *Remark.* Let Y be a closed subspace of X. Then, by restriction of functions, $B = C_k(Y)$ is a $C_k(X)$-module. If we identify $\mathscr{E}_k(X)$ (resp. $\mathscr{E}_k(Y)$) with $\mathscr{P}(A)$ (resp. $\mathscr{P}(B)$), the restriction functor $\mathscr{E}_k(X) \to \mathscr{E}_k(Y)$ may be interpreted as the "extension of scalars" functor $\mathscr{P}(A) \to \mathscr{P}(B)$, defined by $M \mapsto M \otimes_A B$.∗

6.20. Let E be a vector bundle with base X. We now define a Banach space topology on the vector space $\Gamma(X, E)$. If $E = X \times k$, then $\Gamma(X, E) = A$ is actually a Banach algebra under the norm

$$\|s\| = \sup_{x \in X} |s(x)|$$

If E is arbitrary, we may regard $\Gamma(X, E)$ as a module over A. Let $u: A^n \to \Gamma(X, E)$ be a surjective A-module homomorphism and let us give E the quotient topology. Then the topology induced on $\Gamma(X, E)$ by this projection is independent of the choice of u. Suppose $u': A^{n'} \to \Gamma(X, E)$ is another choice. Since $\Gamma(X, E)$ is projective, there exist A-linear maps $v: A^n \to A^{n'}$ and $v': A^{n'} \to A^n$ such that the diagram

$$\begin{array}{ccc} A^n & \underset{v}{\overset{v'}{\leftrightarrows}} & A^{n'} \\ & {\scriptstyle u}\searrow \quad \swarrow{\scriptstyle u'} & \\ & \Gamma(X, E) & \end{array}$$

is commutative. Moreover v and v' are A-linear hence given by matrices, and thus continuous since A is a Banach algebra. Thus the diagram shows that the topology on $\Gamma(X, E)$ is independent of the choice of u.

A vector bundle morphism $f: E \to F$ induces an A-module homomorphism $\Gamma(X,f): \Gamma(X, E) \to \Gamma(X, F)$. If $u: A^n \to \Gamma(X, E)$ and $v: A^m \to \Gamma(X, F)$ are surjective A-linear maps, one can find $\bar{f}: A^n \to A^m$ which makes the diagram

$$\begin{array}{ccc} A^n & \xrightarrow{\bar{f}} & A^m \\ \downarrow & & \downarrow \\ \Gamma(X, E) & \xrightarrow{\Gamma(X,f)} & \Gamma(X, F) \end{array}$$

commutative.

Since $\bar{f}$ is continuous, f is continuous as an A-module homomorphism.

6.21. It is possible to interpret the topology of $\Gamma(X, E)$ in another way. Let (U_i) be a finite cover of X of closed subsets U_i. Each continuous section s of E induces continuous sections s_i of $E_i = E_{U_i}$ such that $s_i|_{U_i \cap U_j} = s_j|_{U_i \cap U_j}$. Conversely, let s_i be continuous sections of E_i such that $s_i|_{U_i \cap U_j} = s_j|_{U_i \cap U_j}$. Then, applying the same method as in 3.1, it can be shown that there exists a unique continuous section s of E, such that $s|_{U_i} = s_i$. In other words, we have the exact sequence

$$0 \to \Gamma(X, E) \longrightarrow \prod_i \Gamma(U_i, E_i) \xrightarrow{r_1 - r_2} \prod_{i,j} \Gamma(U_i \cap U_j, E_{ij})$$

where $E_{ij} = E|_{U_i \cap U_j}$, and r_1 and r_2 are induced by the restrictions $\Gamma(U_i, E_i) \to \Gamma(U_i \cap U_j, E_{ij})$ and $\Gamma(U_j, E_j) \to \Gamma(U_i \cap U_j, E_{ij})$ respectively. According to 6.20, r_1 and r_2 are continuous, thus $\operatorname{Ker}(r_1 - r_2)$ is a closed subspace of the Banach space $\prod_i \Gamma(U_i, E_i)$, and the canonical map $\Gamma(X, E) \to \operatorname{Ker}(r_1 - r_2)$ is bijective. By applying the Banach theorem or using a partition of unity to define a map from $\operatorname{Ker}(r_1 - r_2)$ to $\Gamma(X, E)$, we see that the Banach spaces $\Gamma(X, E)$ and $\operatorname{Ker}(r_1 - r_2)$ are isomorphic. In particular, if the vector bundles E_i are trivial, then $\Gamma(U_i, E_i) \approx C_k(U_i)^{n_i}$ where $E_i \approx U_i \times k^{n_i}$, and $\Gamma(X, E)$ appears as a closed subspace of $\prod_i C_k(U_i)^{n_i}$.

6.22. Let A be an arbitrary Banach algebra, and let M and N be objects of $\mathscr{P}(A)$. Then the vector space $\operatorname{Hom}_A(M, N)$ can be provided with a Banach space topology.

In fact, M and N can be provided with the norms induced by arbitrary A-linear surjections $A^m \to M$, $A^n \to N$. By the argument used in 6.20, the Banach space topology obtained on M and N is independent of the choice of these surjections. Moreover, $\mathrm{Hom}_A(M, N)$ is a closed subspace of $\mathscr{L}(M, N)$, the space of $\mathbb{R}$-linear continuous maps from M to N with the usual topology. Since M and N are projectives and finitely generated, the choice of decompositions $M \oplus M' \approx A^n$ and $N \oplus N' \approx A^m$ enables us to identify $\mathrm{Hom}_A(M, N)$ with a closed subspace of $\mathrm{Hom}_A(A^n, A^m) \approx A^{nm}$. Finally, if P is a third object of $\mathscr{P}(A)$, then the map

$$\mathrm{Hom}_A(M, N) \times \mathrm{Hom}_A(N, P) \longrightarrow \mathrm{Hom}_A(M, P),$$

given by the composition of morphisms, is k-bilinear and continuous (A is a Banach algebra over $k = \mathbb{R}$ or $\mathbb{C}$) since it is induced by the map

$$\mathscr{L}(M, N) \times \mathscr{L}(N, P) \longrightarrow \mathscr{L}(M, P)$$

In particular, let E, F and G be vector bundles with compact base X. If we choose $A = C_k(X)$, we see that $\mathrm{Hom}(E, F) \approx \mathrm{Hom}_A(\Gamma(X, E), \Gamma(X, F))$, $\mathrm{Hom}(F, G) \approx \mathrm{Hom}_A(\Gamma(X, F), \Gamma(X, G))$, and $\mathrm{Hom}(E, G) \approx \mathrm{Hom}_A(\Gamma(X, E), \Gamma(X, G))$. Hence $\mathrm{Hom}(E, F)$, $\mathrm{Hom}(F, G)$, $\mathrm{Hom}(E, G)$ are naturally Banach spaces, and the map

$$\mathrm{Hom}(E, F) \times \mathrm{Hom}(F, G) \longrightarrow \mathrm{Hom}(E, G)$$

is k-bilinear and continuous.

6.23. *Remark.* The isomorphism $\Gamma(X, \mathrm{Hom}(E, F)) \approx \mathrm{Hom}(E, F)$ given in 5.9, is compatible with the natural Banach structure put on the two factors (Hint: write E and F as direct factors of trivial bundles).

6.24. To conclude this section let us consider the pseudo-abelian category $\widetilde{\mathscr{E}_T(X)}$ associated with $\mathscr{E}_T(X)$, for any topological space X. The same ideas as those used in the proof of Theorem 6.12 may be applied to show that $\widetilde{\mathscr{E}_T(X)}$ is equivalent to the full subcategory of $\mathscr{E}(X)$, whose objects are direct factors of trivial bundles. We will denote this category by $\mathscr{E}'(X)$. In general $\mathscr{E}'(X)$ is not equivalent to $\mathscr{E}(X)$. However, we will see that some of the properties of $\mathscr{E}(X)$ are also present in the subcategory $\mathscr{E}'(X)$. More precisely, we prove the following proposition:

6.25. Proposition. *Let (U_i), for $i = 1, \ldots, n$, be a finite open cover of a topological space X, such that there exists a partition of unity associated with (U_i), and let $E_i \in \mathrm{Ob}\,\mathscr{E}'(U_i)$. Let*

$$g_{ji}\colon E_i|_{U_i \cap U_j} \longrightarrow E_j|_{U_i \cap U_j}$$

be isomorphisms such that $g_{ki} = g_{kj} \cdot g_{ji}$ over $U_i \cap U_j \cap U_k$. Then the vector bundle obtained by clutching the E_i using the g_{ji} (cf. 3.2), is an object of $\mathscr{E}'(X)$.

Proof. Let us identify $\mathscr{E}'(Y)$ with $\widetilde{\mathscr{E}_T}(Y)$ for each space Y, so that E_i may be written as (M_i, p_i), where M_i is a trivial bundle and p_i is a projection operator. Then we consider $E=(M,p)$, where $M=M_1 \oplus \cdots \oplus M_n$ and p is the projection operator represented by the matrix (p_{kl}) defined as follows. Let (α_k) be a partition of unity associated with (U_k), and let $\beta_k=\sqrt{\alpha_k}$. Then we define

$$p_{kl}=\beta_k \beta_l g_{kl} p_l,$$

where as usual we use the convention that $g_{kl}=0$ outside $U_k \cap U_l$. We define $g_i: (M_i, p_i) \to (M,p)$ as the column matrix $(g_i)_k=\beta_k g_{ki} p_i$, and $f_i: (M,p) \to (M_i, p_i)$ as the row matrix $(f_i)_k=\beta_k g_{ik} p_k$. Then, a direct computation shows that f_i and g_i are inverse homomorphisms over U_i. We have $g_j^{-1} \cdot g_i = f_j \cdot g_i = \sum_k \beta_k g_{jk} p_k \beta_k g_{ki} p_i = \sum_k (\beta_k)^2 g_{ji} = g_{ji}$ over $U_i \cap U_j$. From 3.3 it follows that (M, p) is the bundle obtained by clutching the bundles E_i. □

6.26. Example. Let X be a locally compact space, let X_1 be open relatively compact, and let $X=X_1 \cup X_2$ where X_2 is open. Let K be the compact space $\overline{X}_1$, and let α_1', α_2' be a partition of unity associated with the cover of K defined by X_1 and $X_2 \cap K$. Then $\alpha_2'|_{K-X_1}=1$ and may be extended by 1 on X outside K. If α_2 denotes this extension, then the pair (α_1', α_2) defines a partition of unity associated with (X_1, X_2). In particular if $E_1 \in \operatorname{Ob}\mathscr{E}'(X_1)$, $E_2 \in \operatorname{Ob}\mathscr{E}'(X_2)$, and $\alpha: E_1|_{X_1 \cap X_2} \to E_2|_{X_1 \cap X_2}$ is an isomorphism, then the clutching of E_1 and E_2 using α is a direct factor of a trivial bundle.

7. Homotopy Theory of Vector Bundles

7.1. Theorem. *Let X be a compact space and let E be a vector bundle over $X \times I$ where $I=[0,1]$. Let $\alpha_t: X \to X \times I$ and $\Pi: X \times I \to X$ be the maps defined by $\alpha_t(x)=(x, t)$ and $\Pi(x, u)=x$, for $x \in X$ and $(t, u) \in I^2$. Then the vector bundles $E_0=\alpha_0^*(E)$ and $E_1=\alpha_1^*(E)$ are isomorphic.*

Proof. Let $E_t=\alpha_t^*(E)$. Then the vector bundles E and $\Pi^*(E_t)$ are isomorphic over the subset $X \times \{t\}$ of $X \times I$. According to 5.11, there exists a neighbourhood V of $X \times \{t\}$ in $X \times I$ such that $E|_V$ and $\Pi^*(E_t)|_V$ are isomorphic. Since X is compact, V must contain a subset of the form $X \times U$ where U is a neighbourhood of t in I. Hence, for t fixed, there is a neighbourhood U of t such that $E_u \approx E_t$ for $u \in U$. Now the connectivity of I implies that E_0 and E_1 must be isomorphic. □

7.2. Theorem. *Let X be a compact space, and let $f_0, f_1: X \to Y$ be two continuous maps which are homotopic. If E is a vector bundle over Y, then the vector bundles $f_0^*(E)$ and $f_1^*(E)$ are isomorphic (see 2.6 for the definition of f^* in general).*

Proof. Let $F: X \times I \to Y$ be a continuous map such that $f_t = F \cdot \alpha_t$ for $t = 0, 1$. Then $f_0^*(E) = (F \cdot \alpha_0)^*(E) = \alpha_0^*(F^*(E)) = (F^*(E))_0$, and $f_1^*(E) = (F \cdot \alpha_1)^*(E) = \alpha_1^*(F^*(E)) = (F^*(E))_1$. Now we apply the above theorem to the bundle $F^*(E)$ over $X \times I$. □

7.3. Theorem. *With the notation of 7.1, the bundles E and $\Pi^* E_0$ are isomorphic.*

Proof. Let $p: X \times I \times I \to X \times I$ be the map defined by $p(x, t, u) = (x, tu)$. Then $E' = p^*(E)$ is a bundle over $X' \times I$, where $X' = X \times I$. But $E'_0 \approx \Pi^* E_0$ and $E'_1 \approx E$. Hence E and $\Pi^* E_0$ are isomorphic by Theorem 7.1 applied to bundles over $X' \times I$. □

7.4. Theorem. *Let X be a contractible compact space. Then every bundle over X is trivial.*

Proof. Let x_0 be a point of X such that if $i: X \to \{x_0\}$ and $j: \{x_0\} \to X$ are the obvious maps, then the map $j \cdot i$ is homotopic to the identity of X. Now $E \approx i^*(j^*(E))$ for any vector bundle E over X. Since the bundle $F = j^*(E)$ is trivial, the bundle $E = i^*(F)$ is also trivial. □

7.5. *Remark.* With much more sophisticated arguments, it is possible to prove the above theorems with the weaker hypothesis, X paraçompact (cf. Husemoller [1]).

7.6. Theorem. *The maps*

$$\pi_{n-1}(\mathrm{GL}_p(k))/\pi_0(\mathrm{GL}_p(k)) \longrightarrow \Phi_p^k(S^n)$$

and

$$[X, \mathrm{GL}_p(k)]'/\pi_0(\mathrm{GL}_p(k)) \longrightarrow \Phi_p^k(S'(X)),$$

defined in 3.10 and 3.14 respectively, are bijective if X is compact.

Proof. Since the second map is a generalization of the first, let us consider only the second map. According to 3.14, it suffices to show that any bundle over $S'(X)$ is isomorphic to a bundle of the form E_f, where $f: X \to \mathrm{GL}_p(k)$ is a continuous map such that $f(e) = 1$. If E is a bundle over $S'(X)$, its restrictions over C^+X and C^-X are trivial since C^+X and C^-X are contractible. Let $E_1 = C^+X \times k^p$ and $E_2 = C^-X \times k^p$, and let $g_1: E_1 \to E|_{C^+X}$ and $g_2: E_2 \to E|_{C^-X}$ be isomorphisms. According to 3.2, E is isomorphic to the bundle obtained by clutching the bundles E_1 and E_2 using the transition function $g_{21}: E_1|_{X \times \{0\}} \to E_2|_{X \times \{0\}}$ defined by $g_{21} = (g_2|_X)^{-1}(g_1|_X)$ where $X \approx X \times \{0\}$. Let $f: X \to \mathrm{GL}_p(k)$ be the map defined by $f(x) = g_{21}(x) \cdot (g_{21}(e))^{-1}$. Then E is isomorphic to E_f by the computation made in 3.9 adapted to $S'(X)$. □

7.7. Proposition. *Let p and p' be two projectors of a vector bundle T of base X such that $\operatorname{Im} p \approx \operatorname{Im} p'$, and let $\bar{p}$ and $\bar{p}'$ be the projectors of $T \oplus T$ defined by $\bar{p} = p \oplus 0_T$ and $\bar{p}' = p' \oplus 0_T$. Then there exists an automorphism δ of $T \oplus T$ which is isotopic to the identity such that $\bar{p}' = \delta \cdot \bar{p} \cdot \delta^{-1}$.*

Proof. Let us put $T_1 = \operatorname{Im} p$, $T_2 = \operatorname{Im}(1-p)$, $T_1' = \operatorname{Im} p'$, and $T_2' = \operatorname{Im}(1-p')$. We write $T \oplus T$ in the form $T_1 \oplus T_2 \oplus T_1' \oplus T_2'$, and notice that any isomorphism $\alpha: T_1 \to T_1'$ induces an automorphism δ of $T \oplus T$, defined by the matrix

$$\delta = \begin{pmatrix} 0 & 0 & -\alpha^{-1} & 0 \\ 0 & 1 & 0 & 0 \\ \alpha & 0 & 0 & 0 \\ 0 & 0 & 0 & 1 \end{pmatrix}$$

Moreover, in this matrix decomposition, $\bar{p}$ and $\bar{p}'$ take the form of matrices

$$\bar{p} = \begin{pmatrix} 1 & 0 & 0 & 0 \\ 0 & 0 & 0 & 0 \\ 0 & 0 & 0 & 0 \\ 0 & 0 & 0 & 0 \end{pmatrix} \qquad \bar{p}' = \begin{pmatrix} 0 & 0 & 0 & 0 \\ 0 & 0 & 0 & 0 \\ 0 & 0 & 1 & 0 \\ 0 & 0 & 0 & 0 \end{pmatrix}$$

and a simple computation shows that $\bar{p}' = \delta \cdot \bar{p} \cdot \delta^{-1}$. On the other hand, on the factor $T_1 \oplus T_1'$ of $T \oplus T$, the automorphism

$$\delta' = \begin{pmatrix} 0 & -\alpha^{-1} \\ \alpha & 0 \end{pmatrix} = \begin{pmatrix} 1 & -\alpha^{-1} \\ 0 & 1 \end{pmatrix} \begin{pmatrix} 1 & 0 \\ \alpha & 1 \end{pmatrix} \begin{pmatrix} 1 & -\alpha^{-1} \\ 0 & 1 \end{pmatrix}$$

is isotopic to the identity: consider the product

$$\begin{pmatrix} 1 & -t\alpha^{-1} \\ 0 & 1 \end{pmatrix} \begin{pmatrix} 1 & 0 \\ t\alpha & 1 \end{pmatrix} \begin{pmatrix} 1 & -t\alpha^{-1} \\ 0 & 1 \end{pmatrix}, \qquad t \in I.$$

Hence $\delta = \delta' \oplus \operatorname{Id}_{T_2 \oplus T_2'}$ is also isotopic to the identity. □

7.8. Let us denote by $\operatorname{Proj}_n(k^N)$, for $N \geqslant n$, the space of projection operators q on k^N, such that $\operatorname{Dim}(\operatorname{Im} q) = n$. A continuous map $g: Y \to \operatorname{Proj}_n(k^N)$ defines a projector $p = \hat{g}$ of the trivial bundle $T = Y \times k^N$, hence a vector bundle $\xi_f = \operatorname{Im} p$ of rank n over Y. If $f: X \to Y$ is a continuous map, we clearly have $\xi_{g \cdot f} = f^*(\xi_g)$. In particular, if we put $Y = \operatorname{Proj}_n(k^N)$, and $g = \operatorname{Id}_Y$, the bundle $\xi_{n,N} = \xi_g$ is called the canonical bundle over $\operatorname{Proj}_n(k^N)$. If $f: X \to \operatorname{Proj}_n(k^N)$, then the vector bundle ξ_f is simply $f^*(\xi_{n,N})$ according to the formula above. Moreover, when X is compact,

Theorem 7.2 shows that up to isomorphism the bundle $\xi_f = f^*(\xi_{n,N})$ only depends on the homotopy class of the map f. Hence, the correspondence $f \mapsto \xi_f$ defines a map

$$C_{n,N}\colon [X, \operatorname{Proj}_n(k^N)] \longrightarrow \Phi_n^k(X).$$

For $N' \geqslant N$, we have the injective map $i_{N',N}\colon \operatorname{Proj}_n(k^N) \to \operatorname{Proj}_n(k^{N'})$ defined by $q \mapsto q \oplus 0$, regarding $k^{N'}$ as $k^N \oplus k^{N'-N}$. Hence, by taking the direct limit, we see that the $C_{n,N}$ induce a map

$$C_n\colon \operatorname{inj\,lim}[X, \operatorname{Proj}_n(k^N)] \longrightarrow \Phi_n^k(X).$$

7.9. Theorem. *For every compact space X, the map C_n defined above is bijective.*

Proof. a) *C_n is surjective.* Let ξ be a vector bundle over X. Since X is compact, $\xi \approx \operatorname{Im} p$, where $p\colon T \to T$ is a projection operator on a trivial bundle $T = X \times k^N$ (6.5). Hence $\xi \approx \xi_{\check{p}}$, where $\check{p}$ is the map from X to $\operatorname{Proj}_n(k^N)$ which is canonically associated with p (1.12).

b) *C_n is injective.* It is enough to verify the following fact: Let $f_0, f_1\colon X \to \operatorname{Proj}_n(k^N)$ be continuous maps such that $\xi_{f_0} \approx \xi_{f_1}$. Then the maps $\bar{f}_\alpha = i_{2N,N} \cdot f_\alpha$ for $\alpha = 0, 1$, are homotopic. To see this, consider the projectors $p_\alpha = \hat{f}_\alpha$ of $T = X \times k^N$ which are associated with the f_α (1.12). According to Proposition 7.7, there exists an automorphism δ of the trivial bundle $T \oplus T$, isotopic to the identity, such that $\bar{p}_1 = \delta \cdot \bar{p}_0 \cdot \delta^{-1}$, where $\bar{p}_\alpha = p_\alpha \oplus 0_T$. Therefore, $\bar{p}_0$ is homotopic to $\bar{p}_1$ among the projectors of $T \oplus T$, and $\bar{f}_0 = \check{\bar{p}}$ is homotopic to $\bar{f}_1 = \check{\bar{p}}_1$ among the maps from X to $\operatorname{Proj}_n(k^{2N})$. □

7.10. Corollary. *Let us define $BGL_n(k) = \operatorname{inj\,lim} \operatorname{Proj}_n(k^N)$. Then the maps C_n induce a functor isomorphism.*

$$[X, BGL_n(k)] \approx \Phi_n^k(X),$$

when X is a compact space.

Proof. Since X is compact, and since $\operatorname{Proj}_n(k^N)$ is closed in $\operatorname{Proj}_n(k^{N'})$ for $N' \geqslant N$, $\operatorname{inj\,lim}[X, \operatorname{Proj}_n(k^N)] \approx [X, \operatorname{inj\,lim} \operatorname{Proj}_n(k^N)] = [X, BGL_n(k)]$ (Karoubi [1]). □

*7.11. *Remark.* The corollary also holds when X is paracompact.*

7.12. Let us put the usual bilinear form (resp. hermitian form) on $\mathbb{R}^n$ (resp. $\mathbb{C}^n$) defined by $\varphi(x, y) = \sum_{i=1}^n x_i y_i$ (resp. $\sum_{i=1}^n x_i \bar{y}_i$). A n-dimensional subspace M of k^N defines a self-adjoint projection operator p on k^N ($k = \mathbb{R}$ or $\mathbb{C}$), given by $\operatorname{Im}(p) = M$ and $\operatorname{Im}(1-p) = M^\perp$. If we let $G_n(k^N)$ denote the set of n-dimensional subspaces of

k^N, we establish a bijective correspondence between $G_n(k^N)$ and the subspace of $\mathrm{Proj}_n(k^N)$ consisting of self-adjoint projection operators. We give $G_n(k^N)$ the topology induced by this bijection (in fact, it is easy to provide $G_n(k^N)$ with the structure of a differentiable manifold: it is the "*Grassmann manifold*").

7.13. Proposition. *The Grassmann manifold $G_n(k^N)$ is a deformation retract of $\mathrm{Proj}_n(k^N)$.*

Proof. If h is a self-adjoint positive operator in k^N it has a unique positive self-adjoint square root $\sqrt{h}$ which depends continuously on h. Let

$$F: \mathrm{Proj}_n(k^N) \times I \longrightarrow \mathrm{Proj}_n(k^N)$$

be the map defined by $F(p, t) = \alpha \cdot p \cdot \alpha^{-1}$, where $\alpha = \sqrt{1 + tJ^* \cdot J}$ for $J = 2p - 1$. Then $F(p, t) = p$ if $p \in G_n(k^N)$, $F(p, 0) = p$, and $F(p, 1) \in G_n(k^N)$ since $\alpha^2 \cdot p \cdot \alpha^{-2} = p^*$ when $t = 1$. □

7.14. Theorem. *Let us define $BO(n) = \text{inj lim}\, G_n(\mathbb{R}^N)$ and $BU(n) = \text{inj lim}\, G_n(\mathbb{C}^N)$. Then the maps C_n induce functor isomorphisms*

$$[X, BO(n)] \approx \Phi_n^{\mathbb{R}}(X) \quad \text{and} \quad [X, BU(n)] \approx \Phi_n^{\mathbb{C}}(X),$$

for every compact space X.

Proof. This is a direct consequence of 7.9 and 7.13. □

7.15. Another description of the spaces $\mathrm{Proj}_n(k^N)$ and $G_n(k^N)$ may be given in terms of homogeneous spaces. We write $\mathrm{O}_r(k)$ for $\mathrm{O}(r)$ if $k = \mathbb{R}$, and for $U(r)$ if $k = \mathbb{C}$. Let p_0 be the projector of $k^N = k^n \oplus k^{N-n}$ defined by $\mathrm{Id}_{k^n} \oplus 0_{k^{N-n}}$. Let

$$\rho: \mathrm{GL}_N(k) \longrightarrow \mathrm{Proj}_n(k^N) \quad \text{and} \quad \sigma: \mathrm{O}_N(k) \longrightarrow G_n(k^N)$$

be the maps defined by $\rho(\alpha) = \alpha \cdot p_0 \cdot \alpha^{-1}$ and $\sigma = \rho|_{\mathrm{O}_N(k)}$.

7.16. Proposition. *The maps ρ and σ induce homeomorphisms*

$$\bar{\rho}: \mathrm{GL}_N(k)/\mathrm{GL}_n(k) \times \mathrm{GL}_{N-n}(k) \xrightarrow{\approx} \mathrm{Proj}_n(k^N) \quad \text{and}$$
$$\bar{\sigma}: \mathrm{O}_N(k)/\mathrm{O}_n(k) \times \mathrm{O}_{N-n}(k) \xrightarrow{\approx} G_n(k^N)$$

Proof. We define a transitive continuous action of $\mathrm{GL}_N(k)$ on $\mathrm{Proj}_n(k^N)$ by the formula $(\alpha, p) \mapsto \alpha p \alpha^{-1}$. Thus $\mathrm{Proj}_n(k^N)$ may be identified (as a $\mathrm{GL}_N(k)$-set) with the homogeneous space $\mathrm{GL}_N(k)/G_0$, where G_0 is the subgroup of $\mathrm{GL}_N(k)$ consisting of matrices α such that $\alpha \cdot p_0 \cdot \alpha^{-1} = p_0$. This subgroup is $\mathrm{GL}_n(k) \times \mathrm{GL}_{N-n}(k)$, imbedded in $\mathrm{GL}_N(k)$ via the map

$$(\alpha_1, \alpha_2) \mapsto \begin{pmatrix} \alpha_1 & 0 \\ 0 & \alpha_2 \end{pmatrix}.$$

From this discussion we obtain a continuous bijective map

$$\bar{\rho}: \mathrm{GL}_N(k)/\mathrm{GL}_n(k) \times \mathrm{GL}_{N-n}(k) \longrightarrow \mathrm{Proj}_n(k^N),$$

and in a completely analogous way, a continuous bijective map

$$\bar{\sigma}: \mathrm{O}_N(k)/\mathrm{O}_n(k) \times \mathrm{O}_{N-n}(k) \longrightarrow G_n(k^N),$$

which is actually the restriction of $\bar{\rho}$. All that remains to be shown is that $\bar{\rho}$ is an open map. For this, we construct a continuous section of ρ on a neighbourhood of each point p of $\mathrm{Proj}_n(k^N)$. If α is an element of $\mathrm{GL}_N(k)$ such that $\rho(\alpha) = p$, then the map s defined by $s(q) = (1 - p - q + 2qp)\,\alpha$ from $\mathrm{Proj}_n(k^N)$ to $\mathrm{GL}_N(k)$ (for q near p) is the required section. In fact,

$$\begin{aligned} s(q) \cdot p_0 &= (1 - p - p + 2qp) \cdot \alpha \cdot p_0 = (1 - p - q + 2qp) \cdot p \cdot \alpha \\ &= q(1 - p - q + 2qp) \cdot \alpha = q \cdot s(q). \quad \square \end{aligned}$$

Exercises (Section I.9) 16, 27, 39, 31–33.

8. Metrics and Forms on Vector Bundles

Let $\lambda \mapsto \bar{\lambda}$ denote a continuous involution of $k = \mathbb{R}$ or $\mathbb{C}$. Classical results show that the only such involutions are the identity and complex conjugation (if $k = \mathbb{C}$).

8.1. Definition. Let E be a k-vector bundle over X. A *sesquilinear form* on E is a continuous map $\varphi: E \times_X E \to k$ which has the following property. The map $\varphi_x: E_x \times E_x \to k$ induced on each fiber is "sesquilinear" with respect to the k-vector space structure of E_x. In other words, φ_x is $\mathbb{R}$-bilinear and $\varphi_x(\lambda e, e') = \varphi_x(e, \bar{\lambda} e') = \lambda \varphi_x(e, e')$ for $\lambda \in k$, $e \in E_x$, and $e' \in E_x$.

8.2. If E is the trivial bundle $X \times k^n$, each sesquilinear form φ induces a continuous map $\check{\varphi}: X \to M_n(k)$ by the formula $\check{\varphi}(x) = (a_{ji}(x))$, where $a_{ji}(x) = \varphi_x(e_i, e_j)$, and (e_i) is the canonical basis of k^n. Conversely, each continuous map $\theta: X \to \mathrm{M}_n(k)$ induces a sesquilinear form $\hat{\theta}$ on E as follows: over each point x of the base, $\hat{\theta}$ is defined by the matrix $\theta(x)=(a_{ji}(x))$. Explicitly, we have the formula

$$\hat{\theta}_x\left(\sum_{i=1}^{n} \lambda_i e_i, \sum_{j=1}^{n} \mu_j e_j\right)=\sum_{i,j} a_{ji}(x)\lambda_i\bar{\mu}_j,$$

showing that $\hat{\theta}$ is continuous if θ is continuous.

8.3. If E is an arbitrary vector bundle, let ${}^tE$ denote its dual bundle E^* if the involution of k is trivial, and its "antidual" $\bar{E}^*$ if $k=\mathbf{C}$ and the involution is complex conjugation (4.8 d) and e)). Then each sesquilinear form φ induces a morphism ψ from E to ${}^tE$ in the following way. On each fiber E_x, the morphism $\psi_x\colon E_x\to({}^tE)_x={}^tE_x$ is induced by the form φ_x by the formula $\psi_x(e)(e')=\varphi_x(e,e')$. In order to prove that ψ is continuous, consider an open subset U of X, over which $E_U\approx U\times k^n$. In this case, ${}^tE$ may also be identified with $U\times k^n$, and on each fiber $E_x=k^n$, ψ induces the linear map defined by the matrix $\check{\varphi}(x)$. Hence ψ is continuous according to 8.2 and 1.12. Conversely, each morphism from E to ${}^tE$ defines a sesquilinear form on E by an analogous argument. The sesquilinear form φ is called non-degenerate if the induced morphism $E\to{}^tE$ is an isomorphism.

8.4. Let $\varepsilon=\pm1$. A sesquilinear form φ on E is called ε-symmetric if $\varphi_x(e',e)=\overline{\varepsilon\varphi_x(e,e')}$, where e and e' are vectors of the same fiber E_x. This is equivalent to having $\psi'=\varepsilon\psi$, where ψ' is the composition $E\approx{}^t({}^tE)\xrightarrow{{}^t\psi}{}^tE$, where ${}^t\psi$ is the transposition of ψ. When $\varepsilon=1$ (resp. $\varepsilon=-1$) and the involution is trivial, such forms will be called *symmetric* (resp. *skew-symmetric*). When $\varepsilon=1$, $k=\mathbf{C}$, and the involution is complex conjugation, such forms will be called *Hermitian* (in this situation one does not generally consider forms with $\varepsilon=-1$, because such forms are obtained from Hermitian forms by multiplication with $i=\sqrt{-1}$).

8.5. Definition. Let E be a real vector bundle (resp. a complex vector bundle). A *metric* on E is a symmetric bilinear form (resp. a Hermitian form) on E such that $\varphi_x(e,e)>0$ for every non-zero vector e of E_x. Two such metrics φ_0 and φ_1 are called *homotopic* if there exists a metric φ on π^*E ($\pi\colon X\times I\to X$), such that $\varphi|_{X\times\{\alpha\}}=\varphi_\alpha$, for $\alpha=0,1$. Finally, two metrics φ_0 and φ_1 are called *isomorphic* if there exists an automorphism f of the vector bundle E such that $\varphi_1(f(e),f(e'))=\varphi_0(e,e')$.

8.6. *Remarks.* It is clear that metrics on vector bundles are always nondegenerate. We will use metrics on vector bundles to split them into direct sums (9.35).

8.7. Theorem. *If the base of the vector bundle E is paracompact, then there exists a metric on E. In particular $E\approx{}^tE$. Moreover, for any base (not necessarily paracompact), any two metrics φ_0 and φ_1 are homotopic.*

Proof. Let us prove the second part of the theorem first. We may identify $\pi^*E\times_{X\times I}\pi^*E$, for $\pi\colon X\times I\to X$, with $(E\times_X E)\times I$, and define a metric φ on π^*E

by the formula $\varphi(e, e', t) = t\varphi_0(e, e') + (1-t)\varphi_1(e, e')$ for e and e' belonging to the same fiber. We have $\varphi(e, e, t) > 0$ for $e \neq 0$, which shows that φ is a metric.

The proof of the first part of the theorem breaks into three parts:

a) $E = X \times k^n$. Then $E \times_X E \approx X \times k^n \times k^n$, and we define a metric φ on E by the formula $\varphi(x, \lambda_1, \ldots, \lambda_n, \mu_1, \ldots, \mu_n) = \sum_{j=1}^{n} \lambda_j \bar{\mu}_j$.

b) *E is isomorphic to $E \times k^n$*. Let $f: E \to T = X \times k^n$ be an arbitrary isomorphism and let $f_1: E \times_X E \to T \times_X T$ be the isomorphism induced by f. If φ is the metric defined in a) on T, it is clear that $\varphi \cdot f_1$ is a metric on E.

c) *E is arbitrary*. Let (U_i), for $i \in I$, be a locally finite open cover of X such that E_{U_i} is trivial, let (α_i) be a partition of unity associated with (U_i), let σ_i be a metric on E_{U_i}, and let $\varphi_i: E \times E \to k$ be the map defined by the formulas

$$\varphi_i(e, e') = \alpha_i(x)\sigma_i(e, e') \quad \text{for } x \in U_i,\ e \text{ and } e' \in E_x,$$

and

$$\varphi_i(e, e') = 0 \quad \text{for } x \notin U_i,\ e \text{ and } e' \in E_x.$$

Then φ_i is continuous because the support of α_i is contained in U_i. Thus we have defined a form on E which is symmetric (resp. Hermitian) if $k = \mathbb{R}$ (resp. $k = \mathbb{C}$). Now let $\varphi: E \times_X E \to k$ be the map defined by $\varphi(e, e') = \sum_{i \in I} \varphi_i(e, e')$ for e and e' belonging to the same fiber. Then this sum is well defined and represents a continuous map, because in a neighbourhood of $\pi^{-1}(\{x\})$ where $\pi: E \to X$, $\varphi_i(e, e') = 0$ except for a finite number of indices. On the other hand φ is a metric, since if e is a non-zero vector of E_x and $i \in I$ such that $\alpha_i(x) > 0$, we have $\varphi(e, e) \geqslant \alpha_i(x)\varphi_i(e, e) > 0$. □

8.8. Theorem. *Let φ_0 and φ_1 be two metrics on a vector bundle E with arbitrary base X. Then φ_0 and φ_1 are isomorphic.*

Proof. Let $\psi_0: E \to {}'E$ and $\psi_1: E \to {}'E$ be the isomorphisms canonically associated with φ_0 and φ_1 (8.3). Then $\psi_0^{-1}\psi_1 = h$ is an automorphism of E which is self-adjoint and positive with respect to the metric φ_0 on each fiber. If f is its self-adjoint positive square root, we have

$$\varphi_0(f(e), f(e')) = \varphi_0(f^2(e), e') = \psi_0(\psi_0^{-1}\psi_1(e))(e') = \psi_1(e)(e') = \varphi_1(e, e'). \quad \square$$

8.9. Corollary. *Let E be a vector bundle over $X \times I$, for X compact, which is provided with a metric φ. Then the vector bundles $E_0 = E|_{X \times \{0\}}$ and $E_1 = E|_{X \times \{1\}}$ are isometric (i.e. there exists an isomorphism between E_0 and E_1 which is compatible with the metric).*

Proof. According to Theorem 7.1, the vector bundles E_0 and E_1 are isomorphic. If $f: E_0 \to E_1$ is such an isomorphism, let E_0' denote the vector bundle E_0 provided with the metric $\varphi_1 \cdot f$. Then f induces an isometry also denoted by f between E_0' and E_1. By 8.8. we have an isometry $g\colon E_0 \to E_0'$, and $g \cdot f$ is an isometry between E_0 and E_1. □

8.10. Let E be a real vector bundle provided with a non-degenerate symmetric bilinear form θ. Over each point x of X, θ induces a form θ_x, and classical theorems on real quadratic forms show that E_x may be written as an orthogonal sum $V^+ \oplus V^-$, where θ_x restricted to V^+(resp. V^-) is positive (resp. negative). The integers $p(x) = \mathrm{Dim}(V^+)$ and $q(x) = \mathrm{Dim}(V^-)$ do not depend on the decomposition and are locally constant functions of x.

8.11. Theorem. *Let E be a real vector bundle on a compact base X, provided with a non-degenerate symmetric bilinear form θ. Then E may be written as an orthogonal sum $E^+ \oplus E^-$, where the restriction of θ to E^+ (resp. E^-) is positive (resp. negative). Moreover, this decomposition is unique up to isomorphism.*

Proof. Let φ be a metric on E (8.7), let $\psi\colon E \xrightarrow{\approx} {}^tE = E^*$ be the isomorphism associated with φ, and let $\chi\colon E \xrightarrow{\approx} {}^tE$ be the isomorphism associated with θ. Then $\omega = \psi^{-1}\chi$ is a self-adjoint automorphism of E with respect to the metric φ. Hence ω may be written as $h \cdot u$, where $h = \sqrt{\omega^2}$ (which is positive) and $u = h^{-1}\omega$. Explicitly, if ω_x is a diagonal matrix $\mathrm{Diag}(\lambda_1, \ldots, \lambda_n)$ in an orthonormal basis, then h_x is the diagonal matrix $\mathrm{Diag}(|\lambda_1|, \ldots, |\lambda_n|)$. The automorphisms h and ω commute, and we have $u^2 = \omega h^{-1} \omega h^{-1} = \omega^2 h^{-2} = 1$. Let $p = (1-u)/2$; then p is a projection operator (cf. 6.3) and we may write $E = E^+ \oplus E^-$, where $E^+ = \mathrm{Ker}\, p$ and $E^- = \mathrm{Ker}(1-p)$. If $e \in E_x^+$, we have $\theta(e, e) = \varphi(\omega(e), e) = \varphi(h(e), e) > 0$ if $e \neq 0$. In the same way $\theta(e, e) < 0$ if $e \in E_x^- - \{0\}$. Moreover E^+ and E^- are orthogonal with respect to both forms, θ and φ.

Conversely, let us suppose that E may be written as $E^+ \oplus E^-$, where θ restricted to E^+ (resp. E^-) is positive (resp. negative), and where E^+ and E^- are orthogonal. Then by the preceding method, this decomposition is associated with the metric φ defined by

$$\varphi_x(e, e') = \theta_x(e, e') \quad \text{if } e \text{ and } e' \text{ belong to } E_x^+,$$

$$\varphi_x(e, e') = -\theta_x(e, e') \quad \text{if } e \text{ and } e' \text{ belong to } E_x^-,$$

and

$$\varphi_x(e, e') = 0 \quad \text{if } e \in E_x^+,\ e' \in E_x^- \text{ or } e \in E_x^-,\ e' \in E_x^+.$$

In fact, h is just the identity in this context.

Finally, let $E = E_0^+ \oplus E_0^-$ and $E = E_1^+ \oplus E_1^-$ be two orthogonal decompositions of E. By what we have just said, they are associated with well defined metrics φ_0 and φ_1. Since the metrics are homotopic (8.7), there exists a bundle F over $X \times I$ provided with a metric φ and the symmetric bilinear form $\Pi^*\theta$, with $\Pi\colon X \times I \to X$, such that $(F, \varphi, \Pi^*\theta)|_{X \times \{0\}} \approx (E, \varphi_0, \theta)$ and $(F, \varphi, \Pi^*\theta)|_{X \times \{1\}} \approx (E, \varphi_1, \theta)$. Hence there exists a bundle F^+ (resp. F^-) over $X \times I$ such that $F^+|_{X \times \{0\}} \approx E_0^+$, and $F^+|_{X \times \{1\}} \approx E_1^+$ (resp. $F^-|_{X \times \{0\}} \approx E_0^-$, and $F^-|_{X \times \{1\}} \approx E_1^-$). According to Theorem 7.1, this implies that the bundles E_0^+ and E_1^+ (resp. E_0^- and E_1^-) are isomorphic. □

8.12. In the case of real bundles provided with skew-symmetric forms or complex bundles provided with symmetric, skew-symmetric or Hermitian forms, we have analogous theorems which are covered in the exercises 9.18–21.

Exercises (Section I.9) 18–26.

9. Exercises

9.1. Prove that TS^1 and TS^3 are trivial bundles.

9.2. A Cayley number is a pair (q_1, q_2) where q_1 and q_2 are elements of $\mathbb{H}$, the field of quaternions. We provide the set C of Cayley numbers with a (non-associative) algebra structure by defining

$$(q_1, q_2)+(q_1', q_2')=(q_1+q_1', q_2+q_2')$$

and $$(q_1, q_2)\cdot(q_1', q_2')=(q_1q_1'-\bar{q}_2'q_2, q_2'q_1+q_2\bar{q}_1'),$$

where $\bar{q}=a-bi-cj-dk$ if we write $q=a+bi+cj+dk$

1) Prove that C has no zero divisors (i.e. the equation $c\cdot c'=0$ implies $c=0$ or $c'=0$).

2) Prove that TS^7 is a trivial bundle.

3) More generally, prove that TS^{n-1} is a trivial bundle if $\mathbb{R}^n$ may be provided with an $\mathbb{R}$-algebra structure without zero divisors.

9.3. Let E'' be the vector bundle considered in 1.9, and let $f: S^1 \to S^1$ be the map defined by $f(z)=z^2$. Prove that $f^*(E)$ is a trivial bundle.

9.4. Prove that $\underbrace{E'' \oplus E'' \oplus \cdots \oplus E''}_{n}$ is trivial if and only if n is even (where E'' is the bundle defined in 1.9).

9.5. Let E and F be vector bundles over X, and let $f: E \to F$ be a morphism.

a) Prove that $\mathrm{Dim}(\mathrm{Ker}\, f_x)$ is a lower-semicontinuous function of x.

b) We assume that f is chosen so that $\mathrm{Dim}(\mathrm{Ker}\, f_x)$ is a continuous function of x. Prove that the quasi-vector bundle defined by $\mathrm{Ker}\, f = \bigsqcup_{x\in X} \mathrm{Ker}(f_x) \subset E$ is a vector bundle. Is $\mathrm{Ker}(f)$ the kernel of f in the category $\mathscr{E}(X)$?

c) Prove the analogous forms of a) and b) for $\mathrm{Coker}(f)$.

∗ **9.6.** Prove that the tangent bundle to a Lie group is trivial.∗

9.7. Make the differential structure on the sphere S^n explicit in such a way that the vector bundle TS^n defined in 1.3 is the tangent bundle to S^n.

9.8. Prove that $\Phi_1^{\mathbb{R}}(X)$ is naturally isomorphic to the set of isomorphism classes of double covers over X if X is paracompact.

9.9. Let (e_1, e_2, e_3) be the canonical basis of $\mathbb{R}^3$, and let x be a vector in $\mathbb{R}^3$ of norm 1. For $\varepsilon = \pm 1$ and $x \neq -\varepsilon e_3$, we write $\mathbb{R}(x, \varepsilon e_3)$ for the rotation in $\mathbb{R}^3$ which transforms εe_3 into x and fixes the vectors orthogonal to x and εe_3.

a) Give the matrix of $R(x, \varepsilon e_3)$, and show that it depends continuously on $x (x \neq -\varepsilon e_3)$.

b) Compute the matrix $R(x, -e_3)^{-1} R(x, e_3)$ when x is a vector of $S^1 \subset S^2$, i.e. a vector whose coordinates (x_1, x_2, x_3) satisfy the relation $(x_1)^2 + (x_2)^2 - 1 = x_3 = 0$.

c) Prove that TS^2 is isomorphic to the nontrivial vector bundle E_f (3.4) associated with the continuous function $f: S^1 \to \mathrm{GL}_2(\mathbb{R})$, defined by

$$f(e^{i\theta}) = \begin{pmatrix} \cos 2\theta & -\sin 2\theta \\ \sin 2\theta & \cos 2\theta \end{pmatrix}.$$

d) Prove that the function $g: S^1 \to \mathrm{GL}_3(\mathbb{R})$, defined by

$$g(e^{i\theta}) = \begin{pmatrix} \cos 2\theta & -\sin 2\theta & 0 \\ \sin 2\theta & \cos 2\theta & 0 \\ 0 & 0 & 1 \end{pmatrix},$$

is homotopic to a constant map (note that $TS^2 \oplus \theta_1$ is a trivial bundle, where θ_1 is the trivial bundle of rank one).

9.10. Let RP_n be the projective space of the real vector space $\mathbb{R}^{n+1}$.

a) Show that $T(RP_n)$ is isomorphic to the quotient of TS^n by the equivalence relation $(x, v) \sim (\varepsilon x, \varepsilon v)$, for $\varepsilon = \pm 1$, $x \in S^n$, and $v \perp x$.

b) If ξ is the canonical line bundle over RP_n, show that $T(RP_n) \oplus \theta_1 \approx \underbrace{\xi^* \oplus \xi^* \oplus \cdots \oplus \xi^*}_{n+1}$, where θ_1 is the trivial bundle of rank 1.

9.11. Let CP_n be the projective space of the complex vector space $\mathbb{C}^{n+1}$.

a) Show that $T(CP_n)$ may be identified with the quotient of TS^{2n+1} by the equivalence relation $(x, v) \sim (\lambda x, \lambda v)$ if $x \in S^{2n+1}$, $v \perp x$, and λ a complex number of norm 1. Provide $T(CP_n)$ with a complex structure.

b) Show that $T(CP_n) \oplus \eta_1$ is isomorphic to $\xi^* \oplus \cdots \oplus \xi^*$, where ξ^* is the dual of the canonical line bundle and η_1 is the *complex* trivial bundle of rank one.

9.12. Let V and W be vector bundles. Prove the formulas

a) $\lambda^n(V\oplus W)\approx \bigoplus_{i+j=n} \lambda^i(V) \otimes \lambda^j(W)$,

b) $S^n(V\oplus W)\approx \bigoplus_{i+j=n} S^i(V)\otimes S^j(W)$,

and c) $V\otimes V\approx\lambda^2(V)\oplus S^2(V)$.

9.13. Let $\pi: O(n)/O(n-k)\to O(n)/O(n-1)$ be the obvious surjective map. Show that the following properties of the pair (n, k) are equivalent:

(i) There exists a continuous map $s: O(n)/O(n-1)\to O(n)/O(n-k)$ such that $\pi\cdot s=\mathrm{Id}$.

(ii) The vector bundle TS^{n-1} admits $(k-1)$ linearly independent sections (note that $S^{n-1}\approx O(n)/O(n-1)$).

***9.14.** Let X be a CW-complex of dimension n, and let E be a real (resp. complex) vector bundle over X of rank $n+p$ (resp. $\geqslant n/2+p$). Show that E admits p linearly independent sections.*

9.15. Let X be a paracompact space and let E be a real vector bundle over X. Show that $\Gamma(X, E)$ may be provided with a Frechet space structure which depends functorially on E.

** **9.16.** Let H be an infinite dimensional Hilbert space over k, and let $\mathrm{Proj}_n(H)$ be the space of continuous endomorphisms q such that $q^2=q$ and $\mathrm{Dim}(\mathrm{Im}\, q)=n$.

a) Show that $\mathrm{Proj}_n(H)$ has the same homotopy type as the space $\Gamma_n(H)$ of endomorphisms D, which have the following two properties:

(i) The spectrum of D does not meet the axis $\mathscr{R}(z)=\frac{1}{2}$.

(ii) The endomorphism

$$p=\frac{1}{2i\pi}\int_\gamma \frac{dz}{z-D},$$

where γ is a differentiable curve in the half plane $\mathscr{R}(z)>\frac{1}{2}$ which contains the spectrum of D in this half plane, belongs to $\mathrm{Proj}_n(H)$.

b) Using Kuiper's theorem [1] ($GL(H)$ contractible), show that $\mathrm{Proj}_n(H)$ has the homotopy type of $BGL_n(k)$.**

9.17. Compute $\Phi_n^{\mathbb{R}}(X)$ and $\Phi_n^{\mathbb{C}}(X)$ for $X=S^1, S^2, S^3$.

9.18. Let E be a real vector bundle with compact base provided with a non-degenerate skew-symmetric form θ. Show that there exists a unique complex structure on E (up to isomorphism) and a metric φ on E, such that $\theta(e, e')=\varphi(ie, e')$.

9.19. Let E be a complex vector bundle with compact base provided with a non-degenerate Hermitian form θ. Show that E may be written as the orthogonal sum

$E^+ \oplus E^-$, where the restriction of θ to E^+ (resp. E^-) is positive (resp. negative). Moreover, show that this decomposition is unique up to isomorphism.

9.20. Let E be a complex vector bundle with compact base provided with a nondegenerate symmetric bilinear form θ. Prove the existence and uniqueness (up to isomorphism) of a real vector bundle $F \subset E_{\mathbb{R}}$ such that

(i) E is the complexification of F, and

(ii) the restriction of θ to F is a real metric.

9.21. Let E be a complex vector bundle with compact base, provided with a nondegenerate skew-symmetric form θ. Show the existence and uniqueness up to isomorphism of a pair (J, φ), where

(i) J is an automorphism of $E_{\mathbb{R}}$ such that $J^2 = -1$ and $iJ = -Ji$, and

(ii) $\theta(e, e') = \varphi(Je, e')$ where φ is a metric on E.

9.22. Let $\Phi^k(X)$ denote the set of isomorphism classes of k-vector bundles over the compact space X ($k = \mathbb{R}$, $\mathbb{C}$ or $\mathbb{H}$).

a) Prove that the Whitney sum of vector bundles provides $\Phi^k(X)$ with the structure of an abelian monoid.

b) Let $k = \mathbb{R}$ or $\mathbb{C}$, and let

1) $\mathrm{Sym}^k_+(X)$ be the set of isomorphism classes of k-vector bundles provided with a nondegenerate symmetric bilinear form.

2) $\mathrm{Sym}^k_-(X)$ be the set of isomorphism classes of k-vector bundles provided with a nondegenerate skew-symmetric form.

3) $\mathrm{Herm}^{\mathbb{C}}(X)$ be the set of isomorphism classes of $\mathbb{C}$-vector bundles provided with a nondegenerate Hermitian form.

With these definitions, prove that the Whitney sum of vector bundles induces an abelian monoid structure on $\mathrm{Sym}^k_+(X)$, $\mathrm{Sym}^k_-(X)$, and $\mathrm{Herm}^{\mathbb{C}}(X)$.

c) Prove the following isomorphisms:

$$\mathrm{Sym}^{\mathbb{R}}_+(X) \approx \Phi^{\mathbb{R}}(X) \times \Phi^{\mathbb{R}}(X) \quad \text{(use 8.11)}$$
$$\mathrm{Sym}^{\mathbb{R}}_-(X) \approx \Phi^{\mathbb{C}}(X) \quad \text{(use 9.18)}$$
$$\mathrm{Sym}^{\mathbb{C}}_+(X) \approx \Phi^{\mathbb{R}}(X) \quad \text{(use 9.20)}$$
$$\mathrm{Sym}^{\mathbb{C}}_-(X) \approx \Phi^{\mathbb{H}}(X) \quad \text{(use 9.21)}$$
$$\mathrm{Herm}^{\mathbb{C}}(X) \approx \Phi^{\mathbb{C}}(X) \times \Phi^{\mathbb{C}}(X) \quad \text{(use 9.19)}$$

9.23. a) Let $\mathrm{O}_{n,p}(k)$ be the subgroup of $\mathrm{GL}_{n+p}(k)$, $k = \mathbb{R}$ or $\mathbb{C}$, which consists of isometries of k^{n+p} provided with the form $\sum_{i=1}^{n} x_i \bar{y}_i - \sum_{i=n+1}^{n+p} x_i \bar{y}_i$. Prove that there exists a deformation retraction of $\mathrm{O}_{n,p}(k)$ onto $\mathrm{O}(n) \times \mathrm{O}(p)$ if $k = \mathbb{R}$, and onto $\mathrm{U}(n) \times \mathrm{U}(p)$ if $k = \mathbb{C}$.

b) Let $\mathrm{Sp}_{2n}(k)$ be the subgroup of $\mathrm{GL}_{2n}(k)$ which consists of isometries of k^{2n}, provided with the skew-symmetric form $\sum_{i=1}^{n} x_i y_{i+n} - \sum_{i=1}^{n} x_{i+n} y_i$. Prove that $\mathrm{Sp}_{2n}(\mathbb{R})$

admits $U(n)$ as a deformation retract. Prove that $Sp_{2n}(\mathbb{C})$ has the same homotopy type as $GL_n(\mathbb{H})$.

c) Let $O_n(\mathbb{C})$ be the subgroup of $GL_n(\mathbb{C})$ which consists of isometries of $\mathbb{C}^n$, provided with the quadratic form $\sum_{i=1}^{n} (x_i)^2$. Prove that $O_n(\mathbb{C})$ admits $O(n)$ as a deformation retract.

9.24. Let E be a vector bundle. Define a bijective correspondence between the vector space of symmetric bilinear forms on E and the space of sections of $S^2(E^*)$. Characterize the sections which correspond to nondegenerate bilinear forms. Do the same work for skew-symmetric forms and sections of $\lambda^2(E^*)$.

9.25. Let E and F be vector bundles provided with a metric, and let $f: E \to F$ be a morphism which induces an epimorphism on each fiber.

a) Show that the map $f^*: F \to E$ defined on each fiber by $(f^*)_x = (f_x^*)$ is a vector bundle morphism.

b) Show that $f \cdot f^*$ is a vector bundle isomorphism which is isotopic to the identity.

* **9.26.** Prove that $\Phi_1^{\mathbb{C}}(X) \approx H^2(X; \mathbb{Z})$ and $\Phi_1^{\mathbb{R}}(X) \approx H^1(X; \mathbb{Z}/2)$, where X is paracompact.*

9.27. (Generalization of 9.9.) Let $(e_1, \ldots, e_{n+1})$ be the canonical basis of $\mathbb{R}^{n+1}$, and let x be a vector of S^n, $x \neq \varepsilon e_{n+1}$ for $\varepsilon = \pm 1$. Let $R(x, \varepsilon e_{n+1})$ be the rotation of $\mathbb{R}^{n+1}$ which transforms x into εe_{n+1}, and leaves the vectors orthogonal to x and e_{n+1} fixed.

a) Let $\rho_{n+1}: \Gamma^0(n+1) \to SO(n+1)$ be the covering of $SO(n+1)$ by the special Clifford group (IV.4). Prove that $R(x, \varepsilon e_{n+1}) = \rho_{n+1}((1 + \varepsilon x e_{n+1})/2$, and deduce from this the continuity of $R(x, \varepsilon e_{n+1})$ as a function of x.

b) Prove the formula $R(x, -e_{n+1})^{-1} R(x, e_{n+1}) = \rho_{n+1}(x e_{n+1})$ for $x \in S^{n-1} = S_+^n \cap S_-^n$.

c) Show that TS^n is isomorphic to the vector bundle E_f associated with the continuous function $f: S^{n-1} \to GL_n(\mathbb{R})$, defined by $f(x) = \rho_n(x e_n)$ where we identify S^{n-1} with a subset of $\Gamma^0(n)$.

d) Explicitly compute the matrix $\rho_n(x e_n)$ for $n = 2, 3, \ldots$.

9.28. Let $\pi: P \to X$ be a surjective continuous map. We say that (P, π, X) is a principal fibration with topological group G, if G acts on the right on P, fiber by fiber, such that $\forall x \in X$, there exists a neighbourhood U of x and an equivariant homeomorphism $\pi^{-1}(U) \to U \times G$ such that the following diagram commutes

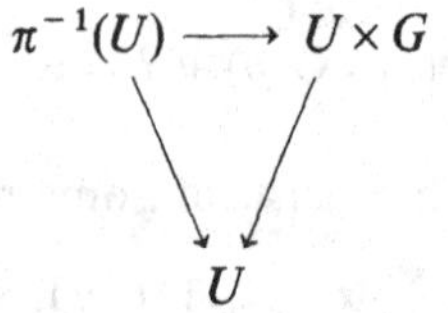

Now let F be a topological space on which G acts continuously on the left. Then we associate with P and F, the space $E=P\times_G F$ which is the quotient of $P\times F$ by the equivalence relation $(p,v)\sim(p\cdot g, g^{-1}\cdot v)$ for $g\in G$. Let $\pi'\colon E\to X$ be the projection defined by $(p,v)\mapsto\pi(p)$.

a) Show that for every point x of X, there exists a neighbourhood U of x, and a homeomorphism $\varphi_U\colon\pi^{-1}(U)\xrightarrow{\approx}U\times F$ which is compatible with the projection on U.

b) More precisely, show that there exists an open cover (U_i) of X, and homeomorphisms $\varphi_{U_i}\colon\pi^{-1}(U_i)\approx U_i\times F$ such that $\varphi_{U_j}^{-1}\cdot\varphi_{U_i}\colon(U_i\cap U_j)\times F\to(U_i\cap U_j)\times F$ is of the form $(x,v)\mapsto(x, g_{ji}(x)\cdot v)$, where the $g_{ji}\colon U_i\cap U_j\to G$ form a G-cocycle.

c) In particular, if $F=k^n$ and G acts on F by linear transformations, show that E is a k-vector bundle. Conversely, if E' is an arbitrary vector bundle of rank n, show the existence of a principal bundle P of group $G=\mathrm{GL}_n(k)$, such that $E'\approx P\times_G k^n$.

d) Show that $\mathrm{O}(n+1)$ is a principal bundle over $\mathrm{O}(n+1)/\mathrm{O}(n)\approx S^n$ with group $G=\mathrm{O}(n)$, and that $TS^n\approx\mathrm{O}(n+1)\times_{\mathrm{O}(n)}\mathbb{R}^n$.

9.29. Show that Theorem 7.2 is also true for X paracompact (Husemoller [1]).

9.30. Let G be a finite group acting on the left on a space X. A G-vector bundle over X is given by a vector bundle E, on which G acts on the left, making the diagram

$$\begin{array}{ccc} G\times E & \xrightarrow{\theta} & E \\ \downarrow & & \downarrow \\ G\times X & \longrightarrow & X \end{array}$$

commutative, and such that the map $e\mapsto\theta(g,e)$ from E_x to $E_{g\cdot x}$ is k-linear. If E and F are G-vector bundles with the same base, then a morphism between E and F is a vector bundle morphism which is equivariant. We write $\mathscr{E}_{G,k}(X)$ or simply $\mathscr{E}_G(X)$ for the category of G-vector bundles over X.

a) If G acts freely on X, show that E/G is a vector bundle over X/G and that $E\approx\pi^*(E/G)$, where $\pi\colon X\to X/G$. Prove that the categories $\mathscr{E}_G(X)$ and $\mathscr{E}(X/G)$ are equivalent under π^*.

* b) Let n be the number of irreducible representations of G ordered from 1 to n (Serre [1]). If the basic field k is $\mathbb{C}$, and if G acts trivially on X, show that every G-vector bundle E may be uniquely decomposed as $E_1\oplus E_2\oplus\cdots\oplus E_n$, where $E_j=T_j\otimes F_j$ for F_j an ordinary vector bundle, and T_j the space of the jth irreducible representation of G.*

c) Give the explicit decomposition of E (as in b)) for $G=\mathbb{Z}/n$.

9.31 (continuing from 9.30). Let E and F be G-vector bundles, and let $f\colon E\to F$ be a morphism between the underlying vector bundles. Let $\bar{f}\colon E\to F$ be the map defined by $\bar{f}(e)=\dfrac{1}{|G|}\sum_{g\in G}g^{-1}f(g\cdot e)$.

a) Show that $\bar{f}(g\cdot e)=g\cdot\bar{f}(e)$ (in other words $\bar{f}$ is a G-vector bundle morphism).

b) If f is a G-vector bundle morphism, then $\bar{f}=f$.

c) Let $h: F\to E$ be a G-vector bundle morphism such that $h\cdot f=\mathrm{Id}_E$. Then $h\cdot\bar{f}=\mathrm{Id}_E$.

d) Deduce from c) that every G-vector bundle is a direct factor of a G-vector bundle of the form $X\times M$, where M is a G-module of finite dimension.

e) Let A be the algebra of continuous functions on X with values in k, and let $B=A[G]$, i.e. the free A-module with basis the elements of G, and with multiplication defined by the rule $(\lambda\cdot g)(\lambda'\cdot g')=(\lambda(g\cdot\lambda'), gg')$, where $(g\cdot\lambda')(x)=\lambda'(g^{-1}x)$. Show that the category of G-vector bundles over X is equivalent to the category of finitely generated projective modules over B.

f) Let E be a G-vector bundle over $X\times I$ where X is compact. Show that $E_0=E|_{X\times\{0\}}$ and $E_1=E|_{X\times\{1\}}$ are isomorphic.

** g) Generalize a)–d) and f) for G a compact group.**

9.32 (Milnor's construction). For each group G, let us consider the subset E'_G of the infinite product $I\times G\times I\times G\times\cdots$, whose elements are sequences $S=(t_0, x_0, t_1, x_1, \ldots)$ where $x_i\in G$ and where $t_i\in[0,1]$, such that $t_i=0$ except for a finite number of indices, and $\sum_{i\in\mathbb{N}} t_i=1$.

In E'_G we consider the following equivalence relation: the sequence S is equivalent to the sequence S' if and only if 1) $t_i=t'_i$ and 2) $x_i=x'_i$ if $t_i=t'_i>0$. The quotient of E'_G by this relation is denoted by E_G, or $G*G*G*\cdots$ ("*infinite join*"), and the class of S is denoted by $(t_0x_0, t_1x_1, \ldots)$.

a) We now suppose that G is a topological group. Show the existence of the coarsest topology on E_G making the maps $t_i: E_G\to[0,1]$ and $x_i: t_i^{-1}(0,1]\to G$ continuous.

b) For this topology, show that G acts freely on E_G under the operation $(t_0x_0, \ldots, t_1x_1, \ldots)\cdot g=(t_0x_0g, \ldots, t_kx_kg, \ldots)$.

c) Prove that E_G is a principal bundle over $B_G=E_G/G$ (9.28).

d) Let X be a paracompact space, and let E be a G-principal bundle over X. Prove the existence of an open cover (U_n), $n\in\mathbb{N}$ of X such that $E|_{U_n}$ is trivial.

e) Using a partition of unity, construct a general morphism (in an obvious sense) between E and E_G which is compatible with the action of G.

* f) If $\mathscr{P}_G(X)$ denotes the set of isomorphism classes of G-principal bundles over the paracompact space X, show that $\mathscr{P}_G(X)\approx[X, B_G]$.*

* g) Give a simple description of the spaces E_G and B_G when $G=\mathbb{Z}/2$, $\mathbb{Z}$, or $U(1)$.*

* h) Prove that B_G has the same homotopy type as $BGL_n(k)$ when $G=\mathrm{GL}_n(k)$ (7.10).*

***9.33.** Show that the main results in the first chapter of this book still hold if we replace the basic field k by a Banach algebra A (where the fibers are finitely generated projective A-modules with the natural topology cf. 6.20). Show that $\mathscr{E}_{C_k(Y)}(X)\sim\mathscr{E}_k(X\times Y)$ if X and Y are compact.*

9.34. Let E be the canonical line bundle over $P(V)$ considered in 2.4 and 2.5.

In the real case, show that $E_p = \underbrace{E \otimes \cdots \otimes E}_{p}$ may be identified with the quotient of $S^n \times \mathbb{R}$ by the equivalence relation $(x, t) \sim (\lambda x, \lambda^p t)$ with $\lambda = \pm 1$. Hence E_p is trivial if p is even and isomorphic to E if p is odd.

In the complex case, show that E_p may be identified with the quotient of $S^{2n+1} \times \mathbb{C}$ by the equivalence relation $(x, t) \sim (\lambda x, \lambda^{-p} t)$, where $\lambda \in U$.

In the same way, compute the dual vector bundle E_p^*.

9.35. Let E be a vector bundle provided with a metric and let $i: E \to F$ be a vector bundle morphism such that $i_x: E_x \to F_x$ is injective. We provide E with the induced metric.

a) Show that the map $i^*: F \to E$, defined by $(i^*)_x = i_x^*$ is a vector bundle morphism such that $i^* \cdot i = \mathrm{Id}_F$.

b) Show that i^* and the quotient map $E \to \mathrm{Coker}(i)$ (9.5) define a direct sum decomposition $E \approx F \oplus \mathrm{Coker}(i)$.

10. Historical Note

Almost all of the material presented in this section is classical; hence we have only selected what we need for topological K-theory from the general theory of bundles. For this general theory, the reader is referred to Steenrod [1] and to Husemoller [1], where more complete results are obtained for bundles over paracompact spaces (for our purposes compact spaces will suffice). The notion of operations on vector bundles is taken from Atiyah [3] and Lang [2]. The proof of the homotopy invariance of $\Phi_n^k(X)$ (Section 7) is also taken from Atiyah [3]. Finally, the proofs we presented of the Serre–Swan theorem (6.18) and of the representability of the functor $\Phi_n^k(X)$, were inspired by the author's thesis [2].

Chapter II

First Notions of *K*-Theory

1. The Grothendieck Group of a Category. The Group $K(X)$

1.1. Let us first consider an abelian monoid M, i.e. a set provided with a composition law (denoted $+$) which satisfies all the properties of an abelian group except possibly the existence of inverses. Then we can associate an abelian group $S(M)$ with M and a homomorphism of the underlying monoids $s: M \to S(M)$, having the following universal property. For any abelian group G, and any homomorphism of the underlying monoids $f: M \to G$, there is a unique group homomorphism $\bar{f}: S(M) \to G$ which makes the following diagram commutative.

$$\begin{array}{ccc} M & \xrightarrow{s} & S(M) \\ & {\scriptstyle f}\searrow \quad \swarrow{\scriptstyle \bar{f}} & \\ & G & \end{array}$$

1.2. There are various possible constructions of s and $S(M)$. Of course they all give the same result up to isomorphism. Consider the free abelian group $\mathscr{F}(M)$ with basis the elements $[m]$ of M. Then the group $S(M)$ is the quotient of $\mathscr{F}(M)$ by the subgroup generated by linear combinations of the form $[m+n]-[m]-[n]$, and the image under s of m in $S(M)$ is the class of $[m]$. We could also consider the product $M \times M$ and form the quotient under the equivalence relation

$$(m, n) \sim (m', n') \Leftrightarrow \exists p \quad \text{such that} \quad m+n'+p = n+m'+p.$$

The quotient monoid is a group and $s(m)$ is the class of the pair $(m, 0)$. Finally, a third construction is to consider the quotient of $M \times M$ by the equivalence relation

$$(m, n) \sim (m', n') \Leftrightarrow \exists p, q \quad \text{such that} \quad (m, n)+(p, p) = (m', n')+(q, q),$$

with $s(m)$ the class of $(m, 0)$ once again. In each of these three constructions, we notice that every element of $S(M)$ can be written as $s(m)-s(n)$ where $m, n \in M$. However, in general, the map s is not injective (see Example 1.5 below).

1.3. Example. One of the most natural examples to consider is $M = \mathbb{N}$. Then, as is well known, $S(M) \approx \mathbb{Z}$.

1.4. Example. Let us take $M=\mathbb{Z}-\{0\}$, an abelian monoid with respect to multiplication. Then $S(M)\approx\mathbb{Q}-\{0\}$.

1.5. Example. Let M be an abelian monoid with the following property: there exists an element ∞ of M such that $m+\infty=\infty$ for every element m of M. Then $S(M)=0$ because each element of $S(M)$ can be written as $s(m)-s(n)=s(m)+s(\infty)-s(n)-s(\infty)=s(\infty)-s(\infty)=0$. Examples of monoids with this property are $\mathbb{Z}$ (with multiplication; then $\infty=0$!), $\mathbb{R}^{+*}\cup\infty$, etc.

1.6. *Remark.* The group $S(M)$ depends "functorially" on M in an obvious way: if $f\colon M\to N$, the universal property enables us to define a unique homomorphism $S(f)\colon S(M)\to S(N)$ which makes the diagram

$$\begin{array}{ccc} M & \xrightarrow{f} & N \\ \downarrow & & \downarrow \\ S(M) & \xrightarrow{s(f)} & S(N) \end{array}$$

commutative. Moreover, $S(g\cdot f)=S(g)\cdot S(f)$ and $S(\mathrm{Id}_M)=\mathrm{Id}_{S(M)}$. The group $S(M)$ is called the symmetrization of the abelian monoid M.

1.7. As a fundamental example, let us now consider an additive category $\mathscr{C}$. If E is an object of $\mathscr{C}$, we denote its isomorphism class by $\dot{E}$. Then the set $\Phi(\mathscr{C})$ of such $\dot{E}$ can be provided with the structure of an abelian monoid if we define $\dot{E}+\dot{F}$ to be $\widehat{E\oplus F}$. This operation is well-defined since the isomorphism class of $E\oplus F$ depends only on the isomorphism classes of E and F. Moreover, the relations $E\oplus(F\oplus G)\approx(E\oplus F)\oplus G$, $E\oplus F\approx F\oplus E$, and $E\oplus 0\approx E$ give the required algebraic identities on $\Phi(\mathscr{C})$. In this situation the group $S(M)$, where M is $\Phi(\mathscr{C})$, is called the *Grothendieck group* of $\mathscr{C}$, and is written $K(\mathscr{C})$. If $\varphi\colon\mathscr{C}\to\mathscr{C}'$ is an additive functor, φ naturally induces a monoid homomorphism $\Phi(\mathscr{C})\to\Phi(\mathscr{C}')$, hence a group homomorphism $K(\mathscr{C})\to K(\mathscr{C}')$ denoted by φ_*. If $\mathscr{C}''$ is a third category and $\psi\colon\mathscr{C}'\to\mathscr{C}''$ an additive functor, we have the formula $(\psi\cdot\varphi)_*=\psi_*\cdot\varphi_*$ from 1.6. Of course if $\mathscr{C}'=\mathscr{C}$ and $\varphi=\mathrm{Id}_{\mathscr{C}}$, then $\varphi_*=\mathrm{Id}_{K(\mathscr{C})}$.

1.8. Example. Let F be an arbitrary field not necessarily commutative and let $\mathscr{C}$ be the category whose objects are finite dimensional F-vector spaces (on the right for example), and whose morphisms are linear maps. Then from the classical theory of dimension of vector spaces we know that $\Phi(\mathscr{C})\approx\mathbb{N}$. Thus Example 1.3 implies that $K(\mathscr{C})\approx\mathbb{Z}$.

1.9. Example. Let $\mathscr{C}$ be an additive category provided with an additive functor $\tau\colon\mathscr{C}\to\mathscr{C}$, and a natural isomorphism $\tau+\mathrm{Id}_{\mathscr{C}}\approx\tau$. Then $K(\mathscr{C})=0$ because the identity above implies $s(\tau(E))+s(E)=s(\tau(E))$, hence $s(E)=0$ (1.5). For example, let $\mathscr{C}$ be the category whose objects are F-vector spaces (not necessarily finite dimensional) and whose morphisms are linear maps. We choose as τ the functor $E\mapsto E\oplus E\oplus\cdots\oplus E\oplus\cdots$. Another example is the category $\mathscr{H}$ with Hilbert spaces as objects, and continuous linear maps as morphisms. Then $\tau(E)=E\oplus E\oplus\cdots\oplus E\oplus\cdots$ (Hilbert sum with the L^2-norm).

1.10. Example (generalization of Example 1.8). Let A be an arbitrary ring with unit, and let $\mathscr{P}(A)$ be the category with finitely generated projective right A-modules as objects, and the A-linear maps as morphisms. By abuse of notation we write $K(A)$ for the Grothendieck group of $\mathscr{P}(A)$. One of the main purposes of "*algebraic K-theory*" is to compute $K(A)$ for interesting rings A (Bass [1]).

1.11. Example. In this book we are mainly concerned with the category $\mathscr{E}(X)$ of vector bundles over a compact space X. We denote the Grothendieck group of $\mathscr{E}(X)$ by $K(X)$. If the basic field k is $\mathbb{R}$ (resp. $\mathbb{C}$) we write $K_{\mathbb{R}}(X)$(resp. $K_{\mathbb{C}}(X)$) for the group $K(X)$ whenever there is some risk of confusion. The object of "*topological K-theory*" is to compute $K(X)$ for interesting spaces X. In this presentation, topological K-theory arises as a special case of algebraic K-theory. Theorem I.6.17 shows that the category $\mathscr{E}(X)$ is equivalent to the category $\mathscr{P}(A)$ where A is the ring of continuous functions on X. Hence the groups $K(\mathscr{E}(X))$ and $K(\mathscr{P}(A))$ are isomorphic. In fact, many of the techniques in algebraic K-theory are inspired by the techniques of topological K-theory which we will develop in this book.

1.12. *Remark.* We have seen in 1.7 that the group $K(\mathscr{C})$ depends *covariantly* on $\mathscr{C}$. In the same way, the group $K(A)$ depends covariantly on the ring A. More precisely, if $u: A\to B$ is a ring homomorphism, there is a functor $\bar{u}: \mathscr{P}(A)\to\mathscr{P}(B)$ associated with u, defined by $E\mapsto E\otimes_A B$; $\bar{u}$ is called the "extension of scalars" functor (regarding B as a left A-module via the homomorphism u). For example, if E is the image of the projection operator $p=(p_{ji})\colon A^n\to A^n$, $\bar{u}(E)$ is the image of the projection operator $q=(u(p_{ji}))\colon B^n\to B^n$. However, the group $K(X)$ depends *contravariantly* on X. More precisely, if $f\colon Y\to X$ is a continuous map, f induces a functor $f^*:\mathscr{E}(X)\to\mathscr{E}(Y)$ (I.2.6), hence a homomorphism $K(X)\to K(Y)$, again called f^*. From I.2.6, we have the identities $(g\cdot f)^*=f^*\cdot g^*$ and $\mathrm{Id}^*=\mathrm{Id}$.

1.13. *Remark about notation.* It is unfortunate that the letter K is simultaneously used to denote the "K-group" of a category, ring, or compact space. So while following conventional notation, to avoid confusion we reserve the first letters of the alphabet $A, B, C,\ldots$ for rings and the last letters $X, Y, Z\ldots$ for spaces. Similarly, categories will be denoted by script letters $\mathscr{A}, \mathscr{B}, \mathscr{C},\ldots$.

1.14. Returning to the definition of the group $K(\mathscr{C})$, we see that we have made very little use of the additive structure of $\mathscr{C}$. Sometimes it is useful to consider categories provided with a *composition law* $\mathscr{C}\times\mathscr{C}\to\mathscr{C}$, denoted by $(E,F)\mapsto E\perp F$, with natural isomorphisms $E\perp(F\perp G)\approx(E\perp F)\perp G$, $E\perp F\approx F\perp E$, $E\perp 0\approx E$ which are "coherent" under iteration. The functor $\perp$ induces an abelian monoid structure on $\Phi(\mathscr{C})$; thus we can define a group $K(\mathscr{C})$ depending on the composition law $\perp$. A typical example is the category of vector bundles provided with nondegenerate symmetric bilinear forms (then $E\perp F$ is induced by the Whitney sum of underlying bundles). The details of this example are dealt with in the exercises.

1.15. Proposition. *Let $\mathscr{C}$ be an additive category, and let $[E]=s(E)$ denote the class of an object E of $\mathscr{C}$ in the Grothendieck group $K(\mathscr{C})$. Then every element of $K(\mathscr{C})$ can be written in the form $[E]-[F]$. Moreover, $[E]-[F]=[E']-[F']$ in $K(\mathscr{C})$, if and only if there exists an object G of $\mathscr{C}$, such that $E \oplus F' \oplus G \approx E' \oplus F \oplus G$.*

Proof. According to 1.2, every element of $K(\mathscr{C})$ is the class of a pair $(\dot{E}, \dot{F})$ which can be written as $s(\dot{E})-s(\dot{F})=[E]-[F]$. Moreover, two such pairs $(\dot{E}, \dot{F})$ and $(\dot{E}', \dot{F}')$ define the same element of $K(\mathscr{C})$ if and only if we can find an object G of $\mathscr{C}$ such that $\dot{E}+\dot{F}'+\dot{G}=\dot{E}'+\dot{F}+\dot{G}$, i.e. $E \oplus F' \oplus G \approx E' \oplus F \oplus G$. □

1.16. Corollary. *Let E and F be objects of $\mathscr{C}$. Then $[E]=[F]$ if and only if there exists an object G of $\mathscr{C}$ such that $E \oplus G \approx F \oplus G$.*

1.17. Proposition. *Let θ_n denote the trivial bundle of rank n over a compact space X. Then every element x of $K(X)$ can be written as $[E]-[\theta_n]$ for some n, and some vector bundle E over X. Moreover, $[E]-[\theta_n]=[F]-[\theta_p]$ if and only if there exists an integer q such that $E \oplus \theta_{p+q} \approx F \oplus \theta_{n+q}$.*

Proof. We will apply the Proposition 1.15 to the category $\mathscr{C}=\mathscr{E}(X)$. By this proposition we already know that each element x of $K(X)$ can be written as E_1-F_1, where E_1 and F_1 are two vector bundles over X. According to I.6.5, there is a vector bundle F_2 such that $F_1 \oplus F_2$ is a trivial bundle, say θ_n. Then $[E_1]-[F_1]=[E_1]+[F_2]-[F_1]-[F_2]=[E]-[\theta_n]$ where $E=E_1 \oplus F_2$. Now suppose that $[E]-[\theta_n]=[F]-[\theta_p]$. By the second part of Proposition 1.15, we can find a vector bundle G such that $E \oplus \theta_p \oplus G \approx F \oplus \theta_n \oplus G$. Let G_1 be a vector bundle such that $G \oplus G_1 \approx \theta_q$ (I.6.5). Then we have $E \oplus \theta_{p+q} \approx E \oplus \theta_p \oplus G \oplus G_1 \approx F \oplus \theta_n \oplus G \oplus G_1 \approx F \oplus \theta_{n+q}$. The converse is obvious. □

1.18. Corollary. *Let E and F be vector bundles. Then $[E]=[F]$ in $K(X)$ if and only if $E \oplus \theta_n \approx F \oplus \theta_n$ for some n.*

1.19. Example. Let E be the tangent bundle of S^p and $F=\theta_p$ (I.2.3). Then, from I.5.5, we have $E \oplus \theta_1 \approx \theta_{p+1}$. Hence $[E]=[\theta_p]$ in $K_{\mathbb{R}}(S^p)$. However, in general E is not trivial (V.2). This gives a nontrivial example of when the map $\Phi(\mathscr{C}) \to K(\mathscr{C})$ is not injective for $\mathscr{C}=\mathscr{E}_{\mathbb{R}}(S^p)$, $p \neq 1, 3, 7$. We will give a homotopic explanation of this phenomena later (1.32).

1.20. Since the functor K is contravariant on the category of compact spaces (1.12), the projection of X onto a point P induces a homomorphism $\alpha: \mathbb{Z} \approx K(P) \to K(X)$, whose cokernel is denoted by $\tilde{K}(X)$ and called the *reduced K-theory* of X. When we want to specify the basic field $k=\mathbb{R}$ or $\mathbb{C}$, we write $\tilde{K}_{\mathbb{R}}(X)$ or $\tilde{K}_{\mathbb{C}}(X)$.

1.21. Proposition. *If $X \neq \varnothing$, we have an exact sequence*

$$0 \longrightarrow \mathbb{Z} \xrightarrow{\alpha} K(X) \xrightarrow{\beta} \tilde{K}(X) \longrightarrow 0.$$

The choice of a point x_0 in X defines a canonical splitting so that

$$\tilde{K}(X) \approx \operatorname{Ker}[K(X) \longrightarrow K(\{x_0\}) \approx \mathbb{Z}] \quad \text{and} \quad K(X) \approx \mathbb{Z} \oplus \tilde{K}(X).$$

Proof. Let us choose $P = \{x_0\}$. Then the inclusion of $\{x_0\}$ in X induces a homomorphism from $K(X)$ to $K(\{x_0\}) \approx \mathbb{Z}$ which is a left-inverse for α. □

1.22. Let $\Phi(X)$ ($\Phi^k(X)$ when we want to specify the basic field k) denote the abelian monoid of isomorphism classes of vector bundles over X, and let γ denote the composition $\Phi(X) \xrightarrow{s} K(X) \xrightarrow{\beta} \tilde{K}(X)$.

1.23. Proposition. *The homomorphism γ is surjective. Moreover, $\gamma(\dot{E}) = \gamma(\dot{F})$ if and only if $E \oplus \theta_n \approx F \oplus \theta_p$ for some trivial bundles θ_n and θ_p.*

Proof. Since the class of θ_n in $\tilde{K}(X)$ is zero, and since any element of $K(X)$ can be written as $E - \theta_n$ (1.17), we have $\beta([E] - [\theta_n]) = \beta([E]) = \gamma(\dot{E})$, thus proving the first part of the proposition. On the other hand, the identity $\gamma(\dot{E}) = \gamma(\dot{F})$ is equivalent to $[E] - [F] = [\theta_q] - [\theta_r]$ for some q and r. By 1.18, this implies $E \oplus \theta_r \oplus \theta_t \approx F \oplus \theta_q \oplus \theta_t$ for some t, hence $E \oplus \theta_n \approx F \oplus \theta_p$, where $n = r + t$ and $p = q + t$. The converse is obvious. □

1.24. *Remark.* This proposition gives a conveniently direct definition of $\tilde{K}(X)$; i.e. $\tilde{K}(X)$ is the quotient of $\Phi(X)$ by the equivalence relation $\dot{E} \sim \dot{F} \Leftrightarrow \exists n, p$ such that $E \oplus \theta_n \approx F \oplus \theta_p$.

1.25. Theorem. *Let X and Y be compact spaces, and let $f_0, f_1 : X \to Y$ be continuous maps which are homotopic. Then f_0 and f_1 induce the same homomorphisms $K(Y) \to K(X)$ and $\tilde{K}(Y) \to \tilde{K}(X)$.*

Proof. According to I.7.2, the maps f_0 and f_1 induce the same homomorphism $\Phi(Y) \to \Phi(X)$, hence the same homomorphism between $K(Y) = S(\Phi(Y))$ and $K(X) = S(\Phi(X))$. Since the diagram

$$\begin{array}{ccc} K(Y) & \xrightarrow{f_\alpha^*} & K(X) \\ & \nwarrow \quad \nearrow & \\ & K(P) & \end{array}$$

(where P is a point and $\alpha = 0, 1$) is commutative, f_0 and f_1 also induce the same homomorphism $\tilde{K}(Y) \to \tilde{K}(X)$. □

1.26. Proposition. *Suppose X is the disjoint union of open subspaces $X_1 \cup X_2 \cup \cdots \cup X_n$. Then the inclusions of the X_i in X induce a decomposition of $K(X)$ as a direct product $K(X_1) \times K(X_2) \times \cdots \times K(X_n) = K(X_1) \oplus K(X_2) \oplus \cdots \oplus K(X_n)$.*

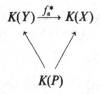

Proof. Since a bundle over X is characterized by its restrictions to each X_i, we have $\Phi(X) \approx \Phi(X_1) \times \Phi(X_2) \times \cdots \times \Phi(X_n)$. Hence $K(X) \approx K(X_1) \times K(X_2) \times \cdots \times K(X_n)$. □

1.27. *Remark.* This last proposition is false for the functor $\tilde{K}$. For example, if X is the disjoint union of two points P_1 and P_2, then $\tilde{K}(X) \approx \mathbb{Z}$, but $\tilde{K}(P_i) = 0$, for $i = 1, 2$.

1.28. Recall that every vector bundle E defines a locally constant function $r: X \to \mathbb{N}$ given by $r(x) = \operatorname{Dim}(E_x)$ (I.2.9). If we let $H^0(X; \mathbb{N})$ denote the abelian monoid of locally constant functions on X with values in $\mathbb{N}$, we see that in fact r defines a monoid homomorphism, also denoted by r, from $\Phi(X)$ to $H^0(X; \mathbb{N})$. It is clear that the symmetrization of $H^0(X; \mathbb{N})$ is the abelian group $H^0(X; \mathbb{Z})$ of locally constant functions over X with values in $\mathbb{Z}$ (* $H^0(X; \mathbb{Z})$ is the first Čech cohomology group of X (Eilenberg-Steenrod [1])*). Therefore, r defines a group homomorphism (denoted again by r)

$$K(X) \longrightarrow H^0(X; \mathbb{Z}).$$

1.29. Proposition. *Letting* $K'(X) = \operatorname{Ker}[K(X) \to H^0(X; \mathbb{Z})]$, *we have an exact sequence*

$$0 \longrightarrow K'(X) \longrightarrow K(X) \xrightarrow{r} H^0(X; \mathbb{Z}) \longrightarrow 0$$

which splits canonically. Moreover, if X is connected, then $K'(X)$ and $\tilde{K}(X)$ are canonically isomorphic.

Proof. Let $f: X \to \mathbb{N}$ be a locally constant function. Since X is compact, f only takes on a finite number of values $n_1, \ldots, n_p$, and $X = X_1 \cup \cdots \cup X_p$ where $X_i = f^{-1}(\{n_i\})$. Let E be the vector bundle defined over X_i by $X_i \times k^{n_i}$. Then the correspondence $f \mapsto E$ defines a monoid homomorphism $t: H^0(X; \mathbb{N}) \to \Phi(X)$ such that $r \cdot t = \operatorname{Id}$. By symmetrization, t induces a group map $H^0(X; \mathbb{Z}) \to K(X)$, which is a right-inverse to $r: K(X) \to H^0(X; \mathbb{Z})$.

If X is connected, we have $H^0(X; \mathbb{Z}) \approx \mathbb{Z}$, and $K'(X) \approx \operatorname{Coker}[\mathbb{Z} \to K(X)]$ following from what we have just said. Now the map $\mathbb{Z} \to K(X)$ is identical to the one induced by the projection of X to a point. Hence $K'(X) \approx \tilde{K}(X)$. □

1.30. Again let $\Phi_n(X)$ ($\Phi_n^k(X)$ when we want to specify the field k) be the set of isomorphism classes of vector bundles of rank n over X. Taking the Whitney sum by trivial bundles enables us to define an inductive system of sets

$$\Phi_0(X) \longrightarrow \Phi_1(X) \longrightarrow \cdots \longrightarrow \Phi_n(X) \longrightarrow \cdots.$$

The direct limit of this system, $\Phi'(X)$, can be provided with an abelian monoid structure, using the maps

$$\Phi_n(X)\times\Phi_p(X)\longrightarrow\Phi_{n+p}(X)$$

induced by the Whitney sum of vector bundles.

If $\dot{E}$, the class of the vector bundle E, is an element of $\Phi_n(X)$, we have $[E]-[\theta_n]\in \operatorname{Ker} r\colon K(X)\to H^0(X;\mathbb{Z})$. The correspondence $\dot{E}\mapsto[E]-[\theta_n]$ for $\dot{E}\in\Phi_n(X)$, induces a monoid homomorphism $\Phi'(X)\to K'(X)$.

1.31. Proposition. *The homomorphism $\Phi'(X)\to K'(X)$ defined above, is an isomorphism. Hence $\Phi'(X)$ is an abelian group.*

Proof. If $[E]-[\theta_n]=[F]-[\theta_p]$ in $K(X)$, by 1.17 there exists an integer q such that $E\oplus\theta_{p+q}\approx F\oplus\theta_{n+q}$. Hence, the map $\Phi'(X)\to K'(X)\subset K(X)$ is injective. Let u be an element of $K'(X)$. Again by 1.17, u can be written as $[E]-[\theta_n]$ with $r([E]-[\theta_n])=0$, i.e. $\operatorname{Dim}(E_x)=n$ for every point $x\in X$; hence the map is surjective. □

1.32. Proposition. *Let BO be the inductive limit of the system of topological spaces*

$$BO(1)\longrightarrow\cdots\longrightarrow BO(n)\longrightarrow\cdots,$$

where the map $BO(n)\to BO(n+1)$ is induced by the map between Grassmannians $G_n(\mathbb{R}^N)\to G_{n+1}(\mathbb{R}^{N+1})$ which consists of adding the subspace generated by the last vector $e_{N+1}=(0,\ldots,1)$. Then we have a natural isomorphism of functors

$$K'_{\mathbb{R}}(X)\approx[X,BO].$$

In the same way, let BU be the inductive limit of the system

$$BU(1)\longrightarrow\cdots\longrightarrow BU(n)\longrightarrow\cdots.$$

Then we have a natural isomorphism of functors

$$K'_{\mathbb{C}}(X)\approx[X,BU].$$

Proof. Since the spaces $BO(n)$ are paracompact [Cartan-Schwarz [1] exposé 5] and hence normal, and since $BO(n)$ is closed in $BO(n+1)$, we have $[X,BO]\approx$ inj lim$[X,BO(n)]$ because X is compact. According to theorem I.7.15, $\Phi_n^{\mathbb{R}}(X)\approx[X,BO(n)]$ and the map $\Phi_n^{\mathbb{R}}(X)\to\Phi_{n+1}^{\mathbb{R}}(X)$, induced by the addition of a trivial bundle of rank one, coincides with the map induced by the inclusion of $BO(n)$ in $BO(n+1)$. Therefore $K'_{\mathbb{R}}(X)\approx\Phi'_{\mathbb{R}}(X)\approx\text{inj lim}\,\Phi_n^{\mathbb{R}}(X)\approx\text{inj lim}[X,BO(n)]\approx[X,BO]$. For $k=\mathbb{C}$, the proof is analogous. □

1.33. Theorem. *For every compact space X we have natural isomorphisms*

$$K_{\mathbb{R}}(X) \approx [X, \mathbb{Z} \times BO] \quad \text{and} \quad K_{\mathbb{C}}(X) \approx [X, \mathbb{Z} \times BU]$$

where $\mathbb{Z}$ is provided with the discrete topology.

Proof. We will only give a proof for the real case, since the complex case is similar. From 1.29, we have $K_{\mathbb{R}}(X) \approx H^0(X; \mathbb{Z}) \oplus K'_{\mathbb{R}}(X)$. Since $H^0(X; \mathbb{Z}) \approx [X, \mathbb{Z}]$, it follows that $K_{\mathbb{R}}(X) \approx [X, \mathbb{Z}] \times [X, BO] \approx [X, \mathbb{Z} \times BO]$. □

1.34. In the case $X = S^n$, it is possible to give a more complete interpretation of $\tilde{K}(X)$ (hence of $K(X) = \mathbb{Z} \oplus \tilde{K}(X)$). It follows from Theorem I.7.6, that $\Phi_p^{\mathbb{C}}(S^n) \approx \pi_{n-1}(\mathrm{GL}_p(\mathbb{C})) \approx \pi_{n-1}(\mathrm{U}(p))$. Therefore $\tilde{K}_{\mathbb{C}}(S^n) \approx \operatorname{inj\,lim} \pi_{n-1}(\mathrm{U}(p)) \approx \pi_{n-1}(\mathrm{U})$, where $\mathrm{U} = \operatorname{inj\,lim} \mathrm{U}(p)$. * In fact, $\pi_{n-1}(\mathrm{U}) \approx \pi_{n-1}(\mathrm{U}(p))$ for $p > n - \frac{1}{2}$ by I.3.13.* In the same way $\Phi_p^{\mathbb{R}}(S^n) \approx \pi_{n-1}(\mathrm{GL}_p(\mathbb{R}))/\pi_0(\mathrm{GL}_p(\mathbb{R})) \approx \pi_{n-1}(\mathrm{O}(p))/\mathbb{Z}/2$. Therefore, $\tilde{K}_{\mathbb{R}}(S^n) \approx \operatorname{inj\,lim} \pi_{n-1}(\mathrm{O}(p))/\mathbb{Z}/2 \approx \operatorname{inj\,lim} \pi_{n-1}(\mathrm{O}(p))$ since the action of $\mathbb{Z}/2$ over $\pi_{n-1}(\mathrm{O}(p))$ is trivial if p is odd. Thus we obtain the isomorphism $\tilde{K}_{\mathbb{R}}(S^n) \approx \pi_{n-1}(\mathrm{O})$, where $\mathrm{O} = \operatorname{inj\,lim} \mathrm{O}(p)$. * In fact, $\pi_{n-1}(\mathrm{O}) \approx \pi_{n-1}(\mathrm{O}(p))$ when $p > n$, by I.3.13.* One of the main purposes of this book is to prove the isomorphisms $K_{\mathbb{R}}(S^n) \approx K_{\mathbb{R}}(S^{n+8})$ and $K_{\mathbb{C}}(S^n) \approx K_{\mathbb{C}}(S^{n+2})$ (see Chapter III). This will prove the theorems stated in I.3.13. More generally, we can prove that $\tilde{K}_{\mathbb{R}}(S'(X)) \approx [X, 0]'$ and $\tilde{K}_{\mathbb{C}}(S'(X)) \approx [X, U]'$ (homotopy classes of maps which preserve base points) in a similar manner.

Exercises (Section II.6) 1–8 and 10.

2. The Grothendieck Group of a Functor. The Group $K(X, Y)$

2.1. Definition. Let $\mathscr{C}$ be an additive category. A *Banach structure* on $\mathscr{C}$ is given by a Banach space structure on all the groups $\mathscr{C}(E, F)$, where E and F run through the objects of $\mathscr{C}$. Moreover, we assume that the map

$$\mathscr{C}(E, F) \times \mathscr{C}(F, G) \longrightarrow \mathscr{C}(E, G)$$

given by the composition of morphisms, is bilinear and continuous (all Banach spaces are over the basic field $k = \mathbb{R}$ or $\mathbb{C}$). A *Banach category* is an additive category provided with a Banach structure.

2.2. Example. For X a compact space, we gave the category $\mathscr{E}(X)$ a Banach structure in I.6.20. More generally, if A is a Banach algebra (with unit but not necessarily commutative), we have given the category $\mathscr{P}(A)$ a Banach structure (I.6.22).

2.3. Example. Let $\mathscr{H}$ be the category considered in 1.9. Then $\mathscr{H}$ becomes a Banach category by giving $\mathscr{H}(E, F)$ the classical norm

$$\|f\| = \operatorname*{Sup}_{x \neq 0} \frac{\|f(x)\|}{\|x\|}$$

2.4. Example. Let $\check{\mathscr{H}}$ be the following category. The objects are the objects of $\mathscr{H}$. The set of morphisms $\check{\mathscr{H}}(E, F)$ is defined by $\check{\mathscr{H}}(E, F) = \mathscr{H}(E, F)/\mathscr{K}(E, F)$, where $\mathscr{K}(E, F)$ is the set of completely continuous operators from E to F (i.e. limits of operators of finite rank). Then $\check{\mathscr{H}}(E, F)$ is a Banach space and the composition of morphisms in $\mathscr{H}$ induces a composition of morphisms in $\check{\mathscr{H}}$.

2.5. Example. If $\mathscr{C}$ is an arbitrary Banach category, then the associated pseudo-abelian category $\tilde{\mathscr{C}}$ (I.6.9) is a Banach category because $\tilde{\mathscr{C}}((E, p), (F, q))$ is a closed subspace of $\mathscr{C}(E, F)$.

2.6. Definitions. Let $\mathscr{C}$ and $\mathscr{C}'$ be additive categories, and $\varphi: \mathscr{C} \to \mathscr{C}'$ be an additive functor. Then φ is called *quasi-surjective* if every object of $\mathscr{C}'$ is a direct factor of an object of the form $\varphi(E)$, where $E \in \mathrm{Ob}(\mathscr{C})$; φ is called *full* if the map $\mathscr{C}(E, F) \to \mathscr{C}'(\varphi(E), \varphi(F))$ is surjective for $E, F \in \mathrm{Ob}(\mathscr{C})$. Finally, if $\mathscr{C}$ and $\mathscr{C}'$ are Banach categories, the functor φ is called a *Banach functor* if the map $\mathscr{C}(E, F) \to \mathscr{C}'(\varphi(E), \varphi(F))$ is linear and continuous.

2.7. Example. Let $\varphi: \mathscr{E}(X) \to \mathscr{E}(Y)$ be the functor defined by $\varphi(E) = E_Y$, where Y is a closed subspace of the compact space X. Then α is a Banach functor which is full and quasi-surjective. By I.5.9 we see that up to isomorphism the "restriction map" $\mathscr{E}(X)(E, F) \to \mathscr{E}(Y)(E_Y, F_Y)$ is identical with the map $\Gamma(X, \mathrm{HOM}(E, F)) \to \Gamma(Y, \mathrm{HOM}(E, F)_Y)$. From I.6.23, it follows that φ is a Banach functor. Moreover, φ is full (I.5.10), and also quasi-surjective because any object of $\mathscr{E}(Y)$ is a direct factor of a trivial bundle $\theta_n = Y \times k^n$ (I.6.5), and clearly $\theta_n = \varphi(E)$ where $E = X \times k^n$.

2.8. Example. More generally, let $f: Y \to X$ be an arbitrary continuous map. Then the inverse image functor $f^*: \mathscr{E}(X) \to \mathscr{E}(Y)$ (I.2.6) is a quasi-surjective Banach functor. To see this, let us consider the map $u_{E,F}: \mathscr{E}(X)(E, F) \to \mathscr{E}(Y)(f^*(E), f^*(F))$, induced by f^*. If $E = X \times k^n$ and $F = X \times k^p$, this is essentially the map $C(X)^{np} \to C(Y)^{np}$ defined by $(\lambda_1, \ldots, \lambda_{np}) \mapsto (\lambda_1 \cdot f, \ldots, \lambda_{np} \cdot f)$, which is clearly linear and continuous. For any E and F, the map $u_{E,F}$ is continuous, since E and F are each a direct factor of some trivial bundle, say $X \times k^n$ and $X \times k^p$ respectively. Finally, f^* is quasi-surjective by the same argument as in 2.7.

2.9. Example. Let A and B be Banach algebras, and $u: A \to B$, a continuous homomorphism. Then u induces a functor $u_*: \mathscr{P}(A) \to \mathscr{P}(B)$, defined by $E \mapsto E \otimes_A B$ (regarding B as a left A-module via u). We claim that u_* is a quasi-surjective Banach functor. To see this, we repeat the argument in 2.8. If $E = A^n$ and $F = A^p$, the map $\mathscr{P}(A)(E, F) \to \mathscr{P}(B)(u_*(E), u_*(F))$ is essentially $A^{np} \to B^{np}$ which is linear and continuous. In general, the map $\mathscr{P}(A)(E, F) \to \mathscr{P}(B)(u_*(E), u_*(F))$ is linear and continuous because E and F can be written as direct factors of free modules. The functor u_* is full if and only if u is surjective.

2.10. Example. The "quotient functor" $\mathscr{H} \to \check{\mathscr{H}}$ (2.4), which is the identity on the objects, is a Banach functor which is full and quasi-surjective.

2.11. Example. Let $\mathscr{C}$ be an arbitrary Banach category, and let $\varphi: \mathscr{C} \to \mathscr{C}$ be the functor defined by $\varphi(E) = \underbrace{E \oplus \cdots \oplus E}_{n}$. Then φ is a Banach functor which is quasi-surjective (but not full unless $\mathscr{C} = 0$ or $n = 1$).

2.12. *Remark.* Let $\varphi: \mathscr{C} \to \mathscr{C}'$ be a Banach functor which is full (resp. faithful, resp. quasi-surjective). Then the functor $\tilde{\varphi}: \tilde{\mathscr{C}} \to \tilde{\mathscr{C}}'$ between the associated pseudo-abelian categories (I.6.9) is full (resp. faithful, resp. quasi-surjective).

2.13. Let $\varphi: \mathscr{C} \to \mathscr{C}'$ be a quasi-surjective Banach functor. We wish to define a "relative" group $K(\varphi)$ (which coincides with $K(\mathscr{C})$ when $\mathscr{C}' = 0$). Let $\Gamma(\varphi)$ denote the set of triples (E, F, α), where E and F are objects of $\mathscr{C}$ and $\alpha: \varphi(E) \to \varphi(F)$ is an isomorphism. Two triples (E, F, α) and (E', F', α') are called *isomorphic* if there exist isomorphisms $f: E \to E'$ and $g: F \to F'$ such that the following diagram commutes.

$$\begin{array}{ccc} \varphi(E) & \xrightarrow{\alpha} & \varphi(F) \\ {\scriptstyle \varphi(f)}\downarrow & & \downarrow{\scriptstyle \varphi(g)} \\ \varphi(E') & \xrightarrow{\alpha'} & \varphi(F') \end{array}$$

A triple (E, F, α) is called *elementary* if $E = F$ and if α is homotopic to $\mathrm{Id}_{\varphi(E)}$ within the automorphisms of $\varphi(E)$. Finally, we define the *sum* of two triples (E, F, α) and (E', F', α') to be $(E \oplus E', F \oplus F', \alpha \oplus \alpha')$.

Then $K(\varphi)$ is the quotient of $\Gamma(\varphi)$ by the following equivalence relation: $\sigma \sim \sigma' \Leftrightarrow \exists$ elementary τ and τ' such that $\sigma + \tau$ is isomorphic to $\sigma' + \tau'$. The sum of triples obviously provides the set $K(\varphi)$ with a monoid structure. We let $d(E, F, \alpha)$ denote the class of (E, F, α) in the monoid $K(\varphi)$. As a direct consequence of the definition, we notice that $d(E, F, \alpha) = 0$ if and only if there exist objects G and H of $\mathscr{C}$, and isomorphisms $u: E \oplus G \to H$ and $v: F \oplus G \to H$, such that $\varphi(v) \cdot (\alpha \oplus \mathrm{Id}_{\varphi(G)}) \cdot \varphi(u^{-1})$ is homotopic to $\mathrm{Id}_{\varphi(H)}$ within the automorphisms of $\varphi(H)$.

2.14. Proposition. *The monoid $K(\varphi)$ is an abelian group which coincides with $K(\mathscr{C})$ when $\mathscr{C}' = 0$. Moreover $d(E, F, \alpha) + d(F, E, \alpha^{-1}) = 0$.*

Proof. Let $d(E, F, \alpha)$ be an arbitrary element of $K(\varphi)$. Then $d(E, F, \alpha) + d(F, E, \alpha^{-1}) = d(E \oplus F, F \oplus E, \alpha \oplus \alpha^{-1})$. The triple $(E \oplus F, F \oplus E, \alpha \oplus \alpha^{-1})$ is isomorphic to

$(E \oplus F, E \oplus F, \beta)$, where β is the automorphism of $\varphi(E) \oplus \varphi(F)$ defined by the matrix

$$\beta = \begin{pmatrix} 0 & -\alpha^{-1} \\ \alpha & 0 \end{pmatrix}.$$

In the group $\mathrm{Aut}(\varphi(E) \oplus \varphi(F))$ we have the identity

$$\begin{pmatrix} 0 & -\alpha^{-1} \\ \alpha & 0 \end{pmatrix} = \begin{pmatrix} 1 & -\alpha^{-1} \\ 0 & 1 \end{pmatrix} \begin{pmatrix} 1 & 0 \\ \alpha & 1 \end{pmatrix} \begin{pmatrix} 1 & -\alpha^{-1} \\ 0 & 1 \end{pmatrix}.$$

Now let $\sigma: I \to \mathrm{Aut}(\varphi(E) \oplus \varphi(F))$ be the continuous map defined by

$$\sigma(t) = \begin{pmatrix} 1 & -t\alpha^{-1} \\ 0 & 1 \end{pmatrix} \begin{pmatrix} 1 & 0 \\ t\alpha & 1 \end{pmatrix} \begin{pmatrix} 1 & -t\alpha^{-1} \\ 0 & 1 \end{pmatrix}.$$

Since $\sigma(0) = 1$ and $\sigma(1) = \beta$, the triple $(E \oplus F, E \oplus F, \beta)$ is elementary, and $d(E, F, \alpha^{-1})$ is the opposite of $d(E, F, \alpha)$ as required. Moreover, let $i: K(\varphi) \to K(\mathscr{C})$ be the homomorphism defined by $i(d(E, F, \alpha)) = [E] - [F]$. It is clear that i is an isomorphism when $\mathscr{C}' = 0$, since the triple (E, F, α) is essentially determined by the pair (E, F) (1.2). □

2.15. Proposition. *Let $d(E, F, \alpha)$ and $d(E, F, \alpha')$ be elements of $K(\varphi)$ such that α and α' are homotopic within the isomorphisms from $\varphi(E)$ to $\varphi(F)$. Then $d(E, F, \alpha) = d(E, F, \alpha')$.*

Proof. By 2.14, we have $d(E, F, \alpha) - d(E, F, \alpha') = d(E, F, \alpha) + d(F, E, \alpha'^{-1}) = d(E \oplus F, F \oplus E, \alpha \oplus \alpha'^{-1}) = d(E \oplus F, E \oplus F, \beta')$, where β' is the automorphism of $\varphi(E) \oplus \varphi(F)$ defined by the matrix

$$\beta' = \begin{pmatrix} 0 & -\alpha'^{-1} \\ \alpha & 0 \end{pmatrix}.$$

Since α' is homotopic to α within the isomorphisms from $\varphi(E)$ to $\varphi(F)$, β' is homotopic to

$$\beta = \begin{pmatrix} 0 & -\alpha^{-1} \\ \alpha & 0 \end{pmatrix},$$

within the automorphisms of $\varphi(E) \oplus \varphi(F)$. As was shown in 2.14, the map β is homotopic to $\mathrm{Id}_{\varphi(E) \oplus \varphi(F)}$. It follows that $(E \oplus F, E \oplus F, \beta')$ is elementary and thus $d(E, F, \alpha) = d(E, F, \alpha')$. □

2.16. Proposition. *Let $d(E, F, \alpha)$ and $d(F, G, \beta)$ be elements of $K(\varphi)$. Then we have the relation*

$$d(E, F, \alpha)+d(F, G, \beta)=d(E, G, \beta\alpha).$$

Proof. The left-hand side of the equation can be written as $d(E\oplus F, F\oplus G, \alpha\oplus\beta)=d(E\oplus F, G\oplus F, \gamma)$, where γ is defined by the matrix

$$\gamma=\begin{pmatrix}0 & -\beta\\ \alpha & 0\end{pmatrix}.$$

On the other hand, $d(E, G, \beta\alpha)=d(E\oplus F, G\oplus F, \gamma')$ where $\gamma'=\beta\alpha\oplus 1$. The automorphism $\gamma\gamma'^{-1}$ is defined by the matrix

$$\begin{pmatrix}0 & -\beta\\ \beta^{-1} & 0\end{pmatrix},$$

which is homotopic to the identity within the automorphisms of $\varphi(G)\oplus\varphi(F)$. Hence γ is homotopic to γ' within the isomorphisms from $\varphi(E)\oplus\varphi(F)$ to $\varphi(G)\oplus\varphi(F)$, and thus $d(E\oplus F, G\oplus F, \gamma)=d(E\oplus F, G\oplus F, \gamma')$ by 2.15. □

2.17. Example. We return to Example 2.11 where $\mathscr{C}=\mathscr{E}_{\mathbf{R}}$, the category of finite dimensional real vector spaces, and where $n=2$. Let $d(E, F, \alpha)$, where $\alpha: E\oplus E\xrightarrow{\approx} F\oplus F$, be an element of $K(\varphi)$. For any isomorphism $u: F\to E$, the sign of the determinant of the composite $E\oplus E\xrightarrow{\alpha} F\oplus F\xrightarrow{u\oplus u} E\oplus E$ is independent of the choice of u. This defines an isomorphism $K(\varphi)\approx\mathbf{Z}/2$.

2.18. Example. Let $d(E, F, \alpha)$ be an element of $K(\varphi)$ with $\varphi:\mathscr{H}\to\hat{\mathscr{H}}$ (2.4). If $\tilde{\alpha}: E\to F$ is a continuous linear map such that $\varphi(\tilde{\alpha})=\alpha$, then the "*index*" of $\tilde{\alpha}$, i.e. $\mathrm{Dim}(\mathrm{Ker}(\tilde{\alpha}))-\mathrm{Dim}(\mathrm{coker}(\tilde{\alpha}))$, depends only on α. This defines an isomorphism $K(\varphi)\approx\mathbf{Z}$ (Exercise 6.12).

2.19. The example that interests us most is Example 2.7. This will be dealt with at the end of this section in terms of $\tilde{K}(X/Y)$, and then $K(\varphi)$ will be called $K(X, Y)$.

2.20. Proposition. *Let $i: K(\varphi)\to K(\mathscr{C})$ (resp. $j: K(\mathscr{C})\to K(\mathscr{C}')$ be the homomorphism defined by $i(d(E, F, \alpha))=[E]-[F]$ (resp. $j([E]-[F])=[\varphi(E)]-[\varphi(F)]$. Then we have an exact sequence*

$$K(\varphi)\xrightarrow{\;i\;}K(\mathscr{C})\xrightarrow{\;j\;}K(\mathscr{C}').$$

Moreover, if there exists a functor $\psi:\mathscr{C}'\to\mathscr{C}$ such that $\varphi\psi\approx\mathrm{Id}_{\mathscr{C}'}$, then we have the split exact sequence

$$0\longrightarrow K(\varphi)\xrightarrow{\;i\;}K(\mathscr{C})\xrightarrow{\;j\;}K(\mathscr{C}')\longrightarrow 0.$$

Proof. We have $(j\cdot i)(d(E,F,\alpha))=[\varphi(E)]-[\varphi(F)]=0$, since $\varphi(E)$ is isomorphic to $\varphi(F)$ by α. On the other hand, let $x=[E]-[F]$ be an element of $K(\mathscr{C})$ such that $j(x)=0$. From 1.16, there exists an object T' of $\mathscr{C}'$ such that $\varphi(E)\oplus T'\approx\varphi(F)\oplus T'$. Since the functor φ is quasi-surjective, we can find an object T of $\mathscr{C}$ and an object T'_1 of $\mathscr{C}'$ such that $\varphi(T)\approx T'\oplus T'_1$. Hence the objects $\varphi(E\oplus T)\approx\varphi(E)\oplus T'\oplus T'_1$ and $\varphi(F\oplus T)\approx\varphi(F)\oplus T'\oplus T'_1$ are isomorphic. Thus we can write $x=[E\oplus T]-[F\oplus T]=i(d(E\oplus T,F\oplus T,\delta))$, where δ is any isomorphism between $\varphi(E\oplus T)$ and $\varphi(F\oplus T)$. Now suppose there exists a functor $\psi:\mathscr{C}'\to\mathscr{C}$ such that $\varphi\psi\approx\mathrm{Id}_{\mathscr{C}'}$. We will prove the injectivity of i, which is the main part of the proof. If $i(d(E,F,\alpha))=[E]-[F]=0$, there exists an object T of $\mathscr{C}$ such that $E\oplus T\approx F\oplus T$. Hence $d(E,F,\alpha)=d(E\oplus T,F\oplus T,\alpha\oplus 1)=d(G,G,\beta)$, where $G=E\oplus T$ and β is the composition of $\alpha\oplus 1$ and the isomorphism from $\varphi(E\oplus T)$ to $\varphi(F\oplus T)$. Thus we have the commutative diagram

$$\begin{array}{ccc}
\varphi(G) & \xrightarrow{\ \beta\ } & \varphi(G)\\
\downarrow{\scriptstyle\gamma} & & \downarrow{\scriptstyle\gamma}\\
(\varphi\psi)(\varphi(G)) & \xrightarrow{(\varphi\psi)(\beta)} & (\varphi\psi)(\varphi(G)),
\end{array}$$

where γ is induced by the isomorphism $\mathrm{Id}_{\mathscr{C}'}\approx\varphi\psi$. If we write $(\varphi\psi)(\varphi(G))$ as $\varphi((\psi\varphi)(G))$, we have $\beta=\gamma^{-1}\cdot(\varphi\psi)(\beta)\cdot\gamma$ and $d(G,G,\beta)=d(G,(\psi\varphi)(G),\gamma)+d((\psi\varphi)(G),(\psi\varphi)(G),(\varphi\psi)(\beta))+d((\psi\varphi)(G),G,\gamma^{-1})$ by applying 2.16 twice. But $d(G,(\psi\varphi)(G),\gamma)+d((\psi\varphi)(G),G,\gamma^{-1})=0$ from 2.14, and $d((\psi\varphi)(G),(\psi\varphi)(G),(\varphi\psi)(\beta))=0$ because the triple $((\psi\varphi)(G),(\psi\varphi)(G),(\varphi\psi)(\beta))$ is isomorphic to $((\psi\varphi)(G),(\psi\varphi)(G),\mathrm{Id})$ thanks to the commutative diagram

$$\begin{array}{ccc}
\varphi(\psi\varphi)(G)=(\varphi\psi)(\varphi(G)) & \xrightarrow{(\varphi\psi)(\beta)} & (\varphi\psi)(\varphi(G))=\varphi(\psi\varphi)(G)\\
\downarrow{\scriptstyle(\varphi\psi)(\beta)} & & \downarrow{\scriptstyle\mathrm{Id}}\\
\varphi(\psi\varphi)(G) & \xrightarrow{\ \mathrm{Id}\ } & \varphi(\psi\varphi)(G).
\end{array}$$ □

2.21. Lemma. *Let A and A' be Banach algebras (with unit but not necessarily commutative), and let $f: A\to A'$ be a continuous surjective ring homomorphism. Let A^*, A'^* denote the group of invertible elements in A, A' respectively. Let $\sigma': I\to A'$ be a continuous map such that $\sigma'(t)\in A'^*$ for every point $t\in I=[0,1]$, and such that $\sigma'(0)=f(\alpha)$ for some $\alpha\in A^*$. Then there is an element β of A^* such that $f(\beta)=\sigma'(1)$.*

Proof. For any Banach algebra C we have a continuous map $\exp: C\to C^*$ defined by $\exp(x)=1+x+x^2/2!+x^3/3!+\cdots$, and its image contains the set V of points y such that $\|y-1\|<1$, since the logarithm function is defined on this set. Since I is compact, we can find a finite sequence of points $t_i\in I$, such that $0=t_0<t_1<t_2<\cdots<t_n=1$ with $\sigma'(t_i)^{-1}\sigma'(t_{i+1})\in V$ (for $C=A'$). Let $\alpha'_i=\mathrm{Log}(\sigma'(t_i)^{-1}\sigma'(t_{i+1}))$, for $i=0,\ldots,n-1$. Then $\alpha'=\sigma'(1)$ can be written as $\sigma'(0)\cdot\exp(\alpha'_0)\cdot\exp(\alpha'_1)\ldots\cdot$

$\exp(\alpha'_{n-1})$. For each $i=0,\ldots,n-1$ choose $\alpha_i \in A$ such that $f(\alpha_i)=\alpha'_i$. Then $\beta=\alpha\cdot\exp(\alpha_0)\cdot\exp(\alpha_1)\ldots\cdot\exp(\alpha_{n-1})$ is the point required. □

The next lemma and proposition will be useful in the following chapters:

2.22. Lemma. *Let $B(X)$ denote, in general, the Banach algebra of continuous functions on the compact space X, with values in the Banach algebra B. Then, if the homomorphism f of Lemma 2.21 induces a surjective ring homomorphism $\tilde{f}: A(I) \to A'(I)$, there exists a continuous map $\sigma: I \to A^*$ such that $\sigma(0)=\alpha$ and $f(\sigma(t))=\sigma'(t)$.*

Proof. We apply Lemma 2.21 to the situation $\tilde{f}: A(I) \to A'(I)$, where α is replaced by the constant function $\tilde{\alpha}$ defined by $\tilde{\alpha}(t)=\alpha$, and σ' is replaced by $\tilde{\sigma}': I \to A'(I)^*$ defined by $\tilde{\sigma}'(t)(u)=\sigma'(tu)$. Therefore, we can find $\sigma_1 \in A(I)^*$ such that $\tilde{f}(\sigma_1(t))=\sigma'(t)$. Then $\sigma(t)=\sigma_1(t)\sigma_1(0)^{-1}\cdot\alpha$ has the required properties. □

2.23. *Remark.* Using more functional analysis it is possible to prove that $A(X) \to A'(X)$ is surjective when f is surjective. We could also prove that $A^* \to A'^*$ is a locally trivial fibration (hence a Serre fibration). However, these results are not necessary for our purposes.

2.24. Proposition. *Let X be a compact space, X' a closed subspace, and E a vector bundle over X. Let $\alpha: E \to E$ be an automorphism, and $\sigma': I \to \operatorname{Aut}(E_{X'})$ a continuous map such that $\sigma'(0)=\alpha|_{X'}$. Then there exists a continuous map $\sigma: I \to \operatorname{Aut}(E)$ such that $\sigma(0)=\alpha$ and $\sigma(t)|_{X'}=\sigma'(t)$.*

Proof. This is a consequence of the previous lemma applied to $A=\operatorname{End}(E)$ and $A'=\operatorname{End}(E_{X'})$ (I.5.10). □

2.25. Proposition. *Let $\varphi: \mathscr{C} \to \mathscr{C}'$ be a full Banach functor and let $\tau=(E, E, \alpha')$ be an elementary triple. Then τ is isomorphic to the triple $(E, E, \mathrm{Id}_{\varphi(E)})$.*

Proof. Let A (resp. A') be the Banach algebra $\operatorname{End}(E)$ (resp. $\operatorname{End}(\varphi(E))$). Then φ induces a continuous ring homomorphism $A \to A'$ which is surjective. Since the triple $(E, E, \mathrm{Id}_{\varphi(E)})$ is elementary, we can find a continuous path $\sigma: I \to A'^*$, such that $\sigma(0)=1$ and $\sigma(1)=\alpha'$. From Lemma 2.21, we can find an automorphism α of E such that $\varphi(\alpha)=\alpha'$. Hence the triples (E, E, α') and $(E, E, \mathrm{Id}_{\varphi(E)})$ are isomorphic, as can be seen from the commutative diagram

$$\begin{array}{ccc} \varphi(E) & \xrightarrow{\alpha'} & \varphi(E) \\ {\scriptstyle \varphi(\alpha)}\downarrow & & \downarrow{\scriptstyle \varphi(\mathrm{Id})} \\ \varphi(E) & \xrightarrow{\mathrm{Id}} & \varphi(E) \end{array}$$

□

2.26. Corollary. *Let $\varphi: \mathscr{C} \to \mathscr{C}'$ be a Banach functor which is full and quasi-surjective, and let $\dot{K}(\varphi)$ be the monoid obtained from the definition of $K(\varphi)$ by replacing ele-*

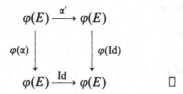

mentary triples by triples of the form $(E, E, \mathrm{Id}_{\varphi(E)})$. Then the natural map from $\dot{K}(\varphi)$ to $K(\varphi)$ is an isomorphism. Hence $\dot{K}(\varphi)$ is an abelian group.

2.27. *Remark.* This corollary provides a purely algebraic description of $K(\varphi)$ when φ is full and quasi-surjective. We identify the groups $\dot{K}(\varphi)$ and $K(\varphi)$ from now on.

2.28. Proposition. *Let $d(E, F, \alpha')$ be an element of $K(\varphi)$ where $\varphi: \mathscr{C} \to \mathscr{C}'$ is a Banach functor which is full and quasi-surjective. Then $d(E, F, \alpha')=0$ if and only if there exists an object G of $\mathscr{C}$ and an isomorphism $\beta: E \oplus G \to F \oplus G$ such that $\varphi(\beta)=\alpha' \oplus \mathrm{Id}_{\varphi(G)}$.*

Proof. According to the previous corollary, we can find two triples $(G, G, \mathrm{Id}_{\varphi(G)})$ and $(H, H, \mathrm{Id}_{\varphi(H)})$ such that the triples $(E \oplus G, F \oplus G, \alpha' \oplus \mathrm{Id}_{\varphi(G)})$ and $(H, H, \mathrm{Id}_{\varphi(H)})$ are isomorphic. If $f: E \oplus G \to H$ and $g: F \oplus G \to H$ are the required isomorphisms, we have the commutative diagram

$$\begin{array}{ccc} \varphi(E \oplus G) & \xrightarrow{\alpha' \oplus \mathrm{Id}_{\varphi(G)}} & \varphi(F \oplus G) \\ \Big\downarrow{\scriptstyle \varphi(f)} & & \Big\downarrow{\scriptstyle \varphi(g)} \\ \varphi(H) & \xrightarrow{\mathrm{Id}_{\varphi(H)}} & \varphi(H) \end{array}$$

Hence, $\alpha' \oplus \mathrm{Id}_{\varphi(G)} = \varphi(\beta)$ where $\beta = g^{-1} \cdot f$. The converse is obvious. $\square$

2.29. By 2.7, we can apply the result above to the case for $\mathscr{C} = \mathscr{E}(X)$, and $\mathscr{C}' = \mathscr{E}(Y)$, where Y is a closed subspace of the compact space X and $\varphi: \mathscr{E}(X) \to \mathscr{E}(Y)$ is the functor induced by the restriction of bundles. In this case we will use the notation $K(X, Y)$ instead of $K(\varphi)$, and more precisely $K_{\mathbb{R}}(X, Y)$ or $K_{\mathbb{C}}(X, Y)$ if we want to specify the basic field. Paraphrasing Corollary 2.26, we may say that every element of $K(X, Y)$ can be written as $d(E, F, \alpha)$, where E and F are vector bundles over X and $\alpha: E_Y \to F_Y$ is an isomorphism. Moreover, $d(E, F, \alpha) = d(E', F', \alpha')$ if and only if there exist triples $(G, G, \mathrm{Id}_{G_Y})$ and $(G', G', \mathrm{Id}_{G'_Y})$, and isomorphisms $f: E \oplus G \to E' \oplus G'$ and $g: F \oplus G \to F' \oplus G'$, such that the following diagram commutes:

$$\begin{array}{ccc} (E \oplus G)|_Y & \xrightarrow{\alpha \oplus \mathrm{Id}_{G_Y}} & (F \oplus G)|_Y \\ \Big\downarrow{\scriptstyle f|_Y} & & \Big\downarrow{\scriptstyle g|_Y} \\ (E' \oplus G')|_Y & \xrightarrow{\alpha' \oplus \mathrm{Id}_{G'_Y}} & (F' \oplus G')|_Y \end{array}$$

From 2.20 we have the exact sequence

$$K(X, Y) \longrightarrow K(X) \longrightarrow K(Y).$$

If Y is a retract of X (i.e. if the inclusion map $i: Y \to X$ admits a left-inverse), there is a functor $\psi: \mathscr{E}(Y) \to \mathscr{E}(X)$ which is a right-inverse to the restriction functor. Then, from the second part of 2.20, we have the split exact sequence

$$0 \longrightarrow K(X, Y) \longrightarrow K(X) \longrightarrow K(Y) \longrightarrow 0.$$

Finally, note that $K(X, \varnothing) \approx K(X)$.

2.30. Example. Let $X = B^2$ (resp. $Y = S^1$) be the unit ball (resp. sphere) in $\mathbb{R}^2$. Then we define an element $d(E, F, \alpha)$ of $K_{\mathbb{C}}(X, Y) = K_{\mathbb{C}}(B^2, S^1)$ by $E = F = B^2 \times \mathbb{C}$, and $\alpha(x, v) = (x, xv)$ for $x \in S^1 \subset B^2 \subset \mathbb{C}$. This element will play a vital role in Chapter III. We shall see later that $d(E, F, \alpha)$ is a generator of $K_{\mathbb{C}}(B^2, S^1) \approx \mathbb{Z}$.

2.31. Example. More generally, let $X = B^2 \times Z$ and $Y = S^1 \times Z$, where Z is an arbitrary compact space. Let G be a complex vector bundle over Z, and let $\pi^*(G)$ be its inverse image by the projection $\pi: B^2 \times Z \to Z$. Then we define an element $d(E, F, \alpha)$ of $K_{\mathbb{C}}(X, Y) = K(B^2 \times Z, S^1 \times Z)$ by $E = F = \pi^*(G) = B^2 \times G$, and $\alpha(x, v) = (x, xv)$ for $x \in S^1 \subset B^2 \subset \mathbb{C}$. In III.1 we will see that the correspondence $G \mapsto d(E, F, \alpha)$ induces an isomorphism $K_{\mathbb{C}}(Z) \approx K_{\mathbb{C}}(B^2 \times Z, S^1 \times Z)$.

2.32. Example. Let $\varphi: \mathscr{H} \to \check{\mathscr{H}}$ be the quotient functor of 2.10. Then we can define an element $d(E, F, \alpha)$ of $K(\varphi)$ by choosing $E = F = H = k \oplus k \oplus \cdots \oplus k \oplus \cdots$ (Hilbert sum of $\aleph_0$ copies of the basic field $k = \mathbb{R}$ or $\mathbb{C}$), and α, the class of the endomorphism $(x_1, x_2, \ldots, x_n, \ldots) \mapsto (x_2, x_3, \ldots, x_n, \ldots)$, which is invertible in $\check{\mathscr{H}}$ (but not in $\mathscr{H}$). It can be proved that $d(E, F, \alpha)$ is a generator of $K(\varphi) \approx \mathbb{Z}$ (Exercise 6.12 and Example 2.18).

2.33. The group $K(X, Y)$ depends "functorially" on the pair (X, Y). More precisely, recall that a morphism between pairs (X, Y) and (X', Y') is a continuous map $f: X \to X'$ such that $f(Y) \subset Y'$. Such a morphism induces a commutative diagram of categories

$$\begin{array}{ccc} \mathscr{E}(X) & \longrightarrow & \mathscr{E}(Y) \\ \uparrow f^* & & \uparrow f_1^* \\ \mathscr{E}(X') & \longrightarrow & \mathscr{E}(Y'). \end{array}$$

where $f_1 = f|_Y$, hence a morphism (again denoted f^*) between $K(X, Y)$ and $K(X', Y')$ given by the formula

$$f^*(d(E', F', \alpha')) = d(f^*(E), f^*(F), f_1^*(\alpha')).$$

2.34. Now consider the "quotient space" X/Y. If Y is non-empty, then X/Y is the compact space obtained by the identification of Y to a single point $\{y\}$. If Y is empty, X/Y is the disjoint union of X and a point outside, again denoted by $\{y\}$.

Notice that X/Y is actually the one point compactification of the locally compact space $X-Y$.

2.35. Theorem (Excision). *The projection $\pi: X \to X/Y$ induces an isomorphism*

$$\pi^*: K(X/Y, \{y\}) \to K(X, Y).$$

Proof. For Y empty, we leave the trivial verification to the reader. For Y non-empty, the proof breaks into two parts:

a) π^* *is surjective.* Let $d(E, F, \alpha)$ be an element of $K(X, Y)$. By adding the same bundle to E and F, we may assume without loss of generality, that F is a trivial bundle, say θ_n. We want to find a triple (E', F', α') defining an element of $K(X/Y, \{y\})$, such that the triples $(\pi^*(E'), \pi^*(F'), \pi_1^*(\alpha'))$ and (E, F, α) are isomorphic with $\pi_1: Y \to \{y\}$. According to I.5.11, there is a closed neighbourhood V of Y and an isomorphism $\beta: E_V \to F_V$ such that $\beta|_Y = \alpha$. Let E' be the vector bundle over X/Y obtained by clutching the bundle $E|_{X-Y}$ and the trivial bundle of rank n over V/Y, using $\beta|_{V-Y}$ (note that $X-Y \approx X/Y - \{y\}$ and $V - Y \approx V/Y - \{y\}$). Let F' be the trivial bundle of rank n over X/Y, and let $\alpha': E'|_{\{y\}} \to F'|_{\{y\}}$ be the isomorphism induced by the above clutching.

Then we can define an isomorphism $f: E \to \pi^*(E')$ by $f|_{X-Y} = \mathrm{Id}$, with the identification $\pi^* E'|_{X-Y} = E'|_{X-Y} = E|_{X-Y}$, and $f|_V = \mathrm{Id}$ with the identification $\pi^*(E')|_V = \theta_n|_V$ (I.3.3). It is now straightforward to check that the diagram

$$\begin{array}{ccc} E|_Y & \xrightarrow{\alpha} & F|_Y \\ {\scriptstyle f|_Y}\downarrow & & \| \\ \pi^*(E')|_Y & \xrightarrow{\pi_1^*(\alpha')} & \pi^*(F')|_Y \end{array}$$

is commutative.

b) π^* *is injective.* Let $d(E', F', \alpha')$ be an element of $K(X/Y, \{y\})$ such that $\pi^*(d(E', F', \alpha')) = d(\pi^*(E'), \pi^*(F'), \pi_1^*(\alpha')) = 0$. According to 2.28, there is a bundle T over X such that $\pi_1^*(\alpha') \oplus \mathrm{Id}_{T|_Y}$ can be extended by an isomorphism $\beta: \pi^*(E') \oplus T \to \pi^*(F') \oplus T$. As before we may assume that T is trivial. Let T' be the trivial bundle over X/Y of the same rank as T. Let $\beta': E' \oplus T' \to F' \oplus T'$ be the isomorphism which is equal to β over $X-Y$, and to α' over $\{y\}$. It is clear that β' is continuous and is an extension of $\alpha' \oplus \mathrm{Id}_{T'|_{\{y\}}}$ over X/Y. Hence $d(E', F', \alpha') = d(E' \oplus T', F' \oplus T', \alpha' \oplus \mathrm{Id}_{T'}) = 0$ by 2.28 again. □

2.36. *Remark.* Since $\{y\}$ is a retract of X/Y, we have $K(X/Y, \{y\}) \approx \mathrm{Ker}(K(X/Y) \to K(\{y\})) \approx \tilde{K}(X/Y)$ from 2.29.

2.37. Corollary. *Let X_1 and X_2 be closed subspaces of X such that $X_1 \cup X_2 = X$. Then the inclusion $(X_1, X_1 \cap X_2) \to (X_1 \cup X_2, X_2)$ induces an isomorphism*

$$K(X_1 \cup X_2, X_2) \xrightarrow{\approx} K(X_1, X_1 \cap X_2)$$

Proof. This follows directly from the commutative diagram

$$\begin{array}{ccc} (X_1, X_1 \cap X_2) & \longrightarrow & (X_1 \cup X_2, X_2) \\ \downarrow & & \downarrow \\ (X_1/(X_1 \cap X_2), \{y\}) & \longrightarrow & ((X_1 \cup X_2)/X_2, \{y\}), \end{array}$$

where $X_1/(X_1 \cap X_2) \approx (X_1 \cup X_2)/X_2$ since the spaces $X_1 - (X_1 \cap X_2)$ and $(X_1 \cup X_2) - X_2$ are homeomorphic. □

2.38. Example. Let B^n (resp. S^{n-1}) be the unit ball (resp. sphere) in $\mathbb{R}^n$. The space B^n/S^{n-1} may be identified with S^n in the following way. By orthogonal projection on $\mathbb{R}^n$, the upper hemisphere S_+^n is homeomorphic to B^n.

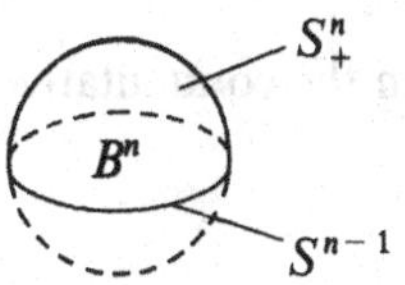

Fig. 7

Every point of S^n may be written as $v \sin\theta + \varepsilon \cos\theta$, $-\pi/2 \leqslant \theta \leqslant \pi/2$, where $v \in S^{n-1}$ and ε is the last basis vector of $\mathbb{R}^{n+1}$. This representation is unique if the point is not $\pm\varepsilon$. Now let $\gamma: S_+^n \to S^n$ be the map defined by $\gamma(v \sin\theta + \varepsilon \cos\theta) = v \sin(2\theta) + \varepsilon \cos(2\theta)$. Then γ is well-defined even for the point $+\varepsilon$, continuous, and induces a homeomorphism from $S_+^n/S^{n-1} \approx B^n/S^{n-1}$ to S^n.

From 2.29. 2.35, and 1.34, it follows that

$$K(B^n, S^{n-1}) \approx K(B^n/S^{n-1}, \{y\}) \approx \tilde{K}(S^n) = \pi_{n-1}(\mathrm{GL}(k)).$$

Hence $K_{\mathbb{R}}(B^n, S^{n-1}) \approx \pi_{n-1}(\mathrm{GL}(\mathbb{R})) \approx \pi_{n-1}(\mathrm{O})$ and $K_{\mathbb{C}}(B^n, S^{n-1}) \approx \pi_{n-1}(\mathrm{U})$. In particular, we have $K_{\mathbb{C}}(B^2, S^1) \approx \pi_1(\mathrm{U}) \approx \pi_1(\mathrm{U}(1)) \approx \mathbb{Z}$.

2.39. Example. More generally, let us consider the sequence

$$K(X \times B^n, X \times S^{n-1}) \xrightarrow{u} K(X \times S^n, X) \xrightarrow{v} \tilde{K}((X \times S^n)/X).$$

By the same argument as above, u is an isomorphism; also v is an isomorphism by 2.35. Hence vu is an isomorphism.

2.40. Example. Let (X, A) and (Y, B) be pairs of compact spaces. Then $(X-A) \times (Y-B)$ is homeomorphic to $X \times Y - C$, where $C = X \times B \cup A \times Y$. On the other hand, if Z and T are compact spaces with base points z_0 and t_0, define $Z_\vee T$ as the subspace of the product $Z \times T$ consisting of points (z, t) such that $z = z_0$ or $t = t_0$, and define $Z_\wedge T$ to be $Z \times T/Z_\vee T$. Then, for $Z = X/A$ and $T = Y/B$, we have $Z \times T - Z_\vee T \approx X \times Y - C$, and so $K(X \times Y, X \times B \cup A \times Y) \approx \tilde{K}(X/A_\wedge Y/B)$.

2.41. Example. If Z is a space with base point z_0, we define its *reduced suspension* $S(Z)$ to be $Z_\wedge S^1 \approx Z \times I/Z \times \{0\} \cup Z \times \{1\} \cup \{z_0\} \times I$. This is the quotient of the suspension $S'(Z)$ considered in I.3.14 by $\{z_0\} \times \mathrm{I}$. It may also be interpreted as the one point compactification of $(Z - \{z_0\}) \times \mathbb{R}$. For $n \geqslant 2$, the nth suspension of Z is inductively defined by $S^n(Z) = S(S^{n-1}(Z))$, and is the one point compactification of $(Z - \{z_0\}) \times \mathbb{R}^n$. Now let us consider the group $K(X \times B^n, X \times S^{n-1} \cup Y \times B^n)$, where Y is a closed subspace of a compact space X. Then $X \times B^n - X \times S^{n-1} \cup Y \times B^n \approx (X - Y) \times \mathbb{R}^n$. Hence $K(X \times B^n, X \times S^{n-1} \cup Y \times B^n) \approx \widetilde{K}(S^n(X/Y))$ by 2.29 and 2.35.

2.42. Theorem. *Let Y be a closed subspace of a compact space X. Then, we have the exact sequence*

$$\widetilde{K}(X/Y) \longrightarrow \widetilde{K}(X) \longrightarrow \widetilde{K}(Y).$$

Proof. From 2.35, 2.36, and 2.29, we have the exact sequence

$$\widetilde{K}(X/Y) \longrightarrow K(X) \longrightarrow K(Y).$$

The exact sequence in the theorem now follows from the commutative diagram

$$\begin{array}{ccccc}
 & & 0 & & 0 \\
 & & \downarrow & & \downarrow \\
0 & & \mathbb{Z} & & \mathbb{Z} \\
\downarrow & & \downarrow & & \downarrow \\
K(X, Y) & \longrightarrow & K(X) & \longrightarrow & K(Y) \\
\approx\downarrow & & \downarrow & & \downarrow \\
\widetilde{K}(X/Y) & \longrightarrow & \widetilde{K}(X) & \longrightarrow & \widetilde{K}(Y) \\
\downarrow & & \downarrow & & \downarrow \\
0 & & 0 & & 0
\end{array} \qquad \square$$

2.43. Theorem. *Let $f_0, f_1 : (X, Y) \to (X', Y')$ be continuous maps between pairs, such that f_0 and f_1 are homotopic. Then f_0 and f_1 induce the same homomorphism $(f_0)^* = (f_1)^* : K(X', Y') \to K(X, Y)$.*

Proof. Let $f : (X \times I, Y \times I) \to (X', Y')$ be the homotopy between f_0 and f_1. Then f induces a homotopy between the quotient maps $\bar{f}_0, \bar{f}_1 : X/Y \to X'/Y'$. Hence $(\bar{f}_0)^*$ and $(\bar{f}_1)^*$ define the same homomorphism from $\widetilde{K}(X'/Y')$ to $\widetilde{K}(X/Y)$. The theorem now follows from the commutative diagram

$$\begin{array}{ccc}
K(X', Y') & \xrightarrow{f_\alpha^*} & K(X, Y) \\
\approx\uparrow & & \uparrow\approx \\
\widetilde{K}(X'/Y') & \xrightarrow{\bar{f}_\alpha^*} & \widetilde{K}(X/Y)
\end{array}$$

$\alpha = 0, 1$, and 2.35–36. $\square$

Exercise II.6.13.

3. The Group K^{-1} of a Banach Category. The Group $K^{-1}(X)$

3.1. Before expanding the theory any further, let us take a brief look at the direction we will follow in the next sections. Our general purpose is to construct groups $K^n(X, Y)$, $n \in \mathbb{Z}$, X compact, and Y closed in X, which depend in a contravariant way on the pair (X, Y). In addition we want to construct the "connecting homomorphisms", i.e. natural transformations

$$\partial^{n-1}: K^{n-1}(Y) \longrightarrow K^n(X, Y).$$

Finally we want the following axioms to be satisfied:

(i) *Exactness.* The sequence

$$K^{n-1}(X) \xrightarrow{j^*} K^{n-1}(Y) \xrightarrow{\partial^{n-1}} K^n(X, Y) \xrightarrow{i^*} K^n(X) \xrightarrow{j^*} K^n(Y)$$

is exact, where in general $K^n(Z) = K^n(Z, \varnothing)$ and the maps j^* and i^* are induced by inclusions $(Y, \varnothing) \subset (X, \varnothing)$ and $(X, \varnothing) \subset (X, Y)$ respectively.

(ii) *Homotopy.* If $f_0, f_1: (X, Y) \to (X', Y')$ are homotopic continuous maps between pairs, they induce the same homomorphisms $(f_0)^* = (f_1)^*: K^n(X', Y') \to K^n(X, Y)$.

(iii) *Excision.* The projection $(X, Y) \to (X/Y, \{y\})$ induces an isomorphism $K^n(X/Y, \{y\}) \xrightarrow{\approx} K^n(X, Y)$.

(iv) *Normalization.* The functor $K^0(X) = K^0(X, \varnothing)$ is the functor $K(X)$ constructed in Section II.1.

So far we have only completed a small part of this task. If we define $K^0(X, Y)$ to be $K(X, Y)$, we have proved the exactness of the sequence

$$K^0(X, Y) \xrightarrow{i^*} K^0(X) \xrightarrow{j^*} K^0(Y)$$

in 2.20 and 2.29. We have also proved the homotopy and excision axioms for $n = 0$ in 2.43 and 2.35 respectively. In the next two sections we will construct "half" of the theory, i.e. the functors $K^n(X, Y)$ for $n < 0$.

The construction of the other half of the theory, i.e. the functors $K^n(X, Y)$ for $n > 0$, is much more difficult. This will be done in chapter III after we prove the "periodicity" of the functors $K^{-n}(X, Y)$ (precisely $K_{\mathbb{R}}^{-n}(X, Y) \approx K_{\mathbb{R}}^{-n-8}(X, Y)$ and $K_{\mathbb{C}}^{-n}(X, Y) \approx K_{\mathbb{C}}^{-n-2}(X, Y)$.

It would be possible to present all these results within the framework of Banach categories as in the author's thesis [2]; however, in the interest of avoiding lengthy developments we will mainly restrict ourselves to the category of vector bundles. The interested reader may consult the reference above or the exercises for many generalizations of the material in the next sections.

3.2. Let $\mathscr{C}$ be a Banach category. In this section we define a group $K^{-1}(\mathscr{C})$ which depends functorially on $\mathscr{C}$ (when $\mathscr{C} = \mathscr{E}(X)$, $K^{-1}(\mathscr{C})$ will be denoted by $K^{-1}(X)$).

If $\varphi: \mathscr{C} \to \mathscr{C}'$ is a quasi-surjective Banach functor, we also construct a "connecting homomorphism"

$$\partial: K^{-1}(\mathscr{C}') \longrightarrow K(\varphi)$$

which is included in an exact sequence

$$K^{-1}(\mathscr{C}) \longrightarrow K^{-1}(\mathscr{C}') \xrightarrow{\partial} K(\varphi) \longrightarrow K(\mathscr{C}) \longrightarrow K(\mathscr{C}').$$

When $\varphi: \mathscr{E}(X) \to \mathscr{E}(Y)$ is the restriction functor, this sequence can be written as

$$K^{-1}(X) \longrightarrow K^{-1}(Y) \xrightarrow{\partial} K^0(X, Y) \longrightarrow K^0(X) \longrightarrow K^0(X),$$

and is the sequence of the exactness axiom for $n=0$.

3.3. More precisely, consider the set of pairs (E, α), where E is an object of $\mathscr{C}$ and α is an automorphism of E. Two pairs (E, α) and (E', α') are *isomorphic* if there is an isomorphism $h: E \to E'$ in the category $\mathscr{C}$, such that the following diagram commutes:

$$\begin{array}{ccc} E & \xrightarrow{h} & E' \\ {\scriptstyle\alpha}\downarrow & & \downarrow{\scriptstyle\alpha'} \\ E & \xrightarrow{h} & E' \end{array}$$

As in the definition of $K(\varphi)$, we define the *sum* of two pairs (E_0, α_0) and (E_1, φ_1) to be $(E_0 \oplus E_1, \alpha_0 \oplus \alpha_1)$. A pair (E, α) is *elementary* if α is homotopic to Id_E within the automorphisms of E, Finally $K^{-1}(\mathscr{C})$ is defined as the quotient of the set of pairs by the following equivalence relation: $\sigma \sim \sigma' \Leftrightarrow \exists \tau$ and τ' elementary such that $\sigma + \tau \approx \sigma' + \tau'$. The sum of pairs induces an abelian monoid structure on $K^{-1}(\mathscr{C})$. We will let $d(E, \alpha)$ denote the class of (E, α) in $K^{-1}(\mathscr{C})$.

3.4. Proposition. *We have the relation* $d(E, \alpha) + d(E, \alpha^{-1}) = 0$. *Thus* $K^{-1}(\mathscr{C})$ *is an abelian group.*

Proof. By definition $d(E, \alpha) + d(E, \alpha^{-1}) = d(E \oplus E, \alpha \oplus \alpha^{-1})$. But $\alpha \oplus \alpha^{-1}$ can be written in the form

$$\begin{pmatrix} \alpha & 0 \\ 0 & \alpha^{-1} \end{pmatrix} = \begin{pmatrix} 0 & -\alpha \\ \alpha^{-1} & 0 \end{pmatrix} \begin{pmatrix} 0 & 1 \\ -1 & 0 \end{pmatrix},$$

and each of the matrices on the right is homotopic to $\mathrm{Id}_{E \oplus E}$ as was shown in 2.14. Therefore, $d(E \oplus E, \alpha \oplus \alpha^{-1}) = 0$. □

3.5. Proposition. *Let* α *and* α' *be automorphisms of* E *which are homotopic within the automorphisms of* E. *Then* $d(E, \alpha) = d(E, \alpha')$.

Proof. We have $d(E,\alpha)-d(E,\alpha')=d(E,\alpha)+d(E,\alpha'^{-1})=d(E\oplus E,\alpha\oplus\alpha'^{-1})$ by 3.4. But the automorphism $\alpha\oplus\alpha'^{-1}$ is homotopic to $\alpha\oplus\alpha^{-1}$, hence to $\mathrm{Id}_{E\oplus E}$. Therefore the pair $(E\oplus E,\alpha\oplus\alpha'^{-1})$ is elementary and $d(E\oplus E,\alpha\oplus\alpha'^{-1})=0$. □

3.6. Proposition. *We have the formula*

$$d(E,\alpha)+d(E,\beta)=d(E,\alpha\beta)=d(E,\beta\alpha).$$

Proof. Consider

$$d(E,\alpha)+d(E,\beta)=d(E\oplus E,\alpha\oplus\beta) \quad \text{and} \quad d(E,\alpha\beta)=d(E\oplus E,\alpha\beta\oplus\mathrm{Id}_E).$$

By 3.5, they are equal if $\alpha\oplus\beta$ and $\alpha\beta\oplus\mathrm{Id}_E$ are homotopic within the automorphisms of $E\oplus E$. But $(\alpha\oplus\beta)^{-1}(\alpha\beta\oplus\mathrm{Id}_E)=\beta\oplus\beta^{-1}$ is homotopic to $\mathrm{Id}_{E\oplus E}$ as was shown in 3.4. Moreover, $d(E,\beta\alpha)=d(E,\beta)+d(E,\alpha)=d(E,\alpha)+d(E,\beta)$. □

3.7. Lemma. *The class $d(E,\alpha)$ is 0 if and only if there is an object G of $\mathscr{C}$, such that $\alpha\oplus\mathrm{Id}_G$ is homotopic to $\mathrm{Id}_{E\oplus G}$ within the automorphisms of $E\oplus G$.*

Proof. Suppose $d(E,\alpha)=0$. By the definition of $K^{-1}(\mathscr{C})$, there exist elementary pairs (G,η) and (G',η') with an isomorphism $h\colon E\oplus G\to G'$, such that the following diagram commutes:

$$\begin{array}{ccc} E\oplus G & \xrightarrow{h} & G' \\ \downarrow{\scriptstyle \alpha\oplus\eta} & & \downarrow{\scriptstyle \eta'} \\ E\oplus G & \xrightarrow{h} & G' \end{array}$$

From this we see that $\alpha\oplus\mathrm{Id}_G$ is homotopic to $\alpha\oplus\eta=h^{-1}\cdot\eta'\cdot h$, which is homotopic to $h^{-1}\cdot\mathrm{Id}_{G'}\cdot h=\mathrm{Id}_{E\oplus G}$. The converse is obvious. □

3.8. Proposition. *We have $d(E,\alpha)=d(F,\beta)$ if and only if there is an object G of $\mathscr{C}$, such that $\alpha\oplus\mathrm{Id}_F\oplus\mathrm{Id}_G$ and $\mathrm{Id}_E\oplus\beta\oplus\mathrm{Id}_G$ are homotopic within the automorphisms of $E\oplus F\oplus G$.*

Proof. Suppose $d(E,\alpha)=d(F,\beta)$. Then $d(E,\alpha)-d(F,\beta)=d(E\oplus F,\alpha\oplus\beta^{-1})=0$ from 3.4. Therefore, there is an object G of $\mathscr{C}$ such that $\alpha\oplus\beta^{-1}\oplus\mathrm{Id}_G$ is homotopic to $\mathrm{Id}_{E\oplus F\oplus G}$ from 3.7. Multiplying this homotopy by $\mathrm{Id}_E\oplus\beta\oplus\mathrm{Id}_G$, we can see that $\alpha\oplus\mathrm{Id}_F\oplus\mathrm{Id}_G$ and $\mathrm{Id}_E\oplus\beta\oplus\mathrm{Id}_G$ are homotopic. Again, the converse is obvious. □

3.9. *Remark.* The proposition above gives an equivalent definition of the group $K^{-1}(\mathscr{C})$.

3.10. Example. Let A be a Banach algebra (with unit but not necessarily commutative), and let $\mathscr{C}=\mathscr{L}(A)$ be the category of finitely generated free right A-modules. Let $\mathrm{GL}_n(A)$ be the group of invertible $n\times n$ matrices with coefficients in A and let $\mathrm{GL}(A)=\mathrm{inj\,lim}\,\mathrm{GL}_n(A)$. We define a map $\gamma: K^{-1}(\mathscr{C})\to\pi_0(\mathrm{GL}(A))=$ inj lim $\pi_0(\mathrm{GL}_n(A))$ by $\gamma(d(A^n,\alpha))=$ class of α in $\pi_0(\mathrm{GL}(A))$. Since the matrices of $\alpha\oplus\mathrm{Id}_{A^n}$ and $\mathrm{Id}_{A^n}\oplus\alpha$ belong to the same connected component in $\mathrm{GL}_{2n}(A)$ (computation of 3.4), Proposition 3.8 shows that this map is well defined. Moreover, by 3.6 we know that γ is a group map if we provide $\pi_0(\mathrm{GL}(A))$ with the (abelian) group structure induced by the product of matrices. Finally γ is clearly surjective and injective by 3.7.

3.11. Example. Let $\mathscr{E}_k$ be the category of finite dimensional k-vector spaces. According to Example 3.10 above, $K^{-1}(\mathscr{E}_{\mathbb{C}})\approx\pi_0(\mathrm{GL}(\mathbb{C}))=0$ and $K^{-1}(\mathscr{E}_{\mathbb{R}})\approx\pi_0(\mathrm{GL}(\mathbb{R}))\approx\mathbb{Z}/2$.

3.12. Example. Let H be an infinite dimensional Hilbert space, and let $\mathscr{K}$ be the closed two-sided ideal of completely continuous operators (i.e. limits of operators of finite rank). Let $A=\mathrm{End}(H)/\mathscr{K}$ be the quotient algebra (called the Calkin algebra in functional analysis). Then $\pi_0(\mathrm{GL}(A))\approx\mathbb{Z}$ by exercise 6.12. Therefore $K^{-1}(\mathscr{L}(A))\approx\mathbb{Z}$.

3.13. Let $\mathscr{C}$ be an arbitrary Banach category, and let $\tilde{\mathscr{C}}$ be the associated pseudo-abelian category (I.6.9). Then the canonical functor $\varphi:\mathscr{C}\to\tilde{\mathscr{C}}$ induces group maps $K(\mathscr{C})\to K(\tilde{\mathscr{C}})$ and $K^{-1}(\mathscr{C})\to K^{-1}(\tilde{\mathscr{C}})$. We know that the first map is not bijective in general (I.3.10 and I.6.16). However, we have the following result for the second map.

3.14. Theorem. *The map*

$$\varphi_*: K^{-1}(\mathscr{C})\longrightarrow K^{-1}(\tilde{\mathscr{C}})$$

is bijective.

Proof.

a) *φ_* is surjective.* Let $d(E,\alpha)\in K^{-1}(\tilde{\mathscr{C}})$, and let T (resp. F) be an object of $\mathscr{C}$ (resp. $\tilde{\mathscr{C}}$) such that $\varphi(T)\xrightarrow[\approx]{h} E\oplus F$. Then $d(E,\alpha)=d(E\oplus F,\alpha\oplus\mathrm{Id}_F)=d(T,\beta)$, where β is the unique automorphism of T such that $\varphi(\beta)=h\cdot(\alpha\oplus\mathrm{Id}_F)\cdot h^{-1}$ (this is possible since φ is fully faithful; cf. I.6.10).

b) *φ_* is injective.* Let $d(E,\alpha)\in K^{-1}(\mathscr{C})$ such that $\varphi_*(d(E,\alpha))=0$. By 3.7 there is an object G of $\tilde{\mathscr{C}}$, which we may assume to be of the form $\varphi(E')$, since φ is quasi-surjective, such that $\varphi(\alpha)\oplus\mathrm{Id}_{\varphi(E')}$ is homotopic to $\mathrm{Id}_{\varphi(E)\oplus\varphi(E')}$ within the automorphisms of $\varphi(E)\oplus\varphi(E')$. Since φ is fully faithful, it follows that $\alpha\oplus\mathrm{Id}_{E'}$ is homotopic to $\mathrm{Id}_{E\oplus E'}$ within the automorphisms of $E\oplus E'$. Hence $d(E,\alpha)=d(E\oplus E',\alpha\oplus\mathrm{Id}_{E'})=0$. □

3.15. Example. Let $\mathscr{C}$ be $\mathscr{L}(A)$, where A is a Banach algebra. Then $\tilde{\mathscr{C}}\sim\mathscr{P}(A)$ by I.6.16. Therefore $K^{-1}(\mathscr{P}(A))\approx\pi_0(\mathrm{GL}(A))$.

3.16. In the proof of Theorem 3.14, we only used the fact that the functor $\varphi: \mathscr{C} \to \tilde{\mathscr{C}}$ is fully faithful and quasi-surjective. Thus, if we replace $\mathscr{C}$ by $\mathscr{E}_T(X)$, the category of trivial bundles over a compact space X, replace $\tilde{\mathscr{C}}$ by $\mathscr{E}(X)$, the category of locally trivial vector bundles over X (I.6.5), and let $\varphi: \mathscr{E}_T(X) \to \mathscr{E}(X)$ be the inclusion factor, the same proof gives an isomorphism

$$K^{-1}(\mathscr{E}_T(X)) \longrightarrow K^{-1}(\mathscr{E}(X)).$$

In order to compute $K^{-1}(\mathscr{E}_T(X))$, and hence $K^{-1}(\mathscr{E}(X))$, we will often make use of the following fact. For E and F vector bundles over X, there exists a bijective correspondence between the continuous maps from Y to $\mathrm{Hom}(E, F)$ and $\mathrm{Hom}(\pi^*(E), \pi^*(F))$, where $\pi: X \times Y \to X$. This is true for trivial bundles because $F(X \times Y, k^n) \approx F(Y, F(X, k^n))$, where F denotes the function space of continuous maps. This is also true for arbitrary vector bundles since they are direct factors of trivial bundles for instance.

3.17. Theorem. *Let X be a compact space and let $[X, \mathrm{GL}(k)] \approx \mathrm{inj\,lim}[X, \mathrm{GL}_n(k)]$, $k = \mathbb{R}$ or $\mathbb{C}$, be the set of homotopy classes of continuous maps from X to $\mathrm{GL}(k)$, provided with the group structure induced by the product of matrices. Let $u: \mathrm{inj\,lim}[X, \mathrm{GL}_n(k)] \to K^{-1}(X) = K^{-1}(\mathscr{E}_k(X))$ be the map defined by $\alpha_n \mapsto d(\theta_n, \hat{\alpha}_n)$, where $\theta_n = X \times k^n$. Then u is an isomorphism.*

Proof. By the above remark, $[X, \mathrm{GL}_n(k)] \approx \pi_0(\mathrm{GL}_n(A))$ where A is the Banach algebra of continuous functions on X with values in k. Moreover, the map u factors into

$$\begin{array}{ccccc} \mathrm{inj\,lim}[X, \mathrm{GL}_n(k)] & \longrightarrow & K^{-1}(\mathscr{E}_T(X)) & \xrightarrow{\approx} & K^{-1}(\mathscr{E}(X)) \\ \wr\wr & & \wr\wr & & \\ \mathrm{inj\,lim}\,\pi_0(\mathrm{GL}_n(A)) & \xrightarrow{\approx} & K^{-1}(\mathscr{L}(A)) & & \end{array}$$

where we use the category equivalence $\mathscr{E}_T(X) \sim \mathscr{L}(A)$ exhibited in the proof of I.6.18. Thus the theorem is essentially a consequence of 3.14. □

3.18. *Remark.* It is easy to give a direct proof of theorem 3.17 (without using Banach algebras) along the lines of 3.14.

3.19. Corollary. *Let* $\mathrm{O} = \mathrm{inj\,lim}\,\mathrm{O}(n)$, *(resp.* $\mathrm{U} = \mathrm{inj\,lim}\,\mathrm{U}(n)$) *be the infinite orthogonal group (resp. the infinite unitary group). Then we have natural isomorphisms*

$$[X, \mathrm{O}] \approx \mathrm{inj\,lim}[X, \mathrm{O}(n)] \xrightarrow{\approx} K_{\mathbb{R}}^{-1}(X)$$

and

$$[X, \mathrm{U}] \approx \mathrm{inj\,lim}[X, \mathrm{U}(n)] \longrightarrow K_{\mathbb{C}}^{-1}(X).$$

3.20. Example. Let X be the sphere S^p with base point $\{e\}$. We define a map $[S^p, \mathrm{GL}(k)] \to \pi_p(\mathrm{GL}(k)) \times \pi_0(\mathrm{GL}(k))$ by $\alpha \mapsto (\alpha', \alpha(e))$ where $\alpha'(x) = \alpha(x)\alpha(e)^{-1}$. It is easy to show that this map is bijective. Hence $K^{-1}(S^p) \approx K^{-1}(\{e\}) \oplus \tilde{K}(S^{p+1})$ (1.34).

3.21. Let $\varphi: \mathscr{C} \to \mathscr{C}'$ be a quasi-surjective Banach functor. Following the outline given in 3.2, we are going to define a "connecting homomorphism"

$$\partial: K^{-1}(\mathscr{C}') \longrightarrow K(\varphi).$$

Let $d(E', \alpha')$ be an element of $K^{-1}(\mathscr{C}')$. Since φ is quasi-surjective, there is an object E of $\mathscr{C}$, an object F' of $\mathscr{C}'$, and an isomorphism $h: \varphi(E) \to E' \oplus F'$. Let $\alpha: \varphi(E) \to \varphi(E)$ be the isomorphism which makes the diagram

$$\begin{array}{ccc} E' \oplus F' & \xrightarrow{h} & \varphi(E) \\ \downarrow{\scriptstyle \alpha' \oplus \mathrm{Id}_{F'}} & & \downarrow{\scriptstyle \alpha} \\ E' \oplus F' & \xrightarrow{h} & \varphi(E) \end{array}$$

commutative. Then $d(E, E, \alpha)$ is the element of $K(\varphi)$ associated with $d(E', \alpha')$ by ∂. We must prove that this element is independent of the choices F', E and h. Let $\bar{F}'$, $\bar{E}$ and $\bar{h}$ be other choices, and let $d(\bar{E}, \bar{E}, \bar{\alpha})$ be the new element of $K(\varphi)$ thus obtained. Then we have $d(E, E, \alpha) = d(E \oplus \bar{E}, E \oplus \bar{E}, \alpha \oplus 1)$ where "$\alpha \oplus 1$" $= \alpha \oplus \mathrm{Id}_{\varphi(\bar{E})}$, is included in the commutative diagram

$$\begin{array}{ccc} (E' \oplus F') \oplus (E' \oplus \bar{F}') & \xrightarrow{h \oplus \bar{h}} & \varphi(E) \oplus \varphi(\bar{E}) \\ \downarrow{\scriptstyle \gamma} & & \downarrow{\scriptstyle \alpha \oplus \mathrm{Id}_{\varphi(\bar{E})}} \\ (E' \oplus F') \oplus (E' \oplus \bar{F}') & \xrightarrow{h \oplus \bar{h}} & \varphi(E) \oplus \varphi(\bar{E}), \end{array}$$

where $\gamma = (\alpha' \oplus \mathrm{Id}_{F'}) \oplus (\mathrm{Id}_{E'} \oplus \mathrm{Id}_{\bar{F}'})$. As was shown in 3.4, γ is homotopic to $\bar{\gamma} = (\mathrm{Id}_{E'} \oplus \mathrm{Id}_{F'}) \oplus (\alpha' \oplus \mathrm{Id}_{\bar{F}'})$ within the automorphisms of $E' \oplus F' \oplus E' \oplus \bar{F}'$. Therefore, $\alpha \oplus \mathrm{Id}_{\varphi(\bar{E})}$ is homotopic to $\mathrm{Id}_{\varphi(E)} \oplus \bar{\alpha}$, as is shown by the analogous commutative diagram

$$\begin{array}{ccc} (E' \oplus F') \oplus (E' \oplus \bar{F}') & \xrightarrow{h \oplus \bar{h}} & \varphi(E) \oplus \varphi(\bar{E}) \\ \downarrow{\scriptstyle \bar{\gamma}} & & \downarrow{\scriptstyle \mathrm{Id}_{\varphi(E)} \oplus \bar{\alpha}} \\ (E' \oplus F') \oplus (E' \oplus \bar{F}') & \xrightarrow{h \oplus \bar{h}} & \varphi(E) \oplus \varphi(\bar{E}). \end{array}$$

According to 2.15, we have the identity $d(E \oplus \bar{E}, E \oplus \bar{E}, \alpha \oplus \mathrm{Id}_{\varphi(\bar{E})}) = d(E \oplus \bar{E}, E \oplus \bar{E}, \mathrm{Id}_{\varphi(E)} \oplus \bar{\alpha}) = d(\bar{E}, \bar{E}, \bar{\alpha})$, showing that ∂ is well-defined and natural.

3.22. Theorem. *The sequence*

$$K^{-1}(\mathscr{C}) \xrightarrow{j_*} K^{-1}(\mathscr{C}') \xrightarrow{\partial} K(\varphi) \xrightarrow{i_*} K(\mathscr{C}) \xrightarrow{j_*} K(\mathscr{C}')$$

is exact.

Proof. From 2.20, we only have to prove exactness at $K(\varphi)$ and $K^{-1}(\mathscr{C}')$.

a) *Exactness at* $K(\varphi)$. Let $d(E', \alpha')$ be an element of $K^{-1}(\mathscr{C}')$. Since φ is quasi-surjective, we may assume without loss of generality that E' is of the form $\varphi(E)$. Then $\partial(d(E', \alpha')) = d(E, E, \alpha')$, and $(i_* \partial)(d(E', \alpha')) = [E] - [E] = 0$.

Conversely, let $d(E, F, \alpha)$ be an element of $K(\varphi)$ such that $i_*(d(E, F, \alpha)) = [E] - [F] = 0$. By 1.16, we can find an object T of $\mathscr{C}$ and an isomorphism $\delta: E \oplus T \to F \oplus T$. Therefore $d(E, F, \alpha) = d(E \oplus T, F \oplus T, \alpha \oplus \mathrm{Id}_{\varphi(T)}) = d(E \oplus T, E \oplus T, \gamma)$ where $\gamma = \varphi(\delta^{-1}) \cdot (\alpha \oplus \mathrm{Id}_{\varphi(T)})$ since the triples $(E \oplus T, E \oplus T, \gamma)$ and $(E \oplus T, F \oplus T, \alpha \oplus \mathrm{Id}_{\varphi(T)})$ are isomorphic, (2.13). Hence $d(E, F, \alpha) = \partial(d(E', \alpha))$ where $E' = \varphi(E) \oplus \varphi(T)$.

b) *Exactness at* $K^{-1}(\mathscr{C})$. Let $d(E, \alpha)$ be an element of $K^{-1}(\mathscr{C})$. Then $(\partial \cdot j_*)(d(E, \alpha)) = d(E, E, \alpha) = 0$ since the triples (E, E, α) and $(E, E, \mathrm{Id}_{\varphi(E)})$ are isomorphic.

Conversely, let $d(E', \alpha')$ be an element of $K^{-1}(\mathscr{C}')$. Since φ is quasi-surjective, we may assume without loss of generality that E' is of the form $\varphi(E)$. If $\partial(d(E', \alpha')) = d(E, E, \alpha') = 0$, we can find two elementary triples (G, G, η) and (H, H, ε) such that $(E, E, \alpha') + (G, G, \eta) \approx (H, H, \varepsilon)$, (2.13). More precisely, we have isomorphisms $u: E \oplus G \to H$ and $v: E \oplus G \to H$ such that the following diagram commutes:

$$\begin{array}{ccc}
\varphi(E) \oplus \varphi(G) & \xrightarrow{\alpha' \oplus \eta} & \varphi(E) \oplus \varphi(G) \\
\downarrow{\scriptstyle \varphi(u)} & & \downarrow{\scriptstyle \varphi(v)} \\
\varphi(H) & \xrightarrow{\quad \varepsilon \quad} & \varphi(H)
\end{array}$$

Therefore $d(E', \alpha') = d(\varphi(E) \oplus \varphi(G), \alpha' \oplus \eta) = d(\varphi(H), \varphi(u)\varphi(v^{-1})\varepsilon) = d(\varphi(H), \varphi(u \cdot v^{-1}))$ by 3.5. Hence $d(E', \alpha') = j_*(d(H, \alpha))$ where $\alpha = u \cdot v^{-1}$. □

3.23. Corollary. *Let X be a compact space and Y be a closed subspace of X. Then we have the exact sequence*

$$K^{-1}(X) \xrightarrow{j^*} K^{-1}(Y) \xrightarrow{\partial} K(X, Y) \xrightarrow{i^*} K(X) \xrightarrow{j^*} K(Y).$$

3.24. Example. Let CP_n be the complex projective space of $\mathbf{C}^{n+1}$. Then $CP_1 \approx S^2$ and $K_{\mathbf{C}}(CP_1) \approx K_{\mathbf{C}}(S^2) \approx \mathbf{Z} \oplus \mathbf{Z}$ by 1.34. Moreover, a nontrivial generator for $\tilde{K}_{\mathbf{C}}(S^2)$ is given by the canonical line bundle over CP_1 (I.2.5). Now let X be CP_2 and Y be CP_1. Since, the canonical line bundle over CP_1 is the restriction of the canonical line bundle over CP_2, the map $K_{\mathbf{C}}(X) \to K_{\mathbf{C}}(Y)$ is surjective. Finally, $K_{\mathbf{C}}^{-1}(S^2) = 0$ because $\pi_2(\mathrm{U}(2)) = 0$ (1.34 and 3.20), and $K(X, Y) \approx K((\overline{X - Y})) \approx \tilde{K}_{\mathbf{C}}(S^4) \approx \mathbf{Z}$ because $\pi_3(\mathrm{U}(2)) \approx \mathbf{Z}$. From this discussion and Corollary 3.23, we obtain the exact sequence

$$\begin{array}{ccccccccc}
0 \longrightarrow & \tilde{K}_{\mathbf{C}}(S^4) & \longrightarrow & K_{\mathbf{C}}(CP_2) & \longrightarrow & K_{\mathbf{C}}(CP_1) & \longrightarrow 0. \\
 & \wr\wr & & & & \wr\wr & \\
 & \mathbf{Z} & & & & \mathbf{Z} \oplus \mathbf{Z} &
\end{array}$$

Hence $K_{\mathbb{C}}(CP_2) \approx \mathbb{Z} \oplus \mathbb{Z} \oplus \mathbb{Z}$. More complete results will be obtained in Chapters III and IV.

3.25. It is possible to generalize the definition of $K^{-1}(X)$ to a "relative" version, $K^{-1}(X, Y)$, for Y closed in X. As in 3.3, we consider the set of pairs (E, α), where E is an object of $\mathscr{E}(X)$, but where α is now an automorphism of E such that $\alpha|_Y = 1$. We make the same definitions as in 3.3, except that we call a pair (E, α) elementary if there exists a continuous map $\sigma: I \to \operatorname{Aut}(E)$, such that $\sigma(0) = 1$, $\sigma(1) = \alpha$ and $\sigma(t)|_Y = \operatorname{Id}_{E_Y}$. We call the group obtained by this procedure $K^{-1}(X, Y)$. Assertions 3.4, 3.5, 3.6, 3.7, and 3.8 can be generalized without difficulty (taking care that the homotopies involved are constant over Y). The proof of Theorem 3.14 can be used again to show that $K^{-1}(X, Y)$ may also be constructed using only trivial bundles (compare with 3.16). It follows that $K^{-1}(X, Y) \approx K^{-1}(X/Y, \{y\}) \approx [X/Y, \operatorname{GL}(k)]'$. where $[\ ,\ ]'$ denotes homotopy classes of maps preserving base points.

3.26. Proposition. *We have the exact sequence*

$$K^{-1}(X, Y) \xrightarrow{i^*} K^{-1}(X) \xrightarrow{j^*} K^{-1}(Y).$$

Moreover, when Y is a retract of X, we have the split exact sequence

$$0 \longrightarrow K^{-1}(X, Y) \xrightarrow{i^*} K^{-1}(X) \xrightarrow{j^*} K^{-1}(Y) \longrightarrow 0.$$

Proof. It is clear that the composition $K^{-1}(X, Y) \to K^{-1}(X) \to K^{-1}(Y)$ is zero. Now let $d(E, \alpha)$ be an element of $K^{-1}(X)$ such that $j^*(d(E, \alpha)) = d(E_Y, \alpha_Y) = 0$. According to 3.7, we can find a vector bundle G over Y, which we may assume to be of the form $F|_Y$, so that $\alpha_Y \oplus \operatorname{Id}_G$ is homotopic to $\operatorname{Id}_{E_Y \oplus G}$ within the automorphisms of $E_Y \oplus G = (E \oplus F)_Y$. According to 2.24, $\alpha \oplus \operatorname{Id}_F$ is homotopic to an automorphism β such that $\beta|_Y = \operatorname{Id}_{(E \oplus F)_Y}$. Hence $d(E, \alpha) = d(E \oplus F, \alpha \oplus \operatorname{Id}_F) = d(E \oplus F, \beta)$ belongs to the image of i^*.

Let us assume now that Y is a retract of X. Then to prove that the exact sequence splits, it is enough to show that i^* is injective. Let x be an element of $K^{-1}(X, Y)$ represented by a continuous map $\gamma: X \to \operatorname{GL}(k)$, such that $\gamma(Y) = \{1\}$ and γ is homotopic to the constant map γ_0 defined by $\gamma_0(X) = \{1\}$. If $\bar{\gamma}: X \times I \to \operatorname{GL}(k)$ is this homotopy, we define $\tilde{\gamma}: X \times I \to \operatorname{GL}(k)$ by the formula $\tilde{\gamma}(x, t) = \bar{\gamma}(x, t) \times \bar{\gamma}(r(x), t)^{-1}$ where $r: X \to Y$ is a retraction. Then $\tilde{\gamma}$ is a homotopy of γ to 1 such that $\tilde{\gamma}(Y \times I) = \{1\}$. □

3.27. Proposition. *Let $f: (X, Y) \to (X', Y')$ be a morphism between compact pairs, which induces homotopy equivalences $X \sim X'$ and $Y \sim Y'$. Then f induces an isomorphism $K(X/Y) \xrightarrow{\approx} K(X/Y)$.*

Proof. Since $X \sim X'$ and $Y \sim Y'$, f induces isomorphisms $K(X') \approx K(X)$, $K^{-1}(X') \approx K^{-1}(X)$, $K(Y') \approx K(Y)$ and $K^{-1}(Y') \approx K^{-1}(Y)$. Therefore, we have the commutative diagram

$$\begin{array}{ccccccccc}
K^{-1}(X') & \longrightarrow & K^{-1}(Y') & \longrightarrow & K(X',Y') & \longrightarrow & K(X') & \longrightarrow & K(Y') \\
\downarrow \approx & & \downarrow \approx & & \downarrow & & \downarrow \approx & & \downarrow \approx \\
K^{-1}(X) & \longrightarrow & K^{-1}(Y) & \longrightarrow & K(X,Y) & \longrightarrow & K(X) & \longrightarrow & K(Y)
\end{array}$$

where four of the vertical maps are isomorphisms. It follows from the five lemma (Northcott [1]) that the map $K(X', Y') \to K(X, Y)$ is also an isomorphism. From the commutative diagram (cf. 2.35)

$$\begin{array}{ccc}
K(X'/Y', \{y'\}) & \xrightarrow{\approx} & K(X', Y') \\
\downarrow & & \downarrow \\
K(X/Y, \{y\}) & \xrightarrow{\approx} & K(X, Y)
\end{array}$$

we obtain the isomorphisms $\tilde{K}(X'/Y') \approx \tilde{K}(X/Y)$ and $K(X'/Y') \approx K(X/Y)$. □

3.28. It is possible to give another interpretation of 3.22 when $\varphi = f^*: \mathscr{E}(X) \to \mathscr{E}(Y)$ is associated with some continuous map $f\colon Y \to X$. Given f, we define its *mapping cylinder* M_f to be the quotient of $X \sqcup Y \times I$ by the equivalence relation, which identifies $(y, 0)$ with $f(y)$ for each y in Y. Then we have the diagram, commutative up to homotopy,

$$\begin{array}{ccc}
Y & \xrightarrow{f} & X \\
\| & & \downarrow u \\
Y & \xrightarrow{i} & M_f,
\end{array}$$

where u is a homotopy equivalence (its homotopy inverse is the "projection" on X), and where $i(y)$ is the class of $(y, 1)$. Therefore, we have the category diagram

$$\begin{array}{ccc}
\mathscr{E}(X) & \xrightarrow{f^*} & \mathscr{E}(Y) \\
u^* \uparrow & & \| \\
\mathscr{E}(M_f) & \xrightarrow{i^*} & \mathscr{E}(Y),
\end{array}$$

and the same argument as used in the proof of 3.27 shows that $K(f^*) \approx K(i^*) \approx \tilde{K}(C'f)$ where $C'f \approx M_f/i(Y)$.

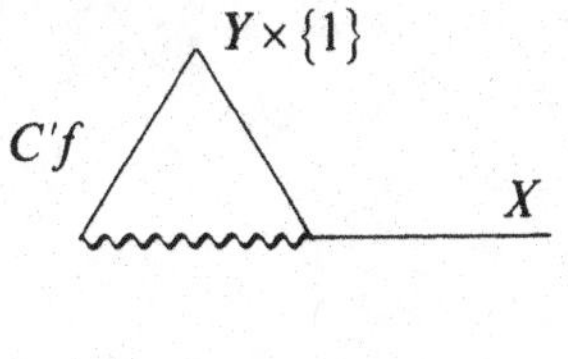

Fig. 8

The *Puppe sequence* associated with f is the sequence

$$Y \xrightarrow{f} X \longrightarrow C'f \longrightarrow S'(Y) \longrightarrow S'(X),$$

where in general $S'(Z)$ denotes the suspension of Z (I.3.14). If x_0 and y_0 are base points in X and Y respectively, and if $f(y_0)=x_0$, we can also consider the "reduced Puppe sequence"

$$Y \xrightarrow{f} X \longrightarrow Cf \longrightarrow S(Y) \longrightarrow S(X),$$

where in general $S(Z)$ denotes the reduced suspension of Z (2.41). The maps $S(Y) \to S(X)$ and $S'(Y) \to S'(X)$ are induced by the functoriality of our construction; the map $C'f \to S'(Y)$ is induced by the identification $C'f/X \approx S'(Y)$. Finally, Cf is the quotient of $C'f$ by the class of $\{x_0\} \times I$; the map $Cf \to S(Y)$ is similarly defined.

3.29. Theorem. *The Puppe sequence and the reduced Puppe sequence induce exact sequences of $\tilde{K}$-groups*

$$\begin{array}{ccccccccc}
\tilde{K}(S'(X)) & \longrightarrow & \tilde{K}(S'(Y)) & \longrightarrow & \tilde{K}(C'f) & \longrightarrow & \tilde{K}(X) & \longrightarrow & \tilde{K}(Y) \\
\uparrow \approx & & \uparrow \approx & & \uparrow \approx & & \| & & \| \\
\tilde{K}(S(X)) & \longrightarrow & \tilde{K}(S(Y)) & \longrightarrow & \tilde{K}(Cf) & \longrightarrow & \tilde{K}(X) & \longrightarrow & \tilde{K}(Y).
\end{array}$$

Proof. Since Cf, $S(Y)$, and $S(X)$, are respectively the quotient of $C'f$, $S'(Y)$, and $S'(X)$, by contractible subsets, Proposition 3.27 shows that proving the first exact sequence will suffice.

a) *Exactness at $\tilde{K}(X)$.* This follows from the diagram

$$\begin{array}{ccccc}
K(C'f) & \longrightarrow & \tilde{K}(X) & \longrightarrow & \tilde{K}(Y) \\
\| & & \uparrow u^* & & \| \\
\tilde{K}(C'f) & \longrightarrow & \tilde{K}(M_f) & \longrightarrow & \tilde{K}(Y),
\end{array}$$

since the last sequence is exact by 2.42.

b) *Exactness at $\tilde{K}(C'f)$.* Since $S'(Y) \approx C'f/X$, exactness follows again from 2.42.

c) *Exactness at $\tilde{K}(S'(Y))$.* We consider the following picture,

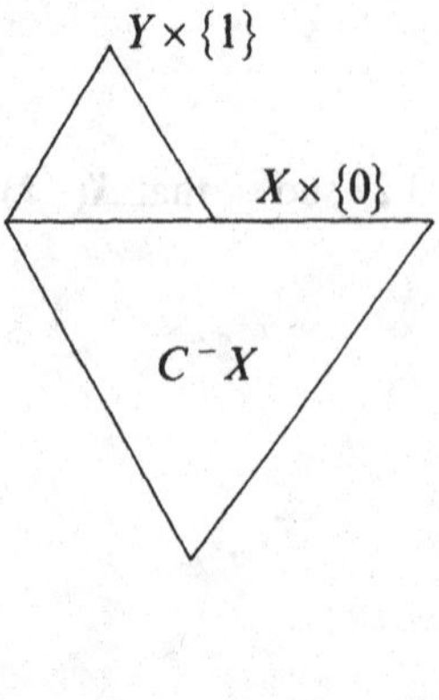

$S'(Y) \cup C^-(X)$ Fig. 9

and the following diagram:

$$\begin{array}{ccccc} C'f & \longrightarrow & S'(Y) & \longrightarrow & S'(X) \\ \| & & \uparrow & & \uparrow \\ C'f & \longrightarrow & S'(Y)\cup C^-(X) & \longrightarrow & S''(X) \\ \| & & \| & & \downarrow \\ C'f & \longrightarrow & S'(Y)\cup C^-(X) & \longrightarrow & S'(X) \end{array}$$

Here $S'(Y)\cup C^-(X)$ is the quotient of $Y\times[0,1]\cup X\times[-1,0]$ by the identification relations $(y,0)\sim(f(y),0)$, $(y,1)\sim(y',1)$, and $(x,-1)\sim(x',-1)$, for $y,y'\in Y$ and $x,x'\in X$. The space $S'(Y)$ is homeomorphic to the quotient of $S'(Y)\cup C^-(X)$ by the contractible subset $C^-(X)$ (I.3.14). Hence, by 3.27, the map $S'(Y)\cup C^-(X)\to S'(Y)$ induces an isomorphism $\tilde{K}(S'(Y))\approx\tilde{K}(S'(Y)\cup C^-(X))$. In the same way, let $S''(X)$ be the quotient of $X\times[-1,1]$ by the relation which identifies $X\times\{1\}$ to a single point, and $X\times\{-1\}$ to another single point as in I.3.14. Then $S'(X)$ may also be identified with the quotient of $S''(X)$ by the contractible subset $C^+(X)$ or $C^-(X)$, and the two vertical quotient maps $S''(X)\to S'(X)$ induce isomorphisms on the $\tilde{K}$-groups. Finally we obtain the commutative diagram

$$\begin{array}{ccccc} \tilde{K}(C'f) & \longleftarrow & \tilde{K}(S'(Y)) & \longleftarrow & \tilde{K}(S'(X)) \\ \| & & \downarrow\approx & & \downarrow\approx \\ \tilde{K}(C'f) & \longleftarrow & \tilde{K}(S'(Y)\cup C^-(X)) & \longleftarrow & \tilde{K}(S''(X)) \\ \| & & \| & & \uparrow\approx \\ \tilde{K}(C'f) & \longleftarrow & \tilde{K}(S'(Y)\cup C^-(X)) & \longleftarrow & \tilde{K}(S'(X)), \end{array}$$

Where all the vertical maps are isomorphisms. Since the bottom horizontal row is an exact sequence by 2.42, the sequence

$$\tilde{K}(S'(X))\longrightarrow\tilde{K}(S'(Y))\longrightarrow\tilde{K}(Cf)$$

is also exact. □

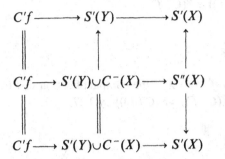

3.30. *Remark*. According to 1.34 and 3.19, $\tilde{K}(S'(\dot{Y}))\approx K^{-1}(Y)$ by the map associating each $\alpha\colon\dot{Y}\to\mathrm{GL}_p(k)$, such that $\alpha(\infty)=1$, with the p-dimensional bundle E_α of I.3.14. Now we claim that the diagram

$$\begin{array}{ccc} \tilde{K}(S'(\dot{Y})) & \longrightarrow & \tilde{K}(C'(\dot{f})) \approx \tilde{K}(C'(f)) \\ \wr\wr & & \swarrow \\ K^{-1}(Y) & \longrightarrow & K(i^*) \end{array}$$

is commutative. In this diagram, $\dot{f}\colon \dot{Y} \to \dot{X}$ is the compactification of f (note that $\dot{Y} = Y \cup \{\infty\}$, $\dot{X} = X \cup \{\infty\}$) and $\tilde{K}(C'(\dot{f})) \approx \tilde{K}(C'(f))$) by 3.27.

Fig. 10

Finally, i^* is the inverse image functor associated with the inclusion $i\colon Y \approx Y \times \{1\} \to M_f$, where M_f is the mapping cylinder of f (note that $C'(f) = M_f / i(Y)$).

The image of the class of E_α by the composition $\tilde{K}(S'(Y)) \to \tilde{K}(C'(f)) \to K(i^*)$ is $d(E, F, \beta)$, where $E = F$ is obtained by clutching the trivial bundles $M_f^+ \times k^n$ and $M_f^- \times k^n$ by the transition function α' over $Y \times \{\frac{1}{2}\}$ (where $\alpha' = \alpha|_Y$; also M_f^+ is the class of $Y \times [\frac{1}{2}, 1]$, and M_f^- is the class of $X \sqcup Y \times [0, \frac{1}{2}]$),

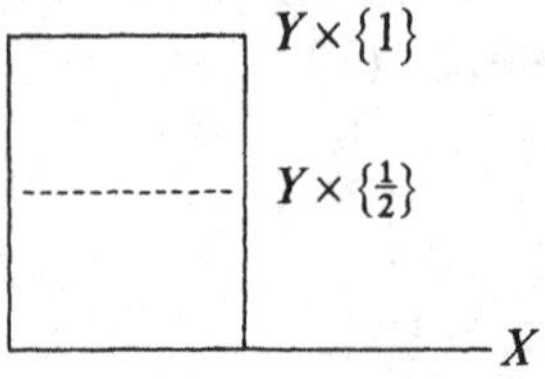

Fig. 11

and where $\beta\colon E|_{Y \times \{1\}} \to F|_{Y \times \{1\}}$ is induced by the identity map of the trivial bundle. Applying I.3.3, it is easy to see that $d(E, F, \beta)$ is also equal to $d(E', F', \beta')$, where $E' = F'$ is obtained by clutching the trivial bundles $M_f \times k^n$ and $Y \times k^n$ by the transition function α over $Y \times \{1\}$, and where β' is induced by the identity map of the trivial bundle. If we identify E' and F' with $M_f \times k^n$, we find the image of E_α in $K(i^*)$ by the connecting homomorphism $K^{-1}(Y) \to K(i^*)$ described in 3.21.

Exercises (Section II.6) 7, 12.

4. The Groups $K^{-n}(X)$ and $K^{-n}(X, Y)$

As promised in 3.1, we are going to define the groups $K^{-n}(X)$ and $K^{-n}(X, Y)$, and prove that they have the desired properties. In fact we will slightly generalize the theory by considering locally compact spaces instead of compact spaces (this generalization will turn out to be very useful).

4.1. As in the preceding sections $\dot{X}$ will denote the one point compactification of a locally compact space X. We define $K(X)$ and $K^{-1}(X)$ as $\mathrm{Ker}[K(\dot{X}) \to K(\{\infty\})]$ and $\mathrm{Ker}[K^{-1}(\dot{X}) \to K^{-1}(\{\infty\})]$ respectively. For X compact, $\dot{X}$ is just the disjoint union of X and $\{\infty\}$; thus by 1.26 and 3.19 this definition agrees with the original definition of $K(X)$ and $K^{-1}(X)$. Let us define a *morphism* between two locally compact spaces X and Y to be a continuous map $f: \dot{X} \to \dot{Y}$, such that $f(\infty) = \infty$. We will write f in the form $f: X \dashrightarrow Y$. It is obvious that locally compact spaces are the objects of a category with morphisms as defined above. Moreover, since the diagrams

$$\begin{array}{ccc} K(\dot{Y}) & \longrightarrow & K(\dot{X}) \\ \downarrow & & \downarrow \\ K(\infty) & = & K(\infty) \end{array} \qquad \begin{array}{ccc} K^{-1}(\dot{Y}) & \longrightarrow & K^{-1}(\dot{X}) \\ \downarrow & & \downarrow \\ K^{-1}(\infty) & = & K^{-1}(\infty) \end{array}$$

are commutative, the functors K and K^{-1} are defined on this category.

4.2. Example. Let $g: X \to Y$ be a continuous *proper* map between X and Y. Then g induces a continuous map $f: \dot{X} \to \dot{Y}$, where $f(x) = g(x)$ if $x \neq \infty$, and $f(\infty) = \infty$. However, not all morphisms are of this form.

4.3. Example. Let Z be a locally compact space, and let T be a closed subspace. Then we define a morphism $f: Z \dashrightarrow Z - T$, or $f: \dot{Z} \to \widehat{Z - T}$, by the formula $f(z) = z$ if $z \notin T$, $f(z) = \infty$ if $z \in T$, and $f(\infty) = \infty$.

4.4. Example. The inclusion of the point $\{\infty\}$ in S^n induces isomorphisms $\tilde{K}(S^n) \approx K(\mathbb{R}^n)$ and $\tilde{K}^{-1}(S^n) \approx K^{-1}(\mathbb{R}^n)$.

4.5. *Remark.* The theory we are developing is similar to *cohomology with compact support*. The reader is warned that in general, Theorems 1.33 and 3.17 are false for locally compact spaces.

4.6. Proposition. *Let X be a compact space, and let Y be a closed subspace. Then we have the exact sequence*

$$K^{-1}(X) \xrightarrow{j^*} K^{-1}(Y) \xrightarrow{\partial} K(X - Y) \xrightarrow{i^*} K(X) \xrightarrow{j^*} K(Y),$$

where i^* and j^* are induced by $i: X \dashrightarrow X - Y$ and $j: Y \dashrightarrow X$ respectively.

Proof. Since the space $X/Y - \{y\}$ may be identified with $X - Y$, we have $(\widehat{X - Y}) \approx X/Y$ and $K(X - Y) \approx \tilde{K}(X/Y)$. The proposition now follows from 2.35 and 3.23. □

4.7. Corollary. *Let X be a locally compact space and let Y be a closed subspace. Then we have the exact sequence*

$$K^{-1}(X) \xrightarrow{j^*} K^{-1}(Y) \xrightarrow{\partial} K(X-Y) \xrightarrow{i^*} K(X) \xrightarrow{j^*} K(Y)$$

where i^* and j^* are induced by $i\colon X \dashrightarrow X-Y$ and $j\colon Y \dashrightarrow X$ respectively.

Proof. We consider the commutative diagram

$$\begin{array}{ccccccccc}
0 & & 0 & & 0 & & 0 & & 0 \\
\uparrow & & \uparrow & & \| & & \uparrow & & \uparrow \\
K^{-1}(\{\infty\}) & \longrightarrow & K^{-1}(\{\infty\}) & \longrightarrow & 0 & \longrightarrow & K(\{\infty\}) & \longrightarrow & K(\{\infty\}) \\
\uparrow & & \uparrow & & \uparrow & & \uparrow & & \uparrow \\
K^{-1}(\dot{X}) & \longrightarrow & K^{-1}(\dot{Y}) & \longrightarrow & K(\dot{X}-\dot{Y}) & \longrightarrow & K(\dot{X}) & \longrightarrow & K(\dot{Y}) \\
\uparrow & & \uparrow & & \wr\wr & & \uparrow & & \uparrow \\
K^{-1}(X) & \longrightarrow & K^{-1}(Y) & \longrightarrow & K(X-Y) & \longrightarrow & K(X) & \longrightarrow & K(Y) \\
\uparrow & & \uparrow & & \uparrow & & \uparrow & & \uparrow \\
0 & & 0 & & 0 & & 0 & & 0
\end{array}$$

where the last row is induced by the others. Since the vertical maps define split exact sequences, elementary diagram chasing shows that the last row is exact when the first two are. Thus the result follows from 4.6. □

4.8. Theorem. *For every locally compact space Y, we have a natural isomorphism $K^{-1}(Y) \approx K(Y \times \mathbb{R})$. Moreover, if Y is a closed subspace of X, we have an exact sequence*

$$K(X \times \mathbb{R}) \longrightarrow K(Y \times \mathbb{R}) \longrightarrow K(X-Y) \longrightarrow K(X) \longrightarrow K(Y).$$

Proof. The second part of this theorem is a consequence of 4.7 and the first part. For the first part, let us consider the space $Z = Y \times \mathbb{R}^+$ where $\mathbb{R}^+ = [0, +\infty[$. Then Z is homeomorphic to $\dot{Y} \times [0, 1] - \dot{Y}_\vee [0, 1]$ (where 1 is the base point of $[0, 1]$). Hence $\dot{Z} \approx Y \times [0, 1]/Y_\vee [0, 1]$. Now we define an homotopy $r\colon \dot{Z} \times [0, 1] \to \dot{Z}$ given by $r(y, t, u) = $ class of $(y, (1+(1-t)u)$ for $(y, t) \in \dot{Y} \times [0, 1]$.

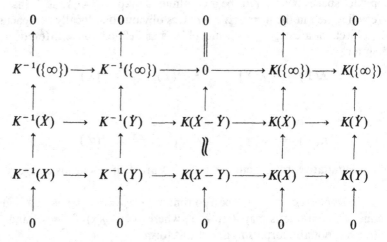

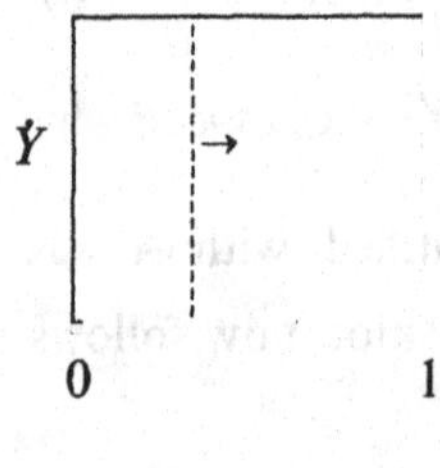

Fig. 12

It follows that $K(Y\times\mathbb{R}^+)=K^{-1}(Y\times\mathbb{R}^+)=0$. Applying 4.7 to the pair $(Y\times\mathbb{R}^+, Y)$ we obtain the exact sequence

$$K^{-1}(Y\times\mathbb{R}^+)\longrightarrow K^{-1}(Y)\longrightarrow K(Y\times\mathbb{R})\longrightarrow K(Y\times\mathbb{R}^+).$$

Therefore $K^{-1}(Y)\approx K(Y\times\mathbb{R})$. □

4.9. The homomorphism

$$\partial=\partial_{X,Y}: K(Y\times\mathbb{R})\longrightarrow K(X-Y)$$

may be described more explicitly as follows. Let Z be now the space $X\times\{0\}\cup Y\times[0, 1]-Y\times\{1\}$.

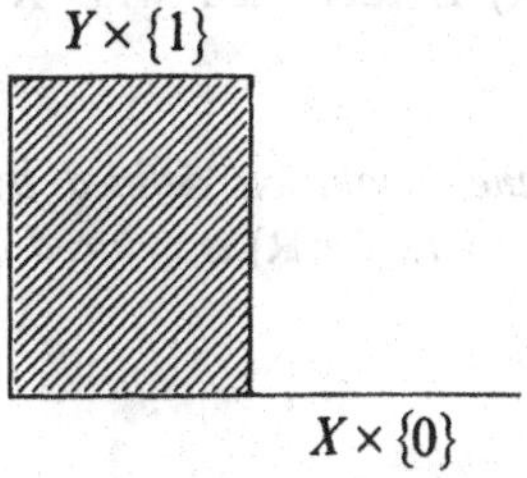

Fig. 13

Since $Z-\overline{Y\times[0, 1)}\approx X-Y$ and $Z-X\times\{0\}\approx Y\times\mathbb{R}$, we have the following diagram

$$\begin{array}{ccc} K(Y\times\mathbb{R}) & \xrightarrow{\partial_{X,Y}} & K(X-Y) \\ & {\scriptstyle\beta^*_{X,Y}}\searrow \quad \swarrow{\scriptstyle\alpha^*_{X,Y}} & \\ & K(Z) & \end{array}$$

In this diagram $\alpha^*_{X,Y}$ is an isomorphism because of the exact sequence

$$K^{-1}(Y\times[0, 1))\longrightarrow K(X-Y)\longrightarrow K(Z)\longrightarrow K(Y\times[0, 1)).$$

In this sequence $K(Y\times[0, 1))=K^{-1}(Y\times[0, 1))=0$ because $\overset{\cdot}{\widehat{Y\times[0, 1)}}\approx\dot{Y}\times I/\dot{Y}\vee I$, which has the K-group of a point (cf. 3.27), hence also the K^{-1}-group of a point by the same argument applied to $Y\times\mathbb{R}$.

Now we claim that the diagram above is commutative (i.e. $\partial_{X,Y}=\alpha^{*-1}_{X,Y}\cdot\beta^*_{X,Y}$). To prove this let us put $\partial'_{X,Y}=\alpha^{*-1}_{X,Y}\cdot\beta^*_{X,Y}$. Then we have two commutative diagrams:

$$\begin{array}{ccc} K(\dot{Y}\times\mathbb{R}) & \xrightarrow{\partial_{\dot{X},\dot{Y}}} & K(\dot{X}-\dot{Y}) \\ \downarrow & & \| \\ K(Y\times\mathbb{R}) & \xrightarrow{\partial_{X,Y}} & K(X-Y) \end{array} \qquad \begin{array}{ccc} K(\dot{Y}\times\mathbb{R}) & \xrightarrow{\partial'_{\dot{X},\dot{Y}}} & K(\dot{X}-\dot{Y}) \\ \downarrow & & \| \\ K(\dot{Y}\times\mathbb{R}) & \xrightarrow{\partial'_{X,Y}} & K(X-Y) \end{array}$$

Therefore, by the substitution $X \mapsto \dot{X}$ and $Y \mapsto \dot{Y}$, we may assume X and Y compact. In that case $\tilde{K}(\dot{Z}) \approx \tilde{K}(C'(f)) \approx \tilde{K}(C'\dot{f}))$ (where $f\colon Y \rightarrowtail X$ is the inclusion map (cf. 3.30)), $\dot{Y} \times \mathbb{R} \approx S(\dot{Y})$, and $\beta^*_{X,Y}\colon K(Y \times \mathbb{R}) \to K(Z)$ coincides modulo isomorphism with the map $\tilde{K}(S(\dot{Y})) \to \tilde{K}(C(\dot{f}))$ in the reduced Puppe sequence 3.29, applied to the map $\dot{f}$. According to 3.30, the diagram

$$\begin{array}{ccc} \tilde{K}(S(\dot{Y})) & \longrightarrow & K(C(\dot{f})) \\ \wr\!\wr & \overset{\partial'_{X,Y}}{\frown} & \wr\!\wr \\ K^{-1}(Y) & \xrightarrow[\partial_{X,Y}]{} & K(X-Y) \end{array}$$

is commutative. Hence $\partial_{X,Y} = \partial'_{X,Y}$ ($K(Y \times \mathbb{R})$ is identified with $K^{-1}(Y)$ by the map $\alpha \mapsto E_\alpha$ described in 3.30).

4.10. Proposition. *Let $T\colon X \times \mathbb{R} \to X \times \mathbb{R}$ be the involution defined by $(x, \lambda) \mapsto (x, -\lambda)$. Then $T^*(u) = -u$, where $T^*\colon K(X \times \mathbb{R}) \to K(X \times \mathbb{R})$ is induced by T.*

Proof. Since we have the split exact sequence

$$0 \longrightarrow K(X \times \mathbb{R}) \longrightarrow K(\dot{X} \times \mathbb{R}) \longrightarrow K(\{\infty\} \times \mathbb{R}) \longrightarrow 0,$$

it suffices to prove the proposition for X compact. Then $K(X \times \mathbb{R}) \approx K(X \times B^1, X \times S^0)$, and T^* can be identified with the involution of $K(X \times B^1, X \times S^0)$ induced by $(x, \lambda) \mapsto (x, -\lambda)$.

Now consider the exact sequence associated with the pair $(X \times B^1, X \times S^0)$:

$$\begin{array}{ccccccccc} K^{-1}(X \times B^1) & \xrightarrow{\Delta_1} & K^{-1}(X \times S^0) & \xrightarrow{\partial} & K(X \times B^1, X \times S^0) & \to & K(X \times B^1) & \xrightarrow{\Delta} & K(X \times S^0) \\ \wr\!\wr & & \wr\!\wr & & & & \wr\!\wr & & \wr\!\wr \\ K^{-1}(X) & & K^{-1}(X) \oplus K^{-1}(X) & & & & K(X) & & K(X) \oplus K(X) \end{array}$$

In this exact sequence $S^0 = \{-1, +1\}$, and the morphisms Δ_1 and Δ are the diagonal homomorphisms with identifications made. Moreover, the involution T induces the involution t^* of $K^{-1}(X \times S^0)$ which switches factors, and we have the commutative diagram

$$\begin{array}{ccc} K^{-1}(X \times S^0) & \xrightarrow{\partial} & K(X \times B^1, X \times S^0) \\ \Big\downarrow{\scriptstyle t^*} & & \Big\downarrow{\scriptstyle T^*} \\ K^{-1}(X \times S^0) & \xrightarrow{\partial} & K(X \times B^1, X \times S^0), \end{array}$$

where ∂ is surjective. If $x = \partial(y) \in K(X \times B^1, X \times S^0)$, it follows that $x + T^*(x) = \partial(y + t^*(y)) = 0$ since $\operatorname{Im}(\partial\Delta_1) = 0$. $\square$

4.11. Definition. If Y is a closed subspace of a locally compact space X, we define $K^{-n}(X, Y)$ to be $K((X-Y)\times\mathbb{R}^n)$.

By Theorem 4.8, this definition agrees (up to isomorphism) with the definition of $K^{-1}(X)=K^{-1}(X, \varnothing)$ given in 3.3. Moreover, since $K^{-1}(X, Y)\approx \operatorname{Ker}\ulcorner K^{-1}(X/Y)\to K^{-1}(\{y\})\urcorner$ for both definitions of $K^{-1}(X, Y)$ (cf. 3.25 and 4.13 below), this definition 4.11 also agrees with the one given in 3.25.

4.12. Proposition. *If X is a compact space and if Y is a closed subspace, we have natural isomorphisms $K^{-n}(X, Y)\approx\tilde{K}(S^n(X/Y))\approx K(X\times B^n, X\times S^{n-1}\cup Y\times B^n)$.*

Proof. We have the homeomorphisms

$$X\times B^n - X\times S^{n-1}\cup Y\times B^n\approx(X-Y)\times(B^n-S^{n-1})\approx(X-Y)\times\mathbb{R}^n \qquad (2.41).$$

Hence

$$K(X\times B^n, X\times S^{n-1}\cup Y\times B^n)\approx K((X-Y)\times(B^n-S^{n-1}))\approx K((X-Y)\times\mathbb{R}^n)$$

Moreover, $S^n(X/Y)\approx B^n/S^{n-1}\wedge X/Y\approx X\times B^n/X\times S^{n-1}\cup Y\times B^n$ and thus the last isomorphism of the proposition also follows. □

We will write $K^{-n}(X)$ instead of $K^{-n}(X, \varnothing)$. Note that the group $K^{-n}(X, Y)$ depends functorially on the pair (X, Y). More precisely, if $f\colon (X, Y)\to(X', Y')$ is a morphism of pairs, it induces a morphism $g\colon X-Y \dashrightarrow X'-Y'$ in the category of locally compact spaces defined by $g(\infty)=\infty$, $g(x)=\infty$ if $f(x)\in Y'$, and $g(x)=f(x)$ if $f(x)\notin Y'$.

4.13. Theorem. *Let X be a locally compact space and let Y be a closed subspace. Then, for any $n\geqslant 0$ we have an exact sequence*

$$K^{-n-1}(X)\longrightarrow K^{-n-1}(Y)\longrightarrow K^{-n}(X, Y)\longrightarrow K^{-n}(X)\longrightarrow K^{-n}(Y)$$

Proof. For $n=0$, the theorem is simply Theorem 4.8. The general case follows by applying 4.8 to the case for $X\times\mathbb{R}^n$ and $Y\times\mathbb{R}^n$. □

4.14. Corollary. *Let X and Y as in 4.13, and let Z be a closed subspace of Y. Then we have an exact sequence*

$$K^{-n-1}(X,Z)\longrightarrow K^{-n-1}(Y,Z)\longrightarrow K^{-n}(X, Y)\longrightarrow K^{-n}(X,Z)\longrightarrow K^{-n}(Y,Z).$$

Proof. We apply 4.13 to the pair $(X-Z, Y-Z)$. □

We adopt the following general convention: the space X/Y will denote the quotient of $\dot{X}$ by the relation which identifies every element of Y with $\{\infty\}$ (for example $X/\varnothing$ is the one point compactification of X). Then we have the following theorem:

4.15. Theorem. *The natural map* $(X, Y) \to (X/Y, \{\infty\})$ *induces an isomorphism* $K^{-n}(X/Y, \{\infty\}) \approx K^{-n}(X, Y)$. *Moreover,* $K^{-n}(X)$ *is naturally isomorphic to* $\operatorname{Ker}[K^{-n}(\dot{X}) \to K^{-n}(\{\infty\})]$.

Proof. Since $X/Y - \{\infty\}$ and $X - Y$ are homeomorphic, $K^{-n}(X/Y, \{\infty\}) \approx K^{-n}(X, Y)$. In particular, $K^{-n}(X) \approx K^{-n}(\dot{X}, \{\infty\})$. Now the exact sequence

$$K^{-n-1}(\dot{X}) \longrightarrow K^{-n-1}(\{\infty\}) \longrightarrow K^{-n}(\dot{X}, \{\infty\}) \longrightarrow K^{-n}(\dot{X}) \longrightarrow K^{n}(\{\infty\})$$

implies the split exact sequence

$$0 \longrightarrow K^{-n}(X) \longrightarrow K^{-n}(\dot{X}) \longrightarrow K^{-n}(\{\infty\}) \longrightarrow 0,$$

since $\{\infty\}$ is a retract of X. □

4.16. Theorem. *Let* $f_0, f_1 : X \dashrightarrow Y$ *be morphisms which are homotopic. Then* f_0 *and* f_1 *induce the same homomorphism*

$$f_0^* = f_1^* : K^{-n}(Y) \longrightarrow K^{-n}(X)$$

Proof. Since f_0 and f_1 are homotopic, there exists a morphism $f: X \times I \dashrightarrow Y$ such that $f_\alpha = f \cdot i_\alpha$, where $i_\alpha : X \dashrightarrow X \times I$ is the morphism associated with the continuous proper map $x \mapsto (x, \alpha)$ for $\alpha = 0, 1$. In order to prove the theorem, it suffices to consider the case $n = 0$, since the general case follows by replacing Y and X by $Y \times \mathbb{R}^n$ and $X \times \mathbb{R}^n$ respectively. Then we have the commutative diagram

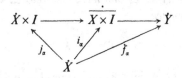

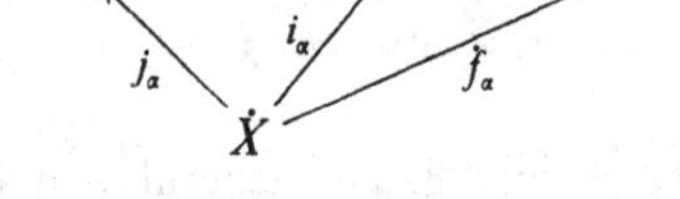

showing that the maps $\dot{f}_0 : \dot{X} \to \dot{Y}$ and $\dot{f}_1 : \dot{X} \to \dot{Y}$ are homotopic. Hence they induce the same homomorphism on the K-groups and we have the commutative diagram:

$$\begin{array}{ccc} K(\dot{Y}) & \longrightarrow & K(\dot{X}) \\ \downarrow & & \downarrow \\ K(\{\infty\}) & = & K(\{\infty\}) \end{array}$$

Therefore $f_0^* = f_1^* : K(Y) \to K(X)$. □

4.17. Let X', X, and X'', be locally compact spaces, and let $i: X' \dashrightarrow X$ and $j: X \dashrightarrow X''$ be morphisms. Then the sequence

$$X' \dashrightarrow X \dashrightarrow X''$$

is called *exact* if it is isomorphic to a sequence of the form

$$T \dashrightarrow X \dashrightarrow X - T,$$

where T is a closed subspace of X. From Theorem 4.13, it follows that we have the exact sequence

$$K^{-n-1}(X) \longrightarrow K^{-n-1}(X') \longrightarrow K^{-n}(X'') \longrightarrow K^{-n}(X) \longrightarrow K^{-n}(X')$$

for $n \geqslant 0$. This formal definition of "exact sequences" may be applied to axiomatically characterize the functors K^{-n} (Karoubi-Villamayor [1]).

4.18. Theorem. *Let X_1 and X_2 be closed subspaces of a locally compact space X such that $X_1 \cup X_2 = X$. Then we have the exact sequence*

$$K^{-n-1}(X_1) \oplus K^{-n-1}(X_2) \xrightarrow{v} K^{-n-1}(X_1 \cap X_2) \xrightarrow{\Delta}$$
$$K^{-n}(X_1 \cup X_2) \xrightarrow{u} K^{-n}(X_1) \oplus K^{-n}(X_2) \xrightarrow{v} K^{-n}(X_1 \cap X_2)$$

for $n \geqslant 0$, where u and v are defined by $u(\alpha) = (\alpha|_{X_1}, \alpha|_{X_2})$ and $v(\alpha_1, \alpha_2) = \alpha_1|_{X_1 \cap X_2} - \alpha_2|_{X_1 \cap X_2}$.

Proof. We have the following commutative diagram

$$\begin{array}{ccccc}
X_1 & \dashrightarrow & X_1 \cup X_2 & \dashrightarrow & X_1 \cup X_2 - X_1 = Z \\
\uparrow & & \uparrow & & \parallel \\
X_1 \cap X_2 & \dashrightarrow & X_2 & \dashrightarrow & X_2 - X_1 \cap X_2 = Z
\end{array}$$

where the horizontal lines are exact. From 4.17, we obtain the following exact sequences and commutative diagram:

$$\begin{array}{ccccccccc}
K^{-n-1}(X_1 \cup X_2) & \longrightarrow & K^{-n-1}(X_1) & \longrightarrow & K^{-n}(Z) & \rightarrow & K^{-n}(X_1 \cup X_2) & \longrightarrow & K^{-n}(X_1) \\
\downarrow & & \downarrow & & \parallel & & \downarrow & & \downarrow \\
K^{-n-1}(X_2) & \longrightarrow & K^{-n-1}(X_1 \cap X_2) & \rightarrow & K^{-n}(Z) & \longrightarrow & K^{-n}(X_2) & \longrightarrow & K^{-n}(X_1 \cap X_2)
\end{array}$$

The zigzag homomorphism $K^{-n-1}(X_1 \cap X_2) \rightarrow K^{-n}(X_1 \cup X_2)$ is the homomorphism Δ written in the text of the theorem. Now the exactness of the sequence of this theorem is a purely formal consequence of the two exact sequences above, except for the sequence

$$K^0(X_1 \cup X_2) \longrightarrow K^0(X_1) \oplus K^0(X_2) \longrightarrow K^0(X_1 \cap X_2).$$

If X_1 and X_2 are compact, let $\alpha_i = [E_i] - [T_i]$ be elements of $K(X_i)$ such that $\alpha_1|_{X_1 \cap X_2} = \alpha_2|_{X_1 \cap X_2}$, and such that T_1 and T_2 are trivial bundles of the same rank n. According to 1.18, by adding trivial bundles, this implies that the vector bundles $E_1|_{X_1 \cap X_2}$ and $E_2|_{X_1 \cap X_2}$ are isomorphic. If $f\colon E_1|_{X_1 \cap X_2} \to E_2|_{X_1 \cap X_2}$ is such an isomorphism, we let E be the vector bundle obtained by clutching E_1 and E_2 using f (I.3.2) and let T be the trivial bundle with rank n. Then $x = [E] - [T]$ is an element of $K(X)$ whose restrictions to $K(X_1)$ and $K(X_2)$ are the elements α_1 and α_2.

In the general case, we consider the diagram

$$\begin{array}{ccccccccc}
 & & 0 & & 0 & & 0 & & \\
 & & \uparrow & & \uparrow & & \uparrow & & \\
0 & \longrightarrow & K^0(\{\infty\}) & \longrightarrow & K^0(\{\infty\}) \oplus K^0(\{\infty\}) & \longrightarrow & K^0(\{\infty\}) & \longrightarrow & 0 \\
 & & \uparrow & & \uparrow & & \uparrow & & \\
 & & K^0(\dot{X}_1 \cup \dot{X}_2) & \longrightarrow & K^0(\dot{X}_1) \oplus K^0(\dot{X}_2) & \longrightarrow & K^0(\dot{X}_1 \cap \dot{X}_2) & & \\
 & & \uparrow & & \uparrow & & \uparrow & & \\
 & & K^0(X_1 \cup X_2) & \longrightarrow & K^0(X_1) \oplus K^0(X_2) & \longrightarrow & K^0(X_1 \cap X_2) & & \\
 & & \uparrow & & \uparrow & & \uparrow & & \\
 & & 0 & & 0 & & 0 & &
\end{array}$$

where the vertical sequences are split exact. The first two horizontal sequences are exact by what we have just proved. It follows that the last horizontal sequence is also exact. □

4.19. Theorem. *Let U_1 and U_2 be two open subspaces of a locally compact space X such that $U_1 \cup U_2 = X$. Then we have the exact sequence*

$$\begin{aligned} K^{-n-1}(U_1) \oplus K^{-n-1}(U_2) &\longrightarrow K^{-n-1}(U_1 \cup U_2) \longrightarrow \\ K^{-n}(U_1 \cap U_2) &\longrightarrow K^{-n}(U_1) \oplus K^{-n}(U_2) \longrightarrow K^{-n}(U_1 \cup U_2). \end{aligned}$$

Proof. We have the following exact sequences of locally compact spaces:

$$U = U_1 - U_1 \cap U_2 \dashrightarrow U_1 \dashrightarrow U_1 \cap U_2$$

$$U = U_1 \cup U_2 - U_2 \dashrightarrow U_1 \cup U_2 \dashrightarrow U_2$$

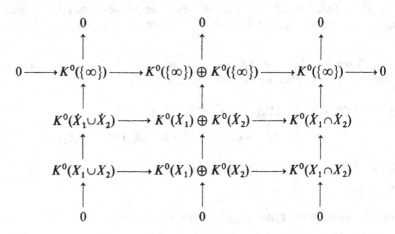

By 4.17 we obtain the following exact sequences

$$\begin{array}{ccccccccc}
K^{-n-1}(U_1) & \longrightarrow & K^{-n-1}(U) & \longrightarrow & K^{-n}(U_1 \cap U_2) & \longrightarrow & K^{-n}(U_1) & \longrightarrow & K^{-n}(U) \\
\downarrow & & \| & & \downarrow & & \downarrow & & \| \\
K^{-n-1}(U_1 \cup U_2) & \longrightarrow & K^{-n-1}(U) & \longrightarrow & K^{-n}(U_2) & \longrightarrow & K^{-n}(U_1 \cup U_2) & \longrightarrow & K^{-n}(U)
\end{array}$$

The homomorphism $K^{-n-1}(U_1 \cup U_2) \to K^{-n}(U_1 \cap U_2)$ is the obvious zigzag homomorphism. The theorem now follows by a simple diagram chasing argument as in 4.18. □

4.20. The sequences of Theorems 4.18 and 4.19 are called *Mayer-Vietoris exact sequences.*

4.21. Proposition. *Let X be a locally compact space and let (U_i) be an inductive family of open subsets in X such that every compact subset of X is contained in at least one U_i. Then*

$$K^{-n}(X) \approx \operatorname{inj\,lim} K^{-n}(U_i).$$

Proof. Replacing X by $X \times \mathbb{R}^n$ when $n > 0$, we see that it suffices to prove the assertion for $n = 0$. Now we have the commutative diagram

$$\begin{array}{ccccc}
K(U_i) & \longrightarrow & K(U_j) & \longrightarrow & K(X) \\
\wr\wr & & \wr\wr & & \wr\wr \\
K(\dot{X}, \dot{X} - U_i) & \longrightarrow & K(\dot{X}, \dot{X} - U_j) & \longrightarrow & K(\dot{X}, \infty)
\end{array}$$

where $\dot{X}$ is the one point compactification of X (note that $\dot{U}_i \approx \dot{X}/\dot{X} - U_i$), and where the map $K(\dot{X}, \dot{X} - U_i) \longrightarrow K(\dot{X}, \dot{X} - U_j)$ is induced by the inclusion $(\dot{X}, \dot{X} - U_j) \subset (\dot{X}, \dot{X} - U_i)$. Since $\dot{X} - U_i$ is a base of closed neighbourhoods for the point $\{\infty\}$, the proposition is a consequence of the following lemma:

4.22. Lemma. *Let Y be a compact space, $y \in Y$, and let (S_i) be a base of closed neighbourhoods for y. Then*

$$K(Y, \{y\}) \approx \operatorname{inj\,lim} K(Y, S_i)$$

Proof. Let us denote the obvious homomorphism from $\operatorname{inj\,lim} K(Y, S_i)$ to $K(Y, \{y\})$ by l.

a) *l is surjective.* Let $d(E, F, \alpha) \in K(Y, \{y\})$ where E and F are vector bundles over Y and $\alpha\colon E|_{\{y\}} \to F|_{\{y\}}$ is an isomorphism. By I.5.11 we can find a neighbourhood S_i of y and an isomorphism $\alpha_i\colon E_{S_i} \to F_{S_i}$ such that $\alpha_i|_{\{y\}} = \alpha|_{\{y\}}$. Then $d(E, F, \alpha_y)$ is the image under l of the class of $d(E, F, \alpha_i)$ in the inductive limit.

b) *l is injective.* Let $d(E, F, \alpha_i)$ be an element of $K(Y, S_i)$ such that

$d(E, F, \alpha_i|_{\{y\}})=0$. According to 2.28 applied to the categories $\mathscr{C}=\mathscr{E}(Y)$ and $\mathscr{C}'=\mathscr{E}(\{y\})$, there is a bundle G over Y such that $\alpha_i|_{\{y\}} \oplus \mathrm{Id}_G|_{\{y\}}$ is the restriction to $\{y\}$ of an isomorphism $\alpha': E \oplus G \to F \oplus G$. Let us put $E'=E \oplus G$, $F'=F \oplus G$, and $\alpha_i'=\alpha_i \oplus \mathrm{Id}_{G_{S_i}}$. Let $\gamma: \pi^*(E')|_{S_i \times I} \to \pi^*(F')|_{S_i \times I}$, where $\pi: Y \times I \to Y$, be the morphism defined on $(x, t) \in S_i \times I$ by the formula $t\alpha_{ix} + (1-t)\alpha'_x$. Since $\gamma|_{\{y\} \times I} = \mathrm{Id}$, there is a neighbourhood V of $\{y\} \times I$ in $S_i \times I$ of the form $S_j \times I$, such that $\gamma|_V$ is an isomorphism (I.5.11). Hence, by 2.15, we have $d(E, F, \alpha_i|_{S_j}) = d(E, F, \alpha'|_{S_j}) = 0$. Therefore, the class of $d(E, F, \alpha_i)$ in the inductive limit is 0. □

Exercises (Section II.6) 14–18.

5. Multiplicative Structures

5.1. Let X and Y be compact spaces, and let E and F be vector bundles with bases X and Y respectively. Then their "external tensor product" $E \boxtimes F$ (I.4.9) is a vector bundle over $X \times Y$. The correspondence $(E, F) \mapsto E \boxtimes F$ induces a functor φ from the category $\mathscr{E}(X) \times \mathscr{E}(Y)$ to $\mathscr{E}(X \times Y)$, such that $\varphi(E \oplus E', F) \approx \varphi(E, F) \oplus \varphi(E', F)$ and $\varphi(E, F \oplus F') \approx \varphi(E, F) \oplus \varphi(E, F')$, with the analogous property for morphisms. From this functor we obtain a bilinear map

$$\varphi_*: K(X) \times K(Y) \longrightarrow K(X \times Y)$$

defined by

$$\varphi_*([E]-[E'], [F]-[F']) = [\varphi(E, F)] + [\varphi(E', F')] - [\varphi(E, F')] - [\varphi(E', F)].$$

Also we denote the *cup-product* $\varphi_*(x, y)$ for $x \in K(X)$ and $y \in K(Y)$, by $x \cup y$.

If Z is another compact space and G is a vector bundle over Z, we have a canonical isomorphism $(E \boxtimes F) \boxtimes G \approx E \boxtimes (F \boxtimes G)$ as an immediate consequence of the associativity of tensor products (I.4.7). This implies the commutativity of the diagram

$$\begin{array}{ccc} K(X) \times K(Y) \times K(Z) & \longrightarrow & K(X \times Y) \times K(Z) \\ \downarrow & & \downarrow \\ K(X) \times K(Y \times Z) & \longrightarrow & K(X \times Y \times Z) \end{array}$$

Similarly we can prove the commutativity of the diagram

$$\begin{array}{ccc} K(X) \times K(Y) & \longrightarrow & K(X \times Y) \\ \downarrow & & \downarrow T^* \\ K(Y) \times K(X) & \longrightarrow & K(Y \times X) \end{array}$$

where T is the permutation $(x, y) \mapsto (y, x)$.

5.2. The diagonal map $X \to X \times X$ induces a homomorphism $K(X \times X) \to K(X)$. If we compose this homomorphism with the homomorphism $K(X) \times K(X) \to K(X \times X)$ defined above, we have a product operation in the group $K(X)$, denoted by $(u, v) \mapsto u \cdot v$ or uv. Thanks to the commutative diagrams in 5.1, it is easy to prove that $K(X)$ is provided with a commutative ring structure. The unit element in this ring is the class of the trivial bundle of rank one (note that the basic field is $k = \mathbb{R}$ or $\mathbb{C}$; these techniques do not work for quaternionic bundles).

5.3. Example. Let $X = CP_n$ be the complex projective space of $\mathbb{C}^{n+1}$ and let ξ be the canonical line bundle over X (I.2.4). In IV.2 we prove that $K(X)$ is isomorphic to the quotient algebra $\mathbb{Z}[u]/(u^{n+1})$ where $u = [\xi] - 1$.

5.4. Proposition. *There is a unique way to naturally extend the "cup-product" defined in 5.1 to the K-groups of locally compact spaces X and Y*

$$K(X) \times K(Y) \longrightarrow K(X \times Y)$$

Moreover, the properties of associativity and commutativity still hold in this case.

Proof. Let $\dot{X}$ and $\dot{Y}$ be the one point compactifications of X and Y respectively. Since $X \times Y \approx \dot{X} \times \dot{Y} - \dot{X} \vee \dot{Y}$, we have the exact sequence

$$K^{-1}(\dot{X} \times \dot{Y}) \longrightarrow K^{-1}(\dot{X} \vee \dot{Y}) \longrightarrow K(X \times Y) \longrightarrow K(\dot{X} \times \dot{Y}) \longrightarrow K(\dot{X} \vee \dot{Y})$$

by 4.6. Let us prove that the first homomorphism in this sequence is surjective. If $\alpha: \dot{X} \vee \dot{Y} \to \mathrm{GL}(k)$ is a continuous map whose class is an element of $K^{-1}(\dot{X} \vee \dot{Y})$ (3.17), then α is the restriction of the continuous map $\tilde{\alpha}: \dot{X} \times \dot{Y} \to \mathrm{GL}(k)$ defined by $\tilde{\alpha}(x, y) = f(x) f(\infty)^{-1} g(y)$, where $f = \alpha|_{\dot{X}}$ and $g = \alpha|_{\dot{Y}}$ (we are identifying $\dot{X}$ with $\dot{X} \times \{\infty\}$ and $\dot{Y}$ with $\dot{Y} \times \{\infty\}$). From this surjectivity we obtain the exact sequence

$$0 \longrightarrow K(X \times Y) \longrightarrow K(\dot{X} \times \dot{Y}) \longrightarrow K(\dot{X} \vee \dot{Y}),$$

and the uniqueness of the cup-product θ by the commutativity of the diagram

$$\begin{array}{ccc}
 & & 0 \\
 & & \downarrow \\
K(X) \times K(Y) & \xrightarrow{\theta} & K(X \times Y) \\
{\scriptstyle i \times j}\downarrow & & \downarrow \\
K(\dot{X}) \times K(\dot{Y}) & \longrightarrow & K(\dot{X} \times \dot{Y}) \\
 & & \downarrow \\
 & & K(\dot{X} \vee \dot{Y}).
\end{array}$$

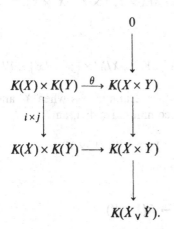

In order to prove existence, we notice first that the map $\gamma: K(\dot{X}_\vee \dot{Y}) \to K(\dot{X}) \times K(\dot{Y})$, induced by projections on the factors, is injective. In fact $\gamma([E]-[T])=0$ implies, at least stably, that $E|_{\dot{X}} \approx T|_{\dot{X}}$ and $E|_{\dot{Y}} \approx T|_{\dot{Y}}$. Since $\dot{X} \cap \dot{Y}$ is a point, $E \approx T$.

Let us now construct the map θ. If x and y are elements of $K(X)$ and $K(Y)$ respectively, $i(x) \cup j(y)$ is an element of $K(\dot{X} \times \dot{Y})$ whose restriction z to $K(\dot{X}_\vee \dot{Y})$ is zero (the image of z by the injective homomorphism $K(\dot{X}_\vee \dot{Y}) \to K(\dot{X}) \times K(\dot{Y})$ is 0 since the restrictions of $i(x)$ and $j(y)$ to $K(\{\infty\})$ are 0). Hence the homomorphism $K(\dot{X}) \times K(\dot{Y}) \to K(\dot{X} \times \dot{Y})$ induces a homomorphism $K(X) \times K(Y) \to K(X \times Y)$ by restriction, with all the required properties. □

5.5. *Remark.* Another way of interpreting Proposition 5.4 is to consider compact spaces with base point. If $\{\infty\}$ denotes the base point of the spaces $X, Y, \ldots$ involved, we have $\widehat{X' \times Y'} \approx X_\wedge Y$, where $X' = X - \{\infty\}$ and $Y' = Y - \{\infty\}$. Hence the cup-product defined in 5.4 induces a cup-product

$$\tilde{K}(X) \times \tilde{K}(Y) \longrightarrow \tilde{K}(X_\wedge Y)$$

with the same formal properties.

5.6. Proposition. *Let (X, X') and (Y, Y') be arbitrary pairs of compact spaces with $X' \subset X$ and $Y' \subset Y$. Then there is a unique natural way to define a bilinear homomorphism*

$$K(X, X') \times K(Y, Y') \longrightarrow K(X \times Y, X \times Y' \cup X' \times Y),$$

which agrees with the one defined in 5.1 when $X' = Y' = \varnothing$.

Proof. Since $X \times Y - X \times Y' \cup X' \times Y \approx (X - X') \times (Y - Y')$, the existence of such a cup-product follows from 5.4. To prove its uniqueness we consider the diagram

$$\begin{array}{ccc} K(X, X') \times K(Y, Y') & \longrightarrow & K(X \times Y, X \times Y' \cup X' \times Y) \\ \uparrow \approx & & \uparrow \approx \\ K(X/X', \{x'\}) \times K(Y/Y', \{y'\}) & \longrightarrow & K(X/X' \times Y/Y', X/X' \times \{y'\} \cup \{x'\} \times Y/Y'). \end{array}$$

This diagram shows that it is enough to prove uniqueness when X' and Y' are reduced to points. In this case we have a commutative diagram

$$\begin{array}{ccc} K(X, X') \times K(Y, Y') & \longrightarrow & K(X \times Y, X \times Y' \cup X' \times Y) \\ \downarrow & & \downarrow \\ K(X) \times K(Y) & \longrightarrow & K(X \times Y), \end{array}$$

where the map $K(X\times Y, X\times Y'\cup X'\times Y)\to K(X\times Y)$ is essentially the map $K(Z\times T)\to K(Z^+\times T^+)$, for $Z=X-X'$ and $T=Y-Y'$. By considering the exact sequences associated with the pairs $(Z^+\times T^+, Z\times T^+)$ and $(Z\times T^+, Z\times T)$, it is easy to prove that the last map is injective as in 4.15. □

5.7. Example. We will prove in III.1.3 that $K_{\mathbb{C}}(B^2, S^1)\approx\mathbb{Z}$ and that the cup-product with a generator induces an isomorphism between $K_{\mathbb{C}}(X, X')$ and $K_{\mathbb{C}}(X\times B^2, X\times S^1\cup X'\times B^2)$. We will also prove in III.5.17 that the cup-product with a generator of $K_{\mathbb{R}}(B^8, S^7)\approx\mathbb{Z}$ induces an isomorphism between $K_{\mathbb{R}}(X, X')$ and $K_{\mathbb{R}}(X\times B^8, X\times S^7\cup X'\times B^8)$. This will prove the classical Bott periodicity theorems.

5.8. Let us now assume that $X=X'$. As in 5.2, the restriction to the diagonal gives rise to another type of product

$$K(X,Y)\times K(X,Y')\longrightarrow K(X, Y\cup Y'),$$

which agrees with the product defined in 5.2 when $Y=Y'=\varnothing$. We denote this product by $(a,b)\mapsto a\cdot b$ or ab.

5.9. Theorem. *Let X be a compact space and let $K'(X)$ be the subgroup of $K(X)$ defined in I.1.29. Then for the ring structure defined in 5.2, the subgroup $K'(X)$ is a nil ideal of $K(X)$ (i.e. every element of $K'(X)$ is nilpotent). In particular, every element of $\tilde{K}(X)\approx\mathrm{Ker}[K(X)\to\mathbb{Z}]$ is nilpotent if X is connected. Moreover, if $X=\bigcup_{i=1}^{n} X_i$, where the X_i are closed contractible subsets, then $(K'(X))^n=0$.*

Proof. For each closed subset Y of X let $K_Y(X)=\mathrm{Ker}[K(X)\to K(Y)]$. From 5.8 it follows that $K_Y(X)\cdot K_{Y'}(X)\subset K_{Y\cup Y'}(X)$ in the ring $K(X)$. More generally, if $Y_1,\ldots,Y_p$ are closed subsets, then $K_{Y_1}(X)\cdot K_{Y_2}(X)\ldots K_{Y_p}(X)\subset K_{Y_1\cup\cdots\cup Y_p}(X)$.

Let $\alpha=[E]-[T]\in K'(X)$. By 1.17 we may assume that T is a trivial bundle, and that $E_x\approx T_x$ for each point x of X. Since X is compact, we can find a finite covering $[X_i^\alpha]$ of closed subsets such that $\alpha|_{X_i^\alpha}=0$. Therefore, $\alpha\in K_{X_i^\alpha}(X)$ for $i=1,\ldots,n$, and by induction on i it follows that $(\alpha)^n\in K_X(X)=0$.

Finally, if $X=\bigcup_{i=1}^{n} X_i$ where the X_i are contractible and closed, we can choose $X_i^\alpha=X_i$ for every element α of $K'(X)$. Therefore if $\alpha_1,\ldots,\alpha_n$ are n elements of $K'(X)$, their product $\alpha_1\ldots\alpha_n$ belongs to $K_X(X)=0$. □

5.10. Examples. If X is the sphere S^p, we choose $X_1=S^p_+$ and $X_2=S^p_-$. Hence, all the products in $K'(X)=\mathrm{Ker}[K(S^p)\to\mathbb{Z}]$ are 0. If $X=CP_n$ or RP_n, we can find $(n+1)$ contractible subsets X_i such that $X=\bigcup_{i=0}^{n} X_i$. In terms of homogeneous coordinates, for each i, X_i is the set of points $(x_0,\ldots,x_n)$ such that $x_i\neq 0$ and $\sum_{j\neq i}\left|\frac{x_j}{x_i}\right|^2\leqslant 1$. Hence $[\tilde{K}(RP^n)]^{n+1}=[\tilde{K}(CP^n)]^{n+1}=0$.

5.11. In the product defined in 5.8 suppose that $Y' = \varnothing$. Then we obtain a third type of product

$$K(X, Y) \times K(X) \longrightarrow K(X, Y),$$

also denoted by $(a, b) \mapsto a \cdot b$ or ab. It is easy to show that $K(X, Y)$ is then provided with a right module structure over the ring $K(X)$. Similarly we may define a left module structure which coincides with the right one.

5.12. As an application of this $K(X)$-module structure which may be defined even when X is not compact, let us consider a vector bundle V over X. If φ is a metric on V (I.8.5), we write $B(V)$ for its "*ball bundle*", i.e. the set of points v of V such that $\varphi(v, v) \leqslant 1$ on each fiber. We write $S(V)$ for its "*sphere bundle*", i.e. the set of points v of V such that $\varphi(v, v) = 1$. Then $B(V) - S(V) \approx V$, and $B(V)$ admits X as a deformation retract. Via the isomorphism $K(X) \xrightarrow{\approx} K(B(V))$ (cf. 4.16), the group $K(V) \approx K(B(V), S(V))$ becomes a $K(X)$-module. This module structure could also be described in the following way: the map $K(X) \times K(V) \to K(V)$ is the composition of the homomorphisms

$$K(X) \times K(V) \longrightarrow K(X \times V) \xrightarrow{\tau^*} K(V),$$

where $\tau \colon V \to X \times V$ is the proper map $v \mapsto (\pi(x), v)$, for $\pi \colon V \to X$ the projection. This follows from the commutative diagram

$$\begin{array}{ccc}
K(X \times V) & \xrightarrow{\quad\tau^*\quad} & K(V) \\
\wr\wr & & \wr\wr \\
K(X \times B(V), X \times S(V)) & \xrightarrow{\tau'^*} & K(B(V), S(V)) \\
\wr\wr & & \| \\
K(B(V) \times B(V), B(V) \times S(V)) & \xrightarrow{d^*} & K(B(V), S(V)),
\end{array}$$

where τ' is defined by the same formula as τ, and where d is the diagonal map [note that $d \colon (B(V), S(V)) \to (B(V) \times B(V), B(V) \times S(V))$ is homotopic to d_0 defined by $d_0(v) = (\pi(v), v)$ where X is considered as a subspace of $B(V)$ via the zero section (I.5.2)].

5.13. Proposition. *Let V be a vector bundle over a locally compact space X, and let x and y be elements of $K(V)$. Let $\Delta : V \to V \times V$ be the diagonal map and let $x' \in K(X)$ be the restriction of x to the zero section. Then $xy = \Delta^*(x \cup y) = x' \cdot y$ in the $K(X)$-module $K(V)$ defined above.*

Proof. The composition $V \xrightarrow{(\pi, id) = \tau} X \times V \xrightarrow{\delta = (i, id)} V \times V$ where i is the zero section, is homotopic within the proper maps to the diagonal map Δ. Since the diagram

$$\begin{array}{ccc} K(V)\times K(V) & \longrightarrow & K(V\times V) \\ \big\downarrow{\scriptstyle (i^*,\, id^*)} & & \big\downarrow{\scriptstyle (i,\, id)^*} \\ K(X)\times K(V) & \longrightarrow & K(X\times V) \end{array}$$

is commutative, $\Delta^*(x\cup y)=(\tau^*\cdot\delta^*)(x\cup y)=\tau^*(x'\cup y)=x'\cdot y$. □

We want to describe the product defined in 5.6 more explicitly. The general description, given in Sections 5.15, 5.16, ..., 5.25, will not be used until Chapter IV.

5.14. For the particular case $X'=\varnothing$, the homomorphism

$$K(X)\times K(Y,Y')\longrightarrow K(X\times Y, X\times Y')$$

may be simply defined in the following way. Let $x=[E]-[F]$ be an element of $K(X)$, and let $y=d(G,H,\alpha)$ be an element of $K(Y,Y')$. Then the cup-product of x and y is $d(E\boxtimes G, E\boxtimes H, 1\boxtimes\alpha)-d(F\boxtimes G, F\boxtimes H, 1\boxtimes\alpha)$. It is easy to see that this product is bilinear and agrees with the product defined in 5.1 when $Y'=\varnothing$. By an argument analogous to the one used in the proof of 5.6, this shows that the above formula is correct.

5.15. In the general case, it is convenient to first give another description of the relative group $K(X,X')$ for X locally compact and X' closed in X. We consider the full subcategory $\mathscr{E}'(X)$ of $\mathscr{E}(X)$ whose objects are the direct factors of trivial bundles (cf. I.6.24; note that by I.6.5 $\mathscr{E}'(X)=\mathscr{E}(X)$ when X is compact). If E is an object of $\mathscr{E}'(X)$ provided with a metric (I.8.5), and if $E_0\oplus E_1$ is an orthogonal decomposition of E with respect to this metric, then an endomorphism D of $E=E_0\oplus E_1$ is said to be *admissible* if:

(i) *it is of degree one and self-adjoint, i.e. it can be written matricially as*

$$\begin{pmatrix} 0 & \alpha^* \\ \alpha & 0 \end{pmatrix}$$

where $\alpha: E_0\to E_1$.

(ii) $D|_Y$ *is an automorphism of* $E|_Y$.

(iii) *There is a compact space* $K\subset X$ *such that* $D|_{X-K}$ *is an automorphism of* $E|_{X-K}$.

5.16. Let $\mathscr{E}$ be the set of pairs (E,D) where D is admissible. An element (E,D) of $\mathscr{E}$ is called *elementary* if D is an automorphism. Two elements (E,D) and (E',D') are called *homotopic* if there exists an isometry $f: E\to E'$ of the form $f_0\oplus f_1$, where $f_i: E_i\to E_i'$, such that $f^{-1}\cdot D'\cdot f$ is homotopic to D within the admissible operators on E. We let $K_0(X,X')$ denote the quotient of $\mathscr{E}$ by the equivalence relation $\sigma\sim\sigma'\Leftrightarrow\exists\tau,\tau'$ elementary, such that $\sigma+\tau$ is homotopic to $\sigma'+\tau'$ (the sum of pairs is defined as usual by the sum of their components; this sum defines a monoid structure on $K_0(X, X')$). We let $\sigma(E, D)$ denote the class of the pair (E, D) in $K_0(X, X')$.

5.17. Proposition. *The monoid $K_0(X, X')$ is an abelian group.*

Proof. Let $\sigma(E, D)$ be an element of $K_0(X, X')$, and let $\bar{E}$ be the vector bundle provided with the "opposite" gradation; i.e. $\bar{E}_0 = E_1$ and $\bar{E}_1 = E_0$. Then $\sigma(E, D) + \sigma(\bar{E}, -D) = \sigma(E \oplus \bar{E}, D \oplus (-D))$ and we have the homotopy

$$\Delta_\theta = \begin{pmatrix} D\cos\theta & \sin\theta \\ \sin\theta & -D\cos\theta \end{pmatrix}, \quad \text{for } \theta \in [0, \pi/2].$$

Since $\Delta_{\pi/2}$ is an automorphism, $\sigma(E, D) + \sigma(\bar{E}, -D) = 0$. □

5.18. Proposition. *If X is a compact space, then the homomorphism*

$$\gamma\colon K_0(X, X') \longrightarrow K(X, X'),$$

defined by $\sigma(E, D) \mapsto d(E_0, E_1, \alpha|_{X'})$, is an isomorphism.

Proof. From the definition of $K(X, X')$ (2.13 and 2.19) we see that γ is well-defined. Let us define a homomorphism $\gamma'\colon K(X, X') \to K_0(X, X')$ in the opposite direction. If $d(E_0, E_1, \beta)$ is an element of $K(X, X')$, we can provide E_0 and E_1 with metrics, which are well-defined up to isomorphism (I.8.8). Let $\alpha\colon E_0 \to E_1$ be an arbitrary extension of β (I.5.10), and let D be the endomorphism of $E_0 \oplus E_1$, defined by the matrix

$$D = \begin{pmatrix} 0 & \alpha^* \\ \alpha & 0 \end{pmatrix}.$$

Then $\sigma(E, D)$ does not depend on which extension is chosen, since any two extensions α and α' are homotopic (consider $t\alpha + (1-t)\alpha'$ for $t \in [0, 1]$, and the corresponding homotopy between the D's). Clearly the homomorphism γ' is inverse to γ. □

5.19. Proposition. *If X is locally compact and X' is closed in X, then the homomorphism*

$$e\colon K_0(X, X') \longrightarrow K_0(X - X'),$$

defined by $\sigma(E, D) \mapsto (E|_{X-X'}, D|_{X-X'})$, is an isomorphism.

Proof. a) *e is surjective.* Let $\sigma(E, D)$ be an element of $K_0(X-X')$, where $E=E_0 \oplus E_1$, and where

$$D=\begin{pmatrix} 0 & \alpha^* \\ \alpha & 0 \end{pmatrix}.$$

By hypothesis, there is a compact K in $X-X'$ such that $\alpha|_{X-X'-K}$ is an isomorphism. On the other hand, by adding the same vector bundle to E_0 and E_1, we may assume without loss of generality, that E_1 is a trivial bundle. Let $\tilde{E}_1$ be the trivial bundle of the same rank as E_1, and let $\tilde{E}_0$ be the vector bundle over X obtained by clutching E_0 and $\tilde{E}_1|_{X-K}$ using $\alpha|_{X-X'-K}$ (I.6.25). Let $\tilde{\alpha}: \tilde{E}_0 \to \tilde{E}_1$ be the morphism obtained by clutching α and $\mathrm{Id}_{\tilde{E}_1|X-K}$ (I.3.2). For any metric of $\tilde{E}_0$ and $\tilde{E}_1$, let $\tilde{D}$ be the endomorphism of $\tilde{E}=\tilde{E}_0 \oplus \tilde{E}_1$ defined by

$$\tilde{D}=\begin{pmatrix} 0 & \tilde{\alpha}^* \\ \tilde{\alpha} & 0 \end{pmatrix}.$$

Then clearly we have $e(\sigma(\tilde{E}, \tilde{D}))=\sigma(E, D)$.

b) *e is injective.* Let $\sigma(E, D)$ be an element of $K_0(X, X')$ such that $e(\sigma(E, D))=0$. From the definition of $K_0(X-X')$, there is an elementary pair (E', D') over $X-X'$, such that $D|_{X-X'} \oplus D'$ is homotopic to an admissible automorphism within the admissible endomorphisms of $E|_{X-X'} \oplus E'$. If D_t denotes this homotopy, there is a relatively compact open subset V of $X-X'$ such that $D|_{X-V}$ and $D_t|_{\overline{V}-V}$ are automorphisms. Therefore the element $\sigma(E_{\overline{V}-V}, D|_{\overline{V}-V})$ is 0. Since $K_0(\overline{V}, \overline{V}-V) \approx K(\overline{V}, \overline{V}-V)$ by γ and γ' from 5.18, it follows from 2.28 that $D|_{\overline{V}-V}$ can be extended to an admissible automorphism $\overline{D}$ over $\overline{V}$ (by adding an elementary pair of the form (F, δ), where $F=F' \oplus F'$ for F' trivial and

$$\delta=\begin{pmatrix} 0 & 1 \\ 1 & 0 \end{pmatrix}).$$

Let Δ: $E \to E$ be the admissible automorphism defined by $\Delta|_{X-V}=D|_{X-V}$ and $\Delta|_{\overline{V}}=\overline{D}$. It is clear that $t\Delta+(1-t)D$ is a homotopy between Δ and D within the admissible endomorphisms of E. Therefore $\sigma(E, D)=\sigma(E, \Delta)=0$ since (E, Δ) is elementary. □

5.20. Corollary. *If X is a locally compact space and X' is a closed subspace, the groups $K_0(X, X')$, $K_0(X-X')$, $K(X, X')$, and $K(X-X')$, are naturally isomorphic.*

Proof. Actually, $K(X-X') \approx K(X, X') \approx K(\dot{X}, \dot{X}') \approx K_0(\dot{X}, \dot{X}')$ by 5.18. Moreover $K_0(\dot{X}, \dot{X}') \approx K_0(\dot{X}-\dot{X}') \approx K_0(X-X')$ by 5.19. □

5.21. Let us return to our original problem of giving explicit formulas for the cup-product

$$K(X, X') \times K(Y, Y') \longrightarrow K(X \times Y, X \times Y' \cup X' \times Y)$$

According to 5.20, we may identify the groups K and K_0, and so equivalently we may attack the same problem for the groups K_0. Let $\sigma(E, \Gamma) \in K_0(X, X')$ and $\sigma(F, \Delta) \in K_0(Y, Y')$. Then $E \boxtimes F$ may be graded by $(E \boxtimes F)_0 = (E_0 \boxtimes F_0) \oplus (E_1 \boxtimes F_1)$ and $(E \boxtimes F)_1 = (E_0 \boxtimes F_1) \oplus (E_1 \boxtimes F_0)$. These decompositions are compatible with the natural metric of $G = E \boxtimes F$. Let $\Omega: G \to G$ be the morphism defined by $\Omega = \Gamma \hat{\boxtimes} 1 + 1 \hat{\boxtimes} \Delta$, that is $\Omega(x_i \otimes y_j) = \Gamma(x_i) \otimes y_j + (-1)^i x_i \otimes \Delta(y_j)$ where x_i and y_j belong to the fibers of E_i and F_j respectively. Then Ω is of degree one, and $\Omega^* = \Gamma^* \hat{\boxtimes} 1 + 1 \hat{\boxtimes} \Delta^* = \Gamma \hat{\boxtimes} 1 + 1 \hat{\boxtimes} \Delta = \Omega$. Moreover, $\Omega^2 = (\Gamma)^2 \hat{\boxtimes} 1 + 1 \hat{\boxtimes} \Delta^2$. Since $(\Gamma_x)^2 \geqslant 0$ and $\Delta_y^2 \geqslant 0$, we have $(\Omega_{x,y})^2 > 0$ if either $(\Gamma_x)^2 > 0$ or $(\Delta_y)^2 > 0$ for $(x, y) \in X \times Y$ (here positivity of operators means positivity of eigenvalues). It follows that $\Omega_{x,y}$ is an automorphism whenever Γ_x *or* Δ_y is an automorphism. Hence $\sigma(G, \Omega)$ is a well-defined element of $K(X \times X', X \times Y' \cup X' \times Y)$. The correspondence

$$[\sigma(E, \Gamma), \sigma(F, \Delta)] \longmapsto \sigma(E \boxtimes F, \Omega)$$

defines the desired bilinear homomorphism.

5.22. Theorem. *We have the commutative diagram*

$$\begin{array}{ccc}
K_0(X, X') \times K_0(Y, Y') & \longrightarrow & K_0(X \times Y, X \times Y' \cup X' \times Y) \\
\downarrow & & \downarrow \\
K_0(X - X') \times K_0(Y - Y') & \longrightarrow & K_0((X - X') \times (Y - Y')) \\
\wr\wr & & \wr\wr \\
K(X - X') \times K(Y - Y') & \longrightarrow & K((X - X') \times (Y - Y')),
\end{array}$$

where the last line is the cup-product defined in 5.4. Moreover, if X and Y are compact, the product

$$\begin{array}{ccc}
K(X, X') \times K(Y, Y') & \longrightarrow & K(X \times Y, X \times Y' \cup X' \times Y) \\
\wr\wr & & \wr\wr \\
K_0(X, X') \times K_0(Y, Y') & \longrightarrow & K_0(X \times Y, X \times Y' \cup X' \times Y)
\end{array}$$

can be directly defined in the following way. Let $d(E_0, E_1, \beta)$ (resp. $d(F_0, F_1, \gamma)$) be an element of $K(X, X')$ (resp. $K(Y, Y')$), and let $\alpha: E_0 \to E_1$ (resp. $\delta: F_0 \to F_1$) be a morphism such that $\alpha|_{X'} = \beta$ (resp. $\delta|_{Y'} = \gamma$). Then the element of $K(X \times Y, X \times Y' \cup X' \times Y)$ associated with $d(E_0, E_1, \beta)$ and $d(F_0, F_1, \gamma)$ is $d(G_0, G_1, \omega)$, where $G_0 =$

$(E_0 \boxtimes F_0) \oplus (E_1 \boxtimes F_1)$, $G_1 = (E_0 \boxtimes F_1) \oplus (E_1 \boxtimes F_0)$, and ω is defined by the matrix

$$\omega = \begin{pmatrix} \alpha \boxtimes 1 & -1 \boxtimes \delta^* \\ \delta \boxtimes 1 & \alpha^* \boxtimes 1 \end{pmatrix}$$

for arbitrary metrics on the vector bundles involved.

Proof. The commutativity of the upper diagram follows directly from the definition of the restriction morphisms $K_0(X, X') \to K_0(X - X')$, etc. By 5.6, it is enough to prove the commutativity of the lower diagram for $X' = Y' = \varnothing$. Then since α and δ are homotopic to 0, we may choose $\alpha = \delta = 0$. In this case, the result follows from the identity

$$E_0 \boxtimes F_0 + E_1 \boxtimes F_1 - E_0 \boxtimes F_1 - E_1 \boxtimes F_0 = (\pi^* E_0 - \pi^* E_1)(p^* F_0 - p^* F_1)$$

in the ring $K(X \times Y)$, where $\pi: X \times Y \to X$ and $p: X \times Y \to Y$ are the canonical projections. □

5.23. If X is compact, the composition of morphisms

$$K_0(X) \times K_0(X - X') \longrightarrow K_0(X - X') \times K_0(X - X') \longrightarrow K_0((X - X') \times (X - X')) \xrightarrow{\Delta^*} K_0(X - X'),$$

where Δ is the diagonal, is defined by

$$([E] - [E'], \sigma(F, D)) \longmapsto \sigma(E \otimes F, 1 \otimes D) - \sigma(E' \otimes F, 1 \otimes D).$$

In the same way, we formally define a product $K_0(X) \times K_0(X, X') \to K_0(X, X')$ by the formula

$$([E] - [E'], \sigma(F, D)) \longmapsto \sigma(E \otimes F, 1 \otimes D) - \sigma(E' \otimes F, 1 \otimes D).$$

We have the commutative diagram (for X compact)

$$\begin{array}{ccc} K(X) \times K(X, X') & \longrightarrow & K(X, X') \\ \uparrow \approx & & \uparrow \approx \\ K_0(X) \times K_0(X, X') & \longrightarrow & K_0(X, X') \\ \downarrow \approx & & \downarrow \approx \\ K_0(X) \times K_0(X - X') & \longrightarrow & K_0(X - X'), \end{array}$$

where the first product is that defined in 5.11. *In other words, the isomorphism $K(X, X') \approx K(X-X')$ is also a $K(X)$-module isomorphism.*

5.24. Example. Let V be a complex vector space of dimension one. Any element v of V defines a homomorphism $d_v\colon \mathbb{C} \to V$ such that $d_v(1)=v$. If V is provided with a positive Hermitian form φ we may define $\partial_v\colon V \to \mathbb{C}$ as the adjoint of d_v (explicitly $\partial_v(w)=\varphi(w, v)$). Now consider the trivial bundle E over V, with $\mathbb{C} \oplus V$ as fiber, so $E=V\times(\mathbb{C}\times V)$. We define $D\colon E \to E$ by the formula $D(v, \lambda, w)=(v, \partial_v(w). d_v(\lambda))$ or in matrix notation

$$D_v = \begin{pmatrix} 0 & \partial_v \\ d_v & 0 \end{pmatrix}.$$

Then $(D_v)^2 = \varphi(v, v)\cdot \mathrm{Id}_E$; hence D_v is an isomorphism for $v \neq 0$. Moreover D is self-adjoint and of degree one with respect to the metric considered. Hence, the pair (E, D) defines an element of $K_{0_{\mathbb{C}}}(V)$. The next proposition formulates this example more precisely.

5.25. Proposition. *Let $B(V)$ (resp. $S(V)$) be the subset of V consisting of points v such that $\varphi(v, v) \leqslant 1$ (resp. $\varphi(v, v)=1$). Then the image of $\sigma(E, D)$ in $K(B(V), S(V))$ under the natural isomorphism $K_{\mathbb{C}}(B(V), S(V)) \approx K_{0_{\mathbb{C}}}(V)$ is $d(F_0, F_1, \alpha)$ where $F_i = E_i|_{B(V)}$ and $\alpha_v = d_v$ for $v \in S(V)$. In particular, for $V=\mathbb{C}$, the image is the element $K_{\mathbb{C}}(B^2, S^2)$ defined in 2.30.*

Proof. The image of $d(F_0, F_1, \alpha)$ in $K_{0_{\mathbb{C}}}(B(V), S(V))$ under the natural isomorphism $K_{0_{\mathbb{C}}}(B(V), S(V)) \approx K_{\mathbb{C}}(B(V), S(V))$ is $\sigma(F, \Delta)$, where $F=F_0 \oplus F_1$ and $\Delta_v = D_v$ for $v \in B(V)$. If we identify $B(V)-S(V)$ with V by the map $v \mapsto \dfrac{v}{1-\|v\|}$ where $\|v\| = \sqrt{\varphi(v, v)}$, then the image of $\sigma(F, \Delta)$ under the isomorphism $K_{0_{\mathbb{C}}}(B(V), S(V)) \approx K_0(V)$ is $\sigma(E, D')$, where $D'_v = D_w$ with $w = \dfrac{\|v\|}{1+\|v\|}\cdot v$. Since $tD_v + (1-t)D'_v$ is an isomorphism for any $t \in I$ and $v \neq 0$, it follows that D' and D are homotopic and $\sigma(E, D') = \sigma(E, D)$ in $K_{0_{\mathbb{C}}}(V)$. □

5.26. Let X and Y be locally compact spaces. The cup-product

$$\begin{array}{ccc} K(X\times \mathbb{R}^n)\times K(Y\times \mathbb{R}^p) & \longrightarrow & K(X\times \mathbb{R}^n \times Y \times \mathbb{R}^p) \\ & & \wr\wr \\ & & K(X\times Y\times \mathbb{R}^{n+p}) \end{array}$$

may be interpreted as a bilinear map

$$K^{-n}(X)\times K^{-p}(Y) \longrightarrow K^{-n-p}(X\times Y)$$

which is characterized axiomatically (Karoubi [5]). This product satisfies the associativity property described in 5.1. However the commutativity property is slightly changed:

5.27. Proposition. *The diagram*

$$\begin{array}{ccc} K^{-n}(X)\times K^{-p}(Y) & \longrightarrow & K^{-n-p}(X\times Y) \\ \downarrow & & \downarrow T^* \\ K^{-p}(Y)\times K^{-n}(X) & \longrightarrow & K^{-n-p}(Y\times X) \end{array}$$

where T^* is induced by $T\colon Y\times X\to X\times Y$, is commutative up to the sign $(-1)^{np}$.

Proof. Clearly we have the commutative diagram

$$\begin{array}{ccc} K(X\times\mathbb{R}^n)\times K(Y\times\mathbb{R}^p) & \longrightarrow & K(X\times\mathbb{R}^n\times Y\times\mathbb{R}^p) \\ & & \downarrow \\ & & K(X\times Y\times\mathbb{R}^n\times\mathbb{R}^p) \\ \downarrow & & \downarrow \\ & & K(X\times Y\times\mathbb{R}^p\times\mathbb{R}^n) \\ & & \downarrow \\ K(Y\times\mathbb{R}^p)\times K(X\times\mathbb{R}^n) & \longrightarrow & K(Y\times\mathbb{R}^p\times X\times\mathbb{R}^n) \end{array}$$

where the homomorphism $K(X\times Y\times\mathbb{R}^n\times\mathbb{R}^p)\to K(X\times Y\times\mathbb{R}^p\times\mathbb{R}^n)$ is induced by the permutation of $\mathbb{R}^{n+p}=\mathbb{R}^n\times\mathbb{R}^p$ which switches the factors. Such a transformation is the product of np transpositions of $\mathbb{R}^{n+p}$ of the form

$$(\lambda_1,\ldots,\lambda_i,\ldots,\lambda_j,\ldots,\lambda_{n+p})\longmapsto(\lambda_1,\ldots,\lambda_j,\ldots,\lambda_i,\ldots,\lambda_{n+p})$$

hence induces $(-1)^{np}$ on the group $K(X\times Y\times\mathbb{R}^{n+p})$ by 4.10. □

5.28. As a corollary of 5.27, let us consider the case where $X=Y$. Then the diagonal map $X\to X\times X$ induces a homomorphism $K^{-n-p}(X\times X)\to K^{-n-p}(X)$. Therefore we obtain a product

$$K^{-n}(X)\times K^{-p}(X)\longrightarrow K^{-n-p}(X)$$

making $K^*(X)=\sum_{n=0}^{\infty}K^{-n}(X)$ a graded algebra. If $x_n\in K^{-n}(X)$ and $x_p\in K^{-p}(X)$ we have $x_nx_p=(-1)^{np}x_px_n$ in $K^*(X)$.

5.29. The product defined in 5.27 also defines a bilinear map

$$K^{-n}(X, X') \times K^{-p}(Y, Y') \longrightarrow K^{-n-p}(X \times Y, X \times Y' \cup X' \times Y)$$

with the aid of the identifications $K^{-n}(X, X') \approx K^{-n}(X - X')$, $K^{-p}(Y, Y') \approx K^{-p}(Y - Y')$, and $K^{-n-p}(X \times Y, X \times Y' \cup X' \times Y) \approx K^{-n-p}((X - X') \times (Y - Y'))$. In particular, if X and Y are compact spaces, we have the product

$$K(X) \times K^{-1}(Y) \longrightarrow K^{-1}(X \times Y),$$

which may be directly defined by the formula

$$([F] - [G]) \cup d(E, \alpha) = d(F \boxtimes E, 1 \boxtimes \alpha) - d(G \boxtimes E, 1 \boxtimes \alpha).$$

To see that this formula is correct, we need only check the commutativity of the diagram

$$\begin{array}{ccc} K(X) \times K^{-1}(Y) & \longrightarrow & K^{-1}(X \times Y) \\ \downarrow{\scriptstyle 1 \times \partial} & & \downarrow{\scriptstyle \partial} \\ K(X) \times K(X \times B^1, Y \times S^0) & \xrightarrow{c} & K(X \times Y \times B^1, X \times Y \times S^0), \end{array}$$

where ∂ is defined as in 3.21, and c is defined as in 5.14. Note that ∂ gives the canonical identification of $K^{-1}(Y)$ with $K(Y \times B^1, Y \times S^0)$, and similarly, of $K^{-1}(X \times Y)$ with $K(X \times Y \times B^1, X \times Y \times S^0)$ (3.30).

5.30. Finally the situation considered in 5.13 may be generalized slightly further to the groups K^{-n}. More precisely, we define the product (denoted by $(\alpha, \beta) \mapsto \alpha \cdot \beta$)

$$K(X \times \mathbb{R}^q) \times K(V \times \mathbb{R}^r) \longrightarrow K(V \times \mathbb{R}^{q+r})$$

or

$$K^{-q}(X) \times K^{-r}(V) \longrightarrow K^{-q-r}(V)$$

as the composition

$$\begin{aligned} K(X \times \mathbb{R}^q) \times K(V \times \mathbb{R}^r) &\longrightarrow K(X \times \mathbb{R}^q \times V \times \mathbb{R}^r) \\ &\approx K(X \times V \times \mathbb{R}^{q+r}) \xrightarrow{\tau^*} K(V \times \mathbb{R}^{q+r}) \end{aligned}$$

where τ is the proper map $(v, \lambda) \mapsto (\pi(v), v, \lambda)$. The following proposition generalizes Proposition 5.13:

5.31. Proposition. *Let V be a vector bundle over a compact space X and let x and y be elements of $K(V \times \mathbb{R}^q) = K^{-q}(V)$ and $K(V \times \mathbb{R}^r) = K^{-r}(V)$ respectively. Then the product of x and y in $K^{-q-r}(V)$ defined by the map obtained in 5.27 is actually $x' \cdot y$, where x' is the restriction of x to $K(X \times \mathbb{R}^q)$ by the zero section.*

Proof. The composition

$$V\times\mathbb{R}^q\times\mathbb{R}^r \xrightarrow{\tau} X\times\mathbb{R}^q\times V\times\mathbb{R}^r \xrightarrow{j} V\times\mathbb{R}^q\times V\times\mathbb{R}^r,$$

where $\tau(v,\mu,\nu)=(\pi(v),\mu,v,\nu)$, $j(x,\mu,v,\nu)=(i(x),\mu,v,\nu)$, and i is the zero section, is homotopic within the proper maps, to the diagonal map Δ defined by $\Delta(v,\mu,v)=(v,\mu,v,\nu)$. Therefore $\Delta^*(x\cup y)=(\tau^*\cdot j^*)(x\cup y)=\tau^*(x'\cup y)=x'\cdot y$. $\square$

Exercises (Section II.6) 9, 11, 19.

6. Exercises

6.1. Let A be an arbitrary ring with unit, and let $\mathscr{A}_n$ be the full subcategory of Mod(A), whose objects M admit a resolution of the type

$$0\longrightarrow P_n\longrightarrow P_{n-1}\longrightarrow\cdots\longrightarrow P_0\longrightarrow M\longrightarrow 0,$$

where the P_i are projective and finitely generated.

a) If N is an object of $\mathscr{A}_{n+1}$, prove the existence of an exact sequence

$$0\longrightarrow Q_1\longrightarrow Q_0\longrightarrow N\longrightarrow 0,$$

where Q_0 and Q_1 are objects of $\mathscr{A}_n$. Moreover, if

$$0\longrightarrow Q_1'\longrightarrow Q_0'\longrightarrow N\longrightarrow 0$$

is another resolution of the same type, show there exists an exact sequence

$$0\longrightarrow Q_1\longrightarrow Q_0\oplus Q_1'\longrightarrow Q_0'\longrightarrow 0.$$

b) We define $G(\mathscr{A}_n)$ as the quotient of the free group generated by the objects of $\mathscr{A}_n$, by the subgroup generated by the relations $[M]=[M']+[M'']$, where

$$0\longrightarrow M'\longrightarrow M\longrightarrow M''\longrightarrow 0$$

is an exact sequence of $\mathscr{A}_n$. Prove that the inclusion functor $\mathscr{A}_n\to\mathscr{A}_{n+1}$ induces an isomorphism $G(\mathscr{A}_n)\approx G(\mathscr{A}_{n+1})$.

c) Let $\mathscr{A}=\cup\mathscr{A}_n$, and let $G(A)$ be the quotient of the free group generated by the objects of $\mathscr{A}$, by the relations $[M]=[M']+[M'']$ where M', M and M'' are related by an exact sequence as above. Prove that $K(A)\approx G(A)$.

6.2. Prove that $K(A)\approx\mathbb{Z}$ if A is a principal ideal domain.

6.3. Let G be a compact Lie group, and let $\mathscr{C}$ be the category of finite dimensional complex representations of G. Let $R(G)$ denote the Grothendieck group $K(\mathscr{C})$ of this category.

a) Compute $R(G)$ when G is the finite group $\mathbb{Z}/n\mathbb{Z}$.

∗ b) In general, prove that $R(G)$ is the free group generated by the irreducible representations of G.∗

6.4. Prove the identities:

$$\begin{array}{ll} \tilde{K}_{\mathbb{C}}(S^1)=0 & \tilde{K}_{\mathbb{R}}(S^1)=\mathbb{Z}/2 \\ \tilde{K}_{\mathbb{C}}(S^2)=\mathbb{Z} & \tilde{K}_{\mathbb{R}}(S^2)=\mathbb{Z}/2 \\ \tilde{K}_{\mathbb{C}}(S^3)=0 & \tilde{K}_{\mathbb{R}}(S^3)=0 \end{array}$$

∗ **6.5.** Let X be a connected CW-complex of dimension $\leqslant 2$. Prove that $\tilde{K}(X)\approx H^2(X;\mathbb{Z})$.∗

6.6. Let $\pi: X\to Y$ be an n-fold covering of Y. If E is a vector bundle over X, we define a vector bundle $F=\pi_*(E)$ over Y by the formula $F_y=\bigoplus_{x\in\pi^{-1}(\{y\})} E_x$. More precisely, if U is an open subset in Y such that $V=\pi^{-1}(U)\approx U\times D$ where D is discrete, we give F_U the topology induced by the bijection $F_U\approx(E_V)^D$.

a) Prove that with the topology defined above, F is a well-defined vector bundle over Y.

b) Prove that the correspondence $E\mapsto F$ induces a group homomorphism $\pi_*: K(X)\to K(Y)$. Moreover, prove the formula

$$\pi_*(\pi^*(y)\cdot x)=y\cdot\pi_*(x)$$

for $y\in K(Y)$ and $x\in K(X)$.

c) We assume that $\pi: X\to Y$ is a principal covering with group G (i.e. the finite group G acts freely on X, and $Y\approx X/G$). Now prove that $(\pi^*\cdot\pi_*)(x)=\sum_{g\in G}\rho(g)^*(x)$, where $\rho(g)^*: K(X)\to K(X)$ is the automorphism of $K(X)$ induced by the action of g. Prove also that $\pi_*(1)=[E]=X\times_G k^n$ (I.9.27), where $n=\mathrm{Card}(G)$ and G acts on k^n by the regular representation.

6.7. Let $f: S^1\to S^1\approx P_1(\mathbb{R})$ be the map defined by $f(z)=z^2$. Now show that $C'(f)$ is homeomorphic to $P_2(\mathbb{R})$ (3.28). From the Puppe sequence

$$S^1\longrightarrow P_1(\mathbb{R})\longrightarrow P_2(\mathbb{R})\longrightarrow S^2\longrightarrow S^2,$$

prove the isomorphisms $K_{\mathbb{C}}(P_2(\mathbb{R}))\approx\mathbb{Z}/2$ and $K_{\mathbb{R}}(P_2(\mathbb{R}))\approx\mathbb{Z}/4$. By the same method compute $K_{\mathbb{C}}(P_3(\mathbb{R}))$ and $K_{\mathbb{R}}(P_3(\mathbb{R}))$.

6.8. Let $\mathscr{E}^Q_{\mathbb{R}}(X)$ be the category of real vector bundles with compact base provided with a nondegenerate symmetric bilinear form (I.8.11). We provide the set M of isomorphism classes of such vector bundles, with the monoid structure induced by the Whitney sum of vector bundles. Show that the symmetrized group of M is $K_{\mathbb{R}}(X) \oplus K_{\mathbb{R}}(X)$. In an analogous way, investigate real vector bundles provided with a nondegenerate skew symmetric form, and complex vector bundles provided with a nondegenerate symmetric or skew symmetric form. Also investigate the case of complex vector bundles provided with a nondegenerate Hermitian form.

6.9. Let X be any finite CW-complex, and let X_n be its nth skeleton. Let $K_{(n)}(X)=\operatorname{Ker}[K(X) \to K(X_{n-1})]$. Show that the cup-product

$$K(X) \times K(Y) \longrightarrow K(X \times Y),$$

where Y is another finite CW-complex, sends $K_{(n)}(X) \times K_{(p)}(Y)$ into $K_{(n+p)}(X \times Y)$; hence the $K_{(n)}(X)$ provide a filtration of the ring $K(X)$, in the sense that $K_{(n)}(X) \cdot K_{(p)}(X) \subset K_{(n+p)}(X)$.

$*$ **6.10.** Let X be a connected finite CW-complex of dimension n, and let E and F be real vector bundles of rank $> n$. Show that $[E]-[F]=0$ in $K_{\mathbb{R}}(X)$ if and only if E and F are isomorphic (in other words, the map $[X, BO(p)] \to [X, BO]$, induced by the inclusion of $O(p)$ in O, is injective for $p>n$). Prove also that each element of $K_{\mathbb{R}}(X)$ may be written as $[E] \oplus [T]-[T']$, where T and T' are trivial vector bundles and E is a vector bundle of rank $\leqslant n$ (in other words the map $[X, BO(p)] \to [X, BO]$ is surjective for $p \geqslant n$). Similarly, for the complex case show that the map $[X, BU(p)] \to [X, BU]$ is injective (resp. surjective) if $p>(n-2)/2$ (resp. $p>(n-1)/2$).$*$

$*$**6.11.** Let S be a multiplicative set in $\mathbb{N}^*$ not equal to $\{1\}$, and let X be a finite connected CW-complex. We let $\mathscr{E}(X)_S$ denote the subcategory of $\mathscr{E}(X)$, whose objects are the vector bundles with rank belonging to S. The tensor product of vector bundles provides the set of isomorphism classes of objects of $\mathscr{E}(X)_S$ with a monoid structure. We let $KP_S(X)$ denote the symmetrized group of this monoid.

a) Let E be an object of $\mathscr{E}(X)_S$. Show the existence of an object F of $\mathscr{E}(X)_S$ such that $E \otimes F$ is a trivial bundle (Hint: write formally $[E]=n\left(1+\dfrac{[E]-n}{n}\right)$, where n is the rank of E).

b) Let $\mathbb{Z}_S^*$ be the multiplicative group of fractions a/b where a and $b \in S$. Prove that $KP_S(X) \approx \mathbb{Z}_S^* \oplus (1+\widetilde{K}(X)_S)_X$, where $\widetilde{K}(X)_S$ denotes the group $\widetilde{K}(X)$ localized at S, and $(1+\widetilde{K}(X)_S)_X$ denotes the *multiplicative group* $1+\widetilde{K}(X)_S$ naturally imbedded in $K(X)_S$.

c) We put $\widetilde{KP}_S(X)=\operatorname{Ker}[KP_S(X) \to KP_S(P_0)]$, where P_0 is a point. Prove that $\widetilde{KP}_S(X)$ is S-divisible. In the case $S=\mathbb{N}-\{0\}$, prove that $\widetilde{KP}_S(X) \approx \widetilde{K}(X) \otimes_{\mathbb{Z}} \mathbb{Q}$.

d) Prove that $\widetilde{KP}_S(X) \approx \operatorname{inj\,lim}_{s \in S} \Phi_s(X)$, where $\Phi_s(X)$ is the set of isomorphism classes of vector bundles of rank s, and where the map $\Phi_s(X) \to \Phi_{st}(X)$ is induced by taking the tensor product with the trivial bundle of rank t.$*$

6.12. Let $\mathscr{H}$ be the category of Hilbert spaces and let $\check{\mathscr{H}}$ be the category with the same objects but with $\check{\mathscr{H}}(E, F) = \mathscr{H}(E, F)/\mathscr{K}(E, F)$, where $\mathscr{K}(E, F)$ denotes the set of completely continuous operators from E to F (i.e. the operators which are limits of operators of finite rank).

a) Prove that a morphism $D: E \to F$ in the category $\mathscr{H}$ is invertible in $\check{\mathscr{H}}$ if and only if D is a *Fredholm operator* (i.e. Ker(D) and Coker(D) are finite dimensional).

b) Let $d: E \to F$ be an invertible morphism of $\check{\mathscr{H}}$, and let D be an operator in $\mathscr{H}$ with class d. Prove that the "*index*" of D (i.e. Dim(Ker(D)) − Dim(Coker(D)) is independent of the choice of D, and is a locally constant function of d.

c) Prove the isomorphism $K^{-1}(\mathscr{H}) \approx \mathbf{Z}$.

Let $\tilde{\check{\mathscr{H}}}$ be the pseudo-abelian category associated with $\check{\mathscr{H}}$ (cf. I.6.10). Prove that $K(\tilde{\check{\mathscr{H}}}) = K(\check{\mathscr{H}}) = 0$.

6.13. Let $\mathscr{C}$ be an additive category. We consider the set of pairs (E, α), where E is an object of $\mathscr{C}$ and α is an automorphism of E. The *Bass group* $K_1(\mathscr{C})$ associated with $\mathscr{C}$, is the quotient of the free group generated by such pairs, by the subgroup generated by the relations

$$(E \oplus F, \alpha \oplus \beta) = (E, \alpha) + (F, \beta)$$

and $$(E, \beta\alpha) = (E, \alpha) + (E, \beta).$$

a) Prove that $K_1(\mathscr{L}(A)) \approx K_1(\mathscr{P}(A))$ (we denote these two groups by $K_1(A)$).

b) Let GL′(A) be the commutator subgroup of GL(A). Prove that $K_1(\mathscr{L}(A)) \approx$ GL(A)/GL′(A), and that GL′(A) is equal to its own commutator subgroup (i.e. GL″(A) = GL′(A)).

c) Let $E_n(A)$ be the subgroup of $GL_n(A)$ generated by the "elementary matrices" (i.e. matrices of the type (a_{ji}) where $a_{ii} = 1$ and $a_{ji} \neq 0$ for at most one pair (i, j) with $i \neq j$). For $n \geqslant 3$, prove that $E_n(A)$ is equal to its own commutator subgroup, and hence $E(A) \subset GL'(A)$ where $E(A) = \text{inj lim } E_n(A)$.

d) Prove that any triangular matrix with 1 on the diagonal belongs to E(A). Using the identities

$$\begin{pmatrix} \alpha\beta\alpha^{-1}\beta^{-1} & 0 & 0 \\ 0 & 1 & 0 \\ 0 & 0 & 1 \end{pmatrix} = \begin{pmatrix} \alpha & 0 & 0 \\ 0 & 1 & 0 \\ 0 & 0 & \alpha^{-1} \end{pmatrix} \begin{pmatrix} \beta & 0 & 0 \\ 0 & \beta^{-1} & 0 \\ 0 & 0 & 1 \end{pmatrix} \begin{pmatrix} \alpha^{-1} & 0 & 0 \\ 0 & 1 & 0 \\ 0 & 0 & \alpha \end{pmatrix} \begin{pmatrix} \beta^{-1} & 0 & 0 \\ 0 & \beta & 0 \\ 0 & 0 & 1 \end{pmatrix}$$

and

$$\begin{pmatrix} u & 0 \\ 0 & u^{-1} \end{pmatrix} = \begin{pmatrix} 1 & -u \\ 0 & 1 \end{pmatrix} \begin{pmatrix} 1 & 0 \\ u^{-1} & 1 \end{pmatrix} \begin{pmatrix} 1 & -u \\ 0 & 1 \end{pmatrix} \begin{pmatrix} 1 & 1 \\ 0 & 1 \end{pmatrix} \begin{pmatrix} 1 & 0 \\ -1 & 1 \end{pmatrix} \begin{pmatrix} 1 & 1 \\ 0 & 1 \end{pmatrix},$$

show that E(A) = GL′(A).

e) We now assume that A is a commutative Banach algebra. Show that $K^{-1}(\mathscr{L}(A)) \approx K^{-1}(\mathscr{P}(A)) \approx \pi_0(\mathrm{GL}(A))$. If we put $SK^{-1}(A) = \pi_0(\mathrm{SL}(A))$ and $SK_1(A) = \mathrm{SL}(A)/\mathrm{E}(A)$, show that these two groups are naturally isomorphic.

∗ f) If A is a commutative Banach algebra, prove that $K_1(A) \approx A^* \oplus SK^{-1}(A)$. Using the results of Chapter III, explicitly compute $K_1(A)$ when A is the ring of continuous complex-valued functions on the torus T^n.∗

6.14. Suppose A is a Banach algebra, and X is a compact space. Let $A(X)$ denote the ring of continuous functions on X with values in A. If Y is a closed subspace of X, we have a Banach functor $\varphi\colon \mathscr{P}(A(X)) \to \mathscr{P}(A(Y))$ associated with the ring map $A(X) \to A(Y)$. We define $K(X, Y; A)$ as the Grothendieck group of the functor φ.

a) Using the material in I.6 show that $K_{\mathbb{R}}(X, Y) \approx K(X, Y; \mathbb{R})$ and $K_{\mathbb{C}}(X, Y) \approx K(X, Y; \mathbb{C})$.

b) We define $K^{-n}(X, Y; A) = K(X \times B^n, X \times S^{n-1} \cup Y \times B^n; A)$. Using the methods of Section II.4 prove the exact sequence

$$K^{-n-1}(X; A) \longrightarrow K^{-n-1}(Y; A) \longrightarrow K^{-n}(X, Y; A) \longrightarrow K^{-n}(X; A) \longrightarrow K^{-n}(Y; A).$$

c) We define $K^{-n}(A)$ to be $K^{-n}(P; A)$ where P is a point. If B is a Banach algebra without unit element, we define $K^{-n}(B)$ as $\mathrm{Ker}[K^{-n}(B^+) \to K^{-n}(\mathbb{R})]$, where B^+ denotes the algebra B augmented by $\mathbb{R}$. If

$$0 \longrightarrow A' \longrightarrow A \longrightarrow A'' \longrightarrow 0$$

is an exact sequence of Banach algebras, prove the exact sequence

$$K^{-n-1}(A) \longrightarrow K^{-n-1}(A'') \longrightarrow K^{-n}(A') \longrightarrow K^{-n}(A) \longrightarrow K^{-n}(A'').$$

e) Let $A(X)$ be the Banach algebra of continuous functions with values in A. Prove that $K^{-n}(X; A) \approx K^{-n}(A(X))$.

d) Prove that $K^{-n-1}(A) \approx \pi_n(\mathrm{GL}(A))$.

6.15. (*Density theorem.*) Let A be a Banach algebra, and let $i\colon B \to A$ be a continuous injection from another Banach algebra B into A, satisfying the following two conditions:

(i) $i(B)$ is dense in A.

(ii) $\mathrm{GL}_n(A) \cap M_n(B) = \mathrm{GL}_n(B)$ for any integer n (considering B as imbedded in A via i).

Under this hypothesis, show that i induces isomorphisms

$$K^{-n}(B) \approx K^{-n}(A)$$

for each $n \geqslant 0$. Generalize this theorem to inductive limits of Banach algebras where $K^{-n}(B)$ is interpreted as $\pi_{n-1}(\mathrm{GL}(B))$ for $n > 0$.

6.16. Let H be a Hilbert space of infinite dimension, let A be the Banach algebra $\mathrm{End}(H)$, and let A'' be the quotient algebra $\mathrm{End}(H)/A'$, where A' is the ideal of completely continuous operators. Show that $K^{-n}(X; A') = \mathrm{Ker}[K^{-n}(X; A'^{+}) \to K^{-n}(X; \mathbb{R})]$ may be identified with $K^{-n}(X)$ (Hint: use the Density theorem above). Prove that $K^{-n}(X; A) = 0$ and that $K^{-n-1}(X; A'') \approx K^{-n}(X; A') \approx K^{-n}(X)$. Prove also that $K(X; A'') \approx [X, \mathscr{F}^1(H)]$ where $\mathscr{F}^1(H)$ denotes the non-trivial connected component of the set of self-adjoint Fredholm operators.

***6.17.** Let (a_{ij}), $(i, j) \in \mathbb{N} \times \mathbb{N}$, be an infinite matrix over A. We say that M is of finite type if there is an integer n such that in each line and each column there are at most n elements $\neq 0$ and if, moreover, the a_{ij} are chosen from at most n distinct elements of A.

a) Show that the set of such infinite matrices is a ring, and that $M = \operatorname{Sup}_j \sum_i \|a_{ij}\|$ is a norm in this ring. Let CA denote the completion of the ring with respect to this norm (this is the "*cone*" of the Banach algebra A). Let $\tilde{A}$ denote the closure of finite matrices in CA, and let SA denote the quotient algebra $CA/\tilde{A}$.

b) Using the same type of arguments as in 6.16, show that $BGL(CA)$ is contractible and that $\Omega(BGL(SA)) \sim K(A) \times BGL(A)$. Deduce that $BGL(A)$ is an infinite loop space for any Banach algebra A.*

6.18. Let $\mathscr{C}$ be a Banach category, and let $\varphi_n : \mathscr{C} \to \mathscr{C}$ be the functor defined by $\varphi_n(E) = E^n$ (2.11). We put $K^{-1}(\mathscr{C}; \mathbb{Z}/n) = K(\varphi_n)$.

a) Prove that $K^{-1}(\mathscr{C}; \mathbb{Z}/n)$ is a group with exponent n if $n \neq 4p+2$ or if $\mathscr{C}$ is a complex Banach category.

b) If $\mathscr{C} = \mathscr{E}(X)$, we put

$$K^{-1}(X; \mathbb{Z}/n) = K^{-1}(\mathscr{C}; \mathbb{Z}/n),$$

$$\tilde{K}^{-1}(X; \mathbb{Z}/n) = \mathrm{Coker}[K^{-1}(P; \mathbb{Z}/n) \longrightarrow K^{-1}(X; \mathbb{Z}/n)],$$

and
$$K^{-p-1}(X, Y; \mathbb{Z}/n) = \tilde{K}^{-1}(S^p(X/Y); \mathbb{Z}/n).$$

Now prove the exact sequences

$$K^{-p-2}(X; \mathbb{Z}/n) \longrightarrow K^{-p-2}(Y; \mathbb{Z}/n) \longrightarrow K^{-p-1}(X, Y; \mathbb{Z}/n) \longrightarrow K^{-p-1}(X; \mathbb{Z}/n) \longrightarrow K^{-p-1}(Y; \mathbb{Z}/n)$$

$$K^{-p-1}(X, Y) \xrightarrow{\times n} K^{-p-1}(X, Y) \longrightarrow K^{-p-1}(X, Y; \mathbb{Z}/n) \longrightarrow K^{-p}(X, Y) \xrightarrow{\times n} K^{-p}(X, Y).$$

c) Compute $K^{-1}(P; \mathbb{Z}/n)$ and $K^{-2}(P; \mathbb{Z}/n)$ in the real and complex cases, where P is a point.

6.19. Let X be a compact space, and let Y be a closed subspace. Let $\mathscr{C}_n(X, Y)$ denote the following category: the objects are the sequences $E_n, E_{n-1}, \ldots, E_0$ of vector bundles together with morphisms $\alpha_i: E_i|_Y \to E_{i-1}|_Y$ such that the sequence

$$0 \longrightarrow E_n|_Y \xrightarrow{\alpha_n} E_{n-1}|_Y \xrightarrow{\alpha_{n-1}} \cdots \xrightarrow{\alpha_1} E_0|_Y \longrightarrow 0$$

is exact. A morphism $\varphi: E \to F$, where $E=(E_i, \alpha_i)$ and $F=(F_i, \beta_i)$, is defined by a sequence of morphisms $\varphi_i: E_i \to F_i$ such that $\beta_i \varphi_i = \varphi_{i-1}\alpha_i$. An object $E=(E_i, \alpha_i)$ is called elementary if it is of the form $(0, \ldots, E_p, E_{p+1}, 0, \ldots, 0)$, where $E_p = E_{p+1}$ and $\alpha_p = \mathrm{Id}$. Let $\mathscr{K}_{(n)}(X, Y)$ denote the quotient of the set of isomorphism classes of objects of $\mathscr{C}_n(X, Y)$ by the equivalence relation generated by the sum of elementary objects.

a) Show that $\mathscr{K}_{(1)}(X, Y) \approx K(X, Y)$.

b) Let (E_i, α_i) be an object of $\mathscr{C}_n(X, Y)$. By adding an elementary object, show that we may assume that α_1 is the restriction to Y of a vector bundle epimorphism over X.

c) Show that the inclusion of $\mathscr{K}_{(i)}(X, Y)$ in $\mathscr{K}_{(i+1)}(X, Y)$ induces isomorphisms $\mathscr{K}_{(i)}(X, Y) \approx \mathscr{K}_{(i+1)}(X, Y)$ (Atiyah [3]).

d) Two objects of $\mathscr{C}_n(X, Y)$ are called homotopic, if there exists an object of $\mathscr{C}_n(X \times I, Y \times I)$, whose restrictions to $X \times \{0\}$ and $X \times \{1\}$ are isomorphic to the two given objects. Show that two homotopic objects have the same class in $\mathscr{K}_{(n)}(X, Y)$ (Hint: notice that an exact sequence over Y can be extended on a neighbourhood of Y).

e) Prove that the tensor product of complexes induces a bilinear map

$$\mathscr{K}_{(n)}(X, Y) \times \mathscr{K}_{(n')}(X', Y') \longrightarrow \mathscr{K}_{(n+n')}(X \times X', X \times Y' \cup X' \times Y),$$

which coincides with the cup-product defined in this chapter modulo the isomorphism $K \approx \mathscr{K}_{(n)}$.

f) Let V be a complex vector space of dimension n provided with a positive Hermitian form and let $B(V)$ (resp. $S(V)$) be the ball (resp. the sphere) associated with V. Let (E_i, α_i) be the complex defined by $E_i = B(V) \times \lambda^{n-i}(V)$, and $\alpha_i: S(V) \times \lambda^{n-i}(V) \to S(V) \times \lambda^{n-i+1}(V)$, where $\alpha_i(v, w) = (v, v \wedge w)$. If $V = \mathbf{C}^n$, show that the element thus defined of $K(B(V), S(V)) = K(B^{2n}, S^{2n-1})$ is the nth cup-product with itself of the element of $K(B^2, S^1)$ defined in 2.30.

7. Historical Note

As stated in the introduction, the definition of the group $K(X)$ was originally due to Grothendieck (cf. Borel-Serre [2]) (in the framework of algebraic geometry), and was studied in the context of algebraic topology by Atiyah and Hirzebruch [3]. The definition of the relative group $K(X, Y)$ is due to Atiyah and Hirzebruch [3] but the presentation adopted here was inspired by the seminar of Cartan-Schwartz [1] and by the author's thesis [2]. Most of the rest of this section is included in the paper of Atiyah and Hirzebruch quoted above; however, some of the investigations on the multiplicative structures stem from the paper of Atiyah, Bott and Shapiro [1].

Chapter III

Bott Periodicity

1. Periodicity in Complex K-Theory

1.1. In this section we will only consider complex K-theory which will be denoted simply by $K(X)$, $K(X, Y)$, etc. instead of $K_{\mathbb{C}}(X)$, $K_{\mathbb{C}}(X, Y)$, ...

1.2. Let us consider again the element u of $K(B^2, S^1)$ (II.2.30) defined by $d(E, F, \alpha)$ where $E=F=B^2 \times \mathbb{C}$ and $\alpha: E|_{S^1} \to F|_{S^1}$ is the isomorphism $(x, v) \mapsto (x, xv)$.

1.3. Theorem. *Let X be a locally compact space, and let Y be a closed subspace. Then the cup-product with u induces isomorphisms*

$$\beta: K^{-n}(X, Y) \xrightarrow{\approx} K^{-n}(X \times B^2, X \times S^{-1} \cup Y \times B^2) = K^{-n-2}(X, Y)$$

(II.4.11).

By replacing the pair (X, Y) by the pair $(X \times B^n, X \times S^{n-1} \cup Y \times B^n)$, we may reduce this theorem to the case $n=0$. Moreover, the commutative diagram

$$\begin{array}{ccc} K(X, Y) & \xrightarrow{\beta} & K(X \times B^2, X \times S^1 \cup Y \times B^2) \\ \approx \uparrow & & \uparrow \approx \\ K(X/Y, \{\infty\}) & \xrightarrow{\beta} & K(X/Y \times B^2, X/Y \times S^1 \cup \{\infty\} \times B^2) \end{array}$$

reduces the theorem to the case X compact and Y a point P. Finally, we have the commutative diagram

$$\begin{array}{ccc} 0 & & 0 \\ \downarrow & & \downarrow \\ K(X, P) & \longrightarrow & K(X \times B^2, X \times S^1 \cup P \times B^2) \approx K((X-P) \times \mathbb{R}^2) \\ \downarrow & & \downarrow \\ K(X) & \longrightarrow & K(X \times B^2, X \times S^1) \approx K(X \times \mathbb{R}^2) \\ \downarrow & & \downarrow \\ K(P) & \longrightarrow & K(P \times B^2, P \times S^1) \approx K(P \times \mathbb{R}^2) \\ \downarrow & & \downarrow \\ 0 & & 0 \end{array}$$

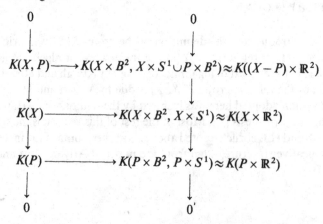

where the vertical sequences are split exact, since P (resp. $P \times \mathbb{R}^2$) is a retract of X (resp. $X \times \mathbb{R}^2$), and by application of II.4.7. Therefore, Theorem 1.3 is actually a consequence of this particular case:

1.4. Theorem. *Let X be a compact space. Then the cup-product with u induces an isomorphism between the groups $K(X)$ and $K(X \times B^2, X \times S^1)$.*

1.5. The proof of this last theorem will take the remainder of this section. Our objective is to reduce 1.4 to a general theorem about Banach algebras (1.11). As we have already seen in I.6.18 and in II.1.11, the group $K(X)$ has an interpretation in terms of the Banach algebra $A = C(X)$ of continuous complex-valued functions on the compact space X. More precisely $K(X) \approx K(A) \approx K(\mathscr{P}(A))$, where $\mathscr{P}(A)$ is the category of finitely generated projective modules over A. The correspondence between $K(X)$ and $K(A)$ is easily seen via projection operators (cf. I.6.17): if E is a vector bundle over X, then E is the image of a projection operator $p: X \times \mathbb{C}^n \to X \times \mathbb{C}^n$ for some n large enough (I.6.5). This projection operator p defines an element $\check{p} \in M_n(A)$ such that $(\check{p})^2 = \check{p}$ (I.1.12), and the projective module associated with E is simply $\operatorname{Im}(\check{p})$. Conversely, if M is a finitely generated projective module over A, then M is the image of $q: A^n \to A^n$ where $(q)^2 = q$. Such a morphism q defines $p = \hat{q}: X \times \mathbb{C}^n \to X \times \mathbb{C}^n$, whose image is a vector bundle (I.6.3).

We are going to give an analogous interpretation of $K(X \times B^2, X \times S^1)$ in terms of the Banach algebra A. The first step is the following lemma:

1.6. Lemma. *Every element of $K(X \times B^2, X \times S^1)$ may be written in the form $d(T, T, \alpha)$ where T is a trivial bundle, and $\alpha: T|_{X \times S^1} \to T|_{X \times S^1}$ is an automorphism such that $\alpha(x, e) = \mathrm{Id}$ for $x \in X$ and $e = (1, 0) \in S^1 \subset \mathbb{R}^2$ (such an automorphism is called normalized). If α and β are normalized, then $d(T, T, \alpha) = d(T, T, \beta)$ if and only if there exists a trivial bundle T' such that $\alpha \oplus \mathrm{Id}_{T'|_{X \times S^1}}$ is homotopic to $\beta \oplus \mathrm{Id}_{T'|_{X \times S^1}}$ within the normalized automorphisms of $(T \oplus T')|_{X \times S^1}$.*

Proof. Let $d(E, F, \gamma)$ be an element of $K(X \times B^2, X \times S^1)$. Since the projection $\pi: X \times B^2 \to X$ is a homotopy equivalence, we may suppose, by I.7.3, that E and F are of the form $\pi^*(E')$ and $\pi^*(F')$ respectively. On the other hand, the isomorphism restricted to the subspace $X \times \{e\}$ of $X \times S^1$ defines an isomorphism $\gamma_e: E' \to F'$, which is itself the restriction of an isomorphism $\pi^*(\gamma_e): \pi^*(E') \to \pi^*(F')$ (identifying X with $X \times \{e\}$). Therefore, we have

$$\begin{aligned} d(\pi^*(E'), \pi^*(F'), \gamma) &= d(\pi^*(E'), \pi^*(F'), \gamma) + d(\pi^*(F'), \pi^*(E'), \pi^*(\gamma_e^{-1})|_{X \times S^1}) \\ &= d(\pi^*(E'), \pi^*(E'), \sigma), \end{aligned}$$

where σ is an automorphism of $\pi^*(E')|_{X \times S^1}$ such that $\sigma(x, e) = \mathrm{Id}$. If E'' is a vector bundle over X such that $E' \oplus E''$ is a trivial bundle T_1, then we have $d(\pi^*(E'), \pi^*(E'), \sigma) = d(\pi^*(E'), \pi^*(E'), \sigma) + d(\pi^*(E''), \pi^*(E''), \mathrm{Id}) = d(T, T, \alpha)$, where $T = \pi^*(T_1)$ and α is normalized.

Let us assume now that $d(T, T, \alpha)=0$, where α is normalized. By II.2.28 there is a trivial vector bundle T' over $X\times B^2$, and an automorphism $\alpha_1: T\oplus T'\to T\oplus T'$, such that $\alpha_1|_{X\times S^1}=\alpha\oplus \mathrm{Id}$. Let $n=\mathrm{Rank}(T\oplus T')$, and let $f: X\times S^1\times I\to \mathrm{GL}_n(\mathbf{C})$ be the continuous map defined by $f(x, z, t)=\check{\alpha}_1(x, zt)\check{\alpha}_1(x, et)^{-1}$. Then f realizes a normalized homotopy between α and $\alpha\oplus \mathrm{Id}_{T\oplus T'|_{X\times S^1}}$.

Finally, let us assume that $d(T, T, \alpha)=d(T, T, \beta)$, or equivalently that $d(T, T, \alpha\beta^{-1})=0$ (II.2.16). According to the discussion above, there is a trivial bundle T' such that $\alpha\beta^{-1}\oplus \mathrm{Id}_{T'|_{X\times S^1}}$ is homotopic to $\mathrm{Id}_{T\oplus T'|_{X\times S^1}}$ within the normalized automorphisms. If we multiply this homotopy on the right by $\beta\oplus \mathrm{Id}_{T'|_{X\times S^1}}$, it follows that $\alpha\oplus \mathrm{Id}_{T'|_{X\times S^1}}$ is homotopic to $\beta\oplus \mathrm{Id}_{T'|_{X\times S^1}}$ within the normalized automorphisms of $(T\oplus T')|_{X\times S^1}$. □

1.7. A triple (T, T, α) where α is normalized, defines a continuous map $\check{\alpha}: X\times S^1\to \mathrm{GL}_n(\mathbf{C})$ such that $\check{\alpha}(x, e)=\mathrm{Id}$ (I.1.12). Hence it also defines a continuous map $\sigma: S^1\to F(X, \mathrm{GL}_n(\mathbf{C}))\approx \mathrm{GL}_n(A)$ with $\sigma(e)=1$. Homotopies between normalized automorphisms may be translated by this correspondence, to homotopies between loops in $\mathrm{GL}_n(A)$ based on the identity element. The next proposition is therefore a more conceptual version of Lemma 1.6.

1.8. Proposition. *The correspondence* $\alpha\mapsto d(T, T, \alpha)$ *defines an isomorphism between* $\pi_1(\mathrm{GL}(A))=\mathrm{inj\,lim}\,\pi_1(\mathrm{GL}_n(A))$ *and the group* $K(X\times B^2, X\times S^1)$, *where* $A=C(X)$.

1.9. Now let A be any complex Banach algebra with unit. We can define a homomorphism

$$K(A)\longrightarrow \pi_1(\mathrm{GL}(A)),$$

generalizing the homomorphism

$$K(X)\longrightarrow K(X\times B^2, X\times S^1)$$

when $A=C(X)$. More precisely, let E be an object of $\mathscr{P}(A)$, and let p be a projector in A^n for some n large enough, such that $E\approx \mathrm{Im}(p)$. Then the projector p defines a loop σ in $\mathrm{GL}_n(A)$, by the formula $\sigma(z)=pz+1-p$ $(z\in S^1\subset \mathbf{C})$. Passing to the inductive limit, we obtain a well-defined element of $\pi_1(\mathrm{GL}(A))=\mathrm{inj\,lim}\,\pi_1(\mathrm{GL}_n(A))$.

1.10. Proposition. *The element of* $\pi_1(\mathrm{GL}(A))$ *defined above is independent of the choice of p. If we let* $\gamma(E)$ *denote this element, we have the relation* $\gamma(E\oplus F)=\gamma(E)\oplus\gamma(F)$. *Therefore the correspondence* $E\mapsto\gamma(E)$ *induces a homomorphism between* $K(A)$ *and* $\pi_1(\mathrm{GL}(A))$, *denoted again by* γ. *Finally, if* $A=C(X)$, *we have the* *commutative diagram*

$$\begin{array}{ccc} K(X) & \xrightarrow{\ \beta\ } & K(X\times B^2, X\times S^1) \\ \approx \Big\downarrow \theta & & \approx \Big\downarrow \varphi \\ K(A) & \xrightarrow{\ \gamma\ } & \pi_1(\mathrm{GL}(A)), \end{array}$$

where the vertical isomorphisms are those defined in 1.5 and 1.8.

Proof. For the moment let us denote the element of $\pi_1(\mathrm{GL}(A))$, associated with E and the projection operator p, by $\gamma(E,p)$. We first have to show that $\gamma(E,p)=\gamma(F,q)$ if $E\approx F$. Since we will pass to the inductive limit, we may assume without loss of generality that p and q are projection operators in A^{2n} for some n large enough. Moreover, we may assume that p and q can be written as $p'\oplus 0$ and $q'\oplus 0$ where p' and q' are projection operators in A^n. In this case, the argument used in the proof of I.7.7 shows the existence of an element δ of $\mathrm{GL}_{2n}(A)$, homotopic to 1, such that $q=\delta\cdot p\cdot\delta^{-1}$. If $\bar{\delta}\colon I\to \mathrm{GL}_{2n}(A)$ denotes a continuous map such that $\bar{\delta}(0)=1$ and $\bar{\delta}(1)=\delta$, we see that the loops associated with p and q are homotopic by the map $(z,t)\mapsto\bar{\delta}(t)(pz+1-p)\bar{\delta}(t)^{-1}$.

If E and F are two objects of $\mathscr{P}(A)$ such that $E\approx \mathrm{Im}(p)$ and $F\approx\mathrm{Im}(q)$, where p and q are projection operators in A^n and A^m respectively, we have $E\oplus F\approx \mathrm{Im}(p\oplus q)$ where $p\oplus q$ is a projector in $A^n\oplus A^m$. Therefore, if we replace p by $p\oplus 0$ and q by $0\oplus q$, we have the formula $(pz+1-p)(qz+1-q)=(p\oplus q)z+1-(p\oplus q)$, which shows that $\gamma(E\oplus F)=\gamma(E)+\gamma(F)$, since the group operation in $\pi_1(\mathrm{GL}(A))$ is induced by the product in the topological group $\mathrm{GL}(A)$ (Godbillon [2]).

Finally let us prove the commutativity of the diagram, i.e. $\varphi\beta=\gamma\theta$. If E is a vector bundle over X and E' is a vector bundle such that $E\oplus E'=T_1$ is trivial, we have $\beta([E])=d(\pi^*(E),\pi^*(E),\alpha)$, where $\pi^*(E)=E\times B^2$ and $\alpha\colon E\times S^1\to E\times S^1$ is defined by $\alpha(e,z)=(ze,z)$. This can also be written as $d(\pi^*(E\oplus E'),\pi^*(E\oplus E'),\alpha')$, where $\alpha'=\alpha\oplus\mathrm{Id}_{\pi^*(E')|X\times S^1}$. If $p\colon T_1\to T_1$ is the projector defining E, we have $\check{\alpha}'(x,z)=z\check{p}(x)+1-\check{p}(x)$ (using the notation of I.1.12). Therefore $(\varphi\beta)([E])=(\gamma\theta)([E])$, and by linearity, $(\varphi\beta)([E]-[F])=(\gamma\theta)([E]-[F])$. □

Proposition 1.10 shows that Theorem 1.4 is a consequence of the following general theorem on Banach algebras:

1.11. Theorem. *Let A be any complex Banach algebra with unit. Then the homomorphism*

$$\gamma\colon K(A)\longrightarrow\pi_1(\mathrm{GL}(A))$$

defined in 1.9 and 1.10 is an isomorphism.

The proof of this theorem requires many steps. We will begin by showing that γ is injective by defining $\gamma'\colon\pi_1(\mathrm{GL}(A))\to K(A)$ left-inverse to γ (1.20). Then we will prove that γ is surjective by a series of reductions (1.21–1.25). Before we proceed any further, we still require some additional definitions and propositions.

1.12. Definition. Let $\sigma: S^1 \to \mathrm{GL}_n(A)$ be a loop in $\mathrm{GL}_n(A)$ based at 1. Then σ is said to be *Laurentian* (resp. *polynomial*, resp. *affine*) if σ can be written as $\sum_{k=-N}^{k=N} a_k z^k$ (resp. $\sum_{k=0}^{k=N} a_k z^k$, resp. $a_0 + a_1 z$) in $M_n(A)$ for some N large enough. We let $\pi_1^L(\mathrm{GL}_n(A))$ denote the set of homotopy classes of Laurentian loops in $\mathrm{GL}_n(A)$ and let $\pi_1^L(\mathrm{GL}(A))$ denote inj lim $\pi_1^L(\mathrm{GL}_n(A))$.

1.13. It is clear that the homomorphism $K(A) \to \pi_1(\mathrm{GL}(A))$ may be decomposed into $K(A) \to \pi_1^L(\mathrm{GL}(A)) \to \pi_1(\mathrm{GL}(A))$. Our first step towards proving Theorem 1.11 is the following proposition:

1.14. Proposition. *The map*

$$\pi_1^L(\mathrm{GL}(A)) \longrightarrow \pi_1(\mathrm{GL}(A))$$

is bijective. In particular $\pi_1^L(\mathrm{GL}(A))$ is an abelian group (with respect to the operation defined by the product in $\mathrm{GL}(A)$).

Proof. Let $\sigma: S^1 \to \mathrm{GL}(A)$ be a continuous map. Since S^1 is compact, there exists a factorization of σ through $\mathrm{GL}_n(A)$ for some n large enough.

We let σ also denote the map of S^1 in $\mathrm{GL}_n(A)$ thus obtained. Since $\mathrm{M}_n(A)$ is a Banach space, Fejer's theorem applied to $\mathrm{M}_n(A)$ shows that σ is a uniform limit of Laurentian loops (of free origin) defined in the following way. Setting

$$a_k = \frac{1}{2\pi}\int_0^{2\pi} \sigma(e^{i\theta})e^{-ik\theta}\,d\theta, \qquad S_k(z) = \sum_{l=-k}^{+k} a_l z^l, \quad \text{and} \quad \sigma_k'(z) = \frac{S_0(z) + \cdots + S_k(z)}{k+1},$$

then for k large enough, $\sigma_k'(e)$ belongs to the open ball of radius 1 centered at the identity element. Moreover, using the local convexity of $\mathrm{GL}_n(A)$ in $\mathrm{M}_n(A)$ again, we have $t\sigma_k'(z) + (1-t)\sigma(z) \in \mathrm{GL}_n(A)$ for k large enough. Therefore, if we let $\sigma_k(z) = \sigma_k'(z) \cdot \sigma_k'(e)^{-1}$, then σ_k is a Laurentian loop which has the same class in $\pi_1(\mathrm{GL}(A))$ as the original loop, thanks to the homotopy $(z, t) \mapsto (t\sigma_k'(z) + (1-t)\sigma(z))(t\sigma'(e) + (1-t)\sigma(e)]^{-1}$. This argument shows then, that the map $\pi_1^L(\mathrm{GL}(A)) \to \pi_1(\mathrm{GL}(A))$ is surjective.

The injectivity of the map is proved by a similar argument applied to the Banach algebra $A(I)$ (ring of continuous functions on I with values in A). More precisely we consider two Laurentian loops σ and τ, which have the same class in $\pi_1(\mathrm{GL}_n(A))$. Then there exists a continuous map $s: S^1 \times I \to \mathrm{GL}_n(A)$, such that $s(z, 0) = \sigma(z)$, $s(z, 1) = \tau(z)$, and $s(e, t) = 1$. Since

$$F(S^1 \times I, \mathrm{GL}_n(A)) \approx F(S^1, \mathrm{GL}_n(A(I))),$$

the argument above applied to the Banach algebra $A(I)$, shows that for every $\varepsilon>0$ there exists a continuous map $s_1: S^1\times I\to \mathrm{GL}_n(A)$, such that $s_1(e,t)=1$, $\|s_1(z,t)-s(z,t)\|<\varepsilon$, and such that $s_1(z,t)$ is a Laurentian function of z. If ε is chosen smaller than $\frac{1}{\|\sigma^{-1}\|}$ and $\frac{1}{\|\tau^{-1}\|}$, we may define a Laurentian homotopy r between σ and τ by the formulas:

$$\begin{aligned} r(z,t)&=3t\sigma(z)+(1-3t)s_1(z,0) && \text{for } 0\leqslant t\leqslant \tfrac{1}{3}\\ r(z,t)&=s_1(z,3t-1) && \text{for } \tfrac{1}{3}\leqslant t\leqslant \tfrac{2}{3}\\ r(z,t)&=(3t-2)s_1(z,1)+(3-3t)\tau(z) && \text{for } \tfrac{2}{3}\leqslant t\leqslant 1\end{aligned}$$

Fig. 14

Notice that the restriction on ε guarantees that the first and third paths lie in $\mathrm{GL}_n(A(I))\subset M_n(A(I))$. □

1.15. Proposition. *Let E be a projective module over $A(I)$, and let E_0 and E_1 be the A-modules obtained by "restriction" to $\{0\}$ and $\{1\}$. Then E_0 and E_1 are isomorphic.*

Proof. According to I.6.16 we should interpret the "restriction" of E to $\{0\}$ and $\{1\}$ in the following way. Let us write E as the image of a projection operator $p(t)\in M_n(A)$, which depends continuously on the parameter $t\in I$. Now we want to prove $\mathrm{Im}(p(0))\approx \mathrm{Im}(p(1))$. For this it suffices to find some $\alpha\in \mathrm{GL}_n(A)$, such that $p(1)=\alpha^{-1}p(0)\alpha$.

For t and $u\in I$, we have the identity $\alpha(t,u)p(u)=p(t)\alpha(t,u)$, where $\alpha(t,u)=1-p(t)-p(u)+2p(t)p(u)$. If u is close to t, then $\alpha(t,u)$ is close to 1, and hence is invertible. Therefore by the compactness of I, there exists a sequence $0=t_0<t_1<\cdots<t_p=1$ such that $\alpha(t_i,t_{i+1})$ is invertible. Now $p(1)=\alpha^{-1}p(0)\alpha$, where $\alpha=\alpha(t_p,t_{p-1})\ldots\alpha(t_1,t_0)$. □

1.16. *Remark.* When $A=C(X)$, by I.6.18 this proposition is simply another interpretation of I.7.2.

1.17. We are now on our way to defining a homomorphism

$$\gamma': \pi_1(\mathrm{GL}(A))\approx \pi_1^L(\mathrm{GL}(A))\longrightarrow K(A)$$

which is left-inverse to γ. For this we consider the Banach algebra $A\langle z\rangle$ (resp. $A\langle z,z^{-1}\rangle$) of formal power series $\sum_{r=0}^{\infty} a_r z^r$ (resp. $\sum_{r=-\infty}^{+\infty} a_r z^r$) such that $\sum_{r=0}^{+\infty}\|a_r\|<+\infty$ (resp. $\sum_{r=-\infty}^{+\infty}\|a_r\|<+\infty$). If $\sigma: S^1\to \mathrm{GL}_n(A)$ is a Laurentian loop, then σ defines an element σ_z of $\mathrm{M}_n(A\langle z, z^{-1}\rangle)$ which is written as $\sum_{r=-\infty}^{+\infty} a_r z^r$. Note that $a_r=0$ except for a finite number of indices r, and that $\mathrm{M}_n(A\langle z, z^{-1}\rangle)$ may be identified with $\mathrm{M}_n(A)\langle z, z^{-1}\rangle$. If $\tau\colon S^1 \to \mathrm{GL}_n(A)$ denotes the "inverse loop" (defined by $\tau(z)=(\sigma(z))^{-1}$), then τ is a differentiable function of class C^2 (actually of class C^∞); hence its Fourier series converges absolutely, and $\tau_z = \sum_{r=-\infty}^{+\infty} b_r z^r \in \mathrm{GL}_n(A\langle z, z^{-1}\rangle)$ is the inverse series of σ_z. For $k \in \mathbb{N}$ we let $\sigma_z^k = \sum_{r=0}^{\infty} a_{r-k} z^r$ and $\tau_z^k = \sum_{r=0}^{\infty} b_{r-k} z^r$. Then we have $\sigma_z^k \cdot \tau_z^k = z^{2k}(1+\varepsilon_k(z))$, where $\varepsilon_k(z) \to 0$ as $k \to +\infty$.

1.18. Lemma. *Let k be chosen large enough so that $\sigma_z^k \in \mathrm{M}_n(A\langle z\rangle)$, and so that $(1+\varepsilon_k(z))$ is invertible. Then the A-module $M_n(\sigma, k) = \mathrm{Coker}(\sigma_z^k)$ is projective of finite type. Moreover, $M_n(\sigma, k+l) = M_n(\sigma, k) \oplus A^{nl}$ for $l \geqslant 0$.*

Proof. The A-module $M_n(\sigma, k)$ is defined by the exact sequence

$$(A\langle z\rangle)^n \xrightarrow{\sigma_z^k} (A\langle z\rangle)^n \longrightarrow M_n(\sigma, k) \longrightarrow 0,$$

where as usual we identify a matrix with its naturally associated linear map. Let $i\colon (A\langle z\rangle)^n \to (A\langle z, z^{-1}\rangle)^n$ (resp. $P\colon (A\langle z, z^{-1}\rangle)^n \to (A\langle z\rangle)^n$ be the natural inclusion (resp. the projection $\sum_{r=-\infty}^{r=+\infty} a_r z^r \mapsto \sum_{r=0}^{+\infty} a_r z^r$). We define $\theta_z^k\colon (A\langle z\rangle)^n \to (A\langle z\rangle)^n$ by the formula $\theta_z^k = P \cdot z^{-k} \cdot \tau_z \cdot i$. Now we have $\theta_z^k \cdot \sigma_z^k = P \cdot z^{-k} \cdot \tau_z \cdot i \cdot z^k \sigma_z = P \cdot z^{-k} \cdot \tau_z \cdot z^k \cdot \sigma_z \cdot i = P \cdot i = \mathrm{Id}$; this shows that σ_z^k is invertible on the left, hence that $M_n(\sigma, k)$ is a direct factor of $(A\langle z\rangle)^n$ as an A-module.

To show that $M_n(\sigma, k)$ is finitely generated and projective, we consider the commutative diagram

$$\begin{array}{ccccccccc}
0 & \longrightarrow & (A\langle z\rangle)^n & \xrightarrow{\sigma_z^k} & (A\langle z\rangle)^n & \rightleftarrows & M_n(\sigma, k) & \longrightarrow & 0 \\
 & & \uparrow{\scriptstyle \tau_z^k \cdot \eta_k} & & \| & & \uparrow & & \\
0 & \longrightarrow & (A\langle z\rangle)^n & \xrightarrow{z^{2k}} & (A\langle z\rangle)^n & \longrightarrow & (A^n)^{2k} & \longrightarrow & 0,
\end{array}$$

where $\eta_k = (1+\varepsilon_k(z))^{-1}$. From this diagram we see that the map $(A^n)^{2k} \to M_n(\sigma, k)$ is surjective and splits; hence $M_n(\sigma, k)$ is finitely generated and projective.

Finally, if $l \geqslant 0$, then the sequence of A-modules and homomorphisms

$$(A\langle z\rangle)^n \xrightarrow{\sigma_z^k} (A\langle z\rangle)^n \xrightarrow{z^l} (A\langle z\rangle)^n$$

induces the exact sequence

$$0 \longrightarrow \mathrm{Coker}(\sigma_z^k) \longrightarrow \mathrm{Coker}(\sigma_z^{k+l}) \longrightarrow \mathrm{Coker}(z^l) \longrightarrow 0$$

i.e. the exact sequence

$$0 \longrightarrow M_n(\sigma, k) \longrightarrow M_n(\sigma, k+l) \longrightarrow A^{nl} \longrightarrow 0.$$

Therefore $M_n(\sigma, k+l) \approx M_n(\sigma, k) \oplus A^{nl}$. □

1.19. Using Lemma 1.18, we define a homomorphism $\gamma': \pi_1^L(\mathrm{GL}(A)) \to K(A)$ in the following way. Let $\sigma: S^1 \to \mathrm{GL}(A)$ be a Laurentian loop and let k and n be two integers such that $z^k\sigma_z \in \mathrm{M}_n(A\langle z\rangle)$, and $1+\varepsilon_k(z)$ is invertible in $\mathrm{M}_n(A\langle z\rangle)$. Then we define $\gamma'(\sigma)$ to be $M_n(\sigma, k) - (A^{nk}) \in K(A)$. Since $M_{n+1}(\sigma, k) = M_n(\sigma, k) \oplus A^k$, we see that $\gamma'(\sigma)$ is independent of the choice of n. Moreover, as soon as $1+\varepsilon_k(z)$ is invertible and $z^k\sigma_z \in \mathrm{M}_n(A\langle z\rangle)$, we see that $\gamma'(\sigma)$ is independent of the choice of k: essentially, $[M_n(\sigma, k+1)] - [A^{n(k+1)}] = [M_n(\sigma, k)] - [A^{nk}]$ by the last part of 1.18.

Let us consider two Laurentian loops σ_0 and σ_1 in $\mathrm{GL}_n(A)$ which are homotopic. Then there exists a Laurentian loop $\sigma: S^1 \to \mathrm{GL}_n(A(I))$, such that $\sigma(z)(0) = \sigma_0(z)$ and $\sigma(z)(1) = \sigma_1(z)$. If k is chosen large enough, the module $M_n(\sigma, k)$ is a finitely generated projective module over $A(I)$, whose restrictions to $\{0\}$ and $\{1\}$ are $M_n(\sigma_0, k)$ and $M_n(\sigma_1, k)$ respectively. Applying Proposition 1.15, we see that $M_n(\sigma_0, k)$ and $M_n(\sigma_1, k)$ are isomorphic. Therefore, the correspondence $\sigma \mapsto M_n(\sigma, k) - [A^{nk}]$ induces a well-defined map γ' from $\pi_1^L(\mathrm{GL}(A)) \approx \pi_1(\mathrm{GL}(A))$ to $K(A)$.

Finally, let us compare $M_n(\sigma, k)$, $M_n(\sigma', k')$, and $M_n(\sigma'\sigma, k+k')$ for k and k' large enough. Since $(\sigma'\sigma)_z^{k+k'} = \sigma_z'^{k'} \cdot \sigma_z^k$, the exact sequence of cokernels

$$0 \longrightarrow \mathrm{Coker}(\sigma_z^k) \longrightarrow \mathrm{Coker}(\sigma_z'^{k'}\sigma_z^k) \longrightarrow \mathrm{Coker}\, \sigma_z'^{k'} \longrightarrow 0$$

can be written as

$$0 \longrightarrow M_n(\sigma, k) \longrightarrow M_n(\sigma'\sigma, k+k') \longrightarrow M_n(\sigma', k') \longrightarrow 0,$$

showing that $M_n(\sigma'\sigma, k+k') \approx M_n(\sigma, k) + M_n(\sigma', k')$. This implies that γ' is a homomorphism, since

$$[M_n(\sigma, k+k')] - [A^{n(k+k')}] = [M_n(\sigma, k)] - [A^{nk}] + [M_n(\sigma', k')] - [A^{nk'}].$$

1.20. Theorem. *The homomorphism*

$$\gamma': \pi_1^L(\mathrm{GL}(A)) \approx \pi_1(\mathrm{GL}(A)) \longrightarrow K(A)$$

is left-inverse to the homomorphism

$$\gamma: K(A) \longrightarrow \pi_1(\mathrm{GL}(A))$$

defined in 1.9 and 1.10. In particular γ is injective.

Proof. For every Banach space E we may define a Banach space $E\langle z\rangle$ as the space of formal power series $\sum_{n=0}^{\infty} e_n z^n$, where $e_n \in E$ and $\sum_{n=0}^{\infty} \|e_n\| < +\infty$. Moreover, if E is an A-module, then $E\langle z\rangle$ is naturally an A-module. In particular $(A\langle z\rangle)^n \approx A^n\langle z\rangle$. Now let E be a projective A-module, and let E' be a supplementary module, such that $E \oplus E' = A^n$. Let p be the projector in A^n associated with this decomposition. By 1.9, the element of $\pi_1(\mathrm{GL}(A))$ associated with E is the class of the loop σ in $\mathrm{GL}_n(A)$, where σ is defined by $\sigma_z = pz + 1 - p$. Therefore by writing $A^n\langle z\rangle$ as $E\langle z\rangle \oplus E'\langle z\rangle$, we obtain the exact sequence

$$\begin{array}{ccccccc} 0 \longrightarrow & (A\langle z\rangle)^n & \xrightarrow{pz+1-p} & (A\langle z\rangle)^n & \longrightarrow E \longrightarrow 0 \\ & \wr\wr & & \wr\wr & \\ & A^n\langle z\rangle & & A^n\langle z\rangle & \end{array}$$

(note that $pz+1-p$ is multiplication by z on the factor $E\langle z\rangle$, and the identity on the factor $E'\langle z\rangle$).

Since γ' is a homomorphism we finally have $(\gamma'\gamma)([E]-[F]) = (\gamma'\gamma)([E]) - (\gamma'\gamma)([F]) = [E]-[F]$. Hence $\gamma'\gamma = \mathrm{Id}_{K(A)}$. □

Following the plan outlined in 1.11, we still must show that γ is surjective. This is the object of the following lemmas.

1.21. Lemma. *Each element of* $\pi_1^L(\mathrm{GL}(A))$ *may be written as* $[\sigma_1]-[\sigma_2]$ *where* σ_1 *and* σ_2 *are polynomial loops* (cf. 1.12).

Proof. In this and the following lemmas, $[\sigma]$ will denote the class of the loop σ in $\pi_1^L(\mathrm{GL}_n(A))$. Now $[z^k\sigma] = [z^k] + [\sigma]$, implying that $[\sigma] = [z^k\sigma] - [z^k]$ and thus we may set $\sigma_1 = [z^k\sigma]$ and $\sigma_2 = [z^k]$, where k is chosen large enough to guarantee $z^k\sigma \in \mathrm{M}_n(A[z])$. □

1.22. Lemma. *Let* $[\sigma]$ *be an element of* $\pi_1^L(\mathrm{GL}(A))$ *with* σ *polynomial. Then* $[\sigma]$ *is the class of an affine loop.*

Proof. Let us write $\sigma(z) = a_0 + a_1 z + \cdots + a_k z^k$ in $\mathrm{GL}_n(A\langle z\rangle)$ where $k > 1$. Then

$$\sigma(t, z) = \begin{pmatrix} 1 & -tz^{k-1} \\ 0 & 1 \end{pmatrix} \begin{pmatrix} \sigma(z) & 0 \\ 0 & 1 \end{pmatrix} \begin{pmatrix} 1 & 0 \\ ta_k z & 1 \end{pmatrix}$$

defines a continuous map from $I \times S^1$ to $\mathrm{GL}_{2n}(A)$, which connects $\sigma(z)$ with a loop of degree $\leqslant k-1$. We "normalize" this homotopy by defining $\tilde{\sigma}(t, z) = \sigma(t, z)\sigma(t, 1)^{-1}$: then $\tilde{\sigma}(t, 1) = 1$, $\tilde{\sigma}(0, z) = \sigma(z)$, and $\tilde{\sigma}(1, z)$ is a Laurentian loop of degree $\leqslant k-1$. The lemma now follows by induction on k. □

1.23. Lemma. *Let $\sigma(z)=a_0+a_1z$ be an affine loop in $\mathrm{GL}_n(A)$, and let $\tau(z)=\sum_{k=-\infty}^{+\infty} b_kz^k\in\mathrm{GL}_n(A\langle z,z^{-1}\rangle)$ be its inverse loop. Then we have the following relations:*

$$\left.\begin{array}{ll}\text{(i)} & b_ia_0b_j=0\\ \text{(ii)} & b_ia_1b_j=0\\ \text{(iii)} & b_ib_j=b_jb_i=0\end{array}\right\}\ \text{if}\ \left\{\begin{array}{c} i<0 \text{ and } j\geqslant 0\\ \text{or}\\ i\geqslant 0 \text{ and } j<0\end{array}\right.$$

$$\begin{array}{ll}\text{(iv)} & b_ib_j=b_jb_i\\ \text{(v)} & a_0b_i=b_ia_0\\ \text{(v')} & a_1b_i=b_ia_1\end{array}$$

Proof. Since τ is a differentiable function of class C^∞, its Fourier series converges absolutely. Thus we already have the relations

$$a_0b_0+a_1b_{-1}=b_0a_0+b_{-1}a_1=1,$$

and $$a_0b_i+a_1b_{i-1}=b_ia_0+b_{i-1}a_1=0 \quad \text{for } i\neq 0.$$

Now let us verify the other relations.

(i) For $i<0$ and $j\geqslant 0$, we write $b_ia_0b_j=-b_{i-1}a_1b_j=b_{i-1}a_0b_{j+1}=\cdots=b_{i-r}a_0b_{j+r}$. As $r\to+\infty$, the last expression converges to 0. Therefore $b_ia_0b_j=0$. In the same way, for $i\geqslant 0$ and $j<0$, we write $b_ia_0b_j=-b_ia_1b_{j-1}=b_{i+1}a_0b_{j-1}=\cdots=b_{i+r}a_0b_{j-r}=0$.

(ii) For $i<0$ and $j\geqslant 0$, we have $b_ia_1b_j=b_ia_0b_{j+1}=0$ by (i).
For $i\geqslant 0$ and $j<0$, we have $b_ia_1b_j=b_{i+1}a_0b_j=0$ by (i).

(iii) In the same range of values for the pair (i,j), since $a_0+a_1=1$, we have $b_ib_j=b_i(a_0+a_1)b_j=b_ia_0b_j+b_ia_1b_j=0$.

(iv) By (iii) it is enough to consider the case where i and j are $\neq 0$ and of the same sign. If $0<i<j$ for instance, we have

$$b_ia_0b_j=-b_ia_1b_{j-1}=b_{i+1}a_0b_{j-1}=\cdots=b_ja_0b_i$$

and $$b_ia_1b_j=-b_{i+1}a_0b_j=b_{i+1}a_1b_{j+1}=\cdots=b_ja_1b_i.$$

By adding these two relations we obtain $b_ib_j=b_jb_i$. For $i<j<0$, the relation $b_ib_j=b_jb_i$ is obtained in an analogous way.

(v) For $i\leqslant 0$, we have $a_0b_i-b_ia_0=-a_1b_{i-1}+b_{i-1}a_1=a_0b_{i-1}-b_{i-1}a_0$ since $a_0+a_1=1$. Therefore $a_0b_i-b_ia_0=a_0b_{i-r}-b_{i-r}a_0=0$. Similarly for $i>0$, we have $a_0b_i-b_ia_0=-a_1b_i+b_ia_1=a_0b_{i+1}-b_{i+1}a_0=a_0b_{i+r}-b_{i+r}a_0=0$.

(v′) By (v) we have $a_1b_i-b_ia_1=-a_0b_i+b_ia_0=0$. □

1.24. Lemma. *With the notation of 1.23 the morphism, $q=a_0b_0$ is a projector, and σ can be written as*

$$\sigma(z)=\sigma^+(z)\sigma^-(z^{-1})(pz+q),$$

where $p=1-q$, $\sigma^+(z)\in\mathrm{GL}_n(A\langle z\rangle)$, and $\sigma^-(z^{-1})\in\mathrm{GL}_n(A\langle z^{-1}\rangle)$.

Proof. We have $q^2 = a_0 b_0 a_0 b_0 = a_0 b_0 (1 - a_1 b_{-1}) = a_0 b_0 - a_0 (b_0 a_1 b_{-1}) = a_0 b_0$ by (ii) of 1.23. Let us define $\sigma^+(z) = p + \sigma(z) q$ and $\sigma^-(z^{-1}) = q + \sigma(z) p z^{-1}$. Then $\sigma^+(z)$ (resp. $\sigma^-(z^{-1})$) is an affine function of z (resp. z^{-1}), and we have

$$\begin{aligned}\sigma^+(z)\sigma^-(z^{-1})(pz+q) &= (p+\sigma(z)q)(q+\sigma(z)pz^{-1})(pz+q)\\ &= (\sigma(z)q+\sigma(z)pz^{-1})(q+pz) = \sigma(z)q+\sigma(z)p = \sigma(z).\end{aligned}$$

Moreover, $p + \sigma(z)q$ (resp. $q + \sigma(z)pz^{-1}$) is invertible in the Banach algebra $\mathrm{M}_n(A\langle z\rangle)$ (resp. $\mathrm{M}_n(A\langle z^{-1}\rangle)$), and its inverse is $p + \tau(z)q$ (resp. $q + \tau(z)pz^{-1}$), since Lemma 1.23 implies that $p + \tau(z)q$ (resp. $q + \tau(z)pz^{-1}$) contains only positive powers of z (resp. negative powers of z). □

1.25. Theorem. *The homomorphism*

$$\gamma: K(A) \longrightarrow \pi_1(\mathrm{GL}(A))$$

is surjective.

Proof. By Lemma 1.22, it suffices to prove that the class of an affine loop $\sigma(z) = a_0 + a_1 z$ is in the image of γ. By Lemma 1.24, we may write $\sigma(z)$ in the form $\sigma^+(z)\sigma^-(z^{-1})(pz+q)$. Let $\theta: I \times S^1 \to \mathrm{GL}(A)$ be the continuous map defined by $\theta(t, z) = \sigma^+(zt)\sigma^-(z^{-1}t)(pz+q)$, and let $\theta_1: I \times S^1 \to \mathrm{GL}(A)$ be the "normalized" homotopy defined by $\theta_1(t, z) = \theta(t, z)\theta(t, 1)^{-1}$. Then θ_1 defines a homotopy between the loop $\sigma(z)$ and the loop $\sigma_0(z) = pz + q$ whose class in $\pi_1(\mathrm{GL}(A))$ is $\gamma([\mathrm{Im}(p)])$. □

1.26. *Remark.* Combining 1.10, 1.20 and 1.25, the proof of 1.4 and hence of 1.3 is now complete. Other types of proofs may be found in the references. It must be pointed out that the basic ideas in this "elementary" proof stem from the work of Atiyah and Bott [1].

1.27. *Remark.* The computation in 1.24 provides more information on the relation between a_0 and the projectors p and q. The element $t\sigma(z) + (1-t)(q+pz) = [ta_0 + (1-t)q] + [ta_1 + (1-t)p]\cdot z$ belongs to $\mathrm{GL}_n(A\langle z, z^{-1}\rangle)$ for $t \in [0, 1]$. In particular, the spectrum of $ta_0 + (1-t)a_1 b_{-1}$ does not meet the axis $\mathscr{R}(z) = \frac{1}{2}$ (where the *spectrum* of an element α of a complex Banach algebra is the set of elements $\lambda \in \mathbf{C}$ such that $\alpha - \lambda$ is not invertible). To show this, we notice that the element above is the product

$$[p + a_0 q + a_1 q(tz + 1 - t)][q + a_1 p + a_0 p(tz^{-1} + 1 - t)]\cdot(p + qz)$$

or equivalently $\sigma^+(tz + 1 - t)\sigma^-(tz^{-1} + 1 - t)(q + pz)$. Now $\sigma^+(tz + 1 - t)$ (resp. $\sigma^-(tz^{-1} + 1 - t)$) belongs to $\mathrm{GL}(A\langle z\rangle)$ (resp. $\mathrm{GL}(A\langle z^{-1}\rangle)$) for $t \in [0, 1]$, since its inverse is obtained by substituting $z \mapsto tz + 1 - t$ (resp. $z^{-1} \mapsto tz^{-1} + 1 - t$) in the expression $p + \beta(z)q$ (resp. $q + \beta(z)pz^{-1}$), which contains only positive (resp.

negative) powers of z. Note that the inverse series converge, since they are inverses of the differentiable functions $z \mapsto \sigma^+(tz+1-t)$ and $z^{-1} \mapsto \sigma^-(tz^{-1}+1-t)$.

This remark will be used at the end of 6.20. Setting

$$g = \frac{2\alpha - 1}{i} \quad \text{where } \alpha = ta_0 + (1-t)a_1 b_{-1},$$

the spectrum of g does not meet the real axis, i.e. $g - \lambda$ is invertible for any real number λ.

Exercises (Section III.7) 1–3, 7, 11, 12.

2. First Applications of Bott Periodicity Theorem in Complex K-Theory

Since $K_{\mathbb{C}}(\mathbb{R}^2) \approx K_{\mathbb{C}}(B^2, S^1)$, Theorem 1.3 may be reformulated in the following way:

2.1. Theorem. *Let X be a locally compact space. Then the cup-product with a generator u of $K_{\mathbb{C}}(\mathbb{R}^2)$ defines an isomorphism $K_{\mathbb{C}}(X) \approx K_{\mathbb{C}}(X \times \mathbb{R}^2)$.*

2.2. Corollary. *Let u be a generator of $\tilde{K}_{\mathbb{C}}(S^2) = K_{\mathbb{C}}(\mathbb{R}^2)$. Then $\tilde{K}_{\mathbb{C}}(S^n) = 0$ if n is odd, and $\tilde{K}_{\mathbb{C}}(S^{2p}) \approx \mathbb{Z}$ with generator u^p* (for the product defined in II.5.4).

2.3. Theorem. *Let $\mathrm{GL}(\mathbb{C}) = \operatorname{inj\,lim} \mathrm{GL}_n(\mathbb{C})$ and $\mathrm{U} = \operatorname{inj\,lim} \mathrm{U}(n)$. Then the inclusion of U in $\mathrm{GL}(\mathbb{C})$ induces a homotopy group isomorphism and the homotopy groups are periodic of period 2. More precisely $\pi_i(\mathrm{U}) \approx \pi_i(\mathrm{GL}(\mathbb{C})) = 0$ when i is even, and $\pi_i(\mathrm{U}) \approx \pi_i(\mathrm{GL}(\mathbb{C})) \approx \mathbb{Z}$ when i is odd.*

Proof. It is well known that the inclusion of $\mathrm{U}(n)$ in $\mathrm{GL}_n(\mathbb{C})$ is a homotopy equivalence (Chevalley [1]). Hence $\pi_i(\mathrm{U}(n)) \approx \pi_i(\mathrm{GL}_n(\mathbb{C}))$ and $\pi_i(\mathrm{U}) \approx \pi_i(\mathrm{GL}(\mathbb{C}))$. Since $K^{-i}(P) \approx K(\mathbb{R}^i) \approx \tilde{K}(S^i) \approx \pi_{i-1}(\mathrm{GL}(\mathbb{C}))$ (II.1.34), the theorem follows from 1.3. □

2.4. Corollary. *For $n > i/2$, we have isomorphisms*

$$\pi_i(\mathrm{U}(n)) \approx \pi_i(\mathrm{GL}_n(\mathbb{C})) \approx \mathbb{Z} \quad \textit{if } i \textit{ is odd},$$

and

$$\pi_i(\mathrm{U}(n)) \approx \pi_i(\mathrm{GL}_n(\mathbb{C})) = 0 \quad \textit{if } i \textit{ is even}.$$

Finally, if i is odd and $n > \frac{i-1}{2}$, the natural map $\pi_i(\mathrm{U}(n-1)) \to \pi_i(\mathrm{U}) \approx \mathbb{Z}$ is surjective.

Proof. We consider the locally trivial fibration

$$U(n) \longrightarrow \mathrm{U}(n+1) \longrightarrow S^{2n+1}.$$

The homotopy exact sequence associated with this fibration may be written as

$$\pi_{i+1}(S^{2n+1}) \longrightarrow \pi_i(\mathrm{U}(n)) \longrightarrow \pi_i(\mathrm{U}(n+1)) \longrightarrow \pi_i(S^{2n+1}).$$

For $n > i/2$, both end terms of this exact sequence are 0. Hence $\pi_i(\mathrm{U}(n)) \approx \pi_i(\mathrm{U}(n+1)) \approx \pi_i(\mathrm{U})$. For $n > \frac{i-1}{2}$, the last term is equal to 0. □

2.5. *Remark.* We will prove later (5.22) that the isomorphism $\pi_i(\mathrm{U}) \approx \pi_{i+2}(\mathrm{U})$ is induced by an explicit weak homotopy equivalence $\mathrm{U} \sim \Omega^2(\mathrm{U})$, where $\Omega^2(\mathrm{U})$ is the iterated loop space of U.

2.6. If E is a complex vector bundle with compact base X, we will denote its "conjugate bundle" (I.4.8.e) by $\overline{E}$: the fiber $\overline{E}_x$ is the conjugate vector space of E_x. Trivially we have $\overline{E \oplus F} \approx \overline{E} \oplus \overline{F}$ and $\overline{\overline{E}} \approx E$. Hence, the correspondence $E \mapsto \overline{E}$ induces an involution on the group $K_{\mathbf{C}}(X)$, which we also denote by $x \mapsto \bar{x}$.

On the other hand, "realification" and "complexification" of bundles induce additive functors

$$r\colon \mathscr{E}_{\mathbf{C}}(X) \longrightarrow \mathscr{E}_{\mathbf{R}}(X)$$

and $\quad c\colon \mathscr{E}_{\mathbf{R}}(X) \longrightarrow \mathscr{E}_{\mathbf{C}}(X)$ respectively.

The first functor associates each complex vector bundle with its underlying real vector bundle. The second one associates each real vector bundle E with the complex vector bundle $E \otimes_{\mathbf{R}} \mathbf{C} = E \oplus E$, where multiplication by i is represented by the matrix

$$I = \begin{pmatrix} 0 & -1 \\ 1 & 0 \end{pmatrix}.$$

These functors induce homomorphisms

$$K_{\mathbf{C}}(X) \longrightarrow K_{\mathbf{R}}(X)$$

and $\quad K_{\mathbf{R}}(X) \longrightarrow K_{\mathbf{C}}(X),$

which we also call r and c.

2.7. Proposition. *The composite homomorphisms*

$$K_{\mathbf{R}}(X) \xrightarrow{c} K_{\mathbf{C}}(X) \xrightarrow{r} K_{\mathbf{R}}(X) \quad \text{and} \quad K_{\mathbf{C}}(X) \xrightarrow{r} K_{\mathbf{R}}(X) \xrightarrow{c} K_{\mathbf{C}}(X)$$

are respectively multiplication by 2 *and the homomorphism* $x \mapsto x + \bar{x}$.

Proof. The first assertion is obvious. To prove the second, it suffices to show that for a complex vector bundle E, the two complex structures on $E \oplus E$ defined by the matrices

$$I=\begin{pmatrix}0 & -1\\ 1 & 0\end{pmatrix} \quad \text{and} \quad J=\begin{pmatrix}i & 0\\ 0 & -i\end{pmatrix}$$

are isomorphic. This is a formal consequence of the identity $J=\alpha I\alpha^{-1}$, where

$$\alpha=\begin{pmatrix}1 & i\\ i & 1\end{pmatrix}. \quad \square$$

2.8. *Remark.* If X is locally compact, we have $K(X)\approx \operatorname{Ker}[K(\dot{X})\to \mathbb{Z}]$. Thus we may also define $c\colon K_{\mathbb{R}}(X)\to K_{\mathbb{C}}(X)$ and $r\colon K_{\mathbb{C}}(X)\to K_{\mathbb{R}}(X)$; once again we have $(rc)(y)=2y$ and $(cr)(x)=x+\bar{x}$. Similar results hold for $K(X,Y)\approx K(X-Y)$.

2.9. Corollary. *Let $K_{\mathbb{C}}(X)^0$ denote the subgroup of $K_{\mathbb{C}}(X)$ consisting of the elements invariant under involution. Then the homomorphism $c\colon K_{\mathbb{R}}(X)\to K_{\mathbb{C}}(X)$ induces an isomorphism $K_{\mathbb{R}}(X)\otimes_{\mathbb{Z}}\mathbb{Z}'\approx K_{\mathbb{C}}(X)^0\otimes_{\mathbb{Z}}\mathbb{Z}'$, where $\mathbb{Z}'=\mathbb{Z}[\frac{1}{2}]$.*

Proof. It is clear that $\operatorname{Im}(c)\subset K_{\mathbb{C}}(X)^0$. If we denote the homomorphism thus defined between $K_{\mathbb{R}}(X)$ and $K_{\mathbb{C}}(X)^0$ by c' and the restriction of r to $K_{\mathbb{C}}(X)^0$ by r', we have $(r'c')(y)=2y$ and $(c'r')(x)=x+\bar{x}=2x$. Hence r' and c' induce an isomorphism $K_{\mathbb{R}}(X)\otimes_{\mathbb{Z}}\mathbb{Z}'\approx K_{\mathbb{C}}(X)^0\otimes_{\mathbb{Z}}\mathbb{Z}'$. $\square$

In the same way, let $K_{\mathbb{C}}(X)^1=\{x\in K_{\mathbb{C}}(X)\mid \bar{x}=-x\}$. Then $K_{\mathbb{C}}(X)\otimes_{\mathbb{Z}}\mathbb{Z}'\approx (K_{\mathbb{C}}(X)^0\otimes_{\mathbb{Z}}\mathbb{Z}')\oplus(K_{\mathbb{C}}(X)^1\otimes_{\mathbb{Z}}\mathbb{Z}')$. Moreover, if X and Y are locally compact spaces, the cup-product

$$K_{\mathbb{C}}(X)\times K_{\mathbb{C}}(Y)\longrightarrow K_{\mathbb{C}}(X\times Y)$$

induces a bilinear map

$$K_{\mathbb{C}}(X)^i\times K_{\mathbb{C}}(Y)^j\longrightarrow K_{\mathbb{C}}(X\times Y)^{i+j},$$

where we regard the indices i,j as elements of $\mathbb{Z}/2$.

2.10. Proposition. *We have $K_{\mathbb{C}}(\mathbb{R}^2)^0=0$ and $K_{\mathbb{C}}(\mathbb{R}^2)^1=K_{\mathbb{C}}(\mathbb{R}^2)\approx\mathbb{Z}$. The cup-product by a generator u of $K_{\mathbb{C}}(\mathbb{R}^2)$ induces an isomorphism*

$$\beta_i\colon K_{\mathbb{C}}(X)^i\longrightarrow K_{\mathbb{C}}(X\times\mathbb{R}^2)^{i+1}.$$

Proof. The generator u of $K_{\mathbb{C}}(\mathbb{R}^2)=K_{\mathbb{C}}(B^2,S^1)$ may be written as $d(E,F,\alpha)$, where $E=F=B^2\times\mathbb{C}$ and $\alpha(z,v)=(z,zv)$. Therefore, $\bar{u}=d(\bar{E},\bar{F},\bar{\alpha})$, where $\bar{\alpha}=\alpha$ on the underlying real vector bundles. Complex conjugation permits us to identify $\bar{E}$ and $\bar{F}$ with $B^2\times\mathbb{C}$ (where $\mathbb{C}$ is provided with the usual complex structure). Under this identification $\bar{\alpha}$ becomes α^{-1} (since $\bar{z}=z^{-1}$ for $z \in S^1 \subset \mathbb{C}$). Hence $\bar{u}=d(F, E, \alpha^{-1})=-u$.

In general, if ε denotes the homomorphism $x \mapsto \bar{x}$ in complex K-theory, we have the commutative diagram

$$\begin{array}{ccc} K_{\mathbb{C}}(X) & \xrightarrow{\beta} & K_{\mathbb{C}}(X \times \mathbb{R}^2) \\ \downarrow{\scriptstyle \varepsilon} & & \downarrow{\scriptstyle -\varepsilon} \\ K_{\mathbb{C}}(X) & \xrightarrow{\beta} & K_{\mathbb{C}}(X \times \mathbb{R}^2). \end{array}$$

Since β is an isomorphism (2.1), it induces an isomorphism $\beta_i\colon K_{\mathbb{C}}(X)^i \approx K_{\mathbb{C}}(X \times \mathbb{R}^2)^{i+1}$. □

2.11. Theorem. *Let v be the image of $u \cup u$ by the homomorphism $K_{\mathbb{C}}(\mathbb{R}^4) \to K_{\mathbb{R}}(\mathbb{R}^4)$. Then the cup-product by v induces an isomorphism $K_{\mathbb{R}}(X) \otimes_{\mathbb{Z}} \mathbb{Z}' \approx K_{\mathbb{R}}(X \times \mathbb{R}^4) \otimes_{\mathbb{Z}} \mathbb{Z}'$ with $\mathbb{Z}'=\mathbb{Z}[\frac{1}{2}]$, for every locally compact space X.*

Proof. By 2.9, $K_{\mathbb{R}}(X) \otimes_{\mathbb{Z}} \mathbb{Z}'$ may be identified with $K_{\mathbb{C}}(X)^0 \otimes_{\mathbb{Z}} \mathbb{Z}'$ by the map c'. We now have the commutative diagram

$$\begin{array}{ccc} K_{\mathbb{R}}(X) \otimes_{\mathbb{Z}} \mathbb{Z}' & \xrightarrow{\beta_{\mathbb{R}}} & K_{\mathbb{R}}(X \times \mathbb{R}^4) \otimes_{\mathbb{Z}} \mathbb{Z}' \\ \downarrow{\scriptstyle c' \otimes 1} & & \downarrow{\scriptstyle c' \otimes 1} \\ K_{\mathbb{C}}(X)^0 \otimes_{\mathbb{Z}} \mathbb{Z}' & \xrightarrow{2\beta_1\beta_0 \otimes 1} & K_{\mathbb{C}}(X \times \mathbb{R}^4)^0 \otimes_{\mathbb{Z}} \mathbb{Z}', \end{array}$$

where $\beta_{\mathbb{R}}$ is the cup-product by v, and $\beta_1\beta_0 \otimes 1$ is the composite $K_{\mathbb{C}}(X)^0 \otimes_{\mathbb{Z}} \mathbb{Z}' \xrightarrow{\beta_0 \otimes 1} K_{\mathbb{C}}(X \times \mathbb{R}^2)^1 \otimes_{\mathbb{Z}} \mathbb{Z}' \xrightarrow{\beta_1 \otimes 1} K_{\mathbb{C}}(X \times \mathbb{R}^4)^0 \otimes_{\mathbb{Z}} \mathbb{Z}'$ (2.10), i.e. is induced by the cup-product with u^2 (note that $c(v)=2u^2$). Since $\beta_1\beta_2$ is an isomorphism (2.10), and since $c' \otimes 1$ is an isomorphism (2.9), $\beta_{\mathbb{R}}$ is an isomorphism as desired. □

2.12. Corollary. *The cup-product by $v \in K_{\mathbb{R}}^{-4}(P)$ induces isomorphisms*

$$K_{\mathbb{R}}^{-n}(X, Y) \otimes_{\mathbb{Z}} \mathbb{Z}' \approx K_{\mathbb{R}}^{-n-4}(X, Y) \otimes_{\mathbb{Z}} \mathbb{Z}'.$$

Moreover, we have natural isomorphisms

$$K_{\mathbb{C}}^{-n}(X, Y) \otimes_{\mathbb{Z}} \mathbb{Z}' \approx [K_{\mathbb{R}}^{-n}(X, Y) \oplus K_{\mathbb{R}}^{-n-2}(X, Y)] \otimes_{\mathbb{Z}} \mathbb{Z}'.$$

Proof. The first part of the corollary is a formal consequence of 2.10 just as 1.3 is a formal consequence of 1.4. For the same reason, it is enough to prove the second part of the corollary for $n=0$ and $Y=\varnothing$. Then $K_{\mathbb{C}}(X) \otimes_{\mathbb{Z}} \mathbb{Z}'$ may be written as

$$\begin{aligned}(K_{\mathbb{C}}(X)^0 \otimes_{\mathbb{Z}} \mathbb{Z}') \oplus (K_{\mathbb{C}}(X)^1 \otimes_{\mathbb{Z}} \mathbb{Z}') &\approx (K_{\mathbb{C}}(X)^0 \otimes_{\mathbb{Z}} \mathbb{Z}') \oplus (K_{\mathbb{C}}(X \times \mathbb{R}^2)^0 \otimes_{\mathbb{Z}} \mathbb{Z}') \\ &\approx (K_{\mathbb{R}}(X) \otimes_{\mathbb{Z}} \mathbb{Z}') \oplus (K_{\mathbb{R}}(X \times \mathbb{R}^2) \otimes_{\mathbb{Z}} \mathbb{Z}'). \quad \square\end{aligned}$$

2.13. From these results in real K-theory we obtain information about the infinite orthogonal group $\mathrm{O} = \operatorname{inj\,lim} \mathrm{O}(n)$ and the infinite general linear group $\mathrm{GL}(\mathbb{R}) = \operatorname{inj\,lim} \mathrm{GL}_n(\mathbb{R})$ (compare with 2.3). In particular $\pi_i(\mathrm{O}) \otimes_{\mathbf{Z}} \mathbf{Z}' \approx \pi_{i+4}(\mathrm{O}) \otimes_{\mathbf{Z}} \mathbf{Z}'$. However, we leave these matters for the moment, since better results will be obtained in the next sections (5.22).

Exercise (III.7.6).

3. Clifford Algebras

3.1. Clifford algebras arise as the solution of the following universal problem. Let k be a (commutative) field, and V a k-vector space provided with a quadratic form Q. We want to find a pair (C, j), where C is a k-algebra (not necessarily commutative), and $j\colon V \to C$ is a homomorphism on the underlying vector spaces with $j(v)^2 = Q(v) \cdot 1$ (1 denotes the unit element in C), which satisfies this condition: for any k-algebra A and any homomorphism on the underlying vector spaces $\varphi\colon V \to A$ such that $(\varphi(v))^2 = Q(v) \cdot 1$, there is a unique algebra homomorphism $\psi\colon C \to A$ making the following diagram commutative

$$\begin{array}{ccc} V & \xrightarrow{\ j\ } & C \\ & {}_{\varphi}\searrow \quad \swarrow_{\psi} & \\ & A & \end{array}$$

3.2. Definition and theorem. The above problem admits a solution (C, j) which is unique up to isomorphism. We will denote it by $C(V, Q)$, or simply $C(V)$ or $C(Q)$. It is the Clifford algebra associated with the pair (V, Q).

Proof. The uniqueness is obvious since we are dealing with a universal problem. Let us prove the existence. For this we consider the tensor algebra $T(V) = \bigoplus_{i=0}^{\infty} T^i(V)$ where $T^0(V) = k$ and $T^i(V) = \underbrace{V \otimes \cdots \otimes V}_{i}$ for $i > 0$. Let $I(Q)$ be the two-sided ideal generated by elements of the form $t(v) = v \otimes v - Q(v) \cdot 1$, where $v \in V$ and 1 is the unit element of $T(V)$. Every element of $I(Q)$ may be written as $\sum \lambda_i t(v_i) \mu_i$, where $\lambda_i, \mu_i \in T(V)$ and $v_i \in V$. Let $C(V) = T(V)/I(Q)$, and let $j\colon V \to C(V)$ be the composition of the isomorphism i from V onto $T^1(V) \subset T(V)$, and the projection p from $T(V)$ to $C(V)$. Then the pair $(C(V), j)$ is the solution of our problem. By the universal property of tensor algebras, φ may be factorized as

$$\begin{array}{ccc} V & \xrightarrow{\ i\ } & T(V) \\ & {}_{\varphi}\searrow \quad \swarrow_{\theta} & \\ & A & \end{array}$$

where θ is an algebra homomorphism. Since $(\varphi(v))^2 = Q(v)\cdot 1$, the map θ is zero on the ideal $I(Q)$ and thus defines the required homomorphism ψ. Since θ is the unique algebra homomorphism that makes the above diagram commutative, the uniqueness of ψ is clear. $\square$

3.3. Example. Let us assume that $Q=0$. Then $C(V, Q)$ is simply the exterior algebra of V (by the universal property of exterior algebras).

3.4. Example. Let $V=k$, and let Q be the quadratic form defined by $Q(x)=dx^2$ for some $d\in k$. Then $T(V)\approx k[X]$ and $I(Q)=(X^2-d)k[X]$. Thus $C(V)\approx k[X]/(X^2-d)$. In particular, for $k=\mathbb{R}$ and $d=-1$ (resp. $d=+1$), we have $C(V, Q)=\mathbb{C}$ (resp. $C(V, Q)=\mathbb{R}\oplus\mathbb{R}$).

3.5. Clearly the Clifford algebra depends "functorially" on the pair (V, Q). More precisely, if $f\colon V\to V'$ is a k-vector space homomorphism such that $Q'(f(v))=Q(v)$, where Q (resp. Q') is a quadratic form on V (resp. V'), then f induces an algebra homomorphism

$$C(f)\colon C(V, Q)\longrightarrow C(V', Q'),$$

and we have the identities $C(g\cdot f)=C(g)\cdot C(f)$ and $C(\mathrm{Id}_V)=\mathrm{Id}_{C(V)}$.

3.6. The tensor algebra $T(V)$ may be considered as $\mathbb{Z}/2$-graded by setting $T^{(0)}(V)=\sum_{i=0}^{\infty} T^{2i}(V)$ and $T^{(1)}(V)=\sum_{i=0}^{\infty} T^{2i+1}(V)$. Letting $I^{(\alpha)}(Q)=I(Q)\cap T^{(\alpha)}(V)$, we have $I(Q)=I^{(0)}(Q)\oplus I^{(1)}(Q)$ (to see this, we decompose the λ_i and μ_i introduced in 3.2, as the sum of homogeneous elements). If we define $C^{(\alpha)}(V, Q)=p(T^{(\alpha)}(V))$, where $p\colon T(V)\to C(V, Q)$ is the canonical projection, we then have $C(V, Q)= C^{(0)}(V, Q)\oplus C^{(1)}(V, Q)$. It follows that the Clifford algebra is also $\mathbb{Z}/2$-graded. We will simply write $C^{(\alpha)}(V)$ or $C^{(\alpha)}(Q)$ instead of $C^{(\alpha)}(V, Q)$. Of course the functoriality described in 3.5 is compatible with the gradation.

3.7. Example. The $\mathbb{Z}/2$-grading for Example 3.3 is given by the even and odd exterior powers: $C^{(0)}(V)=\lambda^{(0)}(V)=\sum_{i=0}^{\infty}\lambda^{2i}(V)$; $C^{(1)}(V)=\lambda^{(1)}(V)=\sum_{i=0}^{\infty}\lambda^{2i+1}(V)$. Similarly in Example 3.4, the algebra $k[X]/(X^2-d)$ is $\mathbb{Z}/2$-graded if we write each element of $C(V)$ in the form $a+bX$, where X is of degree one, and a and b are of degree zero. For instance, if $k=\mathbb{R}$, and $d=-1$, the algebra of complex numbers $\mathbb{C}$ is $\mathbb{Z}/2$-graded by $\mathbb{C}^{(0)}=\mathbb{R}$ and $\mathbb{C}^{(1)}=i\mathbb{R}$.

The following lemma gives an example of how the grading of $C(V, Q)$ may be used.

3.8. Lemma. *Let v and w be vectors of V which are orthogonal with respect to the symmetric bilinear form associated with Q. Then $j(v)j(w)=-j(w)j(v)$. Therefore, if $x=\prod_{i=1}^{n} j(v_i)$ and $y=\prod_{s=1}^{m} j(w_s)$ are elements of $C^{(\alpha)}(V, Q)$ and $C^{(\beta)}(V, Q)$, with*

$\alpha = n \bmod 2$ and $\beta = m \bmod 2$, and if v_i is orthogonal to w_s for each pair (i, s), we have $xy = (-1)^{\alpha\beta} yx$.

Proof. Under the hypothesis of the lemma we have

$$\begin{aligned} Q(v) + Q(w) = Q(v+w) = (j(v+w))^2 &= (j(v))^2 + (j(w))^2 + j(v)j(w) + j(w)j(v) \\ &= Q(v) + Q(w) + j(v)j(w) + j(w)j(v). \end{aligned}$$

Hence $j(v)j(w) = -j(w)j(v)$. □

3.9. If A and B are $\mathbb{Z}/2$-graded algebras, we define their graded tensor product $A \hat{\otimes} B$ as the algebra whose underlying k-vector space is $A \otimes_k B$, and the product of $x \otimes y$ and $z \otimes t$ is defined by the formula

$$(x \otimes y)(z \otimes t) = (-1)^{\alpha\beta} xz \otimes yt$$

where $y \in B^\beta$ and $z \in A^\alpha$.

3.10. Theorem. *Let V and V' be vector spaces over k provided with quadratic forms Q and Q'. Then the Clifford algebra $C(V \oplus V', Q \oplus Q')$ is naturally isomorphic to $C(V, Q) \hat{\otimes} C(V', Q')$* (Chevalley [2]).

Proof. We prove this theorem by explicity defining the desired isomorphism. Let $j: V \to C(V, Q)$ and $j': V' \to C(V', Q')$ denote the canonical maps and let

$$j'': V \oplus V' \longrightarrow C(V, Q) \hat{\otimes} C(V', Q')$$

be the map defined by $j''(v, v') = j(v) \otimes 1 + 1 \otimes j'(v')$. Now we have $(j''(v, v'))^2 = [(j(v))^2 + (j'(v'))^2] \cdot 1 = (Q(v) + Q'(v')) \cdot 1$. Hence, by the universal property of Clifford algebras, j'' induces a homomorphism

$$\psi: C(V \oplus V', Q \oplus Q') \longrightarrow C(V, Q) \hat{\otimes} C(V', Q')$$

In the other direction, we define homomorphisms

$$\gamma: C(V, Q) \longrightarrow C(V \oplus V', Q \oplus Q') \quad \text{and} \quad \gamma': C(V', Q') \longrightarrow C(V \oplus V', Q \oplus Q'),$$

induced by the inclusions of V and V' in $V \oplus V'$ (cf. 3.5). Since V and V' are orthogonal in $V \oplus V'$, we have the identity $\gamma(x)\gamma'(x') = (-1)^{\alpha\alpha'} \gamma'(x')\gamma(x)$, for $x \in C^{(\alpha)}(V, Q)$ and $x' \in C^{(\alpha')}(V', Q')$ by Lemma 3.8. Hence γ and γ' induce a homomorphism

$$\theta: C(V, Q) \hat{\otimes} C(V', Q') \longrightarrow C(V \oplus V', Q \oplus Q')$$

by the formula $\theta(x \otimes x') = \gamma(x) \cdot \gamma'(x')$.

a) $\theta\psi = \mathrm{Id}_{C(V\oplus V', Q\oplus Q')}$. Since $C(V \oplus V', Q \oplus Q')$ is generated by elements of the form $j''(v, v')$, it is enough to compute $(\theta\psi)(j''(v, v'))$. Doing this we find

$$\begin{aligned}(\theta\psi)(j''(v, v')) &= \theta(j(v) \otimes 1 + 1 \otimes j'(v')) = \gamma(j(v)) + \gamma'(j'(v')) \\ &= j''(v, 0) + j''(0, v') = j''(v, v').\end{aligned}$$

b) $\psi\theta = \mathrm{Id}_{C(V,Q)\hat{\otimes} C(V', Q')}$. In the same way, $C(V, Q) \hat{\otimes} C(V', Q')$ is generated by elements of the form $j(v) \otimes 1$ or $1 \otimes j'(v')$. But

$$(\psi\theta)(j(v) \otimes 1) = \psi(\gamma(j(v))) = j(v) \otimes 1,$$

and $$(\psi\theta)(1 \otimes j'(v')) = \psi(\gamma'(j'(v'))) = 1 \otimes j'(v'). \quad \square$$

3.11. Corollary. *Let us assume that V admits an orthogonal basis e_i, for $i = 1, \ldots, n$, with $Q(e_i) = d_i$. Then the Clifford algebra $C(V, Q)$ is of dimension 2^n over k, with basis the products $e_{i_1} e_{i_2} \ldots e_{i_r}$ where $i_1 < i_2 < \cdots < i_r$. The multiplication law is completely determined by the relations*

$$(e_i)^2 = d_i \quad \text{and} \quad e_i e_j = -e_j e_i \quad \text{for } i \neq j.$$

Proof. The vector space V splits into the orthogonal sum $\bigoplus_{i=1}^{n} ke_i$. Hence $C(V, Q) \approx (k \oplus ke_1) \hat{\otimes} (k \oplus ke_2) \hat{\otimes} \cdots \hat{\otimes} (k \oplus ke_n)$ by 3.4 and 3.10 applied $(n-1)$ times. Therefore, the products $e_{i_1} \cdot e_{i_2} \ldots e_{i_r}$ form an additive basis for $C(V, Q)$. The relation $e_i e_j = -e_j e_i$ for $i \neq j$ follows from 3.8. $\square$

3.12. *Remark.* The argument above shows that the map $j: V \to C(V, Q)$ is injective. Thus we may identify e_i with $j(e_i)$. The hypothesis of Corollary 3.11 is satisfied whenever the characteristic of k is $\neq 2$; hence, in this case the map $V \to C(V, Q)$ is injective. A different argument shows that j is injective in general (cf. Bourbaki [2]) but we do not need that here.

3.13. The case we are more interested in is that where k is the field of real numbers and $V = \mathbb{R}^{p+q}$ is provided with the quadratic form $-x_1^2 - \cdots - x_p^2 + \cdots + x_{p+q}^2$. In this case we denote the Clifford algebra by $C^{p,q}$. According to 3.11, the algebra $C^{p,q}$ is generated over $\mathbb{R}$ by the symbols $e_1, e_2, \ldots, e_{p+q}$, subject to the relations

$$\begin{aligned}(e_i)^2 &= -1 \quad \text{for } 1 \leqslant i \leqslant p, \\ (e_i)^2 &= +1 \quad \text{for } p+1 \leqslant i \leqslant p+q,\end{aligned}$$

and $$e_i e_j = -e_j e_i \quad \text{for } i \neq j.$$

One of the purposes of this section is to explicitly compute the algebras $C^{p,q}$ (cf. 3.19, 3.21, 3.22).

3.14. Examples. We have already computed $C^{1,0} \approx \mathbb{C}$ and $C^{0,1} \approx \mathbb{R} \oplus \mathbb{R}$ (3.4). Moreover, $C^{2,0}$ is the skew field $\mathbb{H}$ (the quaternions): put $I = e_1$, $J = e_2$, $K = e_1 e_2$.

We can also verify that $C^{1,1}$ is $M_2(\mathbb{R})$ (ring of 2×2 matrices over $\mathbb{R}$) by letting

$$e_1=\begin{pmatrix}0 & 1\\ -1 & 0\end{pmatrix} \quad\text{and}\quad e_2=\begin{pmatrix}0 & 1\\ 1 & 0\end{pmatrix}.$$

In the same way $C^{0,2}\approx M_2(\mathbb{R})$ by letting

$$e_1=\begin{pmatrix}1 & 0\\ 0 & -1\end{pmatrix} \quad\text{and}\quad e_2=\begin{pmatrix}0 & 1\\ 1 & 0\end{pmatrix}.$$

However, the *graded* algebras $C^{0,2}$ and $C^{1,1}$ are not isomorphic, since the square of an element of degree one always belongs to the center in $C^{0,2}$, but generally does not belong to the center in $C^{1,1}$.

3.15. In the algebra $C^{p,q}$ with $p+q$ even, let us consider the element $\varepsilon=e_1e_2\cdots e_n$ where $n=p+q$. Then

$$(\varepsilon)^2=(-1)^{(n-1)+(n-2)+\cdots+1}(e_1)^2(e_2)^2\ldots(e_n)^2=\pm 1.$$

If $(\varepsilon)^2=+1$, we call the algebra $C^{p,q}$ *positive*. If $(\varepsilon)^2=-1$, we call the algebra $C^{p,q}$ *negative*. In fact $(\varepsilon)^2=(-1)^{n(n-1)/2}(-1)^p$. Therefore, if $p-q\equiv 0, 4 \bmod 8$, then the algebra $C^{p,q}$ is positive; if $p-q\equiv 2, 6 \bmod 8$, then the algebra $C^{p,q}$ is negative (in the case $p-q$ odd, the algebra $C^{p,q}$ has no sign). Moreover, if V is a finite-dimensional real vector space provided with a nondegenerate quadratic form, the choice of a suitable orthogonal basis defines an isomorphism $C(V,Q)\approx C^{p,q}$, where p and q are well-determined by Sylvester's theorem. Hence the notion of positivity of negativity, for a Clifford algebra associated with a finite-dimensional real vector space provided with a nondegenerate quadratic form, is an intrinsic notion.

With the exception of 3.23 all quadratic forms we will now consider are non-degenerate, and will be taken over real vector spaces of finite dimension.

3.16. Proposition. (i) *If* $C(V,Q)>0$ *and* $\mathrm{Dim}(V)$ *even, then*

$$C(V\oplus V', Q\oplus Q')\approx C(V,Q)\otimes C(V',Q').$$

(ii) *If* $C(V,Q)<0$ *and* $\mathrm{Dim}(V)$ *even, then*

$$C(V\oplus V', Q\oplus Q')\approx C(V,Q)\otimes C(V',-Q').$$

Proof. If $\varepsilon=e_1e_2\ldots e_n$, then we have $\varepsilon e_i=(-1)^{n-1}e_i\varepsilon$. Since n is even, we have $\varepsilon e_i=-e_i\varepsilon$ for each i, and therefore $\varepsilon v=-v\varepsilon$ for each vector v of V (identifying V with a subspace of $C(V)$ via the canonical map j; cf. 3.12). In case (i) we define

$$\varphi\colon V\oplus V'\longrightarrow C(V,Q)\otimes C(V',Q')$$

by the formula $\varphi(v, v') = v \otimes 1 + \varepsilon \otimes v'$. Then

$$\begin{aligned}(\varphi(v, v'))^2 &= (v \otimes 1)^2 + (\varepsilon \otimes v)^2 + (v \otimes 1)(\varepsilon \otimes v') + (\varepsilon \otimes v')(v \otimes 1)\\ &= v^2 \otimes 1 + \varepsilon^2 \otimes v'^2 + v\varepsilon \otimes v' + \varepsilon v \otimes v' = Q(v) + Q'(v').\end{aligned}$$

By the universal property of Clifford algebras, we obtain an algebra homomorphism.

$$\psi: C(V \oplus V', Q \oplus Q') \longrightarrow C(V, Q) \otimes C(V', Q')$$

which extends φ. Since the two algebras have the same dimension, we will have that ψ is the required isomorphism, if we show that ψ is surjective. Since $C(V, Q) \otimes C(V', Q')$ is generated by elements of the form $v \otimes 1$ or $1 \otimes v'$, it is enough to show that $1 \otimes v'$ and $v \otimes 1 \in \operatorname{Im}(\psi)$. But $v \otimes 1 = \psi(v, 0)$, and $1 \otimes v' = (\varepsilon \otimes v')(\varepsilon \otimes 1) = \psi(0, v')\psi(\varepsilon, 0)$.

In case (ii), the proof is similar, interpreting φ as a map from $V \oplus V'$ to $C(V, Q) \otimes C(V', -Q')$. □

3.17. Lemma. *Let A be an algebra over k. Then we have algebra isomorphisms:*

(i) $A \otimes_k \mathrm{M}_n(k) \approx \mathrm{M}_n(A)$,

(ii) $\mathrm{M}_p(\mathrm{M}_n(A)) \approx \mathrm{M}_{np}(A)$,

(iii) $\mathrm{M}_n(A) \otimes_k \mathrm{M}_p(k) \approx \mathrm{M}_{np}(A)$.

Proof. For the proof of (i), let $\gamma: A \times \mathrm{M}_n(k) \to \mathrm{M}_n(A)$ be the k-bilinear map

$$\left[a, \begin{pmatrix} \lambda_{11} & \cdots & \lambda_{1n} \\ \lambda_{n1} & \cdots & \lambda_{nn} \end{pmatrix}\right] \longmapsto \begin{pmatrix} a\lambda_{11} & \cdots & a\lambda_{1n} \\ a\lambda_{n1} & \cdots & a\lambda_{nn} \end{pmatrix}.$$

Then γ induces an algebra homomorphism $\bar{\gamma}: A \otimes_k \mathrm{M}_n(k) \to \mathrm{M}_n(A)$. Since $A \otimes_k k^{n^2} \approx A^{n^2}$, $\bar{\gamma}$ is an isomorphism.

The isomorphism $\mathrm{M}_p(\mathrm{M}_n(A)) \approx \mathrm{M}_{np}(A)$ is simply the obvious writing of matrices in blocks.

Finally $\mathrm{M}_n(A) \otimes_k \mathrm{M}_p(k) \approx \mathrm{M}_p(\mathrm{M}_n(A)) \approx \mathrm{M}_{np}(A)$. □

3.18. Proposition. *The algebras $C^{p+n,q+n}$ and $\mathrm{M}_{2^n}(C^{p,q})$ are isomorphic.*

Proof. Since the algebra $C^{1,1}$ is positive, we have $C^{p+1,q+1} \approx C^{p,q} \otimes C^{1,1} \approx C^{p,q} \otimes \mathrm{M}_2(\mathbb{R})$ by 3.16 and 3.14. Therefore

$$C^{p+n,q+n} \approx C^{p,q} \otimes \underbrace{\mathrm{M}_2(\mathbb{R}) \otimes \cdots \otimes \mathrm{M}_2(\mathbb{R})}_{n} \approx C^{p,q} \otimes \mathrm{M}_{2^n}(\mathbb{R}) \approx \mathrm{M}_{2^n}(C^{p,q})$$

by 3.17. □

3.19. Corollary. *We have the following algebra isomorphisms*

$$C^{n,n} \approx M_{2^n}(\mathbb{R}),$$

$$C^{p,q} \approx M_{2^q}(C^{p-q,0}) \quad \text{if } p>q,$$

and $$C^{p,q} \approx M_{2^p}(C^{0,q-p}) \quad \text{if } p<q.$$

3.20. Proposition. *If $C(V,Q)>0$ and the dimension of V is even, then the graded algebras $C(V,Q)$ and $C(V,-Q)$ are isomorphic.*

Proof. Let $h: V \to C(V,Q)$ be the map defined by $h(v)=\varepsilon v$, where ε is defined as in 3.16. Then $(h(v))^2=(\varepsilon v)^2=(\varepsilon v)(\varepsilon v)=-\varepsilon^2 v^2=-v^2=-Q(v)$, so we obtain a homomorphism $\bar{h}: C(V,-Q) \to C(V,Q)$. To prove that $\bar{h}$ is an isomorphism, it is enough to show that $\bar{h}$ is surjective, since the two algebras have the same dimension over k. In fact, it suffices to show that V (considered as a subspace of $C(V,Q)$) is contained in $\operatorname{Im}(\bar{h})$. If $e_1,\ldots,e_n$ is an orthogonal basis in V with $Q(e_i)=\pm 1$, and if ε' denotes the product $e_1 \cdot e_2 \ldots e_n$ in $C(V,-Q)$, we have

$$\bar{h}(\varepsilon' v)=\bar{h}(e_1)\bar{h}(e_2)\ldots\bar{h}(e_n)\bar{h}(v)=(\varepsilon e_1)(\varepsilon e_2)\ldots(\varepsilon e_n)(\varepsilon v)=\pm\varepsilon^2 v=\pm v.$$

Hence $v=\bar{h}(\varepsilon' v)$ or $\bar{h}(-\varepsilon' v)$. □

3.21. Theorem. *The algebras $C^{p+8,q}$, $C^{p,q+8}$, and $M_{16}(C^{p,q})$, are isomorphic.*

Proof. Since $C^{4,0}>0$, we have $C^{8,0} \approx C^{4,0} \otimes C^{4,0}$ by 3.16. For the same reason, $C^{4,0} \otimes C^{4,0} \approx C^{4,0} \otimes C^{0,4} \approx C^{4,4} \approx M_{16}(\mathbb{R})$ by 3.20 and 3.19. Since $C^{8,0}>0$, we also have $C^{p+8,q} \approx C^{p,q} \otimes C^{8,0} \approx C^{p,q} \otimes M_{16}(\mathbb{R}) \approx M_{16}(C^{p,q})$ by 3.17. The proof of the isomorphism $C^{p,q+8} \approx M_{16}(C^{p,q})$ is analogous. □

3.22. By this theorem and Corollary 3.19, we need only compute the algebras $C^{p,0}$ and $C^{0,p}$ for $p<8$, to determine all the others. In fact, since $C^{0,2}$ and $C^{2,0}$ are negative, Proposition 3.16 gives the isomorphisms

$$C^{0,p+2} \approx C^{p,0} \otimes C^{0,2} \approx C^{p,0} \otimes M_2(\mathbb{R}) \approx M_2(C^{p,0}).$$

In the same way, we have the isomorphisms

$$C^{p+2,0} \approx C^{0,p} \otimes C^{2,0} \approx C^{0,p} \otimes \mathbb{H}.$$

Finally, if $p<4$ we have the isomorphisms

$$C^{p+4,0} \approx C^{p,0} \otimes C^{4,0} \approx C^{p,0} \otimes C^{0,4} \approx C^{p,4} \approx M_{2^p}(C^{0,4-p}).$$

Given that $C^{0,0}=\mathbb{R}$, $C^{0,1}=\mathbb{R}\oplus\mathbb{R}$, $C^{0,2}=M_2(\mathbb{R})$, $C^{1,0}=\mathbb{C}$ and $C^{2,0}=\mathbb{H}$, by applying these identities and Lemma 3.17 we obtain the following table (due to

Atiyah, Bott and Shapiro [1]):

p	$C^{p,0}$	$C^{0,p}$
0	$\mathbb{R}$	$\mathbb{R}$
1	$\mathbb{C}$	$\mathbb{R} \oplus \mathbb{R}$
2	$\mathbb{H}$	$M_2(\mathbb{R})$
3	$\mathbb{H} \oplus \mathbb{H}$	$M_2(\mathbb{C})$
4	$M_2(\mathbb{H})$	$M_2(\mathbb{H})$
5	$M_4(\mathbb{C})$	$M_2(\mathbb{H}) \oplus M_2(\mathbb{H})$
6	$M_8(\mathbb{R})$	$M_4(\mathbb{H})$
7	$M_8(\mathbb{R}) \oplus M_8(\mathbb{R})$	$M_8(\mathbb{C})$
8	$M_{16}(\mathbb{R})$	$M_{16}(\mathbb{R})$

3.23. We may also consider Clifford algebras over *complex* vector spaces provided with nondegenerate quadratic forms. Let C'^n be the Clifford algebra of $\mathbb{C}^n$ provided with the quadratic form $\sum_{i=1}^{n} (x_i)^2$. Then by 3.10, we have $C'^n = \underbrace{C'^1 \hat{\otimes} \cdots \hat{\otimes} C'^1}_{n}$ where $C'^1 = \mathbb{C} \otimes_{\mathbb{R}} \mathbb{C} = \mathbb{C} \oplus \mathbb{C}$. Therefore $C'^n \approx C^{n,0} \otimes_{\mathbb{R}} \mathbb{C} \approx C^{0,n} \otimes_{\mathbb{R}} \mathbb{C}$. Moreover, the argument used in 3.16 shows that $C'^{n+2} = C'^n \otimes C'^2$ (choose $\varepsilon = ie_1e_2$ to obtain an isomorphism $C(\mathbb{C}^{n+2}) \approx C(\mathbb{C}^n) \otimes C(\mathbb{C}^2)$). Since $C'^2 = C^{0,2} \otimes_{\mathbb{R}} \mathbb{C} \approx M_2(\mathbb{R}) \otimes_{\mathbb{R}} \mathbb{C} \approx M_2(\mathbb{C})$, we have an isomorphism $C'^{n+2} \approx M_2(C'^n)$. Therefore $C'^{2p} \approx M_{2^p}(\mathbb{C})$ and $C'^{2p+1} \approx M_{2^p}(\mathbb{C}) \oplus M_{2^p}(\mathbb{C})$. This "periodicity" of the Clifford algebras C'^n (period 2) may be compared with the "periodicity" of the Clifford algebras $C^{p,0}$ and $C^{0,p}$ (period 8). These "algebraic" periodicities will be used later to prove the "topological" Bott periodicity in both complex and real K-theory (cf. 5.13).

3.24. In 3.22 we completely determined the algebras $C^{p,0}$ and $C^{0,p}$. However, for our purposes, we require still more information. In particular, we need a description of the inclusions $C^{p,0} \subset C^{p+1,0}$ and $C^{0,p} \subset C^{0,p+1}$. For the first lines of the table in 3.22 it is easy: the inclusions

$$\begin{array}{ccccc} C^{0,0} & \subset & C^{1,0} & \subset & C^{2,0} \\ \| & & \| & & \| \\ \mathbb{R} & & \mathbb{C} & & \mathbb{H} \end{array}$$

are the usual inclusions. Also the inclusions

$$\begin{array}{ccccc} C^{0,0} & \subset & C^{0,1} & \subset & C^{0,2} \\ \wr\wr & & \wr\wr & & \wr\wr \\ \mathbb{R} & & \mathbb{R}\oplus\mathbb{R} & & M_2(\mathbb{R}) \end{array}$$

are defined respectively by $a \mapsto (a, a)$ and $(a, b) \mapsto \begin{pmatrix} a & 0 \\ 0 & b \end{pmatrix}$ (note that the isomorphism $\mathbb{R} \oplus \mathbb{R} \mapsto C^{0,1}$ is defined by $(a, b) \mapsto \frac{a+b}{2} + \frac{a-b}{2} e_1$, and that e_1 is represented in $M_2(\mathbb{R})$ by the matrix $\begin{pmatrix} 1 & 0 \\ 0 & -1 \end{pmatrix}$; cf. 3.14).

Moreover, the formulas given in 3.16 show that we have commutative diagrams

$$\begin{array}{ccc} C^{0,p+2} & \longrightarrow & C^{0,p+3} \\ \wr\wr & & \wr\wr \\ C^{p,0} \otimes C^{0,2} & \longrightarrow & C^{p+1,0} \otimes C^{0,2} \end{array} \quad \text{and} \quad \begin{array}{ccc} C^{p+2,0} & \longrightarrow & C^{p+3,0} \\ \wr\wr & & \wr\wr \\ C^{0,p} \otimes C^{2,0} & \longrightarrow & C^{0,p+1} \otimes C^{2,0} \end{array}$$

where the horizontal maps are the desired inclusions. Therefore the inclusions

$$\begin{array}{ccccc} C^{0,2} & \subset & C^{0,3} & \subset & C^{0,4} \\ \wr\wr & & \wr\wr & & \wr\wr \\ M_2(\mathbb{R}) & & M_2(\mathbb{C}) & & M_2(\mathbb{H}) \end{array} \quad \text{and} \quad \begin{array}{ccccc} C^{2,0} & \subset & C^{3,0} & \subset & C^{4,0} \\ \wr\wr & & \wr\wr & & \wr\wr \\ \mathbb{H} & & \mathbb{H}\oplus\mathbb{H} & & M_2(\mathbb{H}) \end{array}$$

are simply the tensor product of the previous inclusions by $M_2(\mathbb{R})$ and $\mathbb{H}$, respectively.

In the same way, the inclusions

$$\begin{array}{ccccc} C^{0,4} & \subset & C^{0,5} & \subset & C^{0,6} \\ \wr\wr & & \wr\wr & & \wr\wr \\ M_2(\mathbb{H}) & & M_2(\mathbb{H})\oplus M_2(\mathbb{H}) & & M_4(\mathbb{H}) \end{array}$$

are the tensor product by $M_2(\mathbb{R})$ of the inclusions

$$\begin{array}{ccccc} C^{2,0} & \subset & C^{3,0} & \subset & C^{4,0} \\ \wr\wr & & \wr\wr & & \wr\wr \\ \mathbb{H} & & \mathbb{H}\oplus\mathbb{H} & & M_2(\mathbb{H}) \end{array}$$

3.25. To proceed any further, we need a precise description of the inclusions

$$\mathbf{H} \subset \mathbf{H} \otimes_{\mathbf{R}} \mathbf{C} \subset \mathbf{H} \otimes_{\mathbf{R}} \mathbf{H}.$$

We interpret $\mathbf{H}$ as the set of quaternions written in the form $q = \alpha + \beta J$, where $\alpha, \beta \in \mathbf{C}$. Then we set ${}^t q = \bar{\alpha} - \bar{\beta} J$, so that $q \mapsto {}^t q$ is an anti-involution of $\mathbf{H}$. We define an algebra homomorphism

$$\varphi\colon \mathbf{H} \otimes_{\mathbf{R}} \mathbf{H} \longrightarrow \operatorname{End}_{\mathbf{R}}(\mathbf{H}) = \mathrm{M}_4(\mathbf{R})$$

by the formula $\varphi(q \otimes q')(v) = q' v\, {}^t q$. This homomorphism sends $\mathbf{H} \otimes_{\mathbf{R}} \mathbf{C}$ into $\operatorname{End}_{\mathbf{C}}(\mathbf{H}) = \mathrm{M}_2(\mathbf{C})$. In fact direct computation shows that the induced homomorphism

$$\varphi'\colon \mathbf{H} \otimes_{\mathbf{R}} \mathbf{C} \longrightarrow \mathrm{M}_2(\mathbf{C})$$

is a $\mathbf{C}$-algebra homomorphism, and that

$$\varphi'(I \otimes 1) = \begin{pmatrix} -i & 0 \\ 0 & i \end{pmatrix}, \quad \varphi'(J \otimes 1) = \begin{pmatrix} 0 & 1 \\ -1 & 0 \end{pmatrix}, \quad \text{and} \quad \varphi'(K \otimes 1) = \begin{pmatrix} 0 & -i \\ -i & 0 \end{pmatrix}.$$

These formulas show that φ' is in fact an isomorphism. Moreover, $\varphi(1 \otimes J)$ is an anti-automorphism of $\mathbf{H}$ regarded as a complex vector space of dimension 2. Since any element of $\mathrm{M}_4(\mathbf{R}) = \operatorname{End}_{\mathbf{R}}(\mathbf{C}^2)$ may be written as the sum of a $\mathbf{C}$-endomorphism and a $\mathbf{C}$-anti-endomorphism, it follows that φ is surjective, hence bijective. In conclusion, the injections $\mathbf{H} \subset \mathbf{H} \otimes_{\mathbf{R}} \mathbf{C} \subset \mathbf{H} \otimes_{\mathbf{R}} \mathbf{H}$ are therefore

$$\mathbf{H} \overset{u}{\subset} \mathrm{M}_2(\mathbf{C}) \overset{v}{\subset} \mathrm{M}_4(\mathbf{R}),$$

where $u(I)$, $u(J)$, and $u(K)$, are the three matrices above, and where v is induced by the inclusion $\operatorname{End}_{\mathbf{C}}(\mathbf{C}^2) \subset \operatorname{End}_{\mathbf{R}}(\mathbf{C}^2)$, i.e. is the tensor product by $\mathrm{M}_2(\mathbf{R})$ of the inclusion of $\mathbf{C}$ in $\mathrm{M}_2(\mathbf{R})$ defined by

$$a + ib \longmapsto \begin{pmatrix} a & -b \\ b & a \end{pmatrix}.$$

3.26. We are now ready to complete our description of the inclusions $C^{p,0} \subset C^{p+1,0}$ and $C^{0,p} \subset C^{0,p+1}$. Since $C^{p+4,0} \approx C^{p,0} \otimes C^{4,0}$, the inclusions

$$\begin{array}{ccccc} C^{4,0} & \subset & C^{5,0} & \subset & C^{6,0} \\ \wr\wr & & \wr\wr & & \wr\wr \\ \mathrm{M}_2(\mathbf{H}) & & \mathrm{M}_4(\mathbf{C}) & & \mathrm{M}_8(\mathbf{R}) \end{array}$$

are simply the tensor product by $M_2(\mathbb{R})$ of the inclusions $\mathbb{H}\subset M_2(\mathbb{C})\subset M_4(\mathbb{R})$ described above. Moreover, the inclusions

$$\begin{array}{ccccc} C^{6,0} & \overset{u'}{\rightleftarrows} & C^{7,0} & \overset{v'}{\rightleftarrows} & C^{8,0} \\ \wr\wr & & \wr\wr & & \wr\wr \\ M_8(\mathbb{R}) & & M_8(\mathbb{R})\oplus M_8(\mathbb{R}) & & M_{16}(\mathbb{R}) \end{array}$$

are the tensor product by $M_2(\mathbb{H})$ of the inclusions

$$\begin{array}{ccccc} C^{2,0} & \subset & C^{3,0} & \subset & C^{4,0} \\ \wr\wr & & \wr\wr & & \wr\wr \\ \mathbb{H} & & \mathbb{H}\oplus\mathbb{H} & & M_2(\mathbb{H}). \end{array}$$

Therefore we have $u'(a)=(a,a)$ and $v'(a,b)=\begin{pmatrix} a & 0 \\ 0 & b \end{pmatrix}$, where a and b are blocks of 8×8 real matrices.

Finally the inclusions

$$\begin{array}{ccccc} C^{0,6} & \subset & C^{0,7} & \subset & C^{0,8} \\ \wr\wr & & \wr\wr & & \wr\wr \\ M_4(\mathbb{H}) & \subset & M_8(\mathbb{C}) & \subset & M_{16}(\mathbb{R}) \end{array}$$

are simply the tensor product by $M_2(\mathbb{R})$ of the inclusions $M_2(\mathbb{H})\subset M_4(\mathbb{C})\subset M_8(\mathbb{R})$.

4. The Functors $K^{p,q}(\mathscr{C})$ and $K^{p,q}(X)$

4.1. Let $\mathscr{C}$ be a Banach category (for example, the category $\mathscr{E}(X)$ of vector bundles with compact base X), and let A be an $\mathbb{R}$-algebra of finite dimension. Let $\mathscr{C}^A$ denote the category whose objects are the pairs (E,ρ), where $E\in \mathrm{Ob}(\mathscr{C})$ and $\rho: A\to \mathrm{End}(E)$ is an $\mathbb{R}$-algebra homomorphism. A morphism from the pair (E,ρ) to the pair (E',ρ'), is a $\mathscr{C}$-morphism $f: E\to E'$ such that $f\cdot\rho(\lambda)=\rho(\lambda)\cdot f$ for each element λ of A. In particular, if A is the Clifford algebra $C^{p,q}$ (resp. $M_n(\mathbb{R})$), we denote the corresponding category $\mathscr{C}^A$ by $\mathscr{C}^{p,q}$ (resp. $\mathscr{C}(n)$). In general, we notice that when $\mathscr{C}$ is a pseudo-abelian Banach category, $\mathscr{C}^A$ is also (I.6.10).

4.2. Example. Let $\mathscr{C}=\mathscr{E}_{\mathbb{R}}(X)$ and $A=\mathbb{C}$. Then $\mathscr{C}^A\approx\mathscr{E}_{\mathbb{C}}(X)$. Similarly if $A=\mathbb{H}$, then $\mathscr{C}^A\sim\mathscr{E}_{\mathbb{H}}(X)$.

4.3. Example. Let $\mathscr{C}=\mathscr{E}(X)$ and $A=\mathbb{R}[x]/x^2$. Then $\mathscr{C}^A$ is isomorphic to the category of vector bundles provided with an endomorphism whose square is 0 (where the morphisms are compatible with the endomorphism).

4.4. Theorem. *Let $\mathscr{C}$ be a pseudo-abelian Banach category (I.6.7). Then the categories $\mathscr{C}$ and $\mathscr{C}(n)$ are equivalent* [Morita equivalence].

Proof. We want to define a category equivalence

$$\varphi : \mathscr{C} \longrightarrow \mathscr{C}(n).$$

For any object F of $\mathscr{C}$ we define $\varphi(F)$ as (E, ρ), where $E = F^n$, and $\rho : \mathrm{M}_n(\mathbb{R}) \to \mathrm{End}(E) = \mathrm{End}(F^n)$ is the homomorphism which associates each matrix $M = (a_{ji})$ with the endomorphism of F^n defined by the matrix (b_{ji}) with $b_{ji} = a_{ji} \cdot \mathrm{Id}_F$. From now on we will simply write a_{ji} instead of b_{ji}. If $f : F \to F'$ is a morphism in $\mathscr{C}$, we define $\varphi(f) : (E, \rho) \to (E', \rho')$ as the $\mathscr{C}(n)$-morphism whose underlying $\mathscr{C}$-morphism is represented by the diagonal matrix

$$\begin{pmatrix} f & 0 & 0 & \\ 0 & f & & 0 \\ & & \ddots & \\ & 0 & & f \end{pmatrix}.$$

If we write any morphism $g : (E, \rho) \to (E', \rho')$ in the matrix form $g = (g_{ji})$, and require that g must commute with the action of $\mathrm{M}_n(\mathbb{R})$, we obtain relations $\sum_{j=1}^{n} \lambda_{kj} g_{ji} = \sum_{j=1}^{n} g_{kj} \lambda_{ji}$, where the scalars λ belong to $\mathbb{R}$. If we choose all except one of the λ to be 0, we see that g must be of the form $\varphi(f)$. Hence the functor is fully faithful.

Now let (E, ρ) be an arbitrary object of $\mathscr{C}(n)$. Let p_i be the diagonal matrix

$$\begin{array}{cc} & \begin{array}{ccccccc} & & & i & & & \end{array} \\ \begin{array}{c} \\ \\ \\ i \\ \\ \\ \\ \end{array} & \left(\begin{array}{ccccccc} 0 & & & & & & \\ & \ddots & & & 0 & & \\ & & 0 & & & & \\ & & & 1 & & & \\ & & & & 0 & & \\ & 0 & & & & \ddots & \\ & & & & & & 0 \end{array}\right) \end{array}$$

and let τ_{ji} be the transposition matrix $(i \neq j)$

$$\begin{array}{cc} & \begin{array}{ccc} j & i & \end{array} \\ \begin{array}{c} j \\ i \\ \end{array} & \begin{pmatrix} 0 & 1 & 0 \\ 1 & 0 & 0 \\ 0 & 0 & 0 \end{pmatrix}. \end{array}$$

Finally, let $E_i = \operatorname{Im}(p_i) = \operatorname{Ker}(1-p_i)$ (which exists since $\mathscr{C}$ is pseudo-abelian). The relations $p_i p_j = 0$ for $i \neq j$ and $\sum_{i=1}^{n} p_i = 1$, imply $E \approx \bigoplus_{i=1}^{n} E_i$. Moreover, the transposition τ_{ji} enables us to identify E_i with E_j, since $p_j = \tau_{ji} p_i \tau_{ji}^{-1}$. If we set $E_1 = F$, we may therefore assume that $E = F^n$, and that the actions of p_i and τ_{ji} are represented by the two types of matrices above. Since these matrices generate $M_n(\mathbb{R})$ as an $\mathbb{R}$-algebra, it follows that the action of $M_n(\mathbb{R})$ on F^n is the one described at the beginning. Hence the functor φ is essentially surjective. □

4.5. Proposition. *Let A and B be finite dimensional $\mathbb{R}$-algebras and let $\mathscr{C}$ be a Banach category. Then the categories $(\mathscr{C}^A)^B$ and $\mathscr{C}^{A \otimes_{\mathbb{R}} B}$ are isomorphic.*

The proof of this proposition is obvious.

4.6. Corollary. *If $\mathscr{C}$ is a pseudo-abelian Banach category, then up to equivalence the category $\mathscr{C}^{p,q}$ depends only on the difference $p-q$ mod 8.*

Proof. If $p-q = p'-q'$ mod 8 we have proved in 3.18, 3.21 and 3.22 that $C^{p,q} \approx M_n(A)$ and that $C^{p',q'} \approx M_{n'}(A)$ for some algebra A. Therefore $\mathscr{C}^{p,q} \sim \mathscr{C}^{M_n(A)} \sim (\mathscr{C}^A)(n) \sim \mathscr{C}^A$ by 4.4 and 4.5 (note that $M_n(A) \approx A \otimes_{\mathbb{R}} M_n(\mathbb{R})$). By the same argument $\mathscr{C}^{p',q'} \sim \mathscr{C}^A$. □

4.7. Proposition. *Let $\mathscr{C}$ be a pseudo-abelian Banach category, and let A and B be $\mathbb{R}$-algebras. Then the categories $\mathscr{C}^{A \oplus B}$ and $\mathscr{C}^A \times \mathscr{C}^B$ are equivalent.*

Proof. If (E, ρ) and (F, σ) are objects of $\mathscr{C}^A$ and $\mathscr{C}^B$ respectively, we define an object $(E \oplus F, \tau)$ of $\mathscr{C}^{A \oplus B}$ by setting $\tau(a, b) = \rho(a) \oplus \sigma(b)$. In fact, this correspondence defines a functor from $\mathscr{C}^A \times \mathscr{C}^B$ to $\mathscr{C}^{A \oplus B}$ which is fully faithful. Now let (G, τ) be an object of $\mathscr{C}^{A \oplus B}$. Then $p = \tau(1, 0)$ and $q = \tau(0, 1)$ are projectors of G such that $p + q = 1$. If we write G as $E \oplus F$ with $E = \operatorname{Im}(p)$ and $G = \operatorname{Im}(q)$, we see that $G \approx \varphi(E, F)$. This shows that φ is essentially surjective. In particular, when $A = B$ the composition $\mathscr{C}^A \times \mathscr{C}^A \sim \mathscr{C}^{A \oplus A} \to \mathscr{C}^A$, where the functor $\mathscr{C}^{A \oplus A} \to \mathscr{C}^A$ is induced by the algebra map $a \mapsto (a, a)$, is simply $(E, F) \mapsto E \oplus F$. □

4.8. Theorem. *Let $\mathscr{C}$ be a pseudo-abelian Banach category, and let $\mathscr{C}'$ and $\mathscr{C}''$ be the categories $\mathscr{C}^{\mathbb{C}}$ and $\mathscr{C}^{\mathbb{H}}$. Then we have the following table of categories:*

p	$\mathscr{C}^{p,0}$	$\mathscr{C}^{p,1}$	$\mathscr{C}^{0,p}$
0	$\mathscr{C}$	$\mathscr{C}\times\mathscr{C}$	$\mathscr{C}$
1	$\mathscr{C}'$	$\mathscr{C}$	$\mathscr{C}\times\mathscr{C}$
2	$\mathscr{C}''$	$\mathscr{C}'$	$\mathscr{C}$
3	$\mathscr{C}''\times\mathscr{C}''$	$\mathscr{C}''$	$\mathscr{C}'$
4	$\mathscr{C}''$	$\mathscr{C}''\times\mathscr{C}''$	$\mathscr{C}''$
5	$\mathscr{C}'$	$\mathscr{C}''$	$\mathscr{C}''\times\mathscr{C}''$
6	$\mathscr{C}$	$\mathscr{C}'$	$\mathscr{C}''$
7	$\mathscr{C}\times\mathscr{C}$	$\mathscr{C}$	$\mathscr{C}'$
8	$\mathscr{C}$	$\mathscr{C}\times\mathscr{C}$	$\mathscr{C}$

Proof. This theorem follows directly from 3.19, 3.22, 4.7, and 4.6. □

4.9. To understand the categories $\mathscr{C}^{p,q}$ better, we need to compare them by the functors, "extension of scalars" and "restriction of scalars". More precisely, let us consider, for example, the functor $\varphi: \mathscr{C}^{0,q+1} \to \mathscr{C}^{0,q}$, induced by the algebra inclusion $C^{0,q} \subset C^{0,q+1}$ (this is a "restriction of scalars" functor). By the periodicity of Clifford algebras, we need only deal with the cases $0 \leqslant q \leqslant 7$.

$q=0$. Up to equivalence, the functor φ coincides with the functor from $\mathscr{C}\times\mathscr{C}$ to $\mathscr{C}$, defined by $(E, F) \mapsto E \oplus F$.

$q=1$. Again up to equivalence, the functor φ from $\mathscr{C}(2)$ to $\mathscr{C}\times\mathscr{C}$, is induced by the injection of $\mathbb{R}\oplus\mathbb{R}$ into $M_2(\mathbb{R})$ described in 3.14. Therefore, by 4.4 and 4.7, this functor is essentially the "diagonal" functor $E \mapsto (E, E)$.

$q=2$. The functor φ is simply the "restriction of scalars" functor from $\mathscr{C}'$ to $\mathscr{C}$ (we ignore the complex structure).

$q=3$. For the functor $\mathscr{C}'' \to \mathscr{C}'$, we similarly deal with the complex structure underlying the quaternionic structure.

$q=4$. The functor $\mathscr{C}''\times\mathscr{C}'' \to \mathscr{C}''$ is simply $(E, F) \mapsto E \oplus F$.

$q=5$. The functor $\mathscr{C}'' \to \mathscr{C}''\times\mathscr{C}''$ is once again the diagonal functor $E \mapsto (E, E)$.

$q=6$. The functor $\mathscr{C}^{0,7} \to \mathscr{C}^{0,6}$ is induced by the inclusion $M_4(\mathbb{H}) \subset M_8(\mathbb{C})$, described in 3.26. Up to equivalence, the functor φ coincides with the functor

$$\mathscr{C}'(2) = \mathscr{C}^{M_2(\mathbb{C})} \to \mathscr{C}^{\mathbb{H}} = \mathscr{C}''$$

induced by the inclusion of $\mathbb{H}$ in $M_2(\mathbb{C})$, which was described in 3.25. If we examine the composition

$$\mathscr{C}' \xrightarrow{\approx} \mathscr{C}'(2) \longrightarrow \mathscr{C}'',$$

we see that the functor $\mathscr{C}' \to \mathscr{C}''$ is defined by $E \mapsto E \oplus E$, where $\mathbf{H}$ acts on $E \oplus E$ via the embedding of $\mathbf{H}$ in $M_2(\mathbf{C})$. Therefore, the functor φ may be interpreted as an "extension of scalars" functor. For instance, if $\mathscr{C}$ is the category of vector bundles, then φ is isomorphic to the functor $E \mapsto \mathbf{H} \otimes_{\mathbf{C}} E$.
$q=7$. For the same reason as in the case $q=6$, the functor $\mathscr{C} \to \mathscr{C}'$ may be interpreted as an "extension of scalars" functor: it is defined by $E \mapsto E \oplus E$, where $\mathbf{C}$ acts on $E \oplus E$ via the embedding of $\mathbf{C}$ in $M_2(\mathbf{R})$. In the category of vector bundles, φ may be defined as $E \mapsto \mathbf{C} \otimes_{\mathbf{R}} E$.

4.10. From the above list, it is clear that the functors we are considering, are Banach functors [more generally, if B is a sub-algebra of A, it can easily be shown that the functor $\mathscr{C}^A \to \mathscr{C}^B$ is a Banach functor]. This enables us to make the following definition:

4.11. Definition. Let $\mathscr{C}$ be a pseudo-abelian Banach category. Then we define the group $K^{p,q}(\mathscr{C})$ as the Grothendieck group of the functor

$$\mathscr{C}^{p,q+1} \longrightarrow \mathscr{C}^{p,q}$$

in the sense of II.2.13. If $\mathscr{C} = \mathscr{E}(X)$ (or more precisely $\mathscr{E}_{\mathbf{R}}(X)$ or $\mathscr{E}_{\mathbf{C}}(X)$) then $K^{p,q}(X)$ (or $K^{p,q}_{\mathbf{R}}(X)$, $K^{p,q}_{\mathbf{C}}(X)$) will denote the group $K^{p,q}(\mathscr{C})$ obtained.

Up to isomorphism, the group $K^{p,q}(\mathscr{C})$ depends only on the difference $p-q$ mod 8 because of these commutative diagrams

$$\begin{array}{ccc} \mathscr{C}^{p,q+1} & \longrightarrow & \mathscr{C}^{p,q} \\ \wr & & \wr \\ \mathscr{C}^{p+1,q+2} & \longrightarrow & \mathscr{C}^{p+1,q+1} \end{array} \qquad \begin{array}{ccc} \mathscr{C}^{p,q+1} & \longrightarrow & \mathscr{C}^{p,q} \\ \wr & & \wr \\ \mathscr{C}^{p,q+9} & \longrightarrow & \mathscr{C}^{p,q+8} \end{array} \qquad \begin{array}{ccc} \mathscr{C}^{p,q+1} & \longrightarrow & \mathscr{C}^{p,q} \\ \wr & & \wr \\ \mathscr{C}^{p+8,q+1} & \longrightarrow & \mathscr{C}^{p+8,q} \end{array}$$

In the case of complex vector bundles (or more generally, "complex" Banach categories) we can show analogously that the group $K^{p,q}$ depends only on the difference $p-q$ mod 2.

4.12. Theorem. *Let $\mathscr{C}$ be a pseudo-abelian Banach category. Then the groups $K^{0,0}(\mathscr{C})$ and $K^{0,1}(\mathscr{C})$ are canonically isomorphic to the groups $K(\mathscr{C})$ and $K^{-1}(\mathscr{C})$, defined in II.1.7 and II.3.3, respectively. Similarly $K^{0,4}(\mathscr{C}) \approx K(\mathscr{C}'')$ and $K^{0,5}(\mathscr{C}) \approx K^{-1}(\mathscr{C}'')$. Therefore $K^{0,0}(X) \approx K(X)$, $K^{0,1}(X) \approx K^{-1}(X)$, $K^{0,4}_{\mathbf{R}}(X) \approx K_{\mathbf{H}}(X)$, and $K^{0,5}_{\mathbf{R}}(X) \approx K^{-1}_{\mathbf{H}}(X)$.*

Proof. Since $K^{0,0}(\mathscr{C})$ is the Grothendieck group of the functor $\mathscr{C} \times \mathscr{C} \to \mathscr{C}$ by 4.9, the exact sequence proved in II.3.22 may be written as

$$\begin{array}{ccccccccc} K^{-1}(\mathscr{C} \times \mathscr{C}) & \longrightarrow & K^{-1}(\mathscr{C}) & \longrightarrow & K^{0,0}(\mathscr{C}) & \longrightarrow & K(\mathscr{C} \times \mathscr{C}) & \longrightarrow & K(\mathscr{C}) \\ \| & & & & & & \| & & \\ K^{-1}(\mathscr{C}) \oplus K^{-1}(\mathscr{C}) & & & & & & K(\mathscr{C}) \oplus K(\mathscr{C}). & & \end{array}$$

This shows that $K(\mathscr{C}) \approx K^{0,0}(\mathscr{C})$ by the isomorphism which associates the class of an object E with the element $d(E_0, E_1, \alpha)$, where $E_0 = (E, 0)$, $E_1 = (0, E)$, and $\alpha: 0 \oplus E \to 0 \oplus E$ is the canonical isomorphism.

An analogous discussion applied to the exact sequence associated with the "diagonal" functor $\mathscr{C} \to \mathscr{C} \times \mathscr{C}$, i.e.

$$\begin{array}{ccccccccc} K^{-1}(\mathscr{C}) & \longrightarrow & K^{-1}(\mathscr{C} \times \mathscr{C}) & \longrightarrow & K^{0,1}(\mathscr{C}) & \longrightarrow & K(\mathscr{C}) & \longrightarrow & K(\mathscr{C} \times \mathscr{C}) \\ & & \| & & & & & & \| \\ & & K^{-1}(\mathscr{C}) \oplus K^{-1}(\mathscr{C}) & & & & & & K(\mathscr{C}) \oplus K(\mathscr{C}), \end{array}$$

shows that $K^{-1}(\mathscr{C}) \approx K^{0,1}(\mathscr{C})$. Here the isomorphism associates $d(E, \alpha)$ with the element $d(E, E, \beta)$, where $\beta: (E, E) \to (E, E)$ is given by $\beta = (\alpha, 1)$.

Finally, $K^{0,4}(\mathscr{C})$ is the Grothendieck group of the functor $\mathscr{C}^{0,5} \to \mathscr{C}^{0,4}$, hence $\mathscr{C}'' \times \mathscr{C}'' \to \mathscr{C}''$ by 4.9. Therefore $K^{0,4}(\mathscr{C}) \approx K(\mathscr{C}'')$. The group $K^{0,5}(\mathscr{C})$ is the Grothendieck group of the functor $\mathscr{C}^{0,6} \to \mathscr{C}^{0,5}$, hence $\mathscr{C}'' \to \mathscr{C}'' \times \mathscr{C}''$ by 4.9 again. From this we deduce the isomorphism $K^{0,5}(\mathscr{C}) \approx K^{-1}(\mathscr{C}'')$. $\square$

4.13. In the case of the Banach category $\mathscr{C} = \mathscr{E}(X)$, it is important to notice that the groups $K^{p,q}(X) = K^{p,q}(\mathscr{C})$ are naturally modules over the commutative ring $K(X)$: if $x = d(E, F, \alpha)$ is an element of $K^{p,q}(X)$, where E and F are $C^{p,q+1}$ bundles, and α is an isomorphism between their underlying $C^{p,q}$ bundles, we define the product of x by $[G] - [G']$ as $d(E \otimes G, F \otimes G, \alpha \otimes 1) - d(E \otimes G', F \otimes G', \alpha \otimes 1)$. With these structures, the explicit formulas given above show that the isomorphisms $K^{0,0}(X) \approx K(X)$, $K^{0,1}(X) \approx K^{-1}(X)$, $K_{\mathbb{R}}^{0,4}(X) \approx K_{\mathbb{H}}(X)$, and $K_{\mathbb{H}}^{0,5}(X) \approx K_{\mathbb{H}}^{-1}(X)$ are in fact $K(X)$-module isomorphisms. It is also possible to define "external" products

$$K^{p,q}(X) \times K(Y) \longrightarrow K^{p,q}(X \times Y).$$

We leave these to the reader.

4.14. Theorem 4.12 gives some credibility to the conjecture that the groups $K^{p,q}(X)$ and $K^{p-q}(X)$ (defined for $p - q \leqslant 0$ in II.4.11) are isomorphic. In fact this will be the objective of the next sections. This will complete the construction of K^n for $n \in \mathbb{Z}$, and prove Bott periodicity in real and complex K-theory, at the same time.

4.15. In order to work with the groups $K^{p,q}(\mathscr{C})$ and $K^{p,q}(X)$, we describe these groups in a slightly different way (to avoid confusion we temporarily use the notation $K'^{p,q}(\mathscr{C})$ and $K'^{p,q}(X)$).

Let $\mathscr{C}$ be a pseudo-abelian Banach category, and let E be a "$C^{p,q}$-module" (i.e. an object of $\mathscr{C}^{p,q}$). We define a gradation of E to be an endomorphism η of E regarded as an object of $\mathscr{C}$, such that

(i) $\eta^2 = 1$, and

(ii) $\eta\rho(e_i) = -\rho(e_i)\eta$ where the e_i are defined as in 3.13, and $\rho: C^{p,q} \to \mathrm{End}(E)$ defines the $C^{p,q}$-structure.

Equivalently, a gradation of E is a $C^{p,q+1}$-structure of E, extending the initial $C^{p,q}$-structure (put $\eta = \rho(e_{p+q+1})$). The term "gradation" is justified by the splitting of E into the direct sum $E_0 \oplus E_1$, where $E_0 = \operatorname{Ker}\left(\frac{1-\eta}{2}\right)$ and $E_1 = \operatorname{Ker}\left(\frac{1+\eta}{2}\right)$. Then the homomorphism $\rho: C^{p,q} \to \operatorname{End}(E_0 \oplus E_1)$ is a homomorphism of $\mathbb{Z}/2$-graded algebras, where $C^{p,q}$ has the $\mathbb{Z}/2$-grading described in 3.6, and $\operatorname{End}(E_0 \oplus E_1)$ has the $\mathbb{Z}/2$-grading $D_0 \oplus D_1$, with D_0 (resp. D_1) the set of "diagonal" matrices

$$\begin{pmatrix} a & 0 \\ 0 & b \end{pmatrix}, \quad \text{for } a \in \operatorname{End}(E_0) \text{ and } b \in \operatorname{End}(E_1)$$

(resp. "codiagonal" matrices

$$\begin{pmatrix} 0 & a \\ b & 0 \end{pmatrix}$$

for $a \in \mathscr{C}(E_1, E_0)$, and $b \in \mathscr{C}(E_0, E_1)$).

We define $K'^{p,q}(\mathscr{C})$ (or $K'^{p,q}(X)$ if $\mathscr{C} = \mathscr{E}(X)$) to be the quotient of the free group generated by the triples (E, η_1, η_2), where E is a $C^{p,q}$-module, and η_1 and η_2 are gradations, by the subgroup generated by the relations

(i) $(E, \eta_1, \eta_2) + (F, \xi_1, \xi_2) = (E \oplus F, \eta_1 \oplus \xi_1, \eta_2 \oplus \xi_2)$, and

(ii) $(E, \eta_1, \eta_2) = 0$ if η_1 is homotopic to η_2 within the gradations of E.

We let $d(E, \eta_1, \eta_2)$ denote the class of the triple (E, η_1, η_2) in the group $K'^{p,q}(\mathscr{C})$.

4.16. Lemma. *We have the relation*

$$d(E, \eta_1, \eta_2) + d(E, \eta_2, \eta_1) = 0.$$

Moreover, two isomorphic triples (in the obvious sense) have the same class in $K'^{p,q}(\mathscr{C})$.

Proof. We have $d(E, \eta_1, \eta_2) + d(E, \eta_2, \eta_1) = d(E \oplus E, \eta_1 \oplus \eta_2, \eta_2 \oplus \eta_1)$. Then the homotopy

$$\eta(\theta) = \begin{pmatrix} \cos\theta & -\sin\theta \\ \sin\theta & \cos\theta \end{pmatrix} \begin{pmatrix} \eta_1 & 0 \\ 0 & \eta_2 \end{pmatrix} \begin{pmatrix} \cos\theta & \sin\theta \\ -\sin\theta & \cos\theta \end{pmatrix}, \quad \text{for } \theta \in [0, \pi/2]$$

shows that $\eta_1 \oplus \eta_2$ and $\eta_2 \oplus \eta_1$ are homotopic. On the other hand, let $f: (E, \eta_1, \eta_2) \to (E', \eta'_1, \eta'_2)$ be an isomorphism. Therefore, f is an isomorphism between E and E' (denoted again by f) such that $\eta'_i = f \cdot \eta_i \cdot f^{-1}$. Now we have the relation

$$\begin{pmatrix} \eta_2 & 0 \\ 0 & \eta'_1 \end{pmatrix} = \begin{pmatrix} 0 & -f^{-1} \\ f & 0 \end{pmatrix} \begin{pmatrix} \eta_1 & 0 \\ 0 & \eta'_2 \end{pmatrix} \begin{pmatrix} 0 & f^{-1} \\ -f & 0 \end{pmatrix}$$

As before, the matrix

$$\begin{pmatrix} 0 & -f^{-1} \\ f & 0 \end{pmatrix}$$

is homotopic to $\mathrm{Id}_{E \oplus E}$ within the automorphisms of $E \oplus E'$, due to the homotopy

$$\begin{pmatrix} \cos\theta & -f^{-1}\sin\theta \\ f\sin\theta & \cos\theta \end{pmatrix}$$

for $\theta \in [0, \pi/2]$. Therefore

$$\begin{aligned} d(E, \eta_1, \eta_2) - d(E', \eta_1', \eta_2') &= d(E, \eta_1, \eta_2) + d(E', \eta_2', \eta_1') \\ &= d(E \oplus E', \eta_1 \oplus \eta_2', \eta_2 \oplus \eta_1') = 0. \quad \square \end{aligned}$$

4.17. Lemma. *Let η_1, η_2, and η_3 be gradations of E. Then we have the relation*

$$d(E, \eta_1, \eta_2) + d(E, \eta_2, \eta_3) = d(E, \eta_1, \eta_3).$$

Proof. From 4.16 we see that

$$\begin{aligned} d(E, \eta_1, \eta_2) + d(E, \eta_2, \eta_3) - d(E, \eta_1, \eta_3) &= \\ = d(E \oplus E \oplus E, \eta_1 \oplus \eta_2 \oplus \eta_3, \eta_2 \oplus \eta_3 \oplus \eta_1). \end{aligned}$$

We also have the identity $\eta_2 \oplus \eta_3 \oplus \eta_1 = \alpha(\eta_1 \oplus \eta_2 \oplus \eta_3)\alpha^{-1}$, where α is the automorphism of $E \oplus E \oplus E$ represented by the matrix

$$\begin{pmatrix} 0 & 1 & 0 \\ 0 & 0 & 1 \\ 1 & 0 & 0 \end{pmatrix}$$

Since SO(3) is arcwise connected, and since α may be regarded as an element of SO(3) in an obvious manner, this implies that the gradations $\eta_1 \oplus \eta_2 \oplus \eta_3$ and $\eta_2 \oplus \eta_3 \oplus \eta_1$ are homotopic. Therefore

$$d(E \oplus E \oplus E, \eta_1 \oplus \eta_2 \oplus \eta_3, \eta_2 \oplus \eta_3 \oplus \eta_1) = 0. \quad \square$$

4.18. Lemma. *Let η_1, η_2, η_1', η_2' be gradations of E such that η_i is homotopic to η_i' for $i = 1, 2$. Then*

$$d(E, \eta_1, \eta_2) = d(E, \eta_1', \eta_2').$$

Proof. We have $d(E, \eta_1', \eta_1) = d(E, \eta_2, \eta_2') = 0$. Therefore $d(E, \eta_1', \eta_2') = d(E, \eta_1', \eta_1) + d(E, \eta_1, \eta_2) + d(E, \eta_2, \eta_2') = d(E, \eta_1, \eta_2)$ by 4.17. $\square$

4.19. Proposition. *Every element of $K'^{p,q}(\mathscr{C})$ may be written in the form $d(E, \eta_1, \eta_2)$. The identity $d(E, \eta_1, \eta_2) = d(E', \eta'_1, \eta'_2)$ is equivalent to the existence of a triple (T, ζ, ζ), such that the gradations $\eta_1 \oplus \eta'_2 \oplus \zeta$ and $\eta_2 \oplus \eta'_1 \oplus \zeta$ are homotopic.*

Proof. The first assertion follows directly from 4.16. To prove the second, we introduce an auxiliary set, $K''^{p,q}(\mathscr{C})$, which is the quotient of the set of triples (E, η_1, η_2), by the equivalence relation $(E, \eta_1, \eta_2) \sim (E', \eta'_1, \eta'_2) \Leftrightarrow \exists (T, \zeta, \zeta)$ such that $\eta_1 \oplus \eta'_2 \oplus \zeta$ and $\eta_2 \oplus \eta'_1 \oplus \zeta$ are homotopic. The verification that this is an equivalence relation uses the techniques of the previous lemmas. Now $K''^{p,q}(\mathscr{C})$ is an abelian group with respect to the sum of triples.

We define a homomorphism $K'^{p,q}(\mathscr{C}) \to K''^{p,q}(\mathscr{C})$, by $d(E, \eta_1, \eta_2) \mapsto [E, \eta_1, \eta_2]$, where $[E, \eta_1, \eta_2]$ is the class of the triple (E, η_1, η_2); this is well-defined because $(E, \eta_1, \eta_2) \sim 0$ if η_1 is homotopic to η_2. Conversely, we can define a homomorphism $K''^{p,q}(\mathscr{C}) \to K'^{p,q}(\mathscr{C})$ by $[E, \eta_1, \eta_2] \mapsto d(E, \eta_1, \eta_2)$. This is well defined by 4.17 and 4.18. Obviously these two homomorphisms are inverse to each other. $\square$

4.20. Corollary. *We have $d(E, \eta_1, \eta_2) = 0$ in the group $K'^{p,q}(\mathscr{C})$ if and only if there exists a triple (F, ζ, ζ) such that $\eta_1 \oplus \zeta$ is homotopic to $\eta_2 \oplus \zeta$.*

4.21. Lemma. *Let E be a $C^{p,q}$-module, and let* Grad(E) *be the space of gradations of E (provided with the topology induced by* End(E)*). If $\eta: I \to$* Grad(E) *is a continuous map, then there exists a continuous map $\beta: I \to$* Aut(E) *such that $\beta(0) = 1$ and $\eta(t) = \beta(t)\eta(0)\beta(t)^{-1}$.*

Proof. For each pair $(t, u) \in I \times I$, the endomorphism $\beta(t, u) = \dfrac{1 + \eta(t)\eta(u)}{2}$ is a $C^{p,q}$-module endomorphism, which is Id_E for $t = u$, and such that

$$\beta(t, u)\eta(u) = \eta(t)\beta(t, u).$$

Since End(E) is a Banach algebra, there exists $\varepsilon > 0$ such that $\beta(t, u)$ is an automorphism for $|t - u| < \varepsilon$. Let

$$0 = t_0 < t_1 < \cdots < t_n = 1$$

be a partition of the interval $[0, 1]$ such that $|t_{i+1} - t_i| < \varepsilon$. Then, for $t \in [t_i, t_{i+1}]$ we define $\beta(t)$ as $\beta(t, t_i)\beta(t_i, t_{i-1}) \ldots \beta(t_1, t_0)$. $\square$

4.22. Theorem. *The groups $K^{p,q}(\mathscr{C})$ and $K'^{p,q}(\mathscr{C})$ are naturally isomorphic. Hence the groups $K^{p,q}(X)$ and $K'^{p,q}(X)$ are naturally isomorphic.*

Proof. Let $d(E, F, \alpha)$ be an element of $K^{p,q}(\mathscr{C}) = K(\varphi)$, where φ is the functor $\mathscr{C}^{p,q+1} \to \mathscr{C}^{p,q}$. Thus E and F may be regarded as $C^{p,q}$-modules provided with gradations η_E and η_F respectively, and $\alpha: E \to F$ is an isomorphism on the underlying $C^{p,q}$-modules. We associate $d(E, F, \alpha)$ with the class of the triple (E, η_1, η_2), where E is considered as a $C^{p,q}$-module, $\eta_1 = \eta_E$ and $\eta_2 = \alpha^{-1} \cdot \eta_F \cdot \alpha$ (thus η_F induces the gradation η_2 on E via α). If (E, F, α) and (E', F', α') are isomorphic, then the associated triples (E, η_1, η_2) and (E', η'_1, η'_2) are also isomorphic. If $\alpha_0, \alpha_1 : E \to F$ are homotopic, then the associated gradations η_2 are homotopic. Finally, if $E = F$ and $\alpha = \mathrm{Id}$, we have $\eta_1 = \eta_2$. It follows from the definition of $K(\varphi)$ (II.2) and from 4.16, that the correspondence $(E, F, \alpha) \mapsto (E, \eta_1, \eta_2)$ defines a homomorphism from $K^{p,q}(\mathscr{C})$ to $K'^{p,q}(\mathscr{C})$.

Conversely, let $d(E, \eta_1, \eta_2)$ be an element of $K'^{p,q}(\mathscr{C})$. Let E_i be the $C^{p,q+1}$-module (E, η_i), and let $\alpha: E_1 \to E_2$ be the $C^{p,q}$-morphism which is the identity on the underlying $C^{p,q}$-modules. To check that the correspondence $(E, \eta_1, \eta_2) \mapsto (E_1, E_2, \alpha)$ defines a homomorphism $K'^{p,q}(\mathscr{C}) \to K^{p,q}(\mathscr{C})$, we must show that $d(E_1, E_2, \alpha) = 0$ if η_1 and η_2 are two homotopic gradations of E. By Lemma 4.21 there exists a continuous map $\beta: I \to \mathrm{Aut}(E)$, such that $\beta(0) = 1$ and $\beta(1)\eta_1(\beta(1))^{-1} = \eta_2$. Therefore we have the commutative diagram

$$\begin{array}{ccc} E_1 & \xrightarrow{\alpha} & E_2 \\ \| & & \downarrow{\scriptstyle \beta(1)^{-1}} \\ E_1 & \xrightarrow{\gamma} & E_1, \end{array}$$

where $\beta(1)$ is a $C^{p,q+1}$-module isomorphism (since $\beta(1)\eta_1 = \eta_2\beta(1)$), and where $\gamma = \beta(1)^{-1} \cdot \alpha$. Therefore $d(E_1, E_2, \alpha) = d(E_1, E_1, \gamma) = 0$, since γ is homotopic to Id_{E_1} by the homotopy $t \mapsto \beta(t)^{-1}\alpha = \beta(t)^{-1}$.

We leave to the reader the trivial checking that these two homomorphisms are inverse to each other. □

4.23. *From now on, we identify the groups $K^{p,q}(\mathscr{C})$ and $K'^{p,q}(\mathscr{C})$ by the isomorphism defined in 4.22. We also identify $K^{p,q}(X)$ and $K'^{p,q}(X)$.*

4.24. One of the most useful aspects of the definition of $K^{p,q}(X)$ in terms of gradations, is that it gives interesting "classifying spaces" for the functors $K^{p,q}$. More precisely, we already know that $K^{0,0}(X) \approx [X, \mathbb{Z} \times \mathrm{BGL}(k)]$ and $K^{0,1}(X) \approx [X, \mathrm{GL}(k)]$ (4.12, II.1.33 and II.3.17). By the same method, we can prove that $K_{\mathbb{R}}^{0,4}(X) \approx [X, \mathbb{Z} \times \mathrm{BGL}(\mathbb{H})]$ and $K_{\mathbb{R}}^{0,5}(X) \approx [X, \mathrm{GL}(\mathbb{H})]$. By the periodicity of the groups $K^{p,q}$ (4.11), there remains four cases to consider. However, our method will work for all eight cases simultaneously.

4.25. Let A be any ring with unit, and let (M_r) be a sequence of objects of $\mathscr{P}(A)^0$ [3]. We say that (M_r) is cofinal in $\mathscr{P}(A)^0$ if $M_r \oplus M_s = M_{r+s}$, and if every object of $\mathscr{P}(A)^0$ is a direct factor of some M_r. For example, if $A = \mathbb{R} \oplus \mathbb{R}$, the modules $M_r = \mathbb{R}^r \oplus \mathbb{R}^r$ form a cofinal system, but the modules $M_r = \mathbb{R}^r \oplus 0$ do not (notice that $\mathscr{P}(A) \sim \mathscr{P}(\mathbb{R}) \times \mathscr{P}(\mathbb{R})$; cf. 4.7).

---

[3] $\mathscr{P}(A)^0$ denotes the category of finitely generated projective *left* A-modules.

4.26. Proposition. *Let (M_r) be a cofinal system of objects in $\mathscr{P}(C^{p,q+1})^0$. Then every element of $K^{p,q}(X)$ may be written as $d(T_r, \eta_{(r)}, \eta)$, where $(T_r, \eta_{(r)})$ is the "trivial" $C^{p,q+1}$-vector bundle $X \times M_r$, and η is a gradation of the $C^{p,q}$-bundle underlying T_r. The identity $d(T_r, \eta_{(r)}, \eta) = d(T_r, \eta_{(r)}, \zeta)$ is equivalent to the existence of $s \in \mathbb{N}$ such that $\eta \oplus \eta_{(s)}$ is homotopic to $\zeta \oplus \eta_{(s)}$ within the gradations of $T_r \oplus T_s = T_{r+s}$ (regarded as a $C^{p,q}$-bundle).*

Proof. By 4.8, the category $\mathscr{E}(X)^{p,q+1}$ is equivalent to $\mathscr{E}_{\mathbb{R}}(X)$, $\mathscr{E}_{\mathbb{C}}(X)$, $\mathscr{E}_{\mathbb{H}}(X)$, or the product of two of these categories. Therefore, any $C^{p,q+1}$-bundle E is a direct factor of some $X \times M$, where $M \in Ob\mathscr{P}(C^{p,q+1})^0$. Since M is a direct factor of a M_r for some r, E is a direct factor of $X \times M_r$. Therefore, if $d(E, \varepsilon_1, \varepsilon_2) \in K^{p,q}(X)$, we may write it (after addition with a triple of the form (F, ξ, ξ)) as $d(T_r, \eta_{(r)}, \eta)$ for some r.

Let us assume now that $d(T_r, \eta_{(r)}, \eta) = d(T_r, \eta_{(r)}, \zeta)$. According to 4.19, there exists a triple (T, T, ξ), which we may assume to be of the form $(T_u, T_u, \eta_{(u)})$, such that $\eta_{(r)} \oplus \zeta \oplus \eta_{(u)}$ and $\eta \oplus \eta_{(r)} \oplus \eta_{(u)}$ are homotopic gradations of $T_r \oplus T_r \oplus T_u$. If we put $s = r+u$, it follows that $\zeta \oplus \eta_{(s)}$ and $\eta \oplus \eta_{(s)}$ are homotopic. □

4.27. Theorem. *Let (M_r) be a cofinal system of objects in $\mathscr{P}(C^{p,q+1})^0$ and let* $\mathrm{Grad}^{p,q}(M_r)$ *be the space of gradations of the $C^{p,q}$-module underlying M_r. Then for X a compact space, we have natural isomorphisms*

$$K_{\mathbb{R}}^{p,q}(X) \approx \operatorname{inj\,lim}[X, \mathrm{Grad}^{p,q}(M_r)] \approx [X, \mathrm{Grad}^{p,q}(\mathbb{R})],$$

where $\mathrm{Grad}^{p,q}(\mathbb{R}) \approx \operatorname{inj\,lim} \mathrm{Grad}^{p,q}(M_r)$.

Proof. This theorem is simply a reformulation of 4.26. □

4.28. It is possible to describe the spaces $\mathrm{Grad}^{p,q}(\mathbb{R})$ in more familiar terms. To give an idea of the procedure, we give the details for $p=1$ and $q=0$. The other cases can be dealt with in a similar manner. In this case, the Clifford algebra $C^{p,q+1}$ (resp. $C^{p,q}$) is $\mathrm{M}_2(\mathbb{R})$ (resp. $\mathbb{C}$). We may choose $M_r = \mathbb{R}^{2r} = \mathbb{R}^r \oplus \mathbb{R}^r$, regarded as a $C^{p,q+1}$-module by the automorphisms

$$e_1 = \begin{pmatrix} 0 & 1 \\ -1 & 0 \end{pmatrix}, \quad \text{and} \quad e_2 = \begin{pmatrix} 0 & 1 \\ 1 & 0 \end{pmatrix}.$$

A gradation of M_r, regarded as a $C^{p,q}$-module, is simply an automorphism η of $\mathbb{C}^r$, which is antilinear and involutive (i.e. $\eta^2 = 1$ and $\eta(\lambda x) = \bar{\lambda}\eta(x)$ for $\lambda \in \mathbb{C}$). The group $\mathrm{GL}_r(\mathbb{C})$ acts transitively on $\mathrm{Grad}^{1,0}(M_r)$ by the formula $\alpha \cdot \eta = \alpha\eta\alpha^{-1}$, and the isotropy group of e_2 may be identified with $\mathrm{GL}_r(\mathbb{R})$. It follows that the continuous map $\mathrm{GL}_r(\mathbb{C}) \to \mathrm{Grad}^{1,0}(M_r)$, defined by $\alpha \mapsto \alpha e_2 \alpha^{-1}$ induces a continuous bijection $\bar{\varphi}$ from $\mathrm{GL}_r(\mathbb{C})/\mathrm{GL}_r(\mathbb{R})$ to $\mathrm{Grad}^{1,0}(M_r)$. To prove that $\bar{\varphi}$ is bicontinuous, it suffices to construct a local section $s\colon V \to \mathrm{GL}_r(\mathbb{C})$ to φ in a neighbourhood V of each point $\eta_0 \in \mathrm{Grad}^{1,0}(M_r)$. If $\varphi(\alpha_0) = \eta_0$, we simply define $s(\eta) = (1 + \eta\eta_0)/2$ for $\eta \in V$. Taking the inductive limit, we see finally that the spaces $\mathrm{Grad}^{1,0}(\mathbb{R})$ and

GL(ℂ)/GL(ℝ) are homeomorphic. If we treat the other seven cases in the same way, we obtain the following theorem:

4.29. Theorem. *Let X be a compact space. Then we have natural isomorphisms*

$$K_{\mathbb{R}}^{p,q}(X) \approx [X, \mathrm{Grad}^{p,q}(\mathbb{R})],$$

where the spaces $\mathrm{Grad}^{p,q}(\mathbb{R})$ *depend only on the difference* $p-q \bmod 8$, *and are determined by the table:*

$p-q \bmod 8$	$\mathrm{Grad}^{p,q}(\mathbb{R})$
0	$\mathbb{Z} \times BGL(\mathbb{R}) \sim \mathbb{Z} \times \mathrm{GL}(\mathbb{R})/\mathrm{GL}(\mathbb{R}) \times \mathrm{GL}(\mathbb{R})$
-1	$\mathrm{GL}(\mathbb{R}) \sim \mathrm{GL}(\mathbb{R}) \times \mathrm{GL}(\mathbb{R})/\mathrm{GL}(\mathbb{R})$
-2	$\mathrm{GL}(\mathbb{R})/\mathrm{GL}(\mathbb{C})$
-3	$\mathrm{GL}(\mathbb{C})/\mathrm{GL}(\mathbb{H})$
-4	$\mathbb{Z} \times BGL(\mathbb{H}) \sim \mathbb{Z} \times \mathrm{GL}(\mathbb{H})/\mathrm{GL}(\mathbb{H}) \times \mathrm{GL}(\mathbb{H})$
-5	$\mathrm{GL}(\mathbb{H}) \sim \mathrm{GL}(\mathbb{H}) \times \mathrm{GL}(\mathbb{H})/\mathrm{GL}(\mathbb{H})$
-6	$\mathrm{GL}(\mathbb{H})/\mathrm{GL}(\mathbb{C})$
-7	$\mathrm{GL}(\mathbb{C})/\mathrm{GL}(\mathbb{R})$

In this list we must note that the inclusions $\mathrm{GL}(\mathbb{C}) \subset \mathrm{GL}(\mathbb{R})$ and $\mathrm{GL}(\mathbb{H}) \subset \mathrm{GL}(\mathbb{C})$, are induced by $\mathrm{GL}_n(\mathbb{C}) \subset \mathrm{GL}_{2n}(\mathbb{R})$ and $\mathrm{GL}_n(\mathbb{H}) \subset \mathrm{GL}_{2n}(\mathbb{C})$ (3.25). Another observation is that up to homotopy, we could replace the homogeneous spaces in this table by their orthogonal, unitary or symplectic analogs (this follows from the polar decomposition of automorphisms; cf. Chevalley [1]). Then we obtain, in order, the spaces $\mathbb{Z} \times \mathrm{BO}$, O, O/U, U/Sp, $\mathbb{Z} \times \mathrm{Bsp}$, Sp, Sp/U and U/O.

4.30. *Remark.* In complex K-theory one can also prove (in fact more easily) that $K_{\mathbb{C}}^{p,q}(X) \approx [X, \mathrm{Grad}^{p,q}(\mathbb{C})]$ where $\mathrm{Grad}^{p,q}(\mathbb{C}) \sim \mathbb{Z} \times \mathrm{BGL}(\mathbb{C})$ if $p-q$ is even, and $\mathrm{Grad}^{p,q}(\mathbb{C}) \sim \mathrm{GL}(\mathbb{C})$ if $p-q$ is odd.

5. The Functors $K^{p,q}(X, Y)$ and the Isomorphism t. Periodicity in Real K-Theory

5.1. The definition of the group $K^{p,q}(X)$ in terms of gradations (4.15) may be generalized slightly to a "relative" version $K^{p,q}(X, Y)$, where Y is a closed subspace of a compact space X. More precisely, let us consider the set of triples (E, η_1, η_2), where E is a $C^{p,q}$-vector bundle on X, and η_1 and η_2 are gradations of E such that $\eta_1|_Y=\eta_2|_Y$. Then we define the group $K^{p,q}(X, Y)$ to be the quotient of the free group generated by these triples, by the subgroup generated by the relations:

(i) $(E, \eta_1, \eta_2)+(F, \zeta_1, \zeta_2)=(E\oplus F, \eta_1\oplus\zeta_1, \eta_2\oplus\zeta_2)$.

(ii) $(E, \eta_1, \eta_2)=0$ if the gradations η_1 and η_2 are "homotopic", i.e. if there exists a continuous map $\eta: I\to \mathrm{Grad}(E)$ such that $\eta(0)=\eta_1$, $\eta(1)=\eta_2$ and $\eta(t)|_Y=\eta_1|_Y=\eta_2|_Y$.

5.2. Some of the propositions and lemmas proved in section 4 remain valid for the group $K^{p,q}(X, Y)$. For example, it is easy to see that the tensor product of bundles induces a $K(X)$-module structure on $K^{p,q}(X, Y)$ (compare with 4.13), and that $K^{p,q}(X, Y)$ depends only (up to isomorphism) on the difference $p-q \bmod 8$ (mod 2 in the complex case). The proofs of Lemmas 4.16, 4.17, 4.18, and of Proposition 4.19 and its corollary, also carry through for the relative version. Moreover, the proof of Proposition 4.26 shows that we may write each element of $K^{p,q}(X, Y)$ in the form $d(T_r, \eta_{(r)}, \eta)$, and that $d(T_r, \eta_{(r)}, \eta)=d(T_r, \eta_{(r)}, \zeta)$ if and only if $\eta\oplus\eta_{(s)}$ is homotopic to $\zeta\oplus\eta_{(s)}$ within the gradations of $T_r\oplus T_s=R_{r+s}$ (where the homotopy is constant over Y). In other words (compare with 4.27), $K^{p,q}(X, Y)$ may be identified with the set of homotopy classes of continuous maps $\check{\eta}: X\to \mathrm{Grad}^{p,q}(k)$, $k=\mathbb{R}$ or $\mathbb{C}$, such that $\check{\eta}(Y)=\{e_{p+q+1}\}$ where e_{p+q+1} is the base point of $\mathrm{Grad}^{p,q}(k)$.

5.3. Theorem. *The projection $p: (X, Y)\to (X/Y, \{y\})$ induces an isomorphism*

$$K^{p,q}(X/Y, \{y\}) \longrightarrow K^{p,q}(X, Y)$$

Proof. This follows immediately from the observations made in 5.2. □

5.4. Theorem. *We have the exact sequence*

$$K^{p,q}(X, Y) \xrightarrow{i^*} K^{p,q}(X) \xrightarrow{j^*} K^{p,q}(Y).$$

Moreover, if Y is a retract of X, we have the split exact sequence

$$0 \longrightarrow K^{p,q}(X, Y) \xrightarrow{i^*} K^{p,q}(X) \xrightarrow{j^*} K^{p,q}(Y) \longrightarrow 0.$$

Proof. It is clear that the composition $K^{p,q}(X, Y)\xrightarrow{i^*} K^{p,q}(X)\xrightarrow{j^*} K^{p,q}(Y)$ is zero. Now let $d(E, \eta_1, \eta_2)$ be an element of $K^{p,q}(X)$ such that $j^*(d(E, \eta_1, \eta_2))=d(E|_Y, \eta_1|_Y, \eta_2|_Y)=0$. According to 4.20, there exists a triple (F, ζ, ζ) over Y, such that $\eta_1|_Y\oplus\zeta$ and $\eta_2|_Y\oplus\zeta$ are homotopic. Since any $C^{p,q+1}$-bundle is a direct factor of a trivial $C^{p,q+1}$-bundle (compare with the proof of 4.26), we may assume that (F, ζ) is the restriction to Y of a $C^{p,q+1}$-bundle (G, ξ). If we substitute $(E\oplus G, \eta_1\oplus\xi, \eta_2\oplus\xi)$ for (E, η_1, η_2), we may thus assume without loss of generality, that $\eta_1|_Y$ and $\eta_2|_Y$ are homotopic. Hence there exists a continuous map $\eta: I\to \mathrm{Grad}\,(E|_Y)$, such that $\eta(0)=\eta_1|_Y$ and $\eta(1)=\eta_2$. By Lemma 4.21, we may even assume that $\eta(t)=\beta(t)\eta_1\beta(t)^{-1}$, where $\beta: I\to \mathrm{Aut}(E|_Y)$ is a continuous map with $\beta(0)=1$. By 4.8, we may regard E as a real, complex, or quaternionic bundle (or the sum of two of them). Hence, Proposition II.2.24 implies that $\beta(t)=\alpha(t)|_Y$, where $\alpha: I\to \mathrm{Aut}(E)$ is a continuous map with $\alpha(0)=1$. Setting $\eta_2'=\beta(1)\eta_1\beta(1)^{-1}$, we see that $d(E, \eta_1, \eta_2) = d(E, \eta_1, \eta_2')$, which belongs to the image of i^*, since $\eta_1|_Y = \eta_2'|_Y$.

Let us now assume that Y is a retract of X. Then to prove the split exact sequence, it suffices to show that the map $K^{p,q}(X, Y) \to K^{p,q}(X)$ is injective. Let $d(E, \eta_1, \eta_2)$ be an element of $K^{p,q}(X, Y)$, where we may assume E to be of the form $X \times M_r$ and $\eta_1 = \eta_{(r)}$, by 5.2 and 4.26. Hence η_1 and η_2 may be regarded as continuous maps (also denoted η_1 and η_2 instead of $\check{\eta}_1$ and $\check{\eta}_2$ as in I.1.12) from X to $\mathrm{Grad}^{p,q}(M_r)$ (4.27), such that $\eta_1|_Y = \eta_2|_Y$. Assume now that the image of $d(E, \eta_1, \eta_2)$ in $K^{p,q}(X)$ is 0. Therefore, after stabilization, we can find a continuous map $\eta: X \times I \to \mathrm{Grad}^{p,q}(M_r)$, such that $\eta(x, 0) = \eta_1(x)$ and $\eta(x, 1) = \eta_2(x)$ (4.26). By Lemma 4.21, we can find a continuous map $\beta: X \times I \to \mathrm{Aut}(M_r)$ such that $\beta(x, 0) = 1$ and $\eta_2(x) = \beta(x, 1)\eta_1(x)(\beta(x, 1))^{-1}$. Let $r: X \to Y$ denote a retraction, and $\gamma: X \times I \to \mathrm{Aut}(M_r)$ denote the continuous map defined by $\gamma(x, t) = \beta(r(x), t)$. Then, using the obvious notation, we have these successive identities in $K^{p,q}(X, Y)$:

$$\begin{aligned} d(E, \eta_1(x), \eta_2(x)) &= d(E, \eta_1(x), \beta(x, 1)\eta_1(x)\beta(x, 1)^{-1}) \\ &= d(E, \gamma(x, 1)\eta_1(x)\gamma(x, 1)^{-1}, \beta(x, 1)\eta_1(x)\beta(x, 1)^{-1}) \\ &= d(E, \gamma(x, t)\eta_1(x)\gamma(x, t)^{-1}, \beta(x, t)\eta_1(x)\beta(x, t)^{-1}) \\ &= d(E, \eta_1(x), \eta_1(x)) = 0 \end{aligned}$$

(note that $\gamma(x, 1)\eta_1(x)\gamma(x, 1)^{-1} = \eta_1(x)$ since $\eta_1(x)$ is constant). □

5.5. Corollary. *If X is a compact space and Y is a closed subspace, then $K^{p,q}(X/Y) \approx K^{p,q}(X, Y) \oplus K^{p,q}(P)$ where P is a point.*

5.6. Corollary. *If Y is a retract of X, we have the split exact sequence*

$$0 \longrightarrow K^{p,q}(X/Y, P) \longrightarrow K^{p,q}(X, P) \longrightarrow K^{p,q}(Y, P) \longrightarrow 0.$$

Proof. This follows directly from the commutative diagram

$$\begin{array}{ccccccccc} & & 0 & & 0 & & 0 & & \\ & & \downarrow & & \downarrow & & \downarrow & & \\ 0 & \longrightarrow & K^{p,q}(X/Y, P) & \longrightarrow & K^{p,q}(X, P) & \longrightarrow & K^{p,q}(Y, P) & \longrightarrow & 0 \\ & & \approx\downarrow & & \downarrow & & \downarrow & & \\ 0 & \longrightarrow & K^{p,q}(X, Y) & \longrightarrow & K^{p,q}(X) & \longrightarrow & K^{p,q}(Y) & \longrightarrow & 0 \\ & & \downarrow & & \downarrow & & \downarrow & & \\ & & 0 & & K^{p,q}(P) & \xrightarrow{\approx} & K^{p,q}(P) & \longrightarrow & 0 \\ & & & & \downarrow & & \downarrow & & \\ & & & & 0 & & 0 & & \end{array}$$

□

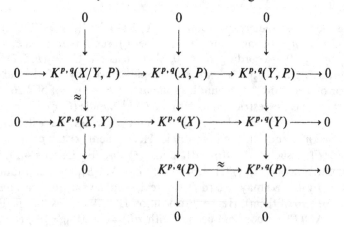

5.7. Proposition. *The groups* $K^{0,0}(X, Y), K^{0,1}(X, Y)$ *(resp.* $K_{\mathbf{R}}^{0,4}(X, Y), K_{\mathbf{R}}^{0,5}(X, Y)$*) are canonically isomorphic to* $K(X, Y)$, $K^{-1}(X, Y)$ *(resp.* $K_{\mathbf{H}}(X, Y)$, $K_{\mathbf{H}}^{-1}(X, Y)$*) as* $K(X)$*-modules (resp.* $K_{\mathbf{R}}(X)$*-modules).*

Proof. We define a homomorphism

$$g: K^{0,0}(X, Y) \longrightarrow K(X, Y)$$

in the following way: $g(d(E, \eta_1, \eta_2)) = d(E_1^0, E_2^0, \alpha)$ where $E_i^0 = \operatorname{Ker}\left(\frac{1-\eta_i}{2}\right)$, and $\alpha: E_1^0|_Y \to E_2^0|_Y$ is the identification isomorphism (note that $\eta_1|_Y = \eta_2|_Y$). When Y is empty, g is the inverse homomorphism to the isomorphism $K^{0,0}(X) \approx K(X)$ (cf. 4.12). Hence, in this case, g is an isomorphism. The morphism g is also an isomorphism when Y is a retract of X (for example when Y is a point). This follows from the diagram

$$\begin{array}{ccccccccc}
0 & \longrightarrow & K^{0,0}(X, Y) & \longrightarrow & K^{0,0}(X) & \longrightarrow & K^{0,0}(Y) & \longrightarrow & 0 \\
 & & \downarrow & & \downarrow \approx & & \downarrow \approx & & \\
0 & \longrightarrow & K(X, Y) & \longrightarrow & K(X) & \longrightarrow & K(Y) & \longrightarrow & 0
\end{array}$$

where the horizontal sequences are exact (5.4 and II.2.29). Finally, for Y arbitrary, we have the commutative diagram

$$\begin{array}{ccc}
K^{0,0}(X/Y, \{y\}) & \xrightarrow{\approx} & K^{0,0}(X, Y) \\
\downarrow g' & & \downarrow g \\
K(X/Y, \{y\}) & \xrightarrow{\approx} & K(X, Y),
\end{array}$$

where the horizontal arrows are isomorphisms (5.3 and II.2.35). Since $\{y\}$ is a retract of X/Y, g' is also an isomorphism. Hence g is an isomorphism as required. Moreover, from the explicit formula given at the beginning, we see immediately that g is a $K(X)$-module map.

Using the same method, we can also define an isomorphism

$$g^{-1}: K^{0,1}(X, Y) \longrightarrow K^{-1}(X, Y).$$

More precisely, every element of $K^{0,1}(X, Y)$ may be written as $d(E, \varepsilon, \eta_1, \eta_2)$, where (E, ε) is a $C^{0,1}$-bundle (thus ε is an involution on E), and η_1 and η_2 are gradations on E (i.e. involutions such that $\eta_i\varepsilon = -\varepsilon\eta_i$). Now we define $g^{-1}(d(E, \varepsilon, \eta_1, \eta_2)) = d(F, \alpha)$, where $F = \operatorname{Ker}\left[\frac{1-\varepsilon}{2}\right]$ and α is the restriction of $\eta_2\eta_1$ to F (cf. II.3.25). If Y is empty, it is clear that g^{-1} is the inverse homomorphism to the isomorphism defined in 4.12. Therefore, using the same type of argument as before (cf. 5.5), it follows that g^{-1} is an isomorphism in general. Moreover, the explicit formula given shows that g^{-1} is a $K(X)$-module isomorphism. The proof of the $K_{\mathbb{R}}(X)$-module isomorphisms $K_{\mathbb{R}}^{0,4}(X, Y) \approx K_{\mathbb{H}}(X, Y)$ and $K_{\mathbb{R}}^{0,5}(X, Y) \approx K_{\mathbb{H}}^{-1}(X, Y)$ follows the same pattern. □

5.8. If we identify the groups $K(Z, T)$ and $K^{0,0}(Z, T)$ in general, it is easy to give another explicit formula for the product

$$K(X, X') \times K(Y, Y') \longrightarrow K(X \times X', X \times Y' \cup X' \times Y)$$

where X and Y are compact spaces, and X' and Y' are closed subsets of X and Y respectively (cf. II.5.6). For this we set

$$d(E, \varepsilon_1, \varepsilon_2) \cup d(F, \eta_1, \eta_2) = d(E \boxtimes F, \zeta_1, \zeta_2),$$

where $$\zeta_1 = \varepsilon_1 \boxtimes \frac{1+\eta_1}{2} + \varepsilon_2 \boxtimes \frac{1-\eta_1}{2}$$

and $$\zeta_2 = \varepsilon_1 \boxtimes \frac{1+\eta_2}{2} + \varepsilon_2 \boxtimes \frac{1-\eta_2}{2}.$$

These formulas are well-defined since $\zeta_1 = \zeta_2$ if $\varepsilon_1 = \varepsilon_2$ or $\eta_1 = \eta_2$, and since they are compatible with the sum of triples. To verify that these formulas give the "right" cup-product, by II.5.6 we need only check the case where $X' = Y' = \varnothing$. In this case we may take $\varepsilon_1 = -\varepsilon_2 = 1$ and $\eta_1 = -\eta_2 = 1$. Then $\zeta_1 = -\zeta_2 = 1$ (cf. 4.12). Therefore, the obvious diagram

$$\begin{array}{ccc} K^{0,0}(X) \times K^{0,0}(Y) & \longrightarrow & K^{0,0}(X \times Y) \\ \wr\wr & & \wr\wr \\ K(X) \times K(Y) & \longrightarrow & K(X \times Y) \end{array}$$

is commutative.

5.9. For each pair (p, q), we define a fundamental homomorphism

$$t\colon K^{p,q+1}(X, Y) \longrightarrow K^{p,q}(X \times B^1, X \times S^0 \cup Y \times B^1).$$

For this, consider an element $d(E, \eta_1(x), \eta_2(x))$ of $K^{p,q+1}(X, Y)$, where x is some point in X. Let $\pi\colon X \times B^1 \longrightarrow X$ denote the first projection, and identify B^1 with the half-circle $B^1 = \{e^{i\theta}$ where $0 \leqslant \theta \leqslant \pi\}$. Finally let E' be the $C^{p,q}$-bundle underlying E, and let $\varepsilon(x)$ be the gradation of E' making it a $C^{p,q+1}$-bundle. On the vector bundle $\pi^*(E') = E' \times B^1$, we can now define two families of gradations, ζ_1 and ζ_2, by

$$\zeta_i(x, \theta) = \varepsilon(x) \cos\theta + \eta_i(x) \sin\theta.$$

Over $X \times S^0 \cup Y \times B^1$, we have $\zeta_1 = \zeta_2$. On the other hand, the correspondence $(E, \eta_1, \eta_2) \mapsto (\pi^*(E'), \zeta_1, \zeta_2)$ is compatible with the sum of triples, and with the homotopy. Hence the homomorphism t is well defined by the formula above.

5.10. Fundamental theorem. *The homomorphism defined above*

$$t: K^{p,q+1}(X, Y) \longrightarrow K^{p,q}(X \times B^1, X \times S^0 \cup Y \times B^1)$$

is an isomorphism.

The proof of this theorem will take all of section 6 of this chapter and is quite technical. Thus we prefer to first derive some of its interesting consequences, especially real Bott periodicity which is our main objective.

5.11. Theorem. *The groups $K^{p,q+n}(X, Y)$ and $K^{p,q}(X \times B^n, X \times S^{n-1} \cup Y \times B^n)$ are canonically isomorphic.*

Proof. Use Theorem 5.10 and induction on n. □

5.12. Theorem. *The groups $K^{p,q}(X, Y)$ and $K^{p-q}(X, Y)$ are canonically isomorphic for $p \leqslant q$.*

Proof. Since the groups $K^{p,q}(X, Y)$ only depend (up to isomorphism) on the difference $p-q$ (4.11 and 5.2), we may assume $p=0$. In this case 5.7 and 5.11 imply $K^{0,q}(X, Y) \approx K^{0,0}(X \times B^q, X \times S^{q-1} \cup Y \times B^q) \approx K^{-q}(X, Y)$ (II.4.11). □

5.13. Theorem (weak Bott periodicity). *The groups $K_{\mathbb{C}}^{-n}(X, Y)$ (resp. $K_{\mathbb{R}}^{-n}(X, Y)$) are periodic with respect to n, of period 2 (resp. 8).*

Proof. This follows from the result that $K_{\mathbb{C}}^{0,n}(X, Y)$ (resp. $K_{\mathbb{R}}^{0,n}(X, Y)$) is periodic with period 2 (resp. 8) by 5.2 and 4.11. □

5.14. Theorem. *The groups $K_{\mathbb{R}}^{-n}(X, Y)$ and $K_{\mathbb{H}}^{-n-4}(X, Y)$ (resp. $K_{\mathbb{H}}^{-n}(X, Y)$ and $K_{\mathbb{R}}^{-n-4}(X, Y)$) are canonically isomorphic.*

Proof. By 5.11 and 5.13, it suffices to show that $K_{\mathbb{H}}(X, Y) \approx K_{\mathbb{R}}^{-4}(X, Y)$. But $K_{\mathbb{H}}(X, Y) \approx K_{\mathbb{R}}^{0,4}(X, Y) \approx K_{\mathbb{R}}^{0,0}(X \times B^4, X \times S^3 \cup Y \times B^4) \approx K_{\mathbb{R}}^{-4}(X, Y)$ by 5.7 and 5.11 again. □

5.15. Theorem. *Let us define $K^n(X, Y)$ to be $K^{n,0}(X, Y)$ for $n > 0$. Then we have the exact sequence*

$$K^{p-1}(X) \longrightarrow K^{p-1}(Y) \longrightarrow K^p(X, Y) \longrightarrow K^p(X) \longrightarrow K^p(Y)$$

for $p \in \mathbb{Z}$.

Proof. Since the groups $K^n(X, Y)$ are periodic with period 8 (at most), and coincide with $K^{n-8r}(X, Y)$ for $r>0$ big enough, we need only prove the theorem for $p\leqslant 0$; however, this was done in II.4.13. □

5.16. *Remark.* This theorem essentially concludes the plan outlined in II.3.1.

5.17. Theorem (strong Bott periodicity). *The group $K_{\mathbb{R}}^{-8}(P)=K_{\mathbb{R}}(B^8, S^7)$ is isomorphic to $\mathbb{Z}$. The cup-product by a generator induces the real periodicity isomorphism*

$$\beta_{\mathbb{R}}: K_{\mathbb{R}}^{-n}(X, Y) \xrightarrow{\approx} K_{\mathbb{R}}^{-n-8}(X, Y)$$

(cf. II.5.26). *In the same way, the group $K_{\mathbb{C}}^{-2}(P)=K_{\mathbb{C}}(B^2, S^1)\approx\mathbb{Z}$, and the cup-product with a generator induces the complex periodicity isomorphism* (compare with 1.3)

$$K_{\mathbb{C}}^{-n}(X, Y) \longrightarrow K_{\mathbb{C}}^{-n-2}(X, Y).$$

Finally, the isomorphism between $K_{\mathbb{R}}^{-n}(X, Y)$ and $K_{\mathbb{H}}^{-n-4}(X, Y)$ is again induced by the cup-product with a generator of $K_{\mathbb{H}}^{-4}(P)\approx\mathbb{Z}$.

Proof. We only prove the first assertion, since the others can be proved in an analogous way. By the Excision theorem (II.4.15), we may assume that Y is empty, and $n=0$. In this case the composition

$$K_{\mathbb{R}}(X) \xrightarrow{\approx} K_{\mathbb{R}}^{0,0}(X) \xrightarrow{\approx} K_{\mathbb{R}}^{0,8}(X) \xrightarrow{\approx} K_{\mathbb{R}}^{0,0}(X\times B^8, X\times S^7)\approx K_{\mathbb{R}}(X\times B^8, X\times S^7)$$

is a $K_{\mathbb{R}}(X)$-module isomorphism (5.7 and 5.9). Therefore it is defined by the cup-product with the image of the unit element of the ring $K_{\mathbb{R}}(X)$, whose image may be considered to be in $K_{\mathbb{R}}(B^8, S^7)\subset K_{\mathbb{R}}(X\times B^8, X\times S^7)$. □

5.18. Theorem. *We have the exact sequence*

$$K_{\mathbb{R}}^{n-1}(X, Y) \xrightarrow{c} K_{\mathbb{C}}^{n-1}(X, Y) \xrightarrow{\gamma} K_{\mathbb{R}}^{n+1}(X, Y) \xrightarrow{r} K_{\mathbb{R}}^{n}(X, Y) \xrightarrow{c} K_{\mathbb{C}}^{n}(X, Y).$$

In this sequence, the homomorphism $c: K_{\mathbb{R}}^n(X, Y)\to K_{\mathbb{C}}^n(X, Y)$ is induced by the complexification of vector bundles (cf. 2.6). *The homomorphism $K_{\mathbb{R}}^{n+1}(X, Y)\to K_{\mathbb{R}}^n(X, Y)$ is induced by the cup-product with the generator of $K_{\mathbb{R}}^{-1}(P)=\pi_0(\mathrm{GL}(\mathbb{R}))=\mathbb{Z}/2$. Finally, the homomorphism $K_{\mathbb{C}}^{n-1}(X, Y)\to K_{\mathbb{R}}^{n+1}(X, Y)$ is the composition of the realification homomorphism (denoted by r) and the periodicity isomorphism $\beta_{\mathbb{C}}: K_{\mathbb{C}}^{n-1}(X, Y)\to K_{\mathbb{C}}^{n+1}(X, Y)$.*

Proof. Since $K_{\mathbb{R}}^{1,0}(X)=K_{\mathbb{R}}^{1}(X)$ is the Grothendieck group of the functor $\varphi: \mathscr{C}^{1,1}\to\mathscr{C}^{1,0}$ where $\mathscr{C}=\mathscr{E}_{\mathbb{R}}(X)$, we have the exact sequence

$$K^{-1}(\mathscr{C}^{1,1}) \longrightarrow K^{-1}(\mathscr{C}^{1,0}) \longrightarrow K(\varphi) \longrightarrow K(\mathscr{C}^{1,1}) \longrightarrow K(\mathscr{C}^{1,0})$$

(cf. II.3.22), i.e.

$$K_{\mathbf{R}}^{-1}(X) \xrightarrow{\bullet} K_{\mathbf{C}}^{-1}(X) \longrightarrow K_{\mathbf{R}}^{1}(X) \longrightarrow K_{\mathbf{R}}(X) \longrightarrow K_{\mathbf{C}}(X)$$

(note that $\mathscr{C} \sim \mathscr{C}^{1,1}$ by 4.8).

If we apply the Excision theorems (II.2.35, II.4.15, and 5.3), we again obtain an exact sequence if we replace X by the pair (X, Y), or better still, by $(X \times B^p, X \times S^{p-1} \cup Y \times B^p)$. If we choose $p = -n \bmod 8$, we obtain the exact sequence required.

Now all that remains is to determine the homomorphisms of the exact sequence; however, this is a much trickier task to accomplish.

The homomorphism $K_{\mathbf{R}}^n(X, Y) \to K_{\mathbf{C}}^n(X, Y)$ is induced by the functor φ, hence is equal to c by 4.9.

The homomorphism $K_{\mathbf{C}}^{-1}(X) \to K_{\mathbf{R}}^1(X)$ in the exact sequence above may be determined by complexifying the situation. More precisely, if we set $\mathscr{C}' = \mathscr{E}_{\mathbf{R}}(X)^{\mathbf{C}} = \mathscr{E}_{\mathbf{C}}(X)$, we have the commutative diagram (up to isomorphism)

$$\begin{array}{ccc} \mathscr{C}^{1,1} & \longrightarrow & \mathscr{C}^{1,0} \\ \uparrow & & \uparrow \\ \mathscr{C}'^{1,1} & \longrightarrow & \mathscr{C}'^{1,0}. \end{array}$$

By the equivalence $\mathscr{C}'^{1,1} \sim \mathscr{C}'$ and $\mathscr{C}'^{1,0} \sim \mathscr{C}' \times \mathscr{C}'$, which follow from the isomorphisms $C^{1,1} \otimes_{\mathbf{R}} \mathbf{C} \approx \mathrm{M}_2(\mathbf{C})$ and $C^{1,0} \otimes_{\mathbf{R}} \mathbf{C} \approx \mathbf{C} \oplus \mathbf{C}$, the diagram can also be read (up to equivalence) as

$$\begin{array}{ccc} \mathscr{C} & \xrightarrow{\varphi} & \mathscr{C}' \\ \mathscr{R} \uparrow & & \uparrow \theta \\ \mathscr{C}' & \xrightarrow{\varphi'} & \mathscr{C}' \times \mathscr{C}', \end{array}$$

where the functors φ', θ, φ, $\mathscr{R}$ are explicitly described as follows. Each object of $\mathscr{C}'^{1,1}$ may be written as $F = E \oplus E$, where $E \in \mathrm{Ob}(\mathscr{C}')$ and the action of $C^{1,1}$ is given by the automorphisms

$$e_1 = \begin{pmatrix} 0 & 1 \\ -1 & 0 \end{pmatrix} \quad \text{and} \quad e_2 = \begin{pmatrix} 0 & 1 \\ 1 & 0 \end{pmatrix}.$$

Therefore, the functor $\mathscr{C}' \to \mathscr{C}'^{1,1} \to \mathscr{C}'^{1,0} \sim \mathscr{C}'^{0,1}$ acts on an object E by the correspondence $E \mapsto (E \oplus E, e_1, e_2) \mapsto (E \oplus E, ie_1)$. On the other hand, the involutions

$$ie_1 = \begin{pmatrix} 0 & -i \\ i & 0 \end{pmatrix} \quad \text{and} \quad e_1 e_2 = \begin{pmatrix} 1 & 0 \\ 0 & -1 \end{pmatrix},$$

are conjugate by the automorphism

$$\alpha=\begin{pmatrix}1 & i\\ i & 1\end{pmatrix}.$$

Therefore $\varphi'(E)\approx E\oplus E$, and thus φ' is isomorphic to the diagonal functor $\mathscr{C}'\to\mathscr{C}'\times\mathscr{C}'$. A similar computation shows that the functor $\theta\colon\mathscr{C}'\times\mathscr{C}'\to\mathscr{C}'$ is defined by $(E,F)\to E\oplus\overline{F}$, since the functors $\mathscr{C}'\times\mathscr{C}'\to\mathscr{C}'^{0,1}\to\mathscr{C}'^{1,0}\to\mathscr{C}^{1,0}$ are defined by $(E,F)\mapsto(E\oplus F,\varepsilon)\mapsto(E\oplus F,i\varepsilon)\mapsto(E_{\mathbb{R}}\oplus F_{\mathbb{R}},i\varepsilon)$, where

$$\varepsilon=\begin{pmatrix}1 & 0\\ 0 & -1\end{pmatrix},$$

$E_{\mathbb{R}}$ and $F_{\mathbb{R}}$ denote the real bundles underlying E and F respectively, and $i\varepsilon$ denotes the action of $C^{1,0}$ on $(E\oplus F)_{\mathbb{R}}$. Finally, by 4.9 the functor φ (resp. $\mathscr{R}$) is induced by complexification (resp. realification). The category diagram given above implies the following commutative K-group diagram (cf. II.3.21)

$$\begin{array}{ccc}
K_{\mathbb{C}}^{-1}(X)\approx K^{-1}(\mathscr{C}') & \xrightarrow{\partial} & K(\varphi)=K_{\mathbb{R}}^{1}(X)\\
\uparrow u & & \uparrow r\\
K_{\mathbb{C}}^{-1}(X)\oplus K_{\mathbb{C}}^{-1}(X)\approx K^{-1}(\mathscr{C}')\oplus K^{-1}(\mathscr{C}') & \xrightarrow{\partial'} & K(\varphi')=K_{\mathbb{C}}^{1}(X)\approx K_{\mathbb{C}}^{-1}(X),
\end{array}$$

where u is the map $(x,y)\mapsto x+\bar{y}$ (i.e. is induced by θ). If x is an element of $K_{\mathbb{C}}^{-1}(X)$, we can write $x=u(x,0)$. Therefore $\gamma(x)=(\partial u)(x,0)=(r\partial')(x)=(r\beta_{\mathbb{C}})(x)$, since the connecting homomorphism ∂' identifies the first factor $K_{\mathbb{C}}^{-1}(X)$ of the sum $K_{\mathbb{C}}^{-1}(X)\oplus K_{\mathbb{C}}^{-1}(X)$ with $K_{\mathbb{C}}^{1}(X)\approx K_{\mathbb{C}}^{-1}(X)$ (cf. 4.12).

Finally we must determine the homomorphism $\sigma\colon K_{\mathbb{R}}^{n+1}(X)\to K_{\mathbb{R}}^{n}(X)$. Since this is a $K_{\mathbb{R}}(X)$-module homomorphism, we have $\sigma(x)=x\cup\sigma(1)$, where $\sigma(1)$ is the image of the unit element under the homomorphism $\sigma\colon K_{\mathbb{R}}(P)\to K_{\mathbb{R}}^{-1}(P)$, obtained by setting $n=-1$ and $X=P$ (a point). The exact sequence

$$K_{\mathbb{C}}(P)\xrightarrow{r}K_{\mathbb{R}}(P)\xrightarrow{\sigma}K_{\mathbb{R}}^{-1}(P),$$

where r is *not* surjective, shows that $\sigma(1)$ is the nontrivial element of $K_{\mathbb{R}}^{-1}(P)=\mathbb{Z}/2$ (II.3.20). This concludes the proof of Theorem 5.18. □

5.19. Theorem. *The groups $K_{\mathbb{C}}^{-n}(P)$, $K_{\mathbb{R}}^{-n}(P)$, and $K_{\mathbb{H}}^{-n}(P)$, are given by the following* table:

n mod 8	$K_{\mathbb{C}}^{-n}(P)$	$K_{\mathbb{R}}^{-n}(P)$	$K_{\mathbb{H}}^{-n}(P)$
0	$\mathbb{Z}$	$\mathbb{Z}$	$\mathbb{Z}$
1	0	$\mathbb{Z}/2$	0
2	$\mathbb{Z}$	$\mathbb{Z}/2$	0
3	0	0	0
4	$\mathbb{Z}$	$\mathbb{Z}$	$\mathbb{Z}$
5	0	0	$\mathbb{Z}/2$
6	$\mathbb{Z}$	0	$\mathbb{Z}/2$
7	0	0	0

Moreover, if r (resp. c) denotes the realification homomorphism (resp. complexification homomorphism), and if u (resp. v) denotes a generator of $K_{\mathbb{C}}^{-2}(P)$ (resp. a suitable generator of $K_{\mathbb{R}}^{-8}(P)$), we have $u^4 = c(v)$, and $w = r(u^2)$ is a generator of $K_{\mathbb{R}}^{-4}(P)$. If η is a generator of $K_{\mathbb{R}}^{-1}(P)$, then $\eta^2 = r(u)$ is a generator of $K_{\mathbb{R}}^{-2}(P)$, and $w^2 = 4v$.

Proof. Obviously we have $K_{\mathbb{R}}^{-1}(P) = \pi_0(\mathrm{GL}(\mathbb{R})) = \mathbb{Z}/2$, $K_{\mathbb{C}}^{-1}(P) = \pi_0(\mathrm{GL}(\mathbb{C})) = 0$, and $K_{\mathbb{H}}^{-1}(P) = K_{\mathbb{R}}^{-5}(P) = \pi_0(\mathrm{GL}(\mathbb{H})) = 0$. On the other hand, elementary algebraic topology shows that $K_{\mathbb{R}}^{-2}(P) \approx \pi_1(\mathrm{GL}(\mathbb{R})) \approx \mathbb{Z}/2$ with generator, the loop

$$\begin{pmatrix} \cos\theta & -\sin\theta & 0 & \cdots \\ \sin\theta & \cos\theta & 0 & \cdots \\ 0 & 0 & 1 & \cdots \\ \vdots & \vdots & \vdots & \ddots \end{pmatrix}$$

It follows that the realification map $K_{\mathbb{C}}^{-2}(P) \to K_{\mathbb{R}}^{-2}(P)$ is surjective. Moreover, the exact sequence

$$\begin{array}{ccccccccc} K_{\mathbb{C}}^{-3}(P) & \xrightarrow{\gamma} & K_{\mathbb{R}}^{-1}(P) & \xrightarrow{\sigma} & K_{\mathbb{R}}^{-2}(P) & \xrightarrow{c} & K_{\mathbb{C}}^{-2}(P) & \xrightarrow{\gamma} & K_{\mathbb{R}}(P) \\ \| & & \| & & & & \| & & \| \\ 0 & & \mathbb{Z}/2 & & & & \mathbb{Z} & \xrightarrow{2} & \mathbb{Z} \end{array}$$

shows that η^2 is the generator of $K_{\mathbb{R}}^{-2}(P)$ since σ is multiplication by η. We also have $K_{\mathbb{H}}^{-2}(P) = \pi_1(\mathrm{GL}(\mathbb{H})) = 0$.

The exact sequences

$$\begin{array}{ccccccc}
 & & & & \mathbf{Z} & & \mathbf{Z} \\
 & & & & \| & & \| \\
K_{\mathbf{C}}^{-5}(P) & \xrightarrow{\gamma} & K_{\mathbf{R}}^{-3}(P) & \xrightarrow{\sigma} & K_{\mathbf{R}}^{-4}(P) & \xrightarrow{c} & K_{\mathbf{C}}^{-4}(P) \\
\| & & & & & & \\
K_{\mathbf{C}}^{-9}(P) & \xrightarrow{\gamma} & K_{\mathbf{R}}^{-7}(P) & \xrightarrow{\sigma} & K_{\mathbf{R}}^{-8}(P) & \xrightarrow{c} & K_{\mathbf{C}}^{-8}(P) \\
\| & & & & \| & & \| \\
0 & & & & \mathbf{Z} & & \mathbf{Z}
\end{array}$$

show that $K_{\mathbf{R}}^{-3}(P)=K_{\mathbf{R}}^{-7}(P)=0$ since cr is multiplication by 2, and since $K_{\mathbf{R}}^{-4}(P)\approx K_{\mathbf{C}}^{-4}(P)\approx\mathbf{Z}$ (2.7 and 5.14).

On the other hand, u^4 is a generator of $K_{\mathbf{C}}^{-8}(P)$, since the periodicity isomorphism in complex K-theory is defined by the cup-product with u (1.3 and 5.17). The exact sequence

$$\begin{array}{ccccccc}
K_{\mathbf{R}}^{-7}(P) & \longrightarrow & K_{\mathbf{R}}^{-8}(P) & \xrightarrow{c} & K_{\mathbf{C}}^{-8}(P) & \longrightarrow & K_{\mathbf{R}}^{-6}(P) \\
\| & & & & & & \| \\
0 & & & & & & 0
\end{array}$$

shows that $u^4=c(v)$, where v is a generator of $K_{\mathbf{R}}^{-8}(P)$.

The exact sequence

$$\begin{array}{ccccc}
K_{\mathbf{C}}^{-4}(P) & \xrightarrow{r} & K_{\mathbf{R}}^{-4}(P) & \xrightarrow{\sigma} & K_{\mathbf{R}}^{-5}(P) \\
 & & & & \| \\
 & & & & 0
\end{array}$$

shows that $w=r(u^2)$ is a generator of $K_{\mathbf{R}}^{-4}(P)$. Since $c(w)=(cr)(u^2)=u^2+u^2=2u^2$ (2.7), we have $c(w^2)=(c(w))^2=4u^4=4c(v)$. Since c is injective on $K_{\mathbf{R}}^{-8}(P)$, we have $w^2=4v$. $\square$

5.20. *Remark*. The exact sequence

$$\begin{array}{ccccccc}
K_{\mathbf{R}}^{-4}(P) & \xrightarrow{c} & K_{\mathbf{C}}^{-4}(P) & \xrightarrow{\gamma} & K_{\mathbf{R}}^{-2}(P) & \longrightarrow & K_{\mathbf{R}}^{-3}(P) \\
 & & & & \wr\wr & & \| \\
 & & & & \mathbf{Z}/2 & & 0
\end{array}$$

shows that the homomorphism $\mathbf{Z}\approx K_{\mathbf{R}}^{-4}(P)\to K_{\mathbf{C}}^{-4}(P)\approx\mathbf{Z}$ is multiplication by 2. Hence the realification homomorphism $K_{\mathbf{C}}^{-4}(P)\to K_{\mathbf{R}}^{-4}(P)$ is an isomorphism. Since the periodicity isomorphism $K^{-n}\approx K^{-n-8}$ is compatible with realification and complexification, we have analogous results for the groups $K_{\mathbb{R}}^{-8p-4}$, $K_{\mathbb{C}}^{-8p-4}$, etc.

5.21. Naturally Theorem 5.19 has a homotopy version (compare with 2.3). However, we can achieve better results by making the iterated loop spaces of GL(ℝ) and GL(ℍ) explicit. To do this, we must slightly alter the definition of the homomorphism t which was given in 5.10. If ε and η are gradations which anti-commute, we have the identity

$$\varepsilon \cos\theta + \eta \sin\theta = (\cos\theta/2 + \eta\varepsilon \sin\theta/2)\varepsilon(\cos\theta/2 - \eta\varepsilon \sin\theta/2).$$

Using the notation of 5.9, we write $t(d(E, \eta_1, \eta_2)) = d(\pi^*(E'), \xi_1, \xi_2)$, where $\xi_1(x, \theta) = \eta_1(x)$ and

$$\begin{aligned}\xi_2(x, \theta) = &(\cos\theta/2 - \eta_1(x)\varepsilon(x)\sin\theta/2)(\varepsilon(x)\cos\theta + \eta_2(x)\sin\theta)\\ &\times(\cos\theta/2 + \eta_1(x)\varepsilon(x)\sin\theta/2)\end{aligned}$$

(for $0 \leqslant \theta \leqslant \pi$). In this form, we see that the homomorphism t is actually induced by a continuous map

$$T\colon \mathrm{Grad}^{p,q+1}(k) \longrightarrow \Omega\, \mathrm{Grad}^{p,q}(k)$$

($k = \mathbb{R}$ or $\mathbb{C}$; cf. 4.28, 4.29, and 5.2). More precisely, if (M_r) is a cofinal system of $C^{p,q+2} \otimes_{\mathbb{R}} k$-modules, it is also a cofinal system of $C^{p,q+1} \otimes_{\mathbb{R}} k$-modules (denoted M_r^0), as can be seen from the table of Clifford algebras (3.24). Now we define a continuous map from $\mathrm{Grad}(M_r)$ to $\Omega\,\mathrm{Grad}(M_r^0)$ by $\eta \mapsto \xi$, where ξ is the loop

$$\xi(e^{i\theta}) = (\cos\theta/4 - \eta_r\varepsilon \sin\theta/4)(\varepsilon\cos\theta/2 + \eta\sin\theta/2)(\cos\theta/4 + \eta_r\varepsilon\sin\theta/4)$$

for $0 \leqslant \theta \leqslant 2\pi$, and where ε is the last generator of the Clifford algebra $C^{p,q+1}$, which is taken as the base point of $\mathrm{Grad}(M_r^0)$. Since t is an isomorphism (5.10), it follows that T is a weak homotopy equivalence,[4)] since $\mathrm{Grad}^{p,q+1}(k) \sim \text{inj lim}\, \mathrm{Grad}(M_r)$ and $\mathrm{Grad}^{p,q}(k) \sim \text{inj lim}\, \mathrm{Grad}(M_r^0)$. Examining the spaces $\mathrm{Grad}^{p,q}$ as in 4.28 and 4.29, we obtain the following theorem:

---

4) We say that a map $T\colon Z \to Z'$ is a weak homotopy equivalence if it induces a bijection $[X, Z] \approx [X, Z']$ for every compact space X.

* In the situation considered here, the map T is a homotopy equivalence, since all the spaces involved have the homotopy type of cw-complexes (cf Milnor [1]).*

5.22. Theorem (Bott [1]). *We have the following weak homotopy equivalences:*[5)]

$$\begin{array}{lll}
\mathrm{GL}(\mathbb{R})\sim\Omega(\mathbb{Z}\times \mathrm{BGL}(\mathbb{R})) & \text{or} & \mathrm{O}\sim\Omega(\mathbb{Z}\times \mathrm{BO})\\
\mathrm{GL}(\mathbb{R})/\mathrm{GL}(\mathbb{C})\sim\Omega(\mathrm{GL}(\mathbb{R})) & \text{or} & \mathrm{O/U}\sim\Omega(\mathrm{O})\\
\mathrm{GL}(\mathbb{C})/\mathrm{GL}(\mathbb{H})\sim\Omega(\mathrm{GL}(\mathbb{R})/\mathrm{GL}(\mathbb{C})) & \text{or} & \mathrm{U/Sp}\sim\Omega(\mathrm{O/U})\\
\mathbb{Z}\times \mathrm{BGL}(\mathbb{H})\sim\Omega(\mathrm{GL}(\mathbb{C})/\mathrm{GL}(\mathbb{H})) & \text{or} & \mathbb{Z}\times \mathrm{Bsp}\sim\Omega(\mathrm{U/Sp})\\
\mathrm{GL}(\mathbb{H})\sim\Omega(\mathbb{Z}\times \mathrm{BGL}(\mathbb{H})) & \text{or} & \mathrm{Sp}\sim\Omega(\mathbb{Z}\times \mathrm{BSp})\\
\mathrm{GL}(\mathbb{H})/\mathrm{GL}(\mathbb{C})\sim\Omega(\mathrm{GL}(\mathbb{H})) & \text{or} & \mathrm{Sp/U}\sim\Omega(\mathrm{Sp})\\
\mathrm{GL}(\mathbb{C})/\mathrm{GL}(\mathbb{R})\sim\Omega(\mathrm{GL}(\mathbb{H})/\mathrm{GL}(\mathbb{C})) & \text{or} & \mathrm{U/O}\sim\Omega(\mathrm{Sp/U})\\
\mathbb{Z}\times \mathrm{BGL}(\mathbb{R})\sim\Omega(\mathrm{GL}(\mathbb{C})/\mathrm{GL}(\mathbb{R})) & \text{or} & \mathbb{Z}\times \mathrm{BO}\sim\Omega(\mathrm{U/O})\\
\mathrm{GL}(\mathbb{C})\sim\Omega(\mathbb{Z}\times \mathrm{BGL}(\mathbb{C})) & \text{or} & U\sim\Omega(\mathbb{Z}\times \mathrm{BU})\\
\mathbb{Z}\times \mathrm{BGL}(\mathbb{C})\sim\Omega(\mathrm{GL}(\mathbb{C})) & \text{or} & \mathbb{Z}\times \mathrm{BU}\sim\Omega(\mathrm{U})
\end{array}$$

In particular, each space of this list has the same weak homotopy type as its 8th iterated loop space (2nd iterated loop space for the last two lines).

5.23. *Remark.* In the preceding list, the weak homotopy equivalences of the type $G\sim\Omega(\mathbb{Z}\times BG)=\Omega(BG)$, with $G=\mathrm{GL}(\mathbb{R})$, $\mathrm{GL}(\mathbb{C})$, $\mathrm{GL}(\mathbb{H})$ or O, U, Sp can be proved in an "elementary" way. The other equivalences are not trivial.

5.24. *Remark.* It is possible to transform these "stable" homotopy equivalences into "non-stable" isomorphisms of the form $\pi_i(\mathrm{Grad}(M_r))\approx\pi_{i+1}(\mathrm{Grad}(M_r^0))$ for r large enough (cf. I.3.13). For example, if $2p>i+1$, we have $\pi_i(\mathrm{O}(2p)/\mathrm{U}(p))\approx \pi_{i+1}(\mathrm{O}(2p))$, etc.

5.25. Example. For the convenience of the reader, we explicitly describe the weak homotopy equivalence between $\mathrm{Grad}^{0,2}(\mathbb{R})$ and $\Omega(\mathrm{Grad}^{0,1}(\mathbb{R}))$, i.e. $\mathrm{GL}(\mathbb{R})/\mathrm{GL}(\mathbb{C})$ and $\Omega(\mathrm{GL}(\mathbb{R}))$. As a cofinal system of $C^{0,3}$-modules, we choose $M_r=\mathbb{C}^r\oplus\mathbb{C}^r$ (considered as a *real* vector space), and the generators e_1, e_2, and e_3, of the Clifford algebra acting as

$$e_1=\begin{pmatrix}1 & 0\\ 0 & -1\end{pmatrix},\quad e_2=\begin{pmatrix}0 & 1\\ 1 & 0\end{pmatrix},\quad\text{and}\quad e_3=\begin{pmatrix}0 & i\\ -i & 0\end{pmatrix}.$$

Then it is easy to check that a gradation of M_r, considered as a $C^{0,2}$-module, must be of the form

$$\begin{pmatrix}0 & J\\ -J & 0\end{pmatrix}$$

where $J^2=-1$. Therefore, the space $\mathrm{Grad}(M_r)$ may be identified with the space of complex structures on $\mathbb{R}^{2r}$. On the other hand, the space $\mathrm{Grad}(M_r^0)$ may be

[5)] Actually homotopy equivalences by the footnote [4)], p. 159.

identified with the space of automorphisms of $\mathbb{C}^r \oplus \mathbb{C}^r$ (considered as a real vector space), which are written in the form

$$\begin{pmatrix} 0 & \alpha \\ \alpha^{-1} & 0 \end{pmatrix}.$$

Hence $\mathrm{Grad}(M_r^0)$ may be identified with $\mathrm{GL}_{2r}(\mathbb{R})$. If we put $\eta_r = e_3$ and $\eta = \begin{pmatrix} 0 & J \\ -J & 0 \end{pmatrix}$, we see by a short computation, that

$$\xi(e^{i\theta}) = \begin{pmatrix} 0 & e^{-i\theta/2}(\cos\theta/2 + J\sin\theta/2) \\ e^{i\theta/2}(\cos\theta/2 - J\sin\theta/2) & 0 \end{pmatrix}.$$

Therefore, the loop in $\mathrm{GL}_{2r}(\mathbb{R})$ *associated with the complex structure* $J \in \mathrm{GL}_{2r}(\mathbb{R})/\mathrm{GL}_r(\mathbb{C})$ *is defined by*

$$e^{i\theta} \longrightarrow e^{-i\theta/2}(\cos\theta/2 + J\sin\theta/2), \quad \text{for } 0 \leqslant \theta \leqslant 2\pi.$$

Taking the limit, we obtain the homotopy equivalence $\mathrm{GL}(\mathbb{R})/\mathrm{GL}(\mathbb{C}) \sim \Omega(\mathrm{GL}(\mathbb{R}))$ as expected. As an exercise the reader may also make the other nine homotopy equivalences of Theorem 5.22 explicit.

Exercises (Section III.7) 4, 5, 13, 14, 15.

6. Proof of the Fundamental Theorem

6.1. The purpose of this section is to prove Theorem 5.10, from which we derived Bott periodicity in III.5. As in Theorem 1.3, Theorem 5.10 is proved using a general theorem on Banach algebras (6.12; compare with 1.11). In order to do this, we need some preliminary lemmas.

6.2. Let us first show that it suffices to prove Theorem 5.10 for Y empty. In fact, the commutative diagram

$$\begin{array}{ccc} K^{p,q+1}(X, Y) & \xrightarrow{\ t\ } & K^{p,q}(X \times B^1, X \times S^0 \cup Y \times B^1) \\ \wr\wr & & \wr\wr \\ K^{p,q+1}(X/Y, \{y\}) & \xrightarrow{\ t\ } & K^{p,q}(X/Y \times B^1, X/Y \times S^0 \cup \{y\} \times B^1) \end{array}$$

enables us to reduce theorem 5.10 to the case where Y is a point (cf. 5.3). Moreover, when Y is a point we have the commutative diagram

$$\begin{array}{ccccccc}
0 \longrightarrow & K^{p,q+1}(X,Y) & \longrightarrow & K^{p,q+1}(X) & \longrightarrow & K^{p,q+1}(Y) & \longrightarrow 0 \\
& \downarrow & & \downarrow & & \downarrow & \\
0 \to & K^{p,q}(X\times B^1, X\times S^0 \cup Y\times B^1) & \to & K^{p,q}(X\times B^1, X\times S^0) & \to & K^{p,q}(Y\times B^1, Y\times S^0) & \to 0
\end{array}$$

The first sequence is exact by 5.4. The second sequence may also be written (up to isomorphism) in the form

$$0 \longrightarrow K^{p,q}(X\times B^1/X\times S^0 \cup Y\times B^1, *) \longrightarrow K^{p,q}(X\times B^1/X\times S^0, *)$$
$$\longrightarrow K^{p,q}(Y\times B^1/Y\times S^0, *) \longrightarrow 0$$

where $*$ is a point; it is exact by 5.6 since $T = Y\times B^1/Y\times S^0$ is a retract of $Z = X\times B^1/X\times S^0$, and since $X\times B^1/X\times S^0 \cup Y\times B^1 \approx Z/T$. Therefore, it suffices to prove Theorem 5.10 for Y empty.

6.3. If E is a $C^{p,q}$-vector bundle provided with a gradation ε, we again let $\varepsilon(\theta)$ denote the gradation on $p^*(E)$, induced by the projection $p\colon X\times B^1 \to X$. If α is an endomorphism of p^*E, we often emphasize the dependence of α as a function of $e^{i\theta} \in B^1$, by writing $\alpha(\theta)$ instead of α(recall that B^1 is identified with the upper half-circle; cf. 5.9).

Now let (M_r) be a cofinal system of $C^{p,q+2}$-modules (over $k = \mathbb{R}$ or $\mathbb{C}$; cf. 4.25). Let $E_r = X\times M_r$, and let E_r^0 be the $C^{p,q}$-bundle underlying E_r. If $e_1, \ldots, e_{p+q+2}$ are the generators of the Clifford algebra $C^{p,q+2}$, we write $\xi_r(\theta)$ for the gradation of $p^*(E_r^0)$, defined by $\xi_r(\theta) = e_{p+q+1}\cos\theta + e_{p+q+2}\sin\theta$ (we always identify the generators of a Clifford algebra with their action on vector bundles).

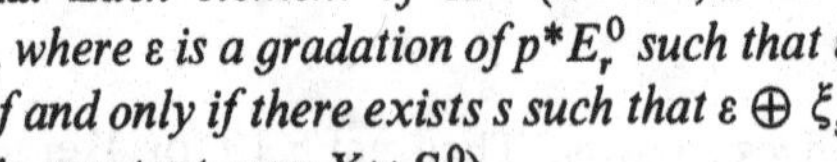

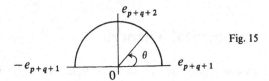

Fig. 15

6.4. Lemma. *Each element of $K^{p,q}(X\times B^1, X\times S^0)$ may be written as $d(p^*E_r^0, \xi_r(\theta), \varepsilon(\theta))$, where ε is a gradation of $p^*E_r^0$ such that $\varepsilon|_{X\times S^0} = \xi_r|_{X\times S^0}$. This element is equal to 0 if and only if there exists s such that $\varepsilon \oplus \xi_s$ is homotopic to ξ_{r+s} (where the homotopy is constant over $X\times S^0$).*

Proof. By 5.2 each element of $K^{p,q}(X\times B^1, X\times S^0)$ may be written as $d(p^*E_r^0, e_{p+q+1}, \varepsilon'(\theta))$, where ε' is a gradation of $p^*E_r^0$ such that $\varepsilon'|_{X\times S^0} = e_{p+q+1}$, or equivalently $\varepsilon'(0) = \varepsilon'(\pi) = e_{p+q+1}$. Applying the inner automorphism defined by $h_r(0) = \cos\theta/2 + e_{p+q+1}e_{p+q+2}\sin\theta/2$, we see that the triple $(p^*E_r^0, e_{p+q+1}, \varepsilon'(\theta))$ is isomorphic to the triple $(p^*E_r^0, e_{p+q+1}\cos\theta + e_{p+q+2}\sin\theta, \varepsilon(\theta))$, where

$$\varepsilon(\theta) = (\cos\theta/2 - e_{p+q+1}e_{p+q+2}\sin\theta/2)\varepsilon'(\theta)(\cos\theta/2 + e_{p+q+1}e_{p+q+2}\sin\theta/2).$$

Therefore, each element of $K^{p,q}(X\times B^1, X\times S^0)$ may be written in the form stated. We remark that the second triple becomes "closer" to the image of t, since $\xi_r(\theta)$ is "$\zeta_1(\theta)$" for some gradation on $p^*E_r^0$ (see the definition of t in 5.9).

To prove the second part of the lemma, let us be more precise by writing e_i^r for the automorphism of E_r or p^*E_r associated with the ith generator e_i of the Clifford algebra $C^{p,q+2}$. By 5.2 again, the equality $d(p^*E_r^0, \xi_r(\theta), \varepsilon(\theta))=0$ implies the existence of an integer s such that $\varepsilon(\theta)\oplus e_{p+q+1}^s$ is homotopic to $\xi_r\oplus e_{p+q+1}^s$, where the homotopy is constant over $X\times S^0$. If we apply the inner automorphism $1\oplus h_s(\theta)$ to this homotopy, we see that $\varepsilon(\theta)\oplus\xi_s$ is homotopic to ξ_{r+s}, where the homotopy is constant over $X\times S^0$. □

6.5. Lemma. *Each element of $K^{p,q}(X\times B^1, X\times S^0)$ may be written as $d(p^*E_r^0, \xi_r(\theta), \varepsilon(\theta))$ where $\varepsilon(\theta)$ may be assumed of the form $f(\theta)e_{p+q+1}f(\theta)^{-1}$, for f an automorphism of p^*E_r such that*

(i) $f(0)=\mathrm{Id}_{E_r^0}$,

and (ii) $e_{p+q+1}\cdot f(\pi)=-f(\pi)\cdot e_{p+q+1}$.

Such an element is zero if and only if there exists an integer s, such that $f\oplus h_s$ is homotopic to h_{r+s} within the automorphisms of $p^*E_{r+s}^0$ which satisfy conditions (i) *and* (ii) *above* (note that the automorphisms $h_s(\theta)$, which were defined in the proof of 6.4, satisfy conditions (i) and (ii)).

Proof. By Lemma 4.21 (where the interval $[0, 1]$ is replaced by $[0, \pi]$) the gradation $\varepsilon(\theta)$ may be written as $f(\theta)e_{p+q+1}(f(\theta))^{-1}$, where $f(0)=1$. Since $\varepsilon(\theta)=\xi_r(\theta)=-e_{p+q+1}$, condition (ii) is automatically satisfied; however, the "lifting" $f(\theta)$ is not unique.

Now let us assume that $d(p^*E_r^0, \xi_r(\theta), \varepsilon(\theta))=0$, where $\varepsilon(\theta)=f(\theta)e_{p+q+1}(f(\theta))^{-1}$. By Lemma 6.4, there exists an integer s such that $\varepsilon\oplus\xi_s$ is homotopic to ξ_{r+s}. Therefore, there exists a continuous map $\eta\colon B^1\times I\to \mathrm{Grad}(E_{r+s}^0)$ such that $\eta(\theta, 0)=\varepsilon(\sigma)\oplus\xi_s(\theta)$, $\eta(\theta, 1)=\xi_{r+s}(\theta)$, $\eta(0, t)=e_{p+q+1}$, and $\eta(\pi, t)=-e_{p+q+1}$. Thus we have the commutative diagram

$$\begin{array}{ccc}
B^1\times\{0, 1\}\cup\{0\}\times I & \xrightarrow{\tilde\eta} & \mathrm{Aut}(E_{r+s}^0) \\
\big\downarrow & \overset{f'}{\nearrow} & \big\downarrow{\scriptstyle\gamma} \\
B^1\times I & \xrightarrow{\eta} & \mathrm{Grad}(E_{r+s}^0),
\end{array}$$

where $\gamma(\alpha)=\alpha e_{p+q+1}\alpha^{-1}$, and $\tilde\eta$ is defined by

$$\tilde\eta(\theta, 0)=f(\theta)\oplus h_s(\theta),$$

$$\tilde\eta(\theta, 1)=h_{r+s}(\theta),$$

and $$\eta(0, t)=1.$$

Since the pair $(B^1 \times I, B^1 \times \{0,1\} \cup \{0\} \times I)$ is homeomorphic to the pair $(B^1 \times I, B^1 \times \{0\})$, Lemma 4.21 applied to a vector bundle over $X \times B^1$, shows the existence of a continuous map $f' \colon B^1 \times I \to \operatorname{Aut}(E^0_{r+s})$, which makes the above diagram commutative. Therefore, we must have the relations

$$f'(\theta, 0) = f(\theta) \oplus h_s(\theta),$$
$$f'(\theta, 1) = h_{r+s}(\theta),$$
$$f'(0, t) = 1,$$

and
$$e_{p+q+1} \cdot f'(\theta, t) = -f'(\theta, t) \cdot e_{p+q+1},$$

and f' realizes the required homotopy between $f \oplus h_s$ and h_{r+s}. □

6.6. We wish to apply the lemma above to a Banach algebra interpretation of the group $K^{p,q}(X \times B^1, X \times S^0)$. From the table of Clifford algebras, we see that the $C^{p,q+2}$ modules $N_r = (C^{p,q+2})^r$ form a cofinal system. If $C(X)$ denotes the algebra of continuous functions on the space X, with values in $k = \mathbb{R}$ or $\mathbb{C}$, then the Banach algebra A which we consider, is the algebra of endomorphisms of $C^{p,q+2} \otimes_k C(X)$, regarded as a left $C^{p,q} \otimes_k C(X)$-module (of rank 4). By I.1.12, a more "geometric" interpretation of A is $\operatorname{Aut}(F_1^0)$, where F_1^0 is the $C^{p,q}$-bundle underlying $X \times N_1$.

We may regard A as a Banach algebra with an involution which is defined by $\alpha \mapsto \bar{\alpha} = e_{p+q+1} \cdot \alpha e_{p+q+1}^{-1}$. Of course this involution induces an involution on the algebra of matrices over A, and hence on $\mathrm{GL}_r(A) = \operatorname{Aut}(F_r^0)$, where F_r^0 is the $C^{p,q}$-bundle underlying $X \times N_r$. We set $\mathrm{GL}_r^-(A) = \{\alpha \in \mathrm{GL}_r(A) \mid \bar{\alpha} = -\alpha\}$. In particular, the element $e = e_{p+q+2} \cdot e_{p+q+1} \in \mathrm{GL}_1^-(A)$. As a base point of $\mathrm{GL}_{2r}^-(A)$, we choose the diagonal matrix

$$e'^r = \begin{pmatrix} e & & & & & & \\ & -e & & & 0 & & \\ & & e & & & & \\ & & & -e & & & \\ & 0 & & & e & & \\ & & & & & -e & \\ & & & & & & \ddots \end{pmatrix}$$

and we set $\mathrm{GL}^-(A) = \operatorname{inj\,lim} \mathrm{GL}_{2r}(A)$ with respect to the maps $\alpha \mapsto \alpha \oplus e'^1$. If $\tilde{N}_r$ denotes the $C^{p,q+1}$-module underlying N_r, provided with the $C^{p,q+2}$-module structure obtained by changing the sign of the action of e_{p+q+2}, then the $C^{p,q+2}$-modules $M_r = N_r \oplus \tilde{N}_r$ also form a cofinal system. Now the Banach algebra $\operatorname{End}(E_r^0)$, where E_r^0 is the $C^{p,q}$-bundle underlying $E_r = X \times M_r$, is also a Banach algebra with involution (i.e. $\alpha \mapsto \bar{\alpha} = e_{p+q+1} \cdot \alpha e_{p+q+1}^{-1}$). The space $\mathrm{GL}_{2r}^-(A)$ may therefore be identified with the space of automorphisms α of E_r^0 such that $\bar{\alpha} = -\alpha$. Note that the action of e_{p+q+2} may be identified with e'^r.

Let $\Omega(\mathrm{GL}_{2r}(A), \mathrm{GL}_{2r}^-(A))$ be the space of paths f in $\mathrm{GL}_{2r}(A)$ parametrized by

$[0, \pi]$, such that $f(0)=1$ and $f(\pi)\in \mathrm{GL}_{2r}^{-}(A)$, provided with the compact-open topology. We set

$$\pi_1(\mathrm{GL}_{2r}(A), \mathrm{GL}_{2r}^{-}(A)) = \pi_0(\Omega(\mathrm{GL}_{2r}(A), \mathrm{GL}_{2r}^{-}(A))$$

and $$\pi_1(\mathrm{GL}(A), \mathrm{GL}^{-}(A)) = \operatorname{inj\,lim} \pi_1(\mathrm{GL}_{2r}(A), \mathrm{GL}_{2r}^{-}(A)),$$

where the inductive limit is taken with respect to the maps $f \mapsto f \oplus h_1$, for $h_1(\theta) = \cos\theta/2 + e'^1 \sin\theta/2$. We set

$$h_r(\theta) = \cos\theta/2 + e'^r \sin\theta/2 = \underbrace{h_1(\theta) \oplus h_1(\theta) \oplus \cdots \oplus h_1(\theta)}_{r}.$$

The matrix direct sum induces a monoid operation on $\pi_1(\mathrm{GL}(A), \mathrm{GL}^{-}(A))$. More precisely, if $\alpha \in \Omega(\mathrm{GL}_{2r}(A), \mathrm{GL}_{2r}^{-}(A))$ and $\beta \in \Omega(\mathrm{GL}_{2s}(A), \mathrm{GL}_{2s}^{-}(A))$, we consider $\alpha \oplus \beta \in \Omega(\mathrm{GL}_{2r+2s}(A), \mathrm{GL}_{2r+2s}^{-}(A))$. Now $\alpha \oplus \beta \oplus h_1 \oplus h_1$ is in the same connected component as $\alpha \oplus h_1 \oplus \beta \oplus h_1$ (apply a permutation of the coordinates, which is homotopic to the identity in $\mathrm{SO}(2r+2s+4)$). Hence, the class of $\alpha \oplus \beta$ in $\pi_1(\mathrm{GL}(A), \mathrm{GL}^{-}(A))$ depends only on the classes of α and β in $\pi_1(\mathrm{GL}(A), \mathrm{GL}^{-}(A))$. The associativity of this operation, and the existence of a zero element (which is the class of h_r), are obvious.

6.7. Proposition. *We have a natural bijection*

$$K^{p,q}(X \times B^1, X \times S^0) \approx \pi_1(\mathrm{GL}(A), \mathrm{GL}^{-}(A)).$$

In particular, the matrix direct sum induces a group operation in $\pi_1(\mathrm{GL}(A), \mathrm{GL}^{-}(A))$.

Proof. Actually, this is a reformulation of Lemma 6.5. More precisely, let us associate a path $f(\theta)$ in $\mathrm{GL}_{2r}(A)$ where $f(0)=1$ and $f(\pi)\in \mathrm{GL}_{2r}^{-}(A)$, with the class of the triple $(p^*E_r^0, \xi_r(\theta), \varepsilon(\theta))$ where $\varepsilon(\theta) = f(\theta)e_{p+q+1}(f(\theta))^{-1}$, identifying $f(\theta)$ with $\hat{f}(\theta)$ and $\varepsilon(\theta)$ with $\hat{\varepsilon}(\theta)$ (cf. I.1.12). Since $f(\pi)\in \mathrm{GL}_{2r}^{-}(A)$, the triple has a well-defined class in the group $K^{p,q}(X \times B^1, X \times S^0)$, which depends only on the class of f in the set $\pi_1(\mathrm{GL}(A), \mathrm{GL}^{-}(A))$. By 6.5 the map we have just defined from $\pi_1(\mathrm{GL}(A), \mathrm{GL}^{-}(A))$ to $K^{p,q}(X \times B^1, X \times S^0)$, is a monoid homomorphism which is surjective and has kernel 0. Hence $\pi_1(\mathrm{GL}(A), \mathrm{GL}^{-}(A))$ is a group, and the homomorphism is an isomorphism. □

6.8. The group $K^{p,q+1}(X)$ may also be identified with an invariant of the Banach algebra with involution A. Let G_r be the space of gradations of E_r^1, where E_r^1 is the $C^{p,q+1}$-bundle underlying $X \times M_r$ (cf. 6.6). Then G_r is the set of elements η in $\mathrm{End}_{C(X)}(M_r \otimes_k C(X))$, such that $\eta^2 = 1$ and $\eta e_i = -e_i \eta$ for $i = 1, \ldots, p+q+1$. In other words, the elements η are antilinear automorphisms of A^{2r} (i.e. $\eta(\lambda x) = \bar{\lambda}\eta(x)$ for $\lambda \in A$ and $x \in A^{2r}$), which anticommute with the action of the e_i, for $i = 1, \ldots, p+q$ (i.e. commute with the action of the $e_i e_{p+q+1} \in GL_{2r}^{-}(A)$). For example

$e_{p+q+2} \in G_r$, and if we choose this element as base point, we set $G = \text{inj lim } G_r$. By 4.27 and 4.29, $K^{p,q+1}(X) \approx \pi_0(G)$.

In order to work "inside" $\mathrm{GL}_{2r}(A)$ and $\mathrm{GL}(A)$, it is more convenient to consider a space $I_r(A) \subset \mathrm{GL}_{2r}(A)$, which is homeomorphic to G_r. $I_r(A)$ is the space of matrices $g \in \mathrm{GL}_{2r}(A)$, such that $\bar{g} = -g$ and $g^2 = -1$. The map $g \mapsto g \cdot e_{p+q+1}$ defines a homeomorphism between $I_r(A)$ and G_r. Therefore $G \approx \text{inj lim } I_r(A)$, where the limit is taken with respect to the maps $g \mapsto g \oplus e'^1$ (where $e'^1 = e_{p+q+2} \cdot e_{p+q+1}$ is acting on A^2; cf. 6.6). From these remarks, we arrive at the following proposition, analogous to 6.7:

6.9. Proposition. *The map defined above is an isomorphism*

$$K^{p,q+1}(X) \approx \pi_0(I(A)).$$

In particular, the matrix direct sum induces a group operation in $\pi_0(I(A))$.

6.10. The discussion above motivates the following definitions: Let A be a Banach algebra with involution, and provided with an element e such that $\bar{e} = -e$ and $e^2 = -1$. We set $\mathrm{GL}_r^-(A) = \{\alpha \in \mathrm{GL}_r(A) \mid \bar{\alpha} = -\alpha\}$. In $\mathrm{GL}_{2r}^-(A)$ we consider the base point defined by the matrix

$$e'^r = \begin{pmatrix} e & & & & & & \\ & -e & & & 0 & & \\ & & e & & & & \\ & & & -e & & & \\ & 0 & & & e & & \\ & & & & & -e & \\ & & & & & & \ddots \end{pmatrix}$$

Let $\Omega(\mathrm{GL}_{2r}(A), \mathrm{GL}_{2r}^-(A))$ denote the space of paths $f(\theta)$, parametrized by $[0, \pi]$, such that $f(0) = 1$ and $f(\pi) \in \mathrm{GL}_{2r}^-(A)$; let $\Omega(\mathrm{GL}(A), \mathrm{GL}^-(A))$ denote the inductive limit $\text{inj lim } \Omega(\mathrm{GL}_{2r}(A), \mathrm{GL}_{2r}^-(A))$ with respect to the maps $f \mapsto f \oplus h^1$, where $h^1(\theta) = \cos\theta/2 + e'^1 \sin\theta/2$. Then we define $\pi_1(\mathrm{GL}(A), \mathrm{GL}^-(A))$ to be

$$\pi_0(\Omega(\mathrm{GL}(A), \mathrm{GL}^-(A))) = \text{inj lim } \pi_0(\Omega(\mathrm{GL}_{2r}(A), \mathrm{GL}_{2r}^-(A))).$$

In the same way, we may consider the subset $I_r(A)$ of elements g in $\mathrm{GL}_{2r}(A)$, such that $\bar{g} = -g$ and $g^2 = -1$. We define $I(A)$ to be $\text{inj lim } I_r(A)$ with respect to the maps $g \mapsto g \oplus e'^1$.

Finally, the sets $I(A)$ and $\Omega(\mathrm{GL}(A), \mathrm{GL}^-(A))$ are connected by a map

$$W: I(A) \to \Omega(\mathrm{GL}(A), \mathrm{GL}^-(A)),$$

which is defined by $g \mapsto f(\theta) = \cos\theta/2 + g \sin\theta/2$ (in fact the map is actually defined from $I_r(A)$ to $\Omega(\mathrm{GL}_{2r}(A), \mathrm{GL}_{2r}^-(A))$ for each r and thus induced on the inductive limit). The map W induces a map

$$w\colon \pi_0(I(A)) \longrightarrow \pi_1(\mathrm{GL}(A), \mathrm{GL}^-(A)),$$

which turns out to be an isomorphism in general (6.12). According to 6.7 and 6.9, this implies that $K^{p,q+1}(X)$ and $K^{p,q}(X\times B^1, X\times S^0)$ are isomorphic. However, we must check that t coincides with w in a particular case (up to isomorphism):

6.11. Proposition. *Let A be the Banach algebra* $\mathrm{End}(C^{p,q+2}\otimes_k C(X))$, *where* $C^{p,q+2}\otimes_k C(X)$ *is regarded as a* $C^{p,q}\otimes_k C(X)$*-module (or* $\mathrm{End}(F_1^0)$ *using notation of 6.6). Then we have the commutative diagram*

$$\begin{array}{ccc} \pi_0(I(A)) & \xrightarrow{w} & \pi_1(\mathrm{GL}(A), \mathrm{GL}^-(A)) \\ \wr\wr & & \wr\wr \\ K^{p,q+1}(X) & \xrightarrow{t} & K^{p,q}(X\times B^1, X\times S^0) \end{array}$$

where the vertical isomorphisms are defined as in 6.7 and 6.8, and t (resp. w) is defined as in 5.9 (resp. 6.10).

Proof. Let g be an element of $I_{2r}(A)$. The element of $K^{p,q+1}(X)$ associated with it is defined by the triple (E_r, η_1, η_2) where $E_r = X\times M_r$ (notation of 6.6), $\eta_1 = e_{p+q+2}$, and $\eta_2 = g\cdot e_{p+q+1}$ (again we identify α with the $\check{\alpha}$ of I.1.12). Therefore, we have $t(d(E_r, \eta_1, \eta_2)) = d(p^*E_r, \zeta_1(\theta), \zeta_2(\theta))$ where

$$\zeta_i(\theta) = e_{p+q+1}\cos\theta + \eta_i \sin\theta = f_i(\theta) e_{p+q+1}(f_i(\theta))^{-1},$$

and where

$$f_1(\theta) = \cos\theta/2 + e'\sin\theta/2$$

and $$f_2(\theta) = \cos\theta/2 + g\sin\theta/2.$$

By 6.7, this shows that the diagram is commutative. □

After this task of "translation", we have thus reduced Theorem 5.10 to the following general theorem on Banach algebras:

6.12. Theorem (Wood [1]). *Let A be a Banach algebra with involution, provided with an element $e\in A$ such that $e^2 = -1$ and $\bar{e} = -e$. Then the map defined in 6.10*

$$w\colon \pi_0(I(A)) \longrightarrow \pi_1(\mathrm{GL}(A), \mathrm{GL}^-(A))$$

is bijective.

6.13. As in the last part of 6.6, the matrix direct sum induces a monoid operation on the sets $\pi_0(I(A))$ and $\pi_1(\mathrm{GL}(A), \mathrm{GL}^-(A))$. It is clear that w is a monoid homomorphism. We check first that $\pi_0(I(A))$ is an abelian group (the reader who is only interested in the Banach algebra $A=\mathrm{End}(C^{p,q+2}\otimes C(X))$ may omit this verification by 6.9).

For $\alpha, \beta \in I_r(A) \subset \mathrm{GL}_{2r}(A)$, the homotopy

$$\begin{pmatrix} \cos t & -\sin t \\ \sin t & \cos t \end{pmatrix}\begin{pmatrix} \alpha & 0 \\ 0 & \beta \end{pmatrix}\begin{pmatrix} \cos t & \sin t \\ -\sin t & \cos t \end{pmatrix} \quad \text{for } 0 \leqslant t \leqslant \pi/2,$$

shows that $\alpha \oplus \beta$ is homotopic to $\beta \oplus \alpha$ in $I_{2r}(A) \subset \mathrm{GL}_{4r}(A)$. Hence, $\pi_0(I(A))$ is an abelian monoid. Moreover, if $g \in I_r(A)$, and if $e' = e''$, we may write $e' \cdot g \cdot e' \oplus g$ as the product of matrices

$$\frac{1}{2}\begin{pmatrix} 1 & \alpha \\ -\alpha^{-1} & 1 \end{pmatrix}\begin{pmatrix} 0 & e' \\ e' & 0 \end{pmatrix}\begin{pmatrix} 1 & -\alpha \\ \alpha^{-1} & 1 \end{pmatrix}, \quad \alpha = e'g,$$

or equivalently as $\beta \cdot e'' \cdot \beta^{-1}$ where

$$e'' = \begin{pmatrix} 0 & e' \\ e' & 0 \end{pmatrix} \quad \text{and} \quad \beta = \begin{pmatrix} 1 & \alpha \\ -\alpha^{-1} & 1 \end{pmatrix}.$$

Now, in $\mathrm{GL}^+_{4r}(A) = \{\gamma \in \mathrm{GL}_{4r}(A) \mid \bar{\gamma} = \gamma\}$, β may be written as

$$\begin{pmatrix} 1 & 0 \\ -\alpha^{-1} & 2 \end{pmatrix}\begin{pmatrix} 1 & \alpha \\ 0 & 1 \end{pmatrix},$$

which is clearly homotopic to

$$\begin{pmatrix} 1 & 0 \\ 0 & 2 \end{pmatrix}.$$

Therefore $e' \cdot g \cdot e' \oplus g$ is homotopic in $I_{2r}(A)$ to

$$\begin{pmatrix} 1 & 0 \\ 0 & 2 \end{pmatrix}\begin{pmatrix} 0 & e' \\ e' & 0 \end{pmatrix}\begin{pmatrix} 1 & 0 \\ 0 & \frac{1}{2} \end{pmatrix},$$

which is independent of g. Hence $e' \cdot g \cdot e' \oplus g$ is homotopic to $e' \cdot e' \cdot e' \oplus e' = (-e') \oplus e'$, which is e'^{2r} up to a permutation of the coordinates. This shows that the class of $e' \cdot g \cdot e'$ is opposite to the class of g in the abelian monoid $\pi_0(I(A))$.

Using the lemmas that follow, we will prove that w is surjective and $\mathrm{Ker}\, w = 0$. This will show that $\pi_1(\mathrm{GL}(A), \mathrm{GL}^-(A))$ is also an abelian group, and that w is an abelian group isomorphism.

6.14. To prove Theorem 6.12, we must interpret the space $\Omega(\mathrm{GL}(A), \mathrm{GL}^-(A))$ in a slightly different way. By making the parameter change $\varphi = \theta/2$, we may assume that the paths are parametrized by $[0, \pi/2]$. The path $h(\varphi)=f(2\varphi)$ where $\varphi \in [0,\pi/2]$, may be uniquely extended to a periodic map on $\mathbb{R}$, with period 2π, via the formulas

$$h(\varphi)=f(2\varphi) \quad \text{for } 0\leqslant\varphi\leqslant\pi/2,$$
$$h(\varphi+\pi)=-h(\varphi) \quad \text{for } 0\leqslant\varphi\leqslant\pi/2,$$

and $$h(-\varphi)=\overline{h(\varphi)}.$$

In general, a continuous function $\alpha\colon \mathbb{R} \to \mathrm{GL}(A)$ is called *adapted* if it is periodic with period 2π, and if

$$\alpha(0)=1,$$
$$\alpha(\varphi+\pi)=-\alpha(\varphi),$$

and $$\alpha(-\varphi)=\overline{\alpha(\varphi)}.$$

If α_0 and α_1 are adapted functions, an adapted homotopy between them is a continuous map

$$\alpha\colon \mathbb{R}\times I \longrightarrow \mathrm{GL}(A),$$

such that $\alpha(\varphi, 0)=\alpha_0(\varphi)$, $\alpha(\varphi, 1)=\alpha_1(\varphi)$, and such that the function $\varphi \mapsto \alpha(\varphi, t)$ is adapted for each $t\in I$. Therefore we obtain a bijection between $\pi_1(\mathrm{GL}(A), \mathrm{GL}^-(A))$ and the homotopy classes of adapted functions between $\mathbb{R}$ and $\mathrm{GL}(A)$. Under this identification, the adapted function associated with $g\in I(A)$ is $\varphi \to \alpha(\varphi)=\cos\varphi+g\sin\varphi$ ($\varphi\in\mathbb{R}$). If we let $\zeta=\cos\varphi+e'\sin\varphi$, then the function above may also be written as

$$\alpha(\varphi)=\frac{1+ge'}{2}\zeta^{-1}+\frac{1-ge'}{2}\zeta$$

(again this expression is meaningful in $\mathrm{GL}_{2r}(A)$ for r large enough; the stabilization is given by the map $\alpha\mapsto\alpha\oplus\zeta_1$). The main idea in the proof of the Theorem 6.12, is to show that each adapted function is homotopic to a function of the type above.

6.15. Definition. An adapted function $\alpha\colon \mathbb{R}\to \mathrm{GL}(A)$ is called *Laurentian* (resp. *quasi-polynomial*, resp. *quasi-affine*) if it can be written as $\sum_{n=-N}^{+N} a_n\zeta^{2n-1}$ (resp. $\sum_{n=0}^{N} a_n\zeta^{2n-1}$, resp. $a_0\zeta^{-1}+a_1\zeta$) in $\mathrm{GL}_{2r}(A)$ for r large enough, with $\bar{a}_n=a_n$. If α_0 and α_1 are adapted functions, a homotopy between them is called Laurentian (resp. quasi-polynomial, resp. quasi-affine) if it can be written as $\sum_{n=-N}^{+N} a_n(t)\zeta^{2n-1}$ (resp. $\sum_{n=0}^{N} a_n(t)\zeta^{2n-1}$, resp. $a_0\zeta^{-1}+a_1\zeta$), where $\overline{a_n(t)}=a_n(t)$ is a continuous function of t. We let $\pi_1^L=\pi_1^L(\mathrm{GL}(A), \mathrm{GL}^-(A))$ (resp. $\pi_1^P=\pi_1^P(\mathrm{GL}(A), \mathrm{GL}^-(A))$, resp. $\pi_1^A=\pi_1^A(\mathrm{GL}(A), \mathrm{GL}^-(A))$ denote the set of homotopy classes of adapted functions which are Laurentian (resp. quasi-polynomial, resp. quasi-affine).

6.16. The sets π_1^L, π_1^P, π_1^A are abelian monoids (with respect to the matrix direct sum). In fact, the proof of Theorem 6.12 will show that they are abelian groups. If α is a Laurentian (resp. quasi-polynomial, resp. quasi-affine) function, we let $d_L(\alpha)$ (resp. $d_P(\alpha)$, resp. $d_A(\alpha)$) denote its class in π_1^L (resp. π_1^P, resp. π_1^A). Since every quasi-affine (resp. quasi-polynomial) function is quasi-polynomial (resp. quasi-affine), we have obvious homomorphisms

$$\pi_1^A \xrightarrow{w_3} \pi_1^P \xrightarrow{w_2} \pi_1^L \xrightarrow{w_1} \pi_1(\mathrm{GL}(A), \mathrm{GL}^-(A)).$$

We define a homomorphism

$$w_4 : \pi_0(I(A)) \longrightarrow \pi_1^A$$

by the formula $g \mapsto \alpha(\varphi) = \dfrac{1+ge'}{2}\zeta^{-1} + \dfrac{1-ge'}{2}\zeta$. Thus we obtain a factorization of w as shown in the commutative diagram

$$\begin{array}{ccccc}
\pi_0(I(A)) & \xrightarrow{\quad w \quad} & & & \pi_1(\mathrm{GL}(A), \mathrm{GL}^-(A)) \\
\Big\downarrow{\scriptstyle w_4} & & & & \Big\uparrow{\scriptstyle w_1} \\
\pi_1^A & \xrightarrow{\quad w_3 \quad} & \pi_1^P & \xrightarrow{\quad w_2 \quad} & \pi_1^L .
\end{array}$$

To prove Theorem 6.12, we show that each morphism w_i, for $i=1, 2, 3, 4$, is surjective and has kernel 0. Since all the sets involved are monoids, and since $\pi_0(I(A))$ is a group (6.13), it will follow that each w_i (hence w) is an isomorphism.

6.17. Lemma. *The homomorphism w_1 is surjective and its kernel is* 0.

Proof. We are going to use the same method as in the proof of 1.14. Let $d(\alpha)$ be an element of $\pi_1(\mathrm{GL}(A), \mathrm{GL}^-(A)) = \operatorname{inj\,lim} \pi_1(\mathrm{GL}_{2r}(A), \mathrm{GL}_{2r}^-(A))$. We may assume α to be the class of a map (denoted again α) from $\mathbb{R}$ to $\mathrm{GL}_{2r}(A)$, which is periodic with period 2π. Let

$$a_m = \frac{1}{\pi}\int_0^{2\pi} \alpha(\varphi)\cos(m\varphi)d\varphi \quad \text{for } m>0,$$

$$b_m = \frac{1}{\pi}\int_0^{2\pi} \alpha(\varphi)\sin(m\varphi)d\varphi \quad \text{for } m>0,$$

and
$$\alpha_n(\varphi) = \sum_{m=1}^{n} (1-m/n)(a_m\cos(m\varphi) + b_m\sin(m\varphi)).$$

By Cesaro's theorem, α is the uniform limit of the α_n. Since α is adapted, we have $a_m = b_m = 0$ when m is even. Moreover, $\alpha_n(\varphi+\pi) = -\alpha_n(\varphi)$ and $\alpha_n(-\varphi) = \overline{\alpha_n(\varphi)}$. Since $M_{2r}(A)$ is a Banach algebra, $\alpha_n(\varphi)$ is invertible for n large enough, and $\alpha(\varphi) = \lim_{n\to\infty} \tilde{\alpha}_n(\varphi)$ where $\tilde{\alpha}_n(\varphi) = \alpha_n(\varphi)\alpha_n(0)^{-1}$, which is adapted. Finally α_n and $\tilde{\alpha}_n$ are Laurentian since $\cos(m\varphi) = \dfrac{\zeta^m + \zeta^{-m}}{2}$ and $\sin(m\varphi) = \dfrac{\zeta^m - \zeta^{-m}}{2} \cdot e'^{-1}$ (note that $\zeta = \cos\varphi + e' \sin\varphi$; cf. 6.13). This shows that w_1 is surjective.

The proof that $\operatorname{Ker}(w_1) = 0$ is based on the same principle applied to the Banach algebra $A(I)$. More precisely, let $d_L(\alpha)$ be an element of $\pi_1^L(GL(A), GL^-(A))$, such that $w_1(d_L(\alpha)) = d(\alpha) = 0$. If we consider α to be a map from $\mathbb{R}$ to $M_{2r}(A)$ for r large enough, there exists a homotopy $\beta(u, \varphi)$ from $\alpha(\varphi)$ to $\zeta(\varphi)$. Let $\beta_L(u, \varphi)$ be a Laurentian approximation to this homotopy. When this approximation is close enough, we have $t\beta(u, \varphi) + (1-t)\beta_L(u, \varphi) \in GL_{2r}(A)$ for $t \in [0, 1]$ (where u is the parameter of the homotopy). If we let

$$\tilde{\beta}_L(u, \varphi) = 3u\beta_L(0, \varphi) + (1-3u)\alpha(\varphi) \qquad \text{for } 0 \leqslant u \leqslant \tfrac{1}{3},$$

$$\tilde{\beta}_L(u, \varphi) = \beta_L(3u-1, \varphi) \qquad \text{for } \tfrac{1}{3} \leqslant u \leqslant \tfrac{2}{3},$$

and

$$\tilde{\beta}_L(u, \varphi) = (3-3u)\beta_L(1, \varphi) + (3u-2)\zeta(\varphi) \qquad \text{for } \tfrac{2}{3} \leqslant u \leqslant 1,$$

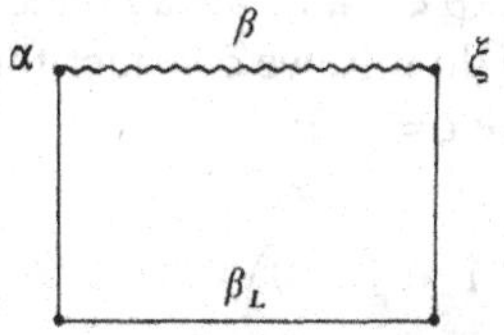

Fig. 16

we obtain a Laurentian homotopy between α and ζ. □

6.18. Lemma. *The homomorphism w_2 is surjective and its kernel is* 0.

Proof. a) *w_2 is surjective.* Let $\alpha = \sum_{n=-N_1}^{N_2} a_{2n-1}\zeta^{2n-1}$ be an adapted function regarded as a map from $\mathbb{R}$ to $GL_{2r}(A)$ (for r large enough), and let $\alpha' = \alpha\zeta^2$. It suffices to prove that $d_L(\alpha) = d_L(\alpha') + d_L(\zeta_{2r}^{-1})$ (we write ζ_{2r} for the adapted function $\cos\varphi + e'^r \sin\varphi$ from $\mathbb{R}$ to $GL_{2r}(A)$, and ζ_{2r}^{-1} for its inverse function; notice that the notation ζ^{-1} has no meaning if we have not fixed r). In order to do this, we consider the following product of matrices in $GL_{4r}(A)$:

$$\begin{pmatrix} \alpha'(\varphi) & 0 \\ 0 & \zeta_{2r} \end{pmatrix} \begin{pmatrix} \cos t & -\sin t \\ \sin t & \cos t \end{pmatrix} \begin{pmatrix} \zeta_{2r}^{-2} & 0 \\ 0 & 1 \end{pmatrix} \begin{pmatrix} \cos t & \sin t \\ -\sin t & \cos t \end{pmatrix}$$

For $t \in [0, \pi/2]$, this defines a Laurentian adapted homotopy between $\alpha \oplus \zeta_{2r}$ and $\alpha' \oplus \zeta_{2r}^{-1}$. Hence we obtain the identity required.

b) $\operatorname{Ker}(w_2) = 0$. Let us assume that $d_L(\alpha) = 0$, where α is quasi-polynomial. Therefore, there exists a Laurentian homotopy $\gamma(\varphi, t)$ between α and ζ. If the order of this homotopy is $-2p-1$ ($p>1$), we show that the order of the homotopy can be "pushed" to $-2p+1$ by increasing the size of the matrix. Assertion b) will then be obtained by downwards induction on p.

The homotopy defined by the product of the four matrices above, actually defines a quasi-polynomial homotopy between $\alpha \oplus \zeta_{2r}$ and $\alpha\zeta^2 \oplus \zeta_{2r}^{-1}$ when α is quasi-polynomial, as is the case. The given Laurentian homotopy of order $-2p-1$ defines a Laurentian homotopy of order $-2p+1$ between $\alpha\zeta^2 \oplus \zeta_{2r}^{-1}$ and $\zeta^3 \oplus \zeta_{2r}^{-1}$. Thus we have a quasi-polynomial homotopy between $\zeta_{2r}^3 \oplus \zeta_{2r}^{-1}$ and $\zeta_{2r} \oplus \zeta_{2r}$. Therefore, we see that

$$\alpha \sim \alpha \oplus \zeta \sim \alpha\zeta^2 \oplus \zeta^{-1} \sim \zeta^3 \oplus \zeta^{-1} \sim \zeta \oplus \zeta = \zeta.$$

The composition of these three homotopies provides the desired Laurentian homotopy of order $-2p+1$ between $\alpha \oplus \zeta$ and $\zeta \oplus \zeta$. □

6.19. Lemma. *The homomorphism w_3 is surjective and its kernel is* 0.

Proof. a) w_3 *is surjective.* Let $d_P(\alpha)$ be an element of $\pi_1^P(\mathrm{GL}(A), \mathrm{GL}^-(A))$ with $\alpha(\varphi) = a_{-1}\zeta^{-1} + a_1\zeta + \cdots + a_{2n-1}\zeta^{2n-1} \in \mathrm{GL}_{2r}(A)$ for $n>1$. We prove that there exists a quasi-polynomial homotopy between $\alpha \oplus \zeta_{2r}$ and an adapted function which is quasi-polynomial of degree $\leqslant 2n-3$. To do this, we consider the product of the following three matrices in $\mathrm{GL}_{4r}(A)$ (where $\zeta = \zeta_{2r}$):

$$\alpha(t, \varphi) = \begin{pmatrix} 1 & t\zeta^{2n-2} \\ 0 & 1 \end{pmatrix} \begin{pmatrix} \alpha & 0 \\ 0 & \zeta^{-1} \end{pmatrix} \begin{pmatrix} 1 & 0 \\ -ta_{2n-1}\zeta^2 & 1 \end{pmatrix}$$

$$= \begin{pmatrix} \alpha - t^2 a_{2n-1}\zeta^{2n-1} & t\zeta^{2n-3} \\ ta_{2n-1}\zeta & \zeta^{-1} \end{pmatrix}.$$

We set $\tilde{\alpha}(t, \varphi) = \alpha(t, \varphi)\alpha(t, 0)^{-1}$. Then $\tilde{\alpha}(t, \varphi)$ is a quasi-polynomial homotopy of degree $\leqslant 2n-1$ between $\alpha \oplus \zeta^{-1}$ and $\alpha' = \tilde{\alpha}(1, \varphi)$, which is of degree $\leqslant 2n-3$. Therefore, we may write

$$d_P(\alpha') = d_P(\alpha') + d_P(\zeta_{2r}^{-1}) = d_P(\alpha) + w_3(d_A((\zeta_{2r}^{-1})).$$

In Lemma 6.20, we prove independently that w_4 is surjective. Therefore $\pi_1^A(\mathrm{GL}(A), \mathrm{GL}^-(A))$ is a group, and $d_P(\alpha) = d_P(\alpha') + w_3(-d_A(\zeta_{2r}^{-1}))$. By decreasing induction on n, this shows that w_3 is surjective.

b) $\mathrm{Ker}(w_3) = 0$. Using the same method as in proving a), we must verify the following assertion: let α be a quasi-polynomial map of degree $\leqslant 2n-3$, for $n>1$, which is homotopic to ζ by a homotopy γ of degree $\leqslant 2n-1$ in $\mathrm{GL}_{2r}(A)$; then $\alpha \oplus \zeta_{2r}^{-1}$ is homotopic to $\zeta_{2r} \oplus \zeta_{2r}^{-1}$ by a homotopy of degree $\leqslant 2n-3$. If we let

$$\gamma(t, \varphi) = a_{-1}(t)\zeta^{-1} + a_1(t)\zeta + \cdots + a_{2n-1}(t)\zeta^{2n-1},$$

then $\gamma(0, \varphi)=\alpha$ and $\gamma(1, \varphi)=\zeta$. In $\mathrm{GL}_{4r}(A)$, let us consider the matrix

$$\delta(t, \varphi)=\begin{pmatrix} \gamma(t, \varphi)-a_{2n-1}(t)\zeta^{2n-1} & \zeta^{2n-3} \\ a_{2n-1}(t)\zeta & \zeta^{-1} \end{pmatrix}.$$

It defines a homotopy of degree $\leqslant 2n-3$, between

$$\delta(0, \varphi)=\begin{pmatrix} \alpha & \zeta^{2n-3} \\ 0 & \zeta^{-1} \end{pmatrix} \quad \text{and} \quad \delta(1, \varphi)=\begin{pmatrix} \zeta & \zeta^{2n-3} \\ 0 & \zeta^{-1} \end{pmatrix},$$

which can themselves be joined by a path of degree $\leqslant 2n-3$ to

$$\begin{pmatrix} \alpha & 0 \\ 0 & \zeta^{-1} \end{pmatrix} \quad \text{and} \quad \begin{pmatrix} \zeta & 0 \\ 0 & \zeta^{-1} \end{pmatrix}$$

respectively. If we replace each of these homotopies by its "normalization" (the normalization of $\sigma(t, \varphi)$ is $\tilde{\sigma}(t, \varphi)=\sigma(t, \varphi)\sigma(t, 0)^{-1}$, we see that the assertion is proved

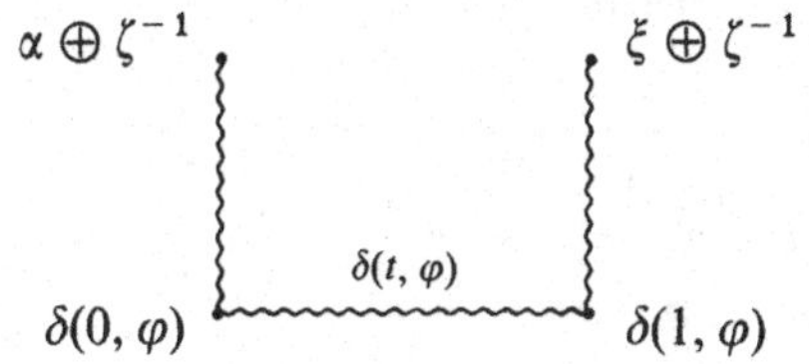

Fig. 17

□

6.20. Lemma. *The homomorphism* w_4 *is surjective and its kernel is* 0.

Proof. Let $d_A(\alpha)$ be an element of $\pi_1^A(\mathrm{GL}(A), \mathrm{GL}^-(A))$. Then α is the class of a path of the form $a_{-1}\zeta^{-1}+a_1\zeta$ where $a_{-1}+a_1=1$. Since $\zeta=\cos\varphi+e'\sin\varphi$, we may write α in the form $\cos\varphi+g\sin\varphi$, where g is an element of $\mathrm{GL}(A)$ such that $\bar{g}=-g$, and such that $g-\lambda$ is invertible for every real number λ. If $J(A)$ denotes the space of such automorphisms g, we see that $\pi_1^A(\mathrm{GL}(A), \mathrm{GL}^-(A))$ may be identified with $\pi_0(J(A))$. On the other hand, $I(A)$ is the subset of $J(A)$ consisting of automorphisms which satisfy the additional condition, $g^2=-1$. So all that remains to be shown, is that the inclusion of $I(A)$ in $J(A)$ induces a bijection on π_0.

If $g \in J(A)$ and if $A'=A \otimes_{\mathbb{R}} \mathbb{C}$, let us consider the element $\beta=\beta_{-1}z^{-1}+\beta_1 z$, where $\beta_1=\dfrac{1+ig}{2}$ and $\beta_{-1}=\dfrac{1-ig}{2}$, regarded as a function of $z \in S^1$. Then, if we write $z=\cos\varphi+i\sin\varphi$, we see that β is invertible, hence may be regarded as an element of $\mathrm{GL}(A'\langle z, z^{-1}\rangle)$ using the notation of 1.17. The computations in 1.24 (with a slight change of notation) show that if we write β^{-1} as $\sum_{n=-\infty}^{+\infty} \gamma_{2n+1}z^{2n+1}$, then $\beta_1\gamma_{-1}$ is a projector. It follows that $\hat{g}=\dfrac{2\beta_1\gamma_{-1}-1}{i}$ is an element of $\mathrm{GL}(A')$

such that $\hat{g}^2 = -1$, and $\bar{\hat{g}} = -\hat{g}$ with respect to the previous involution on A, since applying the involution to β means changing z to z^{-1}, and i to i. Moreover, g is self-conjugate with respect to complex conjugation since applying complex conjugation to β means changing z to z^{-1}, and i to $-i$. Hence $\hat{g} \in \mathrm{GL}(A) \subset \mathrm{GL}(A')$, and the correspondence $g \mapsto \hat{g}$ defines a continuous retraction from $J(A)$ to $I(A)$. It follows that $w_4 : \pi_0(I(A)) \to \pi_0(J(A))$ is injective.

On the other hand, by Remark 1.27, the element $tg + (1-t)\hat{g} \in J(A)$ for each $t \in [0, 1]$. This defines a path between g and $\hat{g}$ in $J(A)$. Therefore w_4 is surjective. □

6.21. *Remark.* The reader who is aware of holomorphic calculus in Banach algebras might perhaps prefer to proceed in the following way for the last proof. Since the spectrum of g does not meet the real axis, we choose two circles γ^+ and γ^- which contain the part of the spectrum located in the half-plane $\mathrm{Im}(z) > 0$, and in the half-plane $\mathrm{Im}(z) < 0$, respectively.

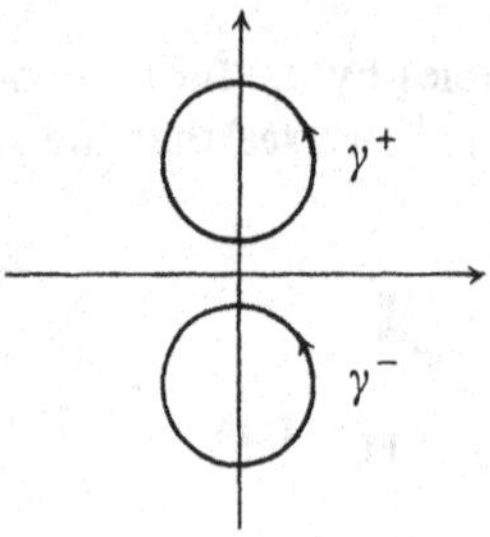

Fig. 18

Then the map

$$g \longrightarrow \hat{g} = \frac{1}{2i\pi} \int_{\gamma^+} \frac{i\,dz}{z-g} + \frac{1}{2i\pi} \int_{\gamma^-} \frac{-i\,dz}{z-g}$$

also defines a retraction from $J(A)$ to $I(A)$, such that $tg + (1-t)\hat{g}$ belongs to $J(A)$. As an exercise, the reader may check that the two definitions of $\hat{g}$ coincide.

Exercises (Section III.7) 8–11.

7. Exercises

7.1. Prove that the external tensor product of bundles (I.4.9) induces an isomorphism

$$K_{\mathbb{C}}(X) \otimes_{\mathbb{Z}} K_{\mathbb{C}}(S^{2n}) \approx K_{\mathbb{C}}(X \times S^{2n}),$$

and deduce that $K_{\mathbb{C}}(X \times S^{2n}) \approx K_{\mathbb{C}}(X) \oplus K_{\mathbb{C}}(X)$ as $K_{\mathbb{C}}(X)$-modules. In the same way, prove that $K_{\mathbb{C}}(X \times S^{2n+1}) \approx K_{\mathbb{C}}(X) \oplus K_{\mathbb{C}}^{-1}(X)$ and compute $K_{\mathbb{R}}(X \times S^n)$.

7.2. Prove that $K_{\mathbb{C}}(P_n(\mathbb{C}))$ is a free group of rank $n+1$. Compute $K_{\mathbb{C}}(X\times P_n(\mathbb{C}))$ as a $K_{\mathbb{C}}(X)$-module.

7.3. Let Y be a compact space and let $\varnothing = Y_0 \subset Y_1 \subset \cdots \subset Y_n = Y$ be a filtration of Y by closed subsets such that $Y_{i+1} - Y_i \approx \mathbb{C}^{p_i}$. Now prove that $K_{\mathbb{C}}(X\times Y)$ is a free $K_{\mathbb{C}}(X)$-module of rank $p = \sum_{i=0}^{n-1} p_i$.

7.4. Let X be a compact differentiable manifold. Prove the existence of a finite open cover (U_i) of X, such that the intersection of p open sets U_i is either empty or homeomorphic to $\mathbb{R}^n$. Apply the Mayer-Vietoris exact sequence (II.4.18) to show that the groups $K^{-n}(X)$ are of finite type.

7.5. Let X be a finite CW-complex of dimension r. Apply a method analogous to 7.4 to prove that $K^{-n}(X)$ is a group of finite type. Prove also that $K'(X)^{r+1}=0$ (cf. II.5.9).

7.6. Compute $K^i_{\mathbb{R}}(P_n(\mathbb{C}))\otimes_{\mathbb{Z}}\mathbb{Z}'$. Show that $\tilde{K}_{\mathbb{R}}(P_2(\mathbb{C})\approx\mathbb{Z}$ and $\tilde{K}^{-1}_{\mathbb{R}}(P_2(\mathbb{C}))=0$. Show that $\tilde{K}_{\mathbb{R}}(P_2(\mathbb{R}))\approx\mathbb{Z}/4$.

7.7. Let A be a complex Banach algebra. Show that $\pi_2(\mathrm{GL}(A))\approx\pi_0(\mathrm{GL}(A))$. Deduce that $\pi_n(\mathrm{GL}(A))\approx K(A)$ if n is odd, and $\pi_n(\mathrm{GL}(A))\approx\pi_0(\mathrm{GL}(A))$ if n is even.

7.8. Let A be a real Banach algebra, and let $A_n = A\otimes_{\mathbb{R}} C^{n,0}$ be provided with the involution $a\otimes b\mapsto a\otimes\bar{b}$, where $b\mapsto\bar{b}$ is the involution on $C^{n,0}$ such that $\bar{e}_i = -e_i$, for $i=1,\ldots,n$.

a) If $n\geqslant 1$, prove that A_{n-1} may be identified with the subalgebra of A_n consisting of elements of zero degree (i.e. such that $\bar{x}=x$), and prove that $\Omega(\mathrm{GL}(A_n),\ \mathrm{GL}^-(A_n))$ has the same homotopy type as $\Omega(\mathrm{GL}(A_n)/\mathrm{GL}^+(A_n))$ where $\mathrm{GL}^+(A_n)=\{\alpha\in\mathrm{GL}(A_n)|\bar{\alpha}=\alpha\}$.

b) If $n\geqslant 2$, prove that $\mathrm{GL}(A_{n-1})$ acts on $I(A_n)$ (notation of 6.10) by inner automorphisms, and that the connected component of $1\otimes e_n$ in $I(A_n)$ is homeomorphic to the connected component of the class of 1 in $\mathrm{GL}(A_{n-1})/\mathrm{GL}(A_{n-2})$.

c) From a), b), and Theorem 6.12, deduce that

$$[\mathrm{GL}(A_{n-1})/\mathrm{GL}(A_{n-2})]^0 \sim [\Omega(\mathrm{GL}(A_n)/\mathrm{GL}(A_{n-1}))]^0 \quad \text{(Wood [1])}$$

(where in general, X^0 denotes a connected component of the space X).

d) Prove that $[\mathrm{GL}(A)]^0\sim[\Omega^8(\mathrm{GL}(A))]^0$.

e) Prove that

$$K(A)\approx\pi_n(\mathrm{GL}(A_n),\ \mathrm{GL}(A_{n-1}))\approx\pi_7(\mathrm{GL}(A)).$$

f) Prove that $K(A\otimes_{\mathbb{R}}\mathbb{H})\approx\pi_3(\mathrm{GL}(A))$.

g) Prove that $\pi_0(\mathrm{GL}(A))\approx\pi_8(\mathrm{GL}(A))$, and show the homotopy equivalence $\mathrm{GL}(A)\sim\Omega^8(\mathrm{GL}(A))$.

7.9. Compute $\pi_n(\mathrm{GL}(A))$ for the following real or complex Banach algebras.

a) $A = C(X)$, the algebra of continuous functions on a compact space X.

b) $A = C_r(X)$, the algebra of differentiable functions of class C^r on the compact manifold X.

c) $A = \mathrm{End}(H)$, where H is a Hilbert space of infinite dimension.

d) $A =$ the subalgebra of $\mathrm{End}(H)$ consisting of operators of the form $\lambda + u$, where λ is a scalar and u is a compact operator (i.e. a limit of operators of finite rank).

e) $A = \mathbb{C}\langle z, z^{-1}\rangle$, the algebra of Laurent series $\sum_{n=-\infty}^{+\infty} a_n z^n$, where $a_n \in \mathbb{C}$ and $\sum_{n=-\infty}^{+\infty} |a_n| < +\infty$.

7.10. Let H be a Hilbert space of infinite dimension over $k = \mathbb{R}$ or $\mathbb{C}$, and let $\mathrm{GL}_c(H)$ be the subgroup of $\mathrm{GL}(H)$ consisting of operators of the form $1 + u$, where u is compact (cf. 7.9).

a) If X is a compact space, prove that $[X, \mathrm{GL}_c(H)] \approx K^{-1}(X) \approx [X, \mathrm{GL}(k)]$.

b) Let $\mathscr{F}(H)$ be the set of Fredholm operators D in H (i.e. such that $\mathrm{Ker}(D)$ and $\mathrm{Coker}(D)$ are finite dimensional). Using the fact that $\mathrm{GL}(H)$ is connected (Kuiper [1]), prove that the map $D \mapsto \mathrm{Dim}(\mathrm{Ker}\, D) - \mathrm{Dim}(\mathrm{Coker}\, D)$ ($=$ *index* of D) induces a bijection $\pi_0(\mathscr{F}(H)) \approx \mathbb{Z}$.

*c) Let $\mathscr{F}(H)^0$ be the subset of $\mathscr{F}(H)$ consisting of operators of index 0. Prove the fibration

$$\mathrm{GL}_c(H) \longrightarrow \mathrm{GL}(H) \longrightarrow \mathscr{F}(H)^0.$$

Using the fact that $\mathrm{GL}(H)$ is contractible (Kuiper [1]), prove that $\Omega\mathscr{F}(H)^0 \sim \mathrm{GL}_c(H)$, and that $K(X) \approx [X, \mathscr{F}(H)]$ (Atiyah [3], Jänich [1]).*

7.11. Let $\sigma: S^1 \to \mathrm{GL}(\mathbb{C})$ be a differentiable function whose Fourier series is $\sigma(z) = \sum_{n=-\infty}^{+\infty} a_n z^n$, where $a_n \in \mathrm{M}_p(\mathbb{C})$ for some p, independent of n.

a) Show that the residue of the function $Tr(\sigma'(z)\sigma(z)^{-1})$ at 0, is an integer which coincides (up to sign) with the integer defined by the periodicity isomorphism $\pi_1(\mathrm{GL}(\mathbb{C})) \approx K(\mathbb{C}) \approx \mathbb{Z}$ (1.11).

b) Show that the infinite matrix

$$\begin{bmatrix} a_0 & a_1 & a_2 & a_3 & \cdots \\ a_{-1} & a_0 & a_1 & a_2 & \cdots \\ a_{-2} & a_{-1} & a_0 & a_1 & \cdots \\ \vdots & \vdots & \vdots & \vdots & \ddots \end{bmatrix}$$

defines a Freholm operator in $H = \mathbb{C} \oplus \mathbb{C} \oplus \cdots \oplus \mathbb{C} \oplus \cdots$ (Hilbert sum), whose index is the integer above (Atiyah [7]).

7.12. Let A be a unitary ring, and let $A[z, z^{-1}]$ be the ring of Laurent polynomials (i.e. the ring of formal power sums $\sum_{n=-\infty}^{+\infty} a_n z^n$ where all but finitely many $a_n = 0$). Applying the methods of section 1, prove the exact sequence (cf.II.6.13)

$$0 \longrightarrow K_1(A) \longrightarrow K_1(A[z]) \oplus K_1(A[z^{-1}]) \longrightarrow K_1(A[z, z^{-1}]) \longrightarrow K(A) \longrightarrow 0$$

(Bass [1], Karoubi [5]).

7.13. (Atiyah [6]). Let X be a compact space provided with an involution. We define a *Real* vector bundle on X (Caution: *Real* is distinct from *real*) to be a complex vector bundle E with base X, provided with an *antilinear* involution $\tau: E \to E$ which commutes with the involution of X. It is easy to see that such vector bundles are the objects of a Banach category $\mathscr{E}\mathscr{R}(X)$, where the morphisms are those morphisms of complex vector bundles which induce the identity on the base X, and commute with the involution. We let $KR(X)$ denote the Grothendieck group of this category. In fact, the tensor product of vector bundles defines a ring structure on $KR(X)$.

a) If the involution of X is trivial, prove that $KR(X) \approx K_{\mathbb{R}}(X)$.

b) We let $S^{p,q}$ (resp. $B^{p,q}$) denote the sphere (resp. the ball) of $\mathbb{R}^{p+q}$, provided with the involution $(x, y) \mapsto (-x, y)$, for $x \in \mathbb{R}^p$ and $y \in \mathbb{R}^q$. If X is provided with the trivial involution, prove that $KR(X \times S^{1,0}) \approx K_{\mathbb{C}}(X)$.

c) Let X be a space provided with the trivial involution. We consider the set of pairs (E, c) where E is a complex vector bundle, and $c: E \to E$ is an antilinear automorphism. Let $\Phi SC(X)$ be the monoid consisting of *homotopy* classes of such pairs (E, c), and let $KSC(X)$ be the symmetrized group. Now prove that $KSC(X) \approx KR(X \times S^{2,0})$.

d) For each pair (X, Y) of compact spaces with involution, we define $KR^{-n}(X, Y) = KR(X \times B^{0,n}, X \times S^{0,n} \cup Y \times B^{0,n})$. Now prove the exact sequence

$$KR^{-n-1}(X) \longrightarrow KR^{-n-1}(Y) \longrightarrow KR^{-n}(X, Y) \longrightarrow KR^{-n}(X) \longrightarrow KR^{-n}(Y),$$

and the excision isomorphism

$$KR^{-n}(X, Y) \approx KR^{-n}(X/Y), \{y\})$$

(compare with II.4.12).

7.14 (7.13 continued). Let $KR^{p,q}(X)$ be the group $K^{p,q}$ of the Banach category $\mathscr{E}\mathscr{R}(X)$ (4.11).

a) Give an interpretation of the group $KR^{p,q}(X)$ in terms of gradations as in 5.1, and from this description obtain a relative definition for $KR^{p,q}(X, Y)$.

b) Applying the ideas of the proof of 5.10, define an isomorphism

$$t: KR^{p,q+1}(X, Y) \longrightarrow KR^{p,q}(X \times B^{0,1}, X \times S^{0,1} \cup Y \times B^{0,1}).$$

c) Applying the ideas of the proof of Theorem 1.3, define an isomorphism

$$KR(X, Y) \longrightarrow KR(X \times B^{1,1}, X \times S^{1,1} \cup Y \times B^{1,1})$$

by the cup-product with a suitable generator of $KR(B^{1,1}, S^{1,1}) \approx \mathbb{Z}$. Conclude that $KR^{n,n}(X, Y) \approx KR(X, Y) \approx KR(X \times B^{n,n}, X \times S^{n,n} \cup Y \times B^{n,n})$.

d) Prove that $KR^{p,q}(X, Y) \approx KR(X \times B^{p,q}, X \times S^{p,q} \cup Y \times B^{p,q})$, and that up to isomorphism the second family of groups depends only on the difference $p-q \bmod 8$. Note that this isomorphism provides a meaning for the concept of a "negative sphere", as the ordinary sphere provided with the antipodal involution.

7.15 (7.14 continued). We set $KR^n(X, Y) \approx KR^{p,q}(X, Y)$ for $n = p-q \in \mathbb{Z}$.

a) Prove the isomorphism $KR^n(X \times S^{p,0}, X \times S^{q,0}) \approx KR^{n+q}(X \times S^{p-q,0})$ for $p \geqslant q$, using the fact that $S^{p,0}/S^{q,0}$ and $S^{p-q,0} \times B^{q,0}/S^{p-q,0} \times S^{q,0}$ are homeomorphic as $\mathbb{Z}/2$-spaces.

b) Let $\eta_q: KR^q(X) \approx KR(X \times B^{q,0}, X \times S^{q,0}) \to KR(X)$ be the morphism induced by the obvious inclusion. Prove that it is a $KR(X)$-module morphism. Show that it is induced by the cup-product with a certain element $\alpha_q \in KR^{-q}(P) = K_{\mathbb{R}}^{-q}(P) = K_{\mathbb{R}}^{-q}(P)$, where P is a point.

c) Show that α_1 and α_2 are not zero. Show that $\eta_q = 0$ for $q \geqslant 3$, and in this case, prove the exact sequences of $KR(X)$-modules

$$0 \longrightarrow KR(X) \longrightarrow KR(X \times S^{q,0}) \longrightarrow KR^{q+1}(X) \longrightarrow 0,$$

$$0 \longrightarrow KR^{-q-1}(X) \longrightarrow KR^{-q-1}(X \times S^{q,0}) \longrightarrow KR(X) \longrightarrow 0.$$

Deduce that $KR(X \times S^{q,0}) \approx KR(X) \oplus KR^{q+1}(X)$.

* d) Show that the exact sequences above split naturally. *

e) Apply the exact sequences associated with the pair $(X \times S^{p,0}, X \times S^{q,0})$ for $(p, q) = (2, 1), (3, 1)$, and $(3, 2)$, to prove the exact sequences

$$\longrightarrow K_{\mathbb{C}}^{n-1}(X) \longrightarrow K_{\mathbb{C}}^{n-1}(X) \longrightarrow KSC^n(X) \longrightarrow K_{\mathbb{C}}^n(X) \longrightarrow K_{\mathbb{C}}^n(X) \longrightarrow,$$

$$\longrightarrow K_{\mathbb{C}}^{n+1}(X) \longrightarrow K_{\mathbb{R}}^n(X) \oplus K_{\mathbb{H}}^n(X) \longrightarrow K_{\mathbb{C}}^n(X) \longrightarrow KSC^{n+2}(X) \longrightarrow,$$

and $\longrightarrow KSC^{n-1}(X) \longrightarrow K_{\mathbb{C}}^n(X) \longrightarrow K_{\mathbb{R}}^n(X) \oplus K_{\mathbb{H}}^n(X) \longrightarrow KSC^n(X) \longrightarrow,$

where X is a compact space (provided with the trivial involution).

* f) Make all the maps explicit in the exact sequences above (Anderson [1]). *

7.16. Let A be a Banach algebra provided with an *anti-involution*, denoted by $\lambda \mapsto \bar{\lambda}$, and let M be a finitely generated projective right A-module. A Hermitian form on M is given by a $\mathbb{Z}$-bilinear map $\Phi: M \times M \to A$, such that $\Phi(x\lambda, y\mu) = \bar{\lambda}\Phi(x, y)\mu$ and $\Phi(y, x) = \overline{\Phi(x, y)}$. The form Φ is called nondegenerate if the homomorphism from M to its antidual, which is induced by Φ, is an isomorphism. We let $L(A)$ denote the symmetrized group of the monoid consisting of isomorphism classes of modules provided with nondegenerate Hermitian forms.

a) We assume that the Banach algebra A satisfies the following condition:

for any matrix $M=(a_{ji})$ with coefficients in A, the matrix $I+M^*M$, where $M^*=(\bar{a}_{ij})$, is an invertible matrix. Now show that $L(A)\approx K(A)\oplus K(A)$.

b) Compute $L(A)$ where $A=\mathrm{M}_2(C_k(X))$, and the anti-involution on $\mathrm{M}_2(C_k(X))$ is given by

$$\begin{pmatrix} a & b \\ c & d \end{pmatrix} \longmapsto \begin{pmatrix} d & -b \\ -c & a \end{pmatrix}$$

(where $C_k(X)$ is the ring of continuous functions on a compact space X with values in $k=\mathbb{R}$ or $\mathbb{C}$).

c) Compute $L(A)$ where A is the ring of continuous functions on X with values in $\mathbb{H}$, provided with the anti-involution

$$a+bI+cJ+dK \longmapsto a-bI-cJ-dK,$$

or

$$a+bI+cJ+dK \longmapsto a+bI+cJ-dK.$$

d) Let X be a space with involution and let A be the (commutative) Banach algebra of complex continuous functions on X, provided with the involution $f\longmapsto \bar{f}$, where $\bar{f}(x)=\overline{f(\bar{x})}$ ($\bar{x}$ being the image of x under the involution). Show that $L(A)\approx KR(X)$.

8. Historical Note

There are now many proofs of the periodicity theorems. The original due to Bott used Morse theory (Milnor [5], Bott [2]). A second proof (in complex K-theory) was given by Atiyah and Bott [1]; this is essentially the one we have presented in III.1 (although with several changes in method of presentation). There is also a proof by Atiyah, which relies on Fredholm operators in Hilbert spaces (cf. Atiyah [7]; see also the author [5] for an algebraic interpretation).

Apart from the original proof by Bott, there are basically two "elementary" ways to prove Bott periodicity in real K-theory. One uses the KR-theory of Atiyah ([6]; cf. also 7.14. The other, which is an adapted proof of 1.3, is due to Wood [1] and the author [2]. Again with changes in presentation, this is the proof we choose to include since it immediately provides the "right" classifying spaces of the groups $K^{-n}(X)$, and the eight homotopy equivalences of Bott (cf. III.5). There is also an "homological" proof of these homotopy equivalences in the Cartan-Moore Seminar 1959/60 [1].

Bott periodicity has been generalized in many directions. First of all there is a deep connection between the periodicity and the methods developed by Atiyah and Singer in the index theorem (Atiyah [7]). In algebraic K-theory, there is an analog to Theorem 1.11, which is due to Bass, Heller, and Swan (cf. 7.12). There are also analogs of Theorems 2.11 and 5.22 (among others) in Hermitian K-theory (cf. the author [4]).

Finally, it must be noted that the introduction of Clifford algebras in real K-theory, which is a key to our proof of the periodicity theorems is due to Atiyah, Bott, and Shapiro [1]. Clifford algebras will play an important role in the next chapter, when we prove Thom isomorphism in real and complex K-theory.

Chapter IV

Computation of Some K-Groups

1. The Thom Isomorphism in Complex K-Theory for Complex Vector Bundles

1.1. The purpose of this section is to define an isomorphism $K_{\mathbb{C}}^q(X) \approx K_{\mathbb{C}}^q(V)$, for any complex vector bundle V over a locally compact space X (note that $K_{\mathbb{C}}^q(V) \approx K_{\mathbb{C}}^q(B(V), S(V))$, with respect to any metric on V; cf. II.5.12). For V trivial, we again obtain Bott periodicity in complex K-theory (cf. III.1.3 and III.2.1); however, Bott periodicity is actually an essential part of our proof. If X is compact, the one point compactification $\dot{V}$ of V is called the *Thom space* of V. Hence, the isomorphism $K_{\mathbb{C}}^q(X) \approx K_{\mathbb{C}}^q(V) \approx \tilde{K}_{\mathbb{C}}^q(\dot{V})$ will enable us to compute the K-theory of the Thom space of a complex vector bundle. Before defining this isomorphism, we will first establish a general theorem (1.3), which will also be useful in next sections (∗ it is the analogous of the Leray-Hirsch-Dold theorem in the framework of K-theory ∗).

1.2. Let $\pi: P \to X$ be a continuous map between two locally compact spaces. Then, using the same technique as in II.5.12, we define a product

$$K(X) \times K(P) \longrightarrow K(P)$$

as the composition $K(X) \times K(P) \to K(X \times P) \xrightarrow{j^*} K(P)$, where $j: P \to X \times P$ is the *proper* map defined by $j(p) = (\pi(p), p)$. More precisely, if we identify the groups K and K_0 of II.5.16 (cf. II.5.20), the product of $\sigma(E, D) \in K(X)$ and $\sigma(F, \Delta) \in K(P)$ is $\sigma(\pi^* E \otimes F, \pi^* D \hat{\otimes} 1 + 1 \hat{\otimes} \Delta)$. Therefore, $K(P)$ may be regarded as a module over the ring $K(X)$ (which may have no unit if X is not compact), using the product above.

More generally, if we set $K^*(Z) = \bigoplus_{q=0}^{\infty} K^{-q}(Z) = \bigoplus_{q=0}^{\infty} K(Z \times \mathbb{R}^q)$ (cf. II.4.11) for any locally compact space Z, then $K^*(P)$ may be regarded as a $K^*(X)$-module with respect to the product

$$K(X \times \mathbb{R}^q) \times K(P \times \mathbb{R}^r) \longrightarrow K(P \times \mathbb{R}^{q+r}),$$

which is defined as the composition

$$K(X \times \mathbb{R}^q) \times K(P \times \mathbb{R}^r) \longrightarrow K(X \times \mathbb{R}^q \times P \times \mathbb{R}^r) \approx K(X \times P \times \mathbb{R}^{q+r}) \xrightarrow{\iota^*} K(P \times \mathbb{R}^{q+r}),$$

where l is the proper map $(p, \lambda) \mapsto (\pi(p), p, \lambda)$. More precisely, if $\sigma(E, D) \in K(X \times \mathbb{R}^q)$ and $\sigma(F, \Delta) \in K(P \times \mathbb{R}^r)$ (cf. II.5.15), their product is $\sigma(\pi_1^* E \otimes \pi_2^* F, \pi_1^* D \,\hat{\otimes}\, 1 + 1 \otimes \pi_2^* \Delta)$, where

$$\pi_1 : P \times \mathbb{R}^{q+r} \longrightarrow X \times \mathbb{R}^q \quad \text{and} \quad \pi_2 : P \times \mathbb{R}^{q+r} \longrightarrow P \times \mathbb{R}^r.$$

1.3. Theorem. *Let $\pi : P \to X$ be a continuous map between two locally compact spaces. Let $e^1, \ldots, e^n$ be elements of $K^0(P)$ such that there exists a finite closed cover (W_i) of X with the following property: for any closed subset Y of W_i, the restrictions of $e^1, \ldots, e^n$ to $K^*(P_Y)$ form a basis of $K^*(P_Y)$ as a $K^*(Y)$-module, where $P_Y = \pi^{-1}(Y)$. Then $K^*(P)$ is a free $K^*(X)$-module with basis $e^1, \ldots, e^n$.*

Proof. Suppose $e^i = \sigma(F^i, \Delta^i)$. Then for *any* locally compact subspace T of X, we define $F_T^i = F^i|_T$, $\Delta_T^i = \Delta^i|_T$, and a fundamental homomorphism

$$\varphi_T^q : K(T \times \mathbb{R}^q)^n \longrightarrow K(P_T \times \mathbb{R}^q),$$

by the formula

$$\varphi_T^q(x_1, \ldots, x_n) = x_1 \cdot e_T^1 + \cdots + x_m \cdot e_T^n,$$

where the expression $x_i \cdot e_T^i$ denotes $\sigma(\pi_T^* E_i \otimes \theta_T^* F_T^i, \pi_T^* D_i \,\hat{\otimes}\, 1 + 1 \,\hat{\otimes}\, \theta_T^* \Delta_T^i)$ with $x_i = \sigma(E_i, D_i) \in K(T \times \mathbb{R}^q)$, $\pi_T : P_T \times \mathbb{R}^q \to T \times \mathbb{R}^q$, and $\theta_T : P_T \times \mathbb{R}^q \to P_T$. If T is closed, then $x_i \cdot e_T^i$ is simply the product of x_i by the restriction of e^i to $\pi^{-1}(T)$ (cf. the second formula given in 1.2). In general (T is not necessarily closed), the formula giving φ_T^q makes sense, since $\pi_T^* D_i \,\hat{\otimes}\, 1 + 1 \,\hat{\otimes}\, \theta_T^* \Delta_T^i$ is admissible in the sense of II.5.15, because it defines a bundle isomorphism outside a compact subset of $P_T \times \mathbb{R}^q$ (cf. the computation made in II.5.21). Thus in order to prove Theorem 1.3, it suffices to show that φ_X^q is an isomorphism.

The homomorphisms φ_T^q have some "natural" properties, which can easily be verified by checking the explicit formulas above. If T' is *closed* in T, we have the commutative diagram

$$\begin{array}{ccc} K(T \times \mathbb{R}^q)^n & \xrightarrow{\varphi_T^q} & K(P_T \times \mathbb{R}^q) \\ \downarrow & & \downarrow \\ K(T' \times \mathbb{R}^q)^n & \xrightarrow{\varphi_{T'}^q} & K(P_{T'} \times \mathbb{R}^q), \end{array}$$

where the vertical maps are defined by restriction.

If T' is *open* in T, we also have the commutative diagram

$$\begin{array}{ccc} K(T \times \mathbb{R}^q)^n & \xrightarrow{\varphi_T^q} & K(P_T \times \mathbb{R}^q) \\ \uparrow & & \uparrow \\ K(T' \times \mathbb{R}^q)^n & \xrightarrow{\varphi_{T'}^q} & K(P_{T'} \times \mathbb{R}^q), \end{array}$$

where the vertical maps are defined by "extension", i.e. as the composition

$$K(T'\times \mathbb{R}^q)\approx K(T\times \mathbb{R}^q, (T-T')\times \mathbb{R}^q)\longrightarrow K(T\times \mathbb{R}^q),$$

or
$$K(P_{T'}\times \mathbb{R}^q)\approx K(P_T\times \mathbb{R}^q, (P_T-P_{T'})\times \mathbb{R}^q)\longrightarrow K(P_T\times \mathbb{R}^q)$$

(cf. II.5.19). Moreover, the diagram

$$\begin{array}{ccc} K(T\times \mathbb{R})^n & \xrightarrow{\varphi_T^1} & K(P_T\times \mathbb{R}) \\ \Big\downarrow \partial_{T,T-T'} & & \Big\downarrow \partial_{P_T,P_T-P_{T'}} \\ K(T')^n & \xrightarrow{\varphi_{T'}^0} & K(P_{T'}), \end{array}$$

where ∂ is the connecting homomorphism described in II.4.9, is also commutative. This follows from the fact that the connecting homorphism ∂ is the composition of two maps of the form α^{*-1} and β^*, where α^* and β^* are induced by extension from open subsets (cf. II.4.9). By the substitutions $T\mapsto T\times \mathbb{R}^q$, $T'\mapsto T'\times \mathbb{R}^q$, and $P\mapsto P\times \mathbb{R}^q$, we also obtain the commutative diagram

$$\begin{array}{ccc} K(T\times \mathbb{R}^{q+1})^n & \xrightarrow{\varphi_T^{q+1}} & K(P_T\times \mathbb{R}^{q+1}) \\ \Big\downarrow \partial & & \Big\downarrow \partial \\ K(T'\times \mathbb{R}^q)^n & \xrightarrow{\varphi_{T'}^q} & K(P_{T'}\times \mathbb{R}^q) \end{array}$$

(up to sign, depending on convention; cf. II.5.27). Finally, if S and T are closed subsets of X, and if $q\leqslant 0$, we have the diagram

$$\begin{array}{ccccc} K^{q-1}(S\times \mathbb{R})^n\oplus K^{q-1}(T\times \mathbb{R})^n & \longrightarrow & K^{q-1}(S\cap T\times \mathbb{R})^n & \xrightarrow{\Delta} & K^q(S\cup T)^n \\ \Big\downarrow \varphi_S^{-q+1}\oplus\varphi_T^{-q+1} & & \Big\downarrow \varphi_{S\cap T}^{-q+1} & & \Big\downarrow \varphi_{S\cup T}^{-q} \\ K^{q-1}(P_S\times \mathbb{R})^n\oplus K^{q-1}(P_T\times \mathbb{R})^n & \longrightarrow & K^{q-1}(P_{S\cap T}\times \mathbb{R})^n & \xrightarrow{\Delta} & K^q(P_{S\cup T})^n \end{array}$$

$$\begin{array}{ccccc} \longrightarrow & K^q(S)^n\oplus K^q(T)^n & \longrightarrow & K^q(S\cap T)^n \\ & \Big\downarrow \varphi_S^{-q}\oplus\varphi_T^{-q} & & \Big\downarrow \varphi_{S\cap T}^{-q} \\ \longrightarrow & K^q(P_S)\oplus K^q(P_T) & \longrightarrow & K^q(P_{S\cap T}). \end{array}$$

In this diagram the horizontal sequences are the Mayer-Vietoris exact sequences (II.4.18). The diagram is commutative by the observations above, and by the fact that the homomorphism Δ, in the Mayer-Vietoris exact sequence, is the composition of restrictions, extensions, and connecting homomorphisms ∂ (cf. II.4.18).

Now let $W_1, \ldots, W_r$ be a finite closed cover of the space X satisfying the hypothesis of the theorem. Let $Z_i = W_1\cup\cdots\cup W_i$. We prove that $\varphi_{Zi}^{-q}\colon K^q(Z_i)\to K^q(P_{Z_i})$ is an isomorphism for $q\leqslant 0$ by induction on i. If we put $S = W_1\cup\cdots\cup W_i$

and $T=W_{i+1}$, we may write the two Mayer-Vietoris exact sequences above, with $S\cup T=Z_{i+1}$. Since φ_Z^{-r} is an isomorphism for $Z=S$, T, or $S\cap T$, by the inductive hypothesis we see that $\varphi_{S\cup T}^{-q}=\varphi_{Z_{i+1}}^{-q}$ is also an isomorphism. This completes the proof of Theorem 1.3. □

1.3.1. *Remark.* In fact the proof above shows that in theorem 1.3 we may restrict ourselves to subspaces $y=w_i$ or obtained by intersection of w_i with the union of w_j, $j<i$.

1.4. We wish to apply Theorem 1.3 to the case where P is the total space of a complex vector bundle V, where $\pi\colon V\to X$ is the canonical projection, and where $K=K_{\mathbf{C}}$. If X is compact, we show that $K_{\mathbf{C}}(V)$ is a $K_{\mathbf{C}}(X)$-module of rank one, generated by an element U_V ($=e^1$ in the notation of the theorem) which belongs to $K_{\mathbf{C}}(V)$, and which will be called the *Thom class* of the complex vector bundle V. This element U_V is constructed using a metric φ on V (I.8.5), and the metric ψ induced on the bundle $\Lambda(V)$ of exterior algebras associated with V(I.4.8.f)). More precisely, if V_x is a fiber of V, then the metric ψ_x on $\Lambda(V_x)=\Lambda(V)_x$, is defined by the formula

$$\psi_x(v_1\wedge\cdots\wedge v_n,\, w_1\wedge\cdots\wedge w_p)=0 \quad \text{if } p\neq n,$$

$$\psi_x(v_1\wedge\cdots\wedge v_n,\, w_1\wedge\cdots\wedge w_n)=\mathrm{Det}(\varphi_x(v_i, w_j)).$$

In particular, the products $e_{i_1}\wedge\cdots\wedge e_{i_p}$ for $i_1<\cdots<i_p$ and $p\leqslant \mathrm{Dim}(V_x)$, are an orthonormal basis of $\Lambda(V_x)$ (which is of complex dimension $2^{\mathrm{Dim}(V_x)}$), if the e_i are an orthonormal basis of V_x. By defining local coordinates, we see that ψ is continuous.

For a vector $v\in V_x$, we let $d_v\colon \Lambda(V_x)\to\Lambda(V_x)$ denote the linear map defined by $d_v(e)=v\wedge e$. Let $\partial_v\colon\Lambda(V_x)\to\Lambda(V_x)$ be the adjoint of d_v with respect to the metric above.

1.5. Lemma. *We have the identity*

$$(d_v+\partial_v)^2=d_v\partial_v+\partial_v d_v=\varphi_x(v,v)=Q_x(v),$$

where Q_x denotes the positive definite quadratic form associated with φ_x.

Proof. Each element e of $\Lambda(V_x)$ may be written as $e=v\wedge w+w'$, where w and w' are orthogonal to v (choose an orthonormal basis (e_i) of V such that $v=\lambda e_1$). Therefore

$$\partial_v(d_v(e))=\partial_v(v\wedge w')=Q_x(v)w',\; d_v(\partial_v(e))=d_v(Q_x(v)w)=Q_x(v)v\wedge w,$$

and $\quad \partial_v(d_v(e))+d_v(\partial_v(e))=Q_x(v)(w'+v\wedge w).$

Hence $\quad (d_v+\partial_v)^2=d_v\partial_v+\partial_v\cdot d_v=Q_x(v).$ □

1.6. By applying Lemma 1.5, the element $U_V\in K_{\mathbf{C}}(V)$ may now be described as follows. We have $U_V=\sigma(\pi^*\Lambda(V),\Delta)$, where $\pi\colon V\to X$, where $\Lambda(V)$ is provided with the $\mathbb{Z}/2$-grading defined in III.3.7 and the metric ψ, and where $\Delta: \pi^*\Lambda(V) \to \pi^*\Lambda(V)$ is defined over the point $(x, v) \in \pi^* V$, for $x \in X$ and $v \in V_x$, by the operator $\Delta_{x,v} = d_v + \partial_v$. Since $(D_{x,v})^2 = Q_x(v)$, Δ is an isomorphism outside the zero section of V, hence Δ is admissible in the sense of II.5.15. Moreover, if V is trivial, say $V = X \times \mathbb{C}^n$, then $\Lambda(V) = X \times \Lambda(\mathbb{C}^n)$ and $U_V = \sigma(E, \Delta)$, where $E = X \times \mathbb{C}^n \times \Delta(\mathbb{C}^n)$, and $\Delta(x, v, e) = (x, v, (d_v + \partial_v)(e))$. Therefore, Δ is continuous.

All that remains to be shown in that $U_V = e^1$ satisfies the hypothesis of Theorem 1.3. To do this, we need the following proposition:

1.7. Proposition. *Let V and V' be complex vector bundles with bases X and X', respectively. Then $U_{V \boxplus V'} = U_V \cup U_{V'}$ (note that $V \boxplus V' = V \times V'$;* cf. I.4.9).

Proof. Let $U_V = \sigma(\pi^*\Lambda(V), \Delta)$ and $U_{V'} = \sigma(\pi'^*\Lambda(V'), \Delta')$ be the Thom classes of V and V', respectively, where $\pi: V \to X$ and $\pi': V' \to X'$. According to the formula given in II.5.21 for the cup-product, we have

$$U_V \cup U_{V'} = \sigma(\pi^*\Lambda(V) \boxtimes \pi'^*\Lambda(V'), \Delta \,\hat{\boxtimes}\, 1 + 1 \,\hat{\boxtimes}\, \Delta').$$

Let $\pi_1, \tilde{\pi}_1, \pi_2, \tilde{\pi}_2, \pi''$ be the obvious projections

$$V \times V' \xrightarrow{\tilde{\pi}_1} V \xrightarrow{\pi} X, \quad \pi_1 = \pi \circ \tilde{\pi}_1, \qquad V \times V' \xrightarrow{\tilde{\pi}_2} V' \xrightarrow{\pi'} X', \quad \pi_2 = \pi' \circ \tilde{\pi}_2$$

$$V \times V' \xrightarrow{\pi''} X \times X'$$

By definition, we have an isomorphism $\pi^*\Lambda(V) \boxtimes \pi'^*\Lambda(V') = \pi_1^*\Lambda(V) \otimes \pi_2^*\Lambda(V')$ (I.4.9). Moreover, we have an isomorphism

$$\pi_1^*\Lambda(V) \otimes \pi_2^*\Lambda(V') \xrightarrow{\theta} \pi''^*\Lambda(V \times V'),$$

defined by the formula

$$(v, v', \sum e_i \otimes e_i') \longmapsto (v, v', \sum e_i \wedge e_i'),$$

where we consider $\Lambda(V_x)$ and $\Lambda(V'_{x'})$ as imbedded in the obvious way in $\Lambda(V \times V')_{x,x'} = \Lambda(V_x \oplus V'_{x'}) \approx \Lambda(V_x) \otimes \Lambda(V'_{x'})$ (cf. III.3.10). Therefore, the proposition is equivalent to the commutativity of the diagram

$$\begin{array}{ccc} \pi_1^*\Lambda(V) \otimes \pi_2^*\Lambda(V') & \xrightarrow{\theta} & \pi''^*\Lambda(V \times V') \\ \downarrow \Gamma & & \downarrow \Delta'' \\ \pi_1^*\Lambda(V) \otimes \pi_2^*\Lambda(V') & \xrightarrow{\theta} & \pi''^*\Lambda(V \times V'), \end{array}$$

where $\Gamma = \tilde{\pi}_1^* \Delta \hat{\otimes} 1 + 1 \otimes \tilde{\pi}_2^* \Delta'$, and where $\sigma(\pi''^* \Lambda(V \times V'), \Delta'')$ is the Thom class of $V \times V' = V \boxplus V'$ as defined in 1.6. The commutativity of this diagram is a matter of linear algebra: we must verify the commutativity of the vector space diagram (where V and V' are now vector spaces)

$$\begin{array}{ccc} \Lambda(V) \otimes \Lambda(V') & \xrightarrow{\theta} & \Lambda(V \oplus V') \\ \Big\downarrow {\scriptstyle (d_v + \partial_v) \hat{\otimes} 1 + 1 \hat{\otimes} (d_{v'} + \partial_{v'})} & & \Big\downarrow {\scriptstyle d_{v+v'} + \partial_{v+v'}} \\ \Lambda(V) \otimes \Lambda(V') & \xrightarrow{\theta} & \Lambda(V \oplus V'), \end{array}$$

where $\quad \theta(\sum e_i \otimes e_i') = \sum e_i \wedge e_i'$, $v \in V \subset V \oplus V'$, and $v' \in V' \subset V \oplus V'$.

Since θ is an isometry, it suffices to verify the commutativity of the diagram

$$\begin{array}{ccc} \Lambda(V) \otimes \Lambda(V') & \xrightarrow{\theta} & \Lambda(V \oplus V') \\ \Big\downarrow {\scriptstyle d_v \hat{\otimes} 1 + 1 \hat{\otimes} d_{v'}} & & \Big\downarrow {\scriptstyle d_{v+v'}} \\ \Lambda(V) \otimes \Lambda(V') & \xrightarrow{\theta} & \Lambda(V \oplus V'). \end{array}$$

However, if e and e' are homogeneous elements of $\Lambda(V)$ and $\Lambda(V')$ respectively, we have

$$\begin{aligned} &\theta((d_v \hat{\otimes} 1 + 1 \hat{\otimes} d_{v'})(e \wedge e')) \\ &\qquad = v \wedge e \wedge e' + (-1)^{\deg(e)} e \wedge v' \wedge e' = v \wedge e \wedge e' + v' \wedge e \wedge e' \\ &\qquad = (v + v') \wedge (e \wedge e') = d_{v+v'}((\theta(e \otimes e')). \quad \square \end{aligned}$$

1.8. Corollary. *If $X = X'$, and if $(\alpha, \beta) \mapsto \alpha \cdot \beta$ denotes the product*

$$K(V) \times K(V') \longrightarrow K(V \oplus V')$$

obtained by the composition $K(V) \times K(V') \to K(V \times V') \xrightarrow{l^*} K(V \oplus V')$, where l is the canonical inclusion $V \oplus V' \subset V \times V'$ (I.4.9), then we have the formula

$$U_{V \oplus V'} = U_V \cdot U_{V'}.$$

1.9. Theorem (Thom isomorphism). *Let V be a complex vector bundle with compact base X. Then $K_{\mathbb{C}}^*(V)$ is a free $K_{\mathbb{C}}^*(X)$-module of rank one, generated by the Thom class U_V.*

Proof. Let (W_i) be a finite closed cover of X such that $V|_{W_i}$ is trivial, say $V|_{W_i} = W_i \times \mathbb{C}^n$. If $Y \subset W_i$, then $V|_Y \approx Y \times \mathbb{C}^n$, and the Thom class $U_V|_Y = U_{V|Y}$ may be written as $1 \cup u_n$, where u_n is the Thom class of $\mathbb{C}^n$ regarded as a bundle over a point. By Bott periodicity in complex K-theory (III.2.1), and 1.2, it is therefore enough to check that u_n is a generator of $\tilde{K}_{\mathbb{C}}(\mathbb{C}^n)=K_{\mathbb{C}}(\mathbb{R}^{2n})\approx\mathbb{Z}$. Since $u_n=(u_1)^n$ by 1.7 (applied to bundles over a point), it actually suffices to check that u_1 is a generator of $K(\mathbb{C})$, but this was shown in II.5.25 and III.1.3. □

1.10. Let us now assume that the base of the vector bundle V provided with a metric, is locally compact. Then even if the Thom class U_V is not defined (see, however, Exercise 8.14), we may define a "Thom homomorphism"

$$\beta_{\mathbb{C}}: K_{\mathbb{C}}(X) \longrightarrow K_{\mathbb{C}}(V)$$

(hence from $K_{\mathbb{C}}^{-r}(X)$ to $K_{\mathbb{C}}^{-r}(V)$) by the formula

$$\sigma(E, D) \longmapsto \sigma(\pi^*E \otimes \pi^*F, \pi^*D \hat{\otimes} 1 + 1 \hat{\otimes} \Delta),$$

where $\pi: V \to X$, $F=\pi^*\Lambda(V)$, and Δ is defined as before (1.6). Since Δ is an isomorphism outside the zero section of V, and since D is an isomorphism outside a compact subset of X, we see that $\pi^*D \hat{\otimes} 1 + 1 \hat{\otimes} \Delta$ is admissible in the sense of II.5.15.

1.11. Theorem (Thom isomorphism for locally compact spaces). *Let V be a complex vector bundle over a locally compact base X. Then the Thom homomorphism defined above,*

$$\beta_{\mathbb{C}}: K_{\mathbb{C}}(X) \longrightarrow K_{\mathbb{C}}(V),$$

is an isomorphism.

Proof. Let us first assume that there exists a pair of compact spaces (Z, T) such that $X=Z-T$, and a complex vector bundle V' over Z such that $V'|_X=V$. Then we have the commutative diagram

$$\begin{array}{ccccccccc}
K_{\mathbb{C}}(Z\times\mathbb{R}) & \longrightarrow & K_{\mathbb{C}}(T\times\mathbb{R}) & \xrightarrow{\partial} & K_{\mathbb{C}}(X) & \longrightarrow & K_{\mathbb{C}}(Z) & \longrightarrow & K_{\mathbb{C}}(T) \\
\downarrow{\scriptstyle \varphi_T^1} & & \downarrow{\scriptstyle \varphi_Z^1} & & \downarrow{\scriptstyle \beta_{\mathbb{C}}} & & \downarrow{\scriptstyle \varphi_Z^0} & & \downarrow{\scriptstyle \varphi_T^0} \\
K_{\mathbb{C}}(V'\times\mathbb{R}) & \longrightarrow & K_{\mathbb{C}}(V'_T\times\mathbb{R}) & \xrightarrow{\partial} & K_{\mathbb{C}}(V) & \longrightarrow & K_{\mathbb{C}}(V') & \longrightarrow & K_{\mathbb{C}}(V'_T),
\end{array}$$

where the horizontal sequences are exact (cf. the "natural" properties of the homomorphism φ_T^q and ∂ proved in 1.3 and II.4.9). Since φ_T^1, φ_Z^1, φ_Z^0, and φ_T^0 are isomorphisms (1.9), $\beta_{\mathbb{C}}$ is also an isomorphism.

For the general case, we have $K_{\mathbb{C}}(X)\approx\text{inj lim } K_{\mathbb{C}}(U_i)$, where (U_i) runs through the set of relatively compact open sets in X (II.4.21). Similarly $K_{\mathbb{C}}(V)\approx\text{inj lim } K_{\mathbb{C}}(V_i)$ where $V_i=V|_{U_i}$. Since $U_i=\bar{U}_i-Fr(U_i)$ (where $Fr(U_i)$ denotes the boundary of U_i), we have the commutative diagram

$$\begin{array}{ccc} \text{inj lim } K_{\mathbb{C}}(U_i) & \xrightarrow{\approx} & K_{\mathbb{C}}(X) \\ \downarrow & & \downarrow \beta_{\mathbb{C}} \\ \text{inj lim } K_{\mathbb{C}}(V_i) & \xrightarrow{\approx} & K_{\mathbb{C}}(V), \end{array}$$

where the first vertical arrow is an isomorphism by what we just proved. Therefore $\beta_{\mathbb{C}}$ is an isomorphism. □

1.12. *Remark.* If we replace X by $X \times \mathbb{R}^q$, and V by $V \times \mathbb{R}^q$, it follows that the groups $K_{\mathbb{C}}^{-q}(X) = K_{\mathbb{C}}(X \times \mathbb{R}^q)$ and $K_{\mathbb{C}}^{-q}(V) = K_{\mathbb{C}}(V \times \mathbb{R}^q)$ are also isomorphic.

1.13. If we choose a metric on V (which is always possible when the base X is paracompact; cf. I.8.7), we have $K_{\mathbb{C}}(V) \approx K_{\mathbb{C}}(B(V) - S(V)) \approx K_{\mathbb{C}}(B(V), S(V))$ (II.5.19). Since $B(V)$ has the homotopy type of X, (it admits X as a deformation retract via the zero section), the exact sequence

$$K_{\mathbb{C}}^{-1}(B(V)) \longrightarrow K_{\mathbb{C}}^{-1}(S(V)) \longrightarrow K_{\mathbb{C}}(B(V), S(V)) \longrightarrow K_{\mathbb{C}}(B(V)) \longrightarrow K_{\mathbb{C}}(S(V))$$

may also be written as

$$K_{\mathbb{C}}^{-1}(X) \xrightarrow{\pi'^*} K_{\mathbb{C}}^{-1}(S(V)) \longrightarrow K_{\mathbb{C}}(X) \xrightarrow{\alpha} K_{\mathbb{C}}(X) \xrightarrow{\pi'^*} K_{\mathbb{C}}(S(V)),$$

and is called the *Gysin exact sequence*. The homomorphism π'^* is induced by the projection $\pi' \colon S(V) \to X$. The homomorphism α is defined by $\sigma(E, D) \mapsto \sum_{i=0}^{\mathrm{rank}(V)} (-1)^i \sigma(E \otimes \lambda^i(V), D \otimes 1)$. When X is compact, α is simply the product with $\chi(V) = \sum_{i=0}^{\mathrm{rank}(V)} (-1)^i [\lambda^i(V)]$. The element $\chi(V)$ is called the *Euler class* (or rather, the analog of the Euler class in $K_{\mathbb{C}}$-theory; cf. V.3) of the bundle V. Finally, if we replace X by $X \times \mathbb{R}^q$, and V by $V \times \mathbb{R}^q$, we also have the exact sequence

$$K_{\mathbb{C}}^{-q-1}(X) \xrightarrow{\pi'^*} K_{\mathbb{C}}^{-q-1}(S(V)) \longrightarrow K_{\mathbb{C}}^{-q}(X) \xrightarrow{\alpha} K_{\mathbb{C}}^{-q}(X) \xrightarrow{\pi'^*} K_{\mathbb{C}}^{-q}(S(V)),$$

where once again, α is defined by the product with the Euler class when X is compact.

1.14. Example. Let CP_n denote the complex projective space of $\mathbb{C}^{n+1}$, and let ξ be the canonical line bundle over CP_n (I.2.4). Then ξ may be identified with the quotient of $S^{2n+1} \times \mathbb{C}$ by the equivalence relation $(x, t) \sim (\lambda x, \lambda^{-1} t)$, for $\lambda \in S^1 \subset \mathbb{C}$. Therefore, $\underbrace{\xi \otimes \cdots \otimes \xi}_{k} = \xi^{\otimes k}$ may be identified with the quotient of $S^{2n+1} \times \mathbb{C}$ by the equivalence relation $(x, t) \sim (\lambda x, \lambda^{-k} t)$. Moreover, we can give

$\xi^{\otimes k}$ the metric defined by $\varphi_x((x,t),(x,t'))=t\bar{t}'$. It follows that $S(\xi^{\otimes k})$ may be identified with the "lens space" $S^{2n+1}/(\mathbb{Z}/k)$, (where $\mathbb{Z}/k$ acts on $S^{2n+1}\subset\mathbb{C}^{n+1}$ via the kth roots of the unity) by the map $(x,t)\mapsto\sqrt[k]{t}\cdot x$. Hence we obtain the exact sequence (cf. 1.13)

$$K_{\mathbb{C}}(P_n(\mathbb{C}))\xrightarrow{\alpha}K_{\mathbb{C}}(P_n(\mathbb{C}))\xrightarrow{\pi'^*}K_{\mathbb{C}}(S^{2n+1}/(\mathbb{Z}/k)\longrightarrow K_{\mathbb{C}}^1(P_n(\mathbb{C})),$$

where π'^* is a ring map and α is multiplication by the Euler class of $\xi^{\otimes k}$, i.e. $1-[\xi^{\otimes k}]$. We will utilize this exact sequence in the computation of $K_{\mathbb{C}}(S^{2n+1}/(\mathbb{Z}/k))$ in the next section.

Exercises (Section IV.8) 2, 11, 13.

2. Complex K-Theory of Complex Projective Spaces and Complex Projective Bundles

2.1. Since we do not consider real K-theory in this section, we denote complex K-theory $K_{\mathbb{C}}$ simply by the letter K. Hence $K(X)=K_{\mathbb{C}}(X)$, $K(X,Y)=K_{\mathbb{C}}(X,Y)$, etc.

2.2 Let V be a complex vector bundle with compact base X. We let $P(V)$ denote the bundle on X, whose fiber over a point x is $P(V_x)$. More precisely, the topology on $P(V)=\bigsqcup_{x\in X}P(V_x)$ is defined by the same procedure as in I.4.5: the functor φ in I.4.3 is replaced by the functor from $\mathscr{E}'_{\mathbb{C}}$ to *Top*, given by $E\mapsto P(E)$, where *Top* is the category of topological spaces, and $\mathscr{E}'_{\mathbb{C}}$ is the category whose objects are the finite dimensional complex vector spaces, and whose morphisms are the isomorphisms between them.

The purpose of this section is to compute $K_{\mathbb{C}}(P(V))$ in terms of $K_{\mathbb{C}}(X)$ (2.16). This is nontrivial even when X is a point, i.e. $P(V)=CP_n$ for some n. In this computation, an important role is played by the *canonical line bundle*, denoted by ξ or ξ_V, over $P(V)$: it is the bundle on X, whose fiber over x is the canonical line bundle on $P(V_x)$. The argument above shows that ξ_V has a well-defined topology. Moreover, ξ_V may be regarded as a line bundle over $P(V)$, since *locally*, we have isomorphisms $V\approx X\times\mathbb{C}^{n+1}$, and $P(V)\approx X\times CP_n$; hence $\xi_V\approx X\times\xi_n$.

2.3. Now let L be a line bundle over X. Then $P(L)$ is isomorphic to X by the canonical projection, and $P(V\oplus L)-P(L)$ may be identified with the line bundle $\xi_V^*\otimes\pi^*L=\mathrm{HOM}(\xi_V,\pi^*L)$, where $\pi: P(V)\to X$, over the space X (I.4.8.d)). More precisely, if $g:\xi_V\to L$ is a

$$\begin{array}{ccc}\xi_V & \xrightarrow{g} & L\\ \downarrow & & \downarrow\\ P(V) & \xrightarrow{\pi} & X\end{array}$$

general morphism over π(I.1.6), and if v is a nonzero vector of ξ_V, then the pair $(v, g(v))$ defines a point of $P(V \oplus L) - P(L)$, which does not depend on the choice of v in a fiber. In particular, the Thom space (cf. 1.1) of $\xi_V^* \otimes \pi^* L$ is homeomorphic to $P(V \oplus L)/P(L) = P(V \oplus L)/X$.

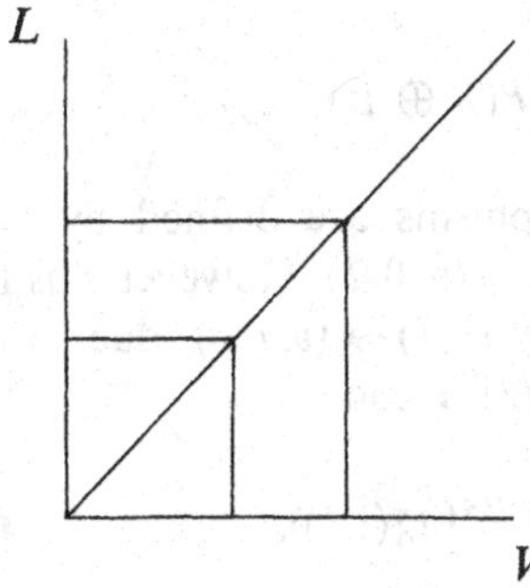

Fig. 19

2.4. Proposition. *We have the split exact sequence* (cf. II.4.13)

$$0 \longrightarrow K^r(P(V \oplus L) - X) \xrightarrow{j_r^*} K^r(P(V \oplus L)) \longrightarrow K^r(X) \longrightarrow 0.$$

If $U \in K^0(P(V \oplus L) - X) \approx K^0(\xi_V^* \otimes \pi^* L)$ denotes the Thom class of the line bundle $\xi_V^* \otimes \pi^*(L)$ over $P(V)$, then $j_0^*(U)$ is the Euler class (1.12) *of the vector bundle $\xi_{V \oplus L}^* \otimes \pi_1^*(L)$ over $P(V \oplus L)$, where $\pi_1 : P(V \oplus L) \to X$. Finally, if $x \in K^r(P(V \oplus L))$ and if x' is the restriction of x to $K^r(P(V))$ (note that $P(V) \subset P(V \oplus L)$), we have the formula $x j_0^*(U) = j_r^*(\Phi(x'))$, where $\Phi : K^r(P(V)) \to K^r(P(V \oplus L) - X)$ is the Thom isomorphism.*

Proof. Since $X \approx P(L)$ is a retract of $P(V \oplus L)$, the cohomology exact sequence II.4.13 implies the first part of the proposition. Let us now consider the commutative diagram

$$\begin{array}{ccc} \xi_V^* \otimes \pi^* L \approx P(V \oplus L) - P(L) & \xrightarrow{s} & P(V \oplus L \oplus L) - P(L) \approx \xi_{V \oplus L}^* \otimes \pi_1^* L \\ \downarrow & & \downarrow \\ P(V) & \longrightarrow & P(V \oplus L). \end{array}$$

Then the map s is a general vector bundle morphism (I.1.6), since

$$\xi_{V \oplus L}^* \otimes \pi_1^* L|_{P(V)} \approx \xi_V^* \otimes \pi^* L.$$

Hence the Thom class of $\xi_V^* \otimes \pi^* L$ is induced from the Thom class U' of $\xi_{V \oplus L}^* \otimes \pi_1^* L$ by s. Moreover, we have the following commutative diagram of

K-groups:

$$\begin{array}{ccc}
K(P(V \oplus L) - P(L)) & \xrightarrow{j_0^*} & K(P(V \oplus L)) \\
\uparrow{\scriptstyle s^*} & & \uparrow{\scriptstyle s'^*} \\
K(P(V \oplus L \oplus L) - P(L)) & \xrightarrow{\tilde{j}_0^*} & K(P(V \oplus L \oplus L))
\end{array}$$

In this diagram the horizontal homomorphisms are defined by "extension" (II.5.19), and s' is induced by the map $(v, l) \mapsto (v, 0, l)$. However s' is homotopic to $i': P(V \oplus L) \to P(V \oplus L \oplus L)$ defined by $(v, l) \to (v, l, 0)$, due to the homotopy $(v, l) \mapsto (v, l \cos \theta, l \sin \theta)$, for $\theta \in [0, \pi/2]$. Hence

$$j_0^*(U) = j_0^*(s^*(U')) = s'^*(\tilde{j}_0^*(U')) = i'^*(\tilde{j}_0^*(U')).$$

If i denotes the zero section of the line bundle $\xi_{V \oplus L}^* \otimes \pi_1^* L$; we also have the commutative diagram of K-groups

$$\begin{array}{ccc}
K(P(V \oplus L \oplus L) - P(L)) & \xrightarrow{\tilde{j}_0^*} & K(P(V \oplus L \oplus L)) \\
 & \searrow{\scriptstyle i^*} & \downarrow{\scriptstyle i'^*} \\
 & & K(P(V \oplus L)),
\end{array}$$

because the image of i is a compact subset of $P(V \oplus L \oplus L) - P(L)$. Since the Euler class is the restriction of the Thom class to the zero section (1.13), $j_0^*(U) = i^*(U')$ is the Euler class of $\xi_{V \oplus L}^* \otimes \pi_1^* L$.

Finally, let us consider an element x of $K^r(P(V \oplus L))$. Since $K^r(P(V \oplus L)) \approx K^r(P(V \oplus L) - X) \oplus K^r(X)$, we must check the formula $x j_0^*(U) = j_r^*(\Phi(x'))$ in two cases:

a) $x \in K^r(X) \approx K^r(P(L)) \subset K^r(P(V \oplus L))$. Since j^* and Φ are $K^*(X)$-module homomorphisms, we have $x j_0^*(U) = x j_0^* \Phi(1) = j_0^* \Phi(x \cdot 1) = j_0^* \Phi(x')$.

b) $x \in K^r(P(V \oplus L) - X)$ or $x = j_r^*(\tilde{x})$. Then $x j_0^*(U) = j_r^*(\tilde{x} U) = j_r^*(x' \cdot U) = j_r^*(\Phi(x'))$ by II.5.31 (where $(\alpha, \beta) \mapsto \alpha \cdot \beta$ is the product defined in II.5.30). $\square$

2.5. Theorem. *Let X be a compact space, and let $P_n = CP_n$ be the complex projective space of $\mathbb{C}^{n+1}$. Then $K^*(X \times P_n)$ is a free $K^*(X)$-module with basis $1, t, \ldots, t^n$, where $t = 1 - [\pi_n^* \xi_n^*]$ is the Euler class of the bundle $\pi_n^* \xi_n^*$, for $\pi_n: X \times P_n \to P_n$ (I.2.4). Moreover, we have $t^{n+1} = 0$, which implies $K^*(X \times P_n) \approx K^*(X)[t]/(t^{n+1})$.*

Proof. We prove the first part of the theorem by induction on n, beginning at $n = 0$. To obtain the $(n+1)$-stage from the n-stage, we consider the split exact sequence

$$0 \longrightarrow K^*(X \times P_{n+1}, X) \xrightarrow{j^*} K^*(X \times P_{n+1}) \longrightarrow K^*(X) \longrightarrow 0.$$

According to 2.4, the Thom isomorphism (denoted by Φ_n) identifies the first group of this sequence with $K^*(X\times P_n)$. To be more precise, let t_r denote the Euler class of $\pi_r^*\xi_r^*$. Since $t_{n+1}|_{K(X\times P_n)}=t_n$, we have $j^*\pi_n^*(t_n^\alpha)=t_{n+1}^{\alpha+1}$ by the last part of Proposition 2.4. Therefore, by the induction hypothesis, $j^*\Phi_n$ is an isomorphism between $K^*(X\times P_n)$ and the free sub-module of $K^*(X\times P_{n+1})$ with basis $t_{n+1}, t_{n+1}^2, \ldots, t_{n+1}^{n+1}$. Since the quotient module is isomorphic to $K^*(X)$, it follows that $1, t_{n+1}, \ldots, t_{n+1}^{n+1}$ is a basis for $K^*(X\times P_{n+1})$ as a $K^*(X)$-module.

Since the restriction of t to any point of P_n is 0, example II.5.10 shows that $t^{n+1}=0$. Hence $K^*(X\times P_n)\approx K^*(X)[t]/(t^{n+1})$. □

2.6. *Remark.* Instead of working with $K^*(Z)=\bigoplus_{r=0}^{\infty} K^{-r}(Z)$ in general, one could as well work with $K^*(Z)=K^0(Z)\oplus K^{-1}(Z)$, where the products $K^i\times K^j\to K^{i+j}$, for i and $j\in\mathbb{Z}/2$, are defined using Bott periodicity.

2.7. Corollary. *The relative group $K^*(X\times P_n, X\times P_k)$ for $k<n$, is the free sub-module of $K^*(X\times P_n)$ generated by $t^{k+1},\ldots,t^n$.*

Proof. This is a direct consequence of the split exact sequence of $K^*(X)$-modules

$$0\longrightarrow K^*(X\times P_n, X\times P_k)\xrightarrow{\alpha} K^*(X\times P_n)\xrightarrow{\beta} K^*(X\times P_k)\longrightarrow 0,$$

where β is surjective (hence α injective) by the previous proposition. □

2.8. Corollary. *We have $K^1(P_n)=0$ and $K^0(P_n)\approx\mathbb{Z}[t]/(t^{n+1})$, where $t=1-[\xi_n^*]$ is the Euler class of ξ_n^*.*

2.9. Corollary. *Let P_n and P_m be complex projective spaces, and let $\eta_1=\pi_1^*\xi_n^*$ and $\eta_2=\pi_2^*\xi_m^*$, where $\pi_1: P_n\times P_m\to P_n$ and $\pi_2: P_n\times P_m\to P_m$. Let x and y be the Euler classes of the line bundles η_1 and η_2. Then $K^1(P_n\times P_m)=0$ and $K^0(P_n\times P_m)\approx \mathbb{Z}[x,y]/(x^{n+1})(y^{m+1})$.*

2.10. Proposition. *Let L_1 and L_2 be line bundles over a compact base X. Then $\chi(L_1\otimes L_2)=\chi(L_1)+\chi(L_2)-\chi(L_1)\chi(L_2)$. Moreover, $z=\chi(L_1)$ is nilpotent, and $\chi(L_1^*)=-z-z^2-\cdots-z^n-\cdots$. Finally, $\chi(L_1^*\otimes L_2)$ may be written as $(\chi(L_2)-\chi(L_1))\cdot h$, where h is a unit element.*

Proof. Since in general $\chi(V)=\sum_{i=0}^{\mathrm{rank}(V)}(-1)^i\lambda^i(V)$ by 1.13, we have $\chi(V)=1-[V]$ if V is a line bundle. Hence,

$$\chi(L_1\otimes L_2)=1-[L_1\otimes L_2]=(1-[L_1])+(1-[L_2])-(1-[L_1])(1-[L_2]),$$

thus proving the first part of the proposition.

Now $\chi(L_1)$ obviously belongs to $K'(X)$, hence is nilpotent by II.5.9. If x is the Euler class of L_1^*, we must have the relation $x+z-zx=0$, since $\chi(L_1\otimes L_1^*)=0$.

Therefore $x = -z(1-z)^{-1} = -z - z^2 - \cdots - z^n - \cdots$ in the ring $K(X)$.

Finally, if we set $t_1 = \chi(L_1)$ and $t_2 = \chi(L_2)$, we have

$$\begin{aligned}\chi(L_1^* \otimes L_2) &= -t_2 + (1-t_2)(-t_1 - t_1^2 - \cdots - t_1^n - \cdots)\\ &= (t_2 - t_1)(1 + t_1 + t_1^2 + \cdots),\end{aligned}$$

which is of the form $(t_2 - t_1)h$, where h is a unit. □

2.11. Corollary. *Let u be the Euler class of $\pi^*\xi_n$, where $\pi: X \times P_n \to X$. Then $K^*(X \times P_n) \approx K^*(X)[u]/(u^{n+1})$.*

2.12. With the aid of 2.11, we now finish the computation begun in 1.14. If $L_{n,k}$ denotes $S^{2n+1}/(\mathbb{Z}/k)$, we have the exact sequence

$$\begin{array}{ccccccc} 0 \longrightarrow & K(P_n) & \xrightarrow{\alpha} & K(P_n) & \longrightarrow & K(L_{n,k}) & \longrightarrow 0, \\ & \wr\wr & & \wr\wr & & & \\ & \mathbb{Z}[u]/(u^{n+1}) & & \mathbb{Z}[u]/(u^{n+1}) & & & \end{array}$$

where α is multiplication by the Euler class of $\xi^{\otimes k}$, i.e. $1-(1-u)^k$. Therefore, $\tilde{K}(L_{n,k})$ is a finite group of at most k^n-torsion. For example, if $k=2$, then $\tilde{K}(L_{n,k}) \approx \tilde{K}(RP_{2n+1}) \approx \mathbb{Z}/2^n\mathbb{Z}$, with generator $[H']-1$, where H' is the complexification of the canonical real line bundle over RP_{2n+1} (I.2.4; cf. also 6.47).

2.13. Proposition. *Let X be a compact space, and let V be a complex vector bundle of rank n over X. Let u be the Euler class of the line bundle ξ_V (2.2). Then $K^*(P(V))$ is a free $K^*(X)$-module with basis $1, u, \ldots, u^{n-1}$. In particular, the homomorphism $K^*(X) \to K^*(P(V))$ is injective.*

Proof. If V is trivial, say $X \times \mathbf{C}^n$, then $\xi_V \approx \pi^*\xi_n$ where $\pi: P(V) \approx X \times P_n \to P_n$. Therefore, by 2.11, the proposition is true in this case. Now let $W_1, \ldots, W_r$ be a finite cover of X such that $V|_{W_i}$ is trivial, and let $e^i = u^{i-1}$. Now we apply Theorem 1.3 since $V|_Y$ is trivial when $Y \subset W_i$. □

2.14.1. Proposition. *Using the notation of 2.13, let us assume that $V = L_1 \oplus L_2 \oplus \cdots \oplus L_n$, where the L_i are line bundles. Then we have the relation $\prod_{i=1}^{n} (u - \chi(L_i)) = 0$, where u is the Euler class of ξ_V.*

Proof. Since the proposition is clear for $n=1$, we proceed by induction on n. Let us consider a line bundle $L_{n+1} = L$ over X, and the product $y = \prod_{i=1}^{n+1} (v - \chi(L_i))$, where v is the Euler class of the line bundle $\xi_{V \oplus L}$ over $P(V \oplus L)$.

Then

$$y = x \cdot \tau, \quad \text{where } x = \prod_{i=1}^{n} (v - \chi(L_i)),$$

and
$$\tau = -\chi(\xi^*_{V \oplus L} \otimes \pi_1^* L) \quad \text{for} \quad \pi_1 : P(V \oplus L) \longrightarrow X,$$

up to a unit element (2.10). Using the notation of 2.4, we have $x \cdot \tau = j_0^* \Phi(x')$, where $x' = \prod_{i=1}^{n} [u - \chi(L_i)] = 0$, by the induction hypothesis. Hence $y = 0$ as required. □

2.14.2. *Remark.* Let $\pi: P(V) \to X$ denote the canonical projection. Then the vector bundle $\xi_V^* \otimes (\pi^* L_1 \oplus \cdots \oplus \pi^* L_n) \approx \xi_V^* \otimes \pi^* V \approx \mathrm{HOM}(\xi_V, \pi^* V)$ has a canonical nonzero section, since ξ_V is a sub-bundle of $\pi^* V$. Therefore its Euler class which is $\prod_{i=1}^{n} (u - \chi(L_i))$ up to a unipotent element by 1.13 and 2.10, must be 0. This provides another proof of 2.14.1.

There still remains the task of determining the ring structure of $K^*(P(V))$, when V is an arbitrary complex vector bundle. For this and many other computations, we use the following theorem called the "*splitting principle*":

2.15. Theorem. *Let V be a complex vector bundle with compact base X. Then we can find a space $F(V)$ and π: $F(V) \to X$, which depend naturally on V, such that*

a) *the homomorphism $\pi^*: K^*(X) \to K^*(F(V))$ is injective,*

b) *the vector bundle $\pi^* V$ splits into the whitney sum of line bundles.*

Proof. We prove the theorem by induction on the rank of V. If the rank is equal to 1, we choose $F(V) = X$, of course. When the rank of V is greater than 1, we consider the projective bundle $P(V)$ associated with V. Now the canonical line bundle $\xi = \xi_V$ on $P(V)$ is a sub-bundle of $V' = p^* V$, where $p: P(V) \to X$. We set $F(V) = F(V'/\xi)$, and π equal to the composition $F(V'/\xi) \xrightarrow{\pi'} P(V) \xrightarrow{p} X$. By 2.13 and the inductive hypothesis, the homomorphism $K^*(X) \xrightarrow{\pi^*} K^*(F(V))$ is injective. Moreover, since $P(V)$ is compact, we may write $V' \approx \xi \oplus V'/\xi$ (I.5.13). Since $\pi'^*(V'/\xi)$ is the sum of line bundles by the inductive hypothesis, we see that $\pi^*(V) = \pi'^* p^*(V)$ is also the sum of line bundles. □

2.16. Theorem. *Let h be the class of the canonical line bundle ξ_V in $K(P(V))$. Then $K^*(P(V))$ is a free $K^*(X)$-module with basis $1, h, \ldots, h^{n-1}$. Moreover, h^n is determined by the relation*

$$h^n - [\lambda^1(V)] h^{n-1} + [\lambda^2(V)] h^{n-2} + \cdots + (-1)^n [\lambda^n(V)] = 0,$$

where $\lambda^i(V)$ is the ith exterior power of V (I.4.8.f)).

Proof. Since $u = 1 - h$, it is clear that $1, h, \ldots, h^{n-1}$ are a basis for $K^*(P(V))$ as a $K^*(X)$-module by 2.14. Now, to prove the relation in the theorem, we may assume that V is the sum of line bundles by the splitting principle (2.15). If $V=\bigoplus_{r=1}^{n} L_r$, we have

$$\lambda^1(V)=\bigoplus_{r=1}^{n} L_r,$$
$$\lambda^2(V)=\bigoplus_{r_1<r_2} L_{r_1}\otimes L_{r_2},$$
$$\vdots$$
$$\lambda^n(V)=L_1\otimes L_2\otimes\cdots\otimes L_n.$$

Hence $\lambda^i(V)$ may be expressed as the ith symmetric polynomial of the $[L_r]$ in the ring $K(X)$, and the relation may be written as

$$\prod_{i=1}^{n}(h-[L_i])=0,$$

which is equivalent to $\prod_{i=1}^{n}(u-\chi(L_i))=0$, since $\chi(L_i)=1-[L_i]$. Therefore, the theorem follows from 2.14. □

The next observations will be very useful in V.3.

2.17. Proposition. *For each vector bundle V of rank n, with compact base X, we define "characteristic classes" $c_i(V)\in K(X)$, for $i=0,\ldots,n$ such that $c_0(V)=1$. These characteristic classes satisfy the following axioms:*

1) The $c_i(V)$ are "natural", i.e. $c_i(V)=f^*(c_i(V'))$ for any general morphism $V\xrightarrow{g}V'$, which induces $f\colon X\to X'$ on the bases, X and X', of V and V' respectively; and which induces an isomorphism on each fiber.

$$\begin{array}{ccc} V & \xrightarrow{g} & V' \\ \downarrow & & \downarrow \\ X & \xrightarrow{f} & X' \end{array}$$

2) If V_1 and V_2 are vector bundles on X, then

$$c_k(V_1\oplus V_2)=\sum_{i+j=k} c_i(V_1)c_j(V_2).$$

3) If the rank of V is one, then $c_1(V)=\chi(V)=1-[V]$, and $c_i(V)=0$ for $i>1$.

Moreover, the characteristic classes $c_i(V)$ are uniquely determined by these axioms.

Proof. Let us first notice that the second axiom may be expressed more briefly as $c_t(V_1\oplus V_2)=c_t(V_1)c_t(V_2)$, where $c_t(V)=\sum_{i=0}^{\infty} t^i c_i(V)\in K(X)[t]$.

We now prove the uniqueness of these classes. If $\pi: F(V) \to X$ is the map described in 2.15, we have $\pi^*V = \bigoplus_{r=1}^{n} L_r$ where the L_r are line bundles. Therefore, we must have

$$\begin{aligned}\pi^*(c_i(V)) &= c_i\left(\bigoplus_{r=1}^{n} L_r\right)\\ &= \sum_{r_1<r_2<\cdots<r_i} c_1(L_{r_1})c_1(L_{r_2})\cdots c_1(L_{r_i})\\ &= \sum_{r_1<r_2<\cdots<r_i} \chi(L_{r_1})\chi(L_{r_2})\cdots\chi(L_{r_i}).\end{aligned}$$

Since π^* is injective (2.13), the classes c_i are determined by the axioms.

To prove existence, we consider the ring $K(P(V))$. By 2.13, u^n is a linear combination of $1, u, \ldots, u^{n-1}$, with coefficients in $K(X)$. Now we define the $c_i(V)$ by the equation

$$u^n - c_1(V)u^{n-1} + \cdots + (-1)^n c_n(V) = 0,$$

and $c_i(V)=0$ for $i > \mathrm{rank}(V)$. With this definition, axioms 1 and 3 are trivial. To verify axiom 2, let us consider the space $F(V_2')$, where $V_2' = \pi_1^* V_2$, with $\pi_1: F(V_1) \to X$. Let π be the composition $F(V_2') \to F(V_1) \to X$. Then $\pi^*: K(X) \to K(F(V_2'))$ is injective, and π^*V_1 and π^*V_2 split into direct sums of line bundles. Therefore, using the homomorphism π^* and the naturality of the characteristic classes, we see that it suffices to verify axiom 2 when V_1 and V_2 are sums of line bundles. By induction on the rank of V_2, it is actually enough to verify the relation $c_t(V \oplus L) = c_t(V)c_t(L)$ for $V = V_1$ a sum of line bundles, and L a line bundle. If we let $V = \bigoplus_{i=1}^{n} L_i$, then Theorem 2.14 enables us to explicitly compute the $c_i(V)$ as

$$\begin{aligned}c_1(V) &= \sum \chi(L_r)\\ c_2(V) &= \sum_{r_1<r_2} \chi(L_{r_1})\chi(L_{r_2})\\ &\vdots\\ c_n(V) &= \chi(L_1)\cdots\chi(L_n).\end{aligned}$$

Therefore,

$$\begin{aligned}c_1(V\oplus L) &= \sum \chi(L_r) + \chi(L) = c_1(V) + c_1(L),\\ c_2(V\oplus L) &= \sum_{r_1<r_2} \chi(L_{r_1})\chi(L_{r_2}) + \left(\sum_{r=1}^{n} \chi(L_r)\right)\chi(L) = c_2(V) + c_1(V)c_1(L),\end{aligned}$$

$$
\begin{aligned}
c_i(V \oplus L) = & \sum_{r_1 < r_2 < \cdots < r_i} \chi(L_{r_1}) \cdots \chi(L_{r_i}) \\
& + \sum_{r_1 < r_2 < \cdots < r_{i-1}} \chi(L_{r_1}) \cdots \chi(L_{r_{i-1}}) \chi(L) \\
& = c_i(V) + c_{i-1}(V) c_1(L), \quad \text{for } i \leqslant n,
\end{aligned}
$$

$$c_{n+1}(V \oplus L) = \chi(L_1)\chi(L_2) \cdots \chi(L_n)\chi(L) = c_n(V)c_1(L),$$

and $\quad c_i(V \oplus L) = 0 \quad \text{for } i > n+1.$

These relations, which may be simplified to $c_t(V \oplus L) = c_t(V)c_t(L)$, are the ones we wished to verify. □

2.18. We may in fact determine the $c_i(V)$ in terms of the exterior powers $\lambda^i(V)$ by the following method. By 2.16, we have the equation

$$h^n - [\lambda^1(V)]h^{n-1} + \cdots + (-1)^n[\lambda^n(V)] = 0.$$

If we replace h by $1-u$, we obtain the equation

$$(u-1)^n - [\lambda^1(V)](u-1)^{n-1} + [\lambda^2(V)](u-1)^{n-2} + \cdots + [\lambda^n(V)] = 0.$$

Therefore, if we identify the coefficients of u^i as $(-1)^{n-i}c_{n-i}(V)$, we must have

$$
\begin{aligned}
c_1(V) &= \tbinom{n}{1}[\lambda^0(V)] - \tbinom{n-1}{0}[\lambda^1(V)], \\
c_2(V) &= \tbinom{n}{2}[\lambda^0(V)] - \tbinom{n-1}{1}[\lambda^1(V)] + \tbinom{n-2}{0}[\lambda^2(V)], \\
c_3(V) &= \tbinom{n}{3}[\lambda^0(V)] - \tbinom{n-1}{2}[\lambda^1(V)] + \tbinom{n-2}{1}[\lambda^2(V)] - \tbinom{n-3}{0}[\lambda^3(V)], \\
& \cdots\cdots\cdots\cdots\cdots\cdots
\end{aligned}
$$

and $\quad c_n(V) = [\lambda^0(V)] - [\lambda^1(V)] + [\lambda^2(V)] + \cdots + (-1)^n[\lambda^n(V)].$

Another interpretation of these results will be given in IV.7 (in the framework of real and complex K-theory).

Exercises (Section IV.8) 3, 10.

3. Complex K-Theory of Flag Bundles and Grassmann Bundles. K-Theory of a Product

3.1. As in Section IV.2, the letter K will again denote complex K-theory, $K_{\mathbb{C}}$.

3.2. Let E be a complex vector space of dimension n. A *flag* in E is a sequence of subspaces $0 = E_0 \subset E_1 \subset \cdots \subset E_n = E$, where E_i is of dimension i. We denote the set of flags in E by $F(E)$; it may be provided with a topology in the following way. If we choose a basis in E, i.e. an isomorphism $\mathbb{C}^n \approx E$, we see that the group $\mathrm{GL}_n(\mathbb{C})$

acts transitively in $F(E) \approx F(\mathbb{C}^n)$, and that the subgroup leaving the canonical flag $0 \subset \mathbb{C} \subset \mathbb{C}^2 \subset \cdots \subset \mathbb{C}^n$ fixed, is the subgroup τ_n^+ of upper triangular matrices. Hence $F(E) \approx \mathrm{GL}_n(\mathbb{C})/\tau_n^+$ may be provided with the quotient topology of $\mathrm{GL}_n(\mathbb{C})$; this topology is independent of the choice of basis. Moreover, $F(E)$ is a compact space, since $\mathrm{GL}_n(\mathbb{C})/\tau_n^+ \approx \mathrm{U}(n)/T^n$, where T^n is the group of diagonal matrices with elements of norm 1 on the diagonal.

3.3. Now let V be a complex vector bundle over a compact space X. We define the "flag bundle" $F(V)$ as the bundle on X whose fiber over $x \in X$ is $F(V_x)$. More precisely, $F(V) = \coprod_{x \in X} F(V_x)$, and the topology on $F(V)$ is defined by the same procedure as in I.4.5 (compare with 2.2).

The space $F(V)$ may also be constructed by induction on the rank of V, by the following procedure. Let $P(V)$ be the projective bundle of V, and let $V' = p^* V$ where $p: P(V) \to X$. Then $F(V) \approx F(V'/\xi)$, where ξ is the canonical line bundle over $P(V)$. More precisely, we have a bijection $F(V) \to F(V'/\xi)$: it associates the flag $0 \subset E_1 \subset \cdots \subset E_{n-1} \subset V_x$ in V_x, with the flag over $\{E_1\} \in P(V)$ defined by $0 \subset E_2/E_1 \subset \cdots \subset V_x/E_1$. This is clearly a continuous map, hence a homeomorphism since $F(V)$ and $F(V'/\xi)$ are compact (one may also define a continuous map in the opposite direction). This construction also shows that the space $F(V)$, introduced in 2.15 to prove the splitting principle, is the same (up to isomorphism) as the one considered here.

3.4. Over the space $F(V)$, we have a sequence of bundles $0 \subset V_1 \subset V_2 \subset \cdots \subset V_n = \pi^* V$, where $\pi: F(V) \to X$, and where the fiber of V_i over the flag $\Delta = \{0 \subset E_1 \subset \cdots \subset E_n = V_x\} \in F(V_x)$ is the set of vectors v belonging to E_i (with the topology induced by the inclusion $V_i \subset \pi^* V$). According to I.5.14 applied $(n-1)$ times, the quotients V_i/V_{i-1} are well-defined line bundles L_i over $F(V)$, with $\bigoplus_{i=1}^{n} L_i \approx \pi^* V$. We denote the class of L_i in $K(F(V))$ by h_i.

3.5. Theorem. *Let X be a compact space, and let V be a complex vector bundle over X of rank n. Then $K^*(F(V))$ is a free $K^*(X)$-module of rank $n!$, and with basis, the products $h_1^{r_1} h_2^{r_2} \cdots h_{n-1}^{r_{n-1}}$ for $r_i \leqslant n-i$.*

Proof. We prove this proposition by induction on n. Assume that $K^*(F(V'/\xi))$ is a free $K^*(P(V))$-module with basis, the products $h_2^{r_2} h_3^{r_3} \cdots h_{n-1}^{r_{n-1}}$ for $r_i \leqslant n-i$. Since $K^*(P(V))$ is a free $K^*(X)$-module with basis $h_1^{r_1}$ for $r_1 \leqslant n-1$ (2.13), the theorem is proved. □

To avoid the assymmetric role played by the h_i, we also prove the following:

3.6. Theorem. *Let*

$$\varphi: K^*(X)[x_1, \ldots, x_n] \longrightarrow K^*(F(V))$$

be the $K^*(X)$-algebra homomorphism sending x_i to h_i. Then φ is surjective, and its kernel is the ideal I generated by the elements $\sigma_i - [\lambda^i(V)]$, where σ_i denotes the ith elementary symmetric function of the x_i. Hence φ induces an isomorphism

$$\varphi' : K^*(X)[x_1, \ldots, x_n]/I \approx K^*(F(V))$$

Proof. By 3.5, φ is surjective. On the other hand, since $\bigoplus_{i=1}^{n} L_i = V$, it is clear that $\sigma_i - [\lambda^i(V)]$ belongs to the kernel of φ. By a well-known theorem in algebra (3.28), $K^*(X)[x_1, \ldots, x_n]$ is a free $K^*(X)[\sigma_1, \ldots, \sigma_n]$-module with basis $x_1^{r_1} \cdots x_{n-1}^{r_{n-1}}$, where $r_i \leqslant n-i$. Hence, the quotient M of $K^*(X)[x_1, \ldots, x_n]$, by the ideal generated by $\sigma_i - [\lambda^i(V)]$, is a free $K^*(X)$-module with basis, the products $x_1^{r_1} \cdots x_{n-1}^{r_{n-1}}$. Since $K^*(F(V))$ is a free $K^*(X)$-module, and since φ induces a homomorphism $\varphi' : M \to K^*(F(V))$ which sends the basis of M onto the basis of $K^*(F(V))$, φ' is an isomorphism. □

3.7. If E is a complex vector space of dimension n, we call the set of q-dimensional subspaces of E, the *Grassmannian* of q-planes in E. If we denote this set by $G_q(E)$, we saw in I.7.16 that $G_q(E)$ may be identified with $\mathrm{U}(n)/\mathrm{U}(q) \times \mathrm{U}(n-q)$, hence may be provided with the topology of a compact space. Moreover, we have a continuous map $F(E) \to G_q(E)$: it associates each flag $E_1 \subset E_2 \subset \cdots \subset E_n = E$ with the qth element E_q. Up to isomorphism, this map coincides with the map $\mathrm{U}(n)/T^n \to \mathrm{U}(n)/\mathrm{U}(q) \times \mathrm{U}(n-q)$, induced by the inclusion of T^n in $\mathrm{U}(q) \times \mathrm{U}(n-q)$.

3.8. Lemma. *The map $\pi : F(\mathbf{C}^n) \to G_q(\mathbf{C}^n)$ is a fibration with fiber $F(\mathbf{C}^q) \times F(\mathbf{C}^{n-q})$.*

Proof. By "fibration", we mean here that for each point S^0 of $G_q(\mathbf{C}^n)$, we can find a neighbourhood W of S^0, such that $\pi^{-1}(W) \approx W \times F(\mathbf{C}^q) \times F(\mathbf{C}^{n-q})$. To prove this, we first give a slightly different description of the space $F(E)$, as the set of sequences $(L_1, \ldots, L_n)$ of linearly independent one-dimensional subspaces of E, which are mutually orthogonal (take $E_i = L_1 \oplus \cdots \oplus L_i$ or, conversely, $L_i = E_{i+1} \cap E_i^{\perp}$). Next we choose W to be the set of elements S of $G_q(\mathbf{C}^n)$, such that the orthogonal projection of S^0 on S is an isomorphism. If we fix an orthonormal basis $B^0 = \{e_1^0, \ldots, e_q^0\}$ of S^0, and an orthonormal basis $C^0 = \{e_{q+1}^0, \ldots, e_n^0\}$ of $S^{0\perp}$, then the orthogonal projection B (resp. C) of S^0 (resp. $S^{0\perp}$) is a basis of S (resp. $S^{\perp}$). The Gram-Schmidt orthonormalization process, applied to $B \cup C$, gives a new orthonormal basis of $\mathbf{C}^n$, hence a unitary isomorphism $\alpha_S : \mathbf{C}^n \to \mathbf{C}^n$ which depends continuously on $S \in W$, and such that $\alpha_S(S^0) = S$. The isomorphism

$$\theta : \pi^{-1}(W) \longrightarrow W \times F(S^0) \times F(S^{0\perp})$$

is now defined by the formula

$$\theta(L_1, \ldots, L_n) = (S, \Delta, \Gamma),$$

where

$$S=L_1\oplus\cdots\oplus L_q,$$
$$\Delta=(\alpha_S^{-1}(L_1),\ldots,\alpha_S^{-1}(L_q)),$$
and
$$\Gamma=(\alpha_S^{-1}(L_{q+1}),\ldots,\alpha_S^{-1}(L_n)).$$

The inverse isomorphism θ^{-1} is defined by the formula

$$\theta^{-1}(S,\Delta^0,\Gamma^0)=(\alpha_S(L_1^0),\ldots,\alpha_S(L_q^0),\ldots,\alpha_S(L_n^0)),$$
for
$$\Delta^0=(L_1^0,\ldots,L_q^0)\quad\text{and}\quad\Gamma^0=(L_{q+1}^0,\ldots,L_n^0).\ \square$$

3.9. Theorem. *Let X be a compact space, and let*

$$\beta\colon X\times F(\mathbb{C}^n)\longrightarrow X\times G_p(\mathbb{C}^n)$$

be the continuous map defined by $(x,e)\mapsto(x,\pi(e))$. Then $\beta^*\colon K^*(X\times G_p(\mathbb{C}^n))\to K^*(X\times F(\mathbb{C}^n))$ is injective, and its image is the invariant subgroup $K^*(X\times F(\mathbb{C}^n))^G$, where $G=S_p\times S_{n-p}$ acts on $F(\mathbb{C}^n)$ by permutation of the L_i (cf. the description of $F(\mathbb{C}^n)$ given in the proof of 3.8).

Proof. By 3.8 we have the fibration

$$F(\mathbb{C}^p)\times F(\mathbb{C}^{n-p})\longrightarrow X\times F(\mathbb{C}^n)\xrightarrow{\ \beta\ }X\times G_p(\mathbb{C}^n).$$

If we denote the classes of the canonical line bundles on $X\times F(\mathbb{C}^n)$ by $h_1,\ldots,h_n$, then the restrictions of the products

$$h_1^{r_1}\cdots h_{p-1}^{r_{p-1}}h_{p+1}^{s_{p+1}}\cdots h_{n-1}^{s_{n-1}},\ \text{for } r_i\leqslant p-i \text{ and } s_j<n-j+1,\ \text{to } \beta^{-1}(X\times Y),$$

where Y is chosen so that $\pi^{-1}(Y)\approx Y\times F(\mathbb{C}^p)\times F(\mathbb{C}^{n-p})$, are a basis of $K^*(\beta^{-1}(X\times Y))$ as a $K^*(X\times Y)$-module, by 3.5 applied twice. Therefore, by 1.3 $K^*(X\times F(\mathbb{C}^n))$ is a free $K^*(X\times G_p(\mathbb{C}^n))$-module with basis, the products above. In particular, the homomorphism $K^*(X\times G_p(\mathbb{C}^n))\to K^*(X\times F(\mathbb{C}^n))$ is injective.

Now the map $F(\mathbb{C}^n)\xrightarrow{\pi}G_p(\mathbb{C}^n)$ is equivariant with respect to the trivial action of $S_p\times S_{n-p}$ on the Grassmannian. It follows that β^* sends $K^*(X\times G_p(\mathbb{C}^n))$ into the invariant part $K^*(X\times F(\mathbb{C}^n))^G$, under the action of $G=S_p\times S_{n-p}$. To compute $K^*(X\times F(\mathbb{C}^n))^G$, we write $K^*(X\times F(\mathbb{C}^n))$ as the quotient of the polynomial algebra $K^*(X)[x_1,\ldots,x_n]$, by the ideal generated by the elementary symmetric polynomials $\sigma_1,\ldots,\sigma_n$ (3.6). Let τ_i (resp. γ_j) be the ith elementary symmetric polynomial of $x_1,\ldots,x_p$ (resp. the jth elementary symmetric polynomial of $x_{p+1},\ldots,x_n$). Then an elementary computation shows that $K^*(X\times F(\mathbb{C}^n))^G$ may be identified with the quotient of $K^*(X)[\tau_1,\ldots,\tau_p,\gamma_1,\ldots,\gamma_{n-p}]$ by the ideal generated by $\sigma_r=\sum_{i=0}^{r}\tau_i\gamma_{r-i}$, for $r=1,\ldots,n$ (with the convention $\tau_0=\gamma_0=1$). This is the formula expressing the rth elementary symmetric polynomial of

$x_1, \ldots, x_n$ in terms of the τ's and the γ's, and which is obtained by computing the product $\prod_{r=1}^{n} (1+tx_r) = \prod_{i=1}^{p} (1+tx_i) \prod_{j=1}^{n-p} (1+tx_{p+j})$, in two different ways.

Finally, let us consider the canonical vector bundle ξ of rank p on $G_p(\mathbb{C}^n)$ (cf. I.7.8). It may be identified with the subspace of $G_p(\mathbb{C}^n) \times \mathbb{C}^n$, consisting of pairs (X, x), where X is a p-plane in $\mathbb{C}^n$ and x is a vector of X. We define $\xi^\perp$ to be its "orthogonal", i.e. the subset of $G_p(\mathbb{C}^n) \times \mathbb{C}^n$ consisting of pairs (X, x) where $X \in G_p(\mathbb{C}^n)$ and $x \in X^\perp$. We set $T = \delta^* \xi$ and $T^\perp = \delta^* \xi^\perp$, where $\delta: X \times G_p(\mathbb{C}^n) \to G_p(\mathbb{C}^n)$, $r_i = [\lambda^i(T)]$ for $i \leqslant p$, and $s_j = [\lambda^j(T^\perp)]$ for $j \leqslant n-p$. Let

$$g: K^*(X)[\tau_1, \ldots, \tau_p, \gamma_1, \ldots, \gamma_{n-p}] \longrightarrow K^*(X \times G_p(\mathbb{C}^n))$$

be the $K^*(X)$-algebra homomorphism sending τ_i to r_i and γ_j to s_j. Since $T \oplus T^\perp$ is trivial, this homomorphism is zero on the ideal generated by $\sigma_r = \sum_{i=0}^{r} \tau_i \gamma_{r-i}$. Therefore, it defines a homomorphism

$$K^*(X \times F(\mathbb{C}^n))^G \approx K^*(X) \otimes K(F(\mathbb{C}^n))^G \longrightarrow K^*(X \times G_p(\mathbb{C}^n)).$$

Because $T = \bigoplus_{i=1}^{p} L_i$ and $T^\perp = \bigoplus_{j=p+1}^{n} L_j$, the composition

$$K^*(X) \otimes K(F(\mathbb{C}^n))^G \longrightarrow K^*(X \times G_p(\mathbb{C}^n)) \longrightarrow K^*(X) \otimes K(F(\mathbb{C}^n))^G$$

is the identity. Since $K^*(X \times G_p(\mathbb{C}^n)) \to K^*(X) \otimes K(F(\mathbb{C}^n))^G$ is injective, the theorem is proved. □

3.10. Corollary. *Let $d = \binom{n}{p}$. Then there exist integral polynomials $P_1, \ldots, P_d$ of the classes r_i and s_j, such that $K^*(X \times G_p(\mathbb{C}^n))$ is a free $K^*(X)$-module with basis $P_1, \ldots, P_d$.*

Proof. We have the inclusions as $\mathbb{Z}$-modules

$$K(F(\mathbb{C}^n)) \supset K(F(\mathbb{C}^n))^G \supset \mathbb{Z},$$

where $K(F(\mathbb{C}^n))$ is a free $\mathbb{Z}$-module of rank $n!$, and $K(F(\mathbb{C}^n))$ is a free $K(F(\mathbb{C}^n))^G$-module of rank $p!\,(n-p)!$ by the first part of the proof of 3.9. Therefore, $K(F(\mathbb{C}^n))^G$ is a free $\mathbb{Z}$-module of rank d with basis, classes of suitable polynomials $P_1, \ldots, P_d$ of the τ_i and γ_j. It follows that $K^*(X \times G_p(\mathbb{C}^n)) \approx K^*(X) \otimes K(F(\mathbb{C}^n))^G$ is a free $K^*(X)$-module with basis, the polynomials P_α evaluated on the r_i and s_j. □

3.11. Corollary. *Let $d = \binom{n}{p}$. Then $K^{-1}(G_p(\mathbb{C}^n)) = 0$, and $K(G_p(\mathbb{C}^n))$ is a free group of rank d.*

Now let V be a vector bundle of rank n, with compact base X. Since the functor $E \mapsto G_p(E)$ from the category of n-dimensional complex vector spaces to the category of compact spaces, is "continuous" in an obvious sense, the method of I.4.5 enables us to construct a compact space $G_p(V)$, fibered over X with fiber, a Grassmannian (when $p=1$ this reduces to the definition of $P(V)$ cf. 2.2). We call T the canonical p dimensional vector bundle over $G_p(V)$, and $T^\perp$, its "orthogonal" with respect to an arbitrary metric on V.

3.12. Theorem. *Let*

$$\psi: K^*(X)[\tau_1,\ldots,\tau_p,\gamma_1,\ldots,\gamma_{n-p}] \longrightarrow K^*(G_p(V))$$

be the $K^*(X)$-algebra homomorphism sending τ_i to $r_i=[\lambda^i(T)]$ and γ_j to $s_j=[\lambda^j(T^\perp)]$. Then ψ is surjective with kernel the ideal J generated by the $\sigma_r-\lambda^r(V)$ where $\sigma_r=\sum_{i+j=r}\tau_i\gamma_j$. Hence ψ induces an isomorphism

$$\psi': K^*(X)[\tau_1,\ldots,\tau_p,\gamma_1,\ldots,\gamma_{n-p}]/J \longrightarrow K^*(G_p(V)).$$

Proof. Since $T\oplus T^\perp$ is isomorphic to π^*V, where $\pi: G_p(V)\to X$, it is clear that the kernel of ψ contains the ideal generated by the $\sigma_r-\lambda^r(V)$. By Theorem 1.3, $K^*(G_p(V))$ is a free $K^*(X)$-module with basis, the polynomials $P_1,\ldots,P_d$ defined in 3.10. Hence ψ' is well-defined and surjective.

On the other hand, we have the commutative diagram

$$\begin{array}{ccc} K^*(X)[\tau_1,\ldots,\tau_p,\gamma_1,\ldots,\gamma_{n-p}]/J & \xrightarrow{\psi'} & K^*(G_p(V)) \\ \theta\downarrow & & \downarrow \\ K^*(X)[x_1,\ldots,x_n]/I & \xrightarrow{\varphi'} & K^*(F(V)), \end{array}$$

where φ' is the isomorphism defined in 3.6, and where θ sends τ_i (resp. γ_j) to the ith (resp. j^{th}) elementary symmetric polynomial of $x_1,\ldots,x_p$ (resp. $x_{p+1},\ldots,x_n$). Since θ is clearly injective, ψ' must also be injective. □

3.13. The observations above may be extended to "generalized flag bundles". More precisely, if $p_1,\ldots,p_s$ are integers such that $p_1+p_2+\cdots+p_s=n$, we consider the bundle $F_{p_1,\ldots,p_s}(V)$ over X, whose fiber over $x\in X$ is the set of orthogonal subspaces $L_1,\ldots,L_s$, such that $L_1\oplus\cdots\oplus L_s=V$ with $\mathrm{Dim}(L_i)=p_i$. On $F_{p_1,\ldots,p_s}(V)$, we have canonical bundles $T_1,\ldots,T_s$ of respective ranks $p_1,\ldots,p_s$. The following theorem may be proved in the same way as Theorem 3.12:

3.14. Theorem. *Let*

$$\psi: K^*(X)[\tau_i^j] \longrightarrow K^*(F_{p_1,\ldots,p_s}(V)), \quad \text{for } 1\leqslant i\leqslant p_j,$$

be the $K^*(X)$-algebra homomorphism sending τ_i^j to $[\lambda^i(T_j)]$. Then ψ is surjective with kernel, the ideal generated by $\sigma_r-[\lambda^r(V)]$ where $\sigma_r=\sum_{j_1+\cdots+j_s=r}\tau_{j_1}^1\cdot\tau_{j_2}^2\cdots\tau_{j_s}^s$.

3.15. *Example.* Let X be the space $U(n)/U(p_1)\times\cdots\times U(p_s)$ where $p_1+\cdots+p_s=n$. Then $K^{-1}(X)=0$, and $K(X)$ is a free group of rank $n!/p_1!p_2!\cdots p_s!$.

3.16. The above results have been expressed in terms of exterior powers of the vector bundles involved. Sometimes it is more convenient to work with the characteristic classes of V, $c_i(V)$, rather than $\lambda^i(V)$. For example, Theorem 3.14 (which implies all the others) may be expressed in the following form:

3.17. Theorem. *Let*

$$\psi: K^*(X)[\tau_i^j] \longrightarrow K^*(F_{p_1,\ldots,p_s}(V)), \quad \text{for} \quad 1 \leqslant i \leqslant p_j,$$

be the $K^*(X)$-algebra homomorphism sending τ_i^j to $c_i(T_j)$. Then ψ is surjective with kernel, the ideal generated by $\sigma_r - c_r(V)$ where $\sigma_r = \sum_{j_1+\cdots+j_s=r} c_{j_1}^1\cdot c_{j_2}^2\cdots c_{j_s}^s$.

3.18. *Example.* Assume $s=2$ and $p_1=1$. If we set $\tau_1^1=\tau_1$ and $\tau_j^2=\gamma_j$, then we obtain the relations

$$\begin{aligned}
&\gamma_1+\tau_1=c_1(V),\\
&\gamma_2+\gamma_1\tau_1=c_2(V),\\
&\cdot\quad\cdot\quad\cdot\quad\cdot\quad\cdot\quad\cdot\quad\cdot\\
&\gamma_{n-1}+\gamma_{n-2}\gamma_1=c_{n-1}(V)\\
&\gamma_{n-1}\gamma_1=c_n(V).
\end{aligned}$$

Therefore, if we set $u=c_1(T_1^*)$, then $K^*(F_{1,n-1}(V))\approx K^*(P(V))$ is a free $K^*(X)$-module, with basis $1, u, \ldots, u^{n-1}$. Moreover, we have the relation

$$u^n - c_1(V)u^{n-1} + \cdots + (-1)^n c_n(V) = 0.$$

Hence, we recover some of the observations in IV.2. If we set $h=[T_1]$, we can also prove Theorem 2.16 by the same method.

3.19. Corollary. *Let $c_i=c_i(T)$ and $d_j=c_j(T^\perp)$, for $i\leqslant p$ and $j\leqslant n-p$, be the characteristic classes of the vector bundles T and $T^\perp$ on $G_p(\mathbb{C}^n)$. Let*

$$\psi: \mathbb{Z}[\tau_1,\ldots,\tau_p,\gamma_1,\ldots,\gamma_{n-p}] \longrightarrow K(G_p(\mathbb{C}^n))$$

be the $\mathbb{Z}$-algebra homomorphism sending τ_i to c_i and γ_j to d_j. Then ψ is surjective with kernel the ideal generated by products of the form $\sigma_r = \sum_{i+j=r} \tau_i\gamma_j$.

3.20. *Example.* Let $X=U(4)/U(2)\times U(2)=G_2(\mathbb{C}^4)$. If we let $x=c_1$ and $y=c_2$, a trivial computation shows that $K(X)$ is a free group of rank 6, with basis $1, x, x^2, y, y^2, xy$. The multiplication operation in $K(X)$ is given by the relations $x^3=2xy$ and $x^2y=y^2$ (these relations imply $xy^2=0$ and $y^3=0$).

3.21. An advantage of the formulation using characteristic classes, c_i and d_j, is that these classes are nilpotent (2.10 and 2.17). This enables us to compute proj lim $K(G_p(\mathbf{C}^n))$. More precisely, if X is any topological space, we write $\mathscr{K}(X)$ for proj lim $K(X_\alpha)$, where X_α runs through the set of compact subsets of X. If (X_n) is a cofinal system of compact spaces, we have $\mathscr{K}(X) = \text{proj lim } K(X_n)$. In particular, if p is fixed and if $X = \text{inj lim } G_p(\mathbf{C}^n)$ is provided with the inductive limit topology, then $\mathscr{K}(X) \approx \text{proj lim } K(G_p(\mathbf{C}^n))$. Note that X has the same homotopy type as the space $BU(p)$ considered in I.7.14.

3.22. Theorem (Atiyah [3]). *Let*

$$\Gamma : \mathbf{Z}[[\tau_1, \ldots, \tau_p]] \longrightarrow \text{proj lim } K(G_p(\mathbf{C}^n))$$

be the $\mathbf{Z}$*-algebra homomorphism sending the formal sum*

$$\sum a_{i_1 \cdots i_p} (\tau_1)^{i_1} \cdots (\tau_p)^{i_p} \quad \text{to} \quad \sum a_{i_1 \cdots i_p} (c_1)^{i_1} \cdots (c_p)^{i_p}.$$

Then Γ *is an isomorphism. In particular,* $\mathscr{K}(BU(p)) \approx \mathbf{Z}[[c_1, \ldots, c_p]]$, *where the* c_i *are the characteristic classes of the canonical vector bundle over* $BU(p)$ *(in an obvious sense).*

Proof. The relations of Corollary 3.19 may be explicitly written as

$$\begin{cases} \tau_1 + \gamma_1 = 0 \\ \tau_2 + \tau_1 \gamma_1 + \gamma_2 = 0 \\ \cdot \quad \cdot \quad \cdot \quad \cdot \quad \cdot \quad \cdot \\ \tau_p \gamma_{n-2p} + \tau_{p-1} \gamma_{n-2p+1} + \cdots + \gamma_{n-p} = 0 \end{cases}$$

$$\begin{cases} \tau_p \gamma_{n-2p+1} + \tau_{p-1} \gamma_{n-2p+2} + \cdots + \tau_1 \gamma_{n-p} = 0 \\ \cdot \quad \cdot \quad \cdot \quad \cdot \quad \cdot \quad \cdot \quad \cdot \quad \cdot \quad \cdot \quad \cdot \\ \tau_p \gamma_{n-p} = 0. \end{cases}$$

From the first $n-p$ relations, it follows that the γ's are functions of the τ's. More precisely, γ_j is a polynomial of weight j in the τ's. The last p relations may be written as $P_i(\tau_1, \ldots, \tau_p) = 0$, where P_i is isobaric of weight $n-p+i$. It follows that $K(G_p(\mathbf{C}^n)) \approx \mathbf{Z}[\tau_1, \ldots, \tau_p]/I_{p,n}$, where $I_{p,n}$ is an ideal generated by polynomials of order $> \dfrac{n-p}{p}$. If we denote the ideal generated by *all* polynomials of order $\geqslant r$ by I_r, we therefore have $I_{p,n} \subset I_r$ when $n > rp + p$.

On the other hand, let us denote the Euler classes of the canonical line bundles over $F(\mathbf{C}^n)$ by $u_1, \ldots, u_n$. They are induced from the Euler class of the canonical line bundle over $P(\mathbf{C}^n)$, by the canonical maps $F(\mathbf{C}^n) \to P(\mathbf{C}^n)$. Therefore $(u_i)^n = 0$ by 2.11. Moreover, Theorem 3.6, written in terms of characteristic classes as in 3.14, shows that $K(F(\mathbf{C}^n))$ is a free $\mathbf{Z}$-module, with basis $u_1^{r_1} \cdots u_{n-1}^{r_{n-1}}$ where $r_i \leqslant n-i$. Hence, if $\alpha \in \widetilde{K}(F(\mathbf{C}^n))$, then α^m is a sum of monomials $u_1^{r_1} \cdots u_{n-1}^{r_{n-1}}$, where $r_1+\cdots+r_{n-1}\geqslant m$. If we choose $m>(n-1)^2$, at least one of the r_i is $\geqslant n$, and this shows that $(\tilde{K}(F(\mathbb{C}^n))^m=0$. Therefore, every element of $\tilde{K}(G_p(\mathbb{C}^n))\subset \tilde{K}(F(\mathbb{C}^n))$ (3.9) is nilpotent of order $\leqslant(n-1)^2+1$, a fact which may be expressed in the form $I_r\subset I_{p,n}$ when $r>(n-1)^2$. This shows that the filtrations of $\mathbb{Z}[\tau_1,\ldots,\tau_p]$ by the ideals $I_{p,n}$ and I_n, are equivalent. Hence

$$\begin{aligned}\mathcal{K}(BU(p))&\approx \operatorname{Proj\,lim} K(G_p(\mathbb{C}^n))\approx \operatorname{Proj\,lim} \mathbb{Z}[\tau_1,\ldots,\tau_p]/I_{p,n}\\ &\approx \operatorname{Proj\,lim} \mathbb{Z}[\tau_1,\ldots,\tau_p]/I_n\approx \mathbb{Z}[[\tau_1,\ldots,\tau_p]].\ \square\end{aligned}$$

3.23. Let X be a locally compact space, and let (X_i), $i=1,\ldots,n$, be a finite closed cover of X. The cover (X_i) is said to be *adapted* if any intersection $X_{i_1}\cap X_{i_2}\cap\cdots\cap X_{i_r}$ is either empty or homeomorphic to $Z\times\mathbb{R}^p$, where Z is a contractible compact space. The space X is said to be of *finite type* if there exists a finite closed cover of X which is adapted. For example, compact differentiable manifolds and finite CW-complexes are spaces of finite type. If X is of finite type, the product $X\times\mathbb{R}^p$ is of finite type. More generally, the product of any two spaces of finite type is of finite type.

3.24. Proposition. *Let X be a locally compact space of finite type, and let Y be a locally compact space such that $K(Y)$ and $K^{-1}(Y)$ are free abelian groups. Then, if we define in general $K^*(Z)=K^0(Z)\oplus K^{-1}(Z)$, the cup-product induces an isomorphism*

$$K^*(X)\otimes K^*(Y)\xrightarrow{\approx} K^*(X\times Y);$$

i.e. more explicitly,

$$[K^0(X)\otimes K^0(Y)]\oplus[K^{-1}(X)\otimes K^{-1}(Y)]\approx K^0(X\times Y),$$

and

$$[K^0(X)\otimes K^{-1}(Y)]\oplus[K^{-1}(X)\otimes K^0(Y)]\approx K^{-1}(X\times Y).$$

Proof. For any closed subspace Z of X, let Φ_Z denote the homomorphism between $K^*(Z)\otimes K^*(Y)$ and $K^*(Z\times Y)$, induced by the cup-product. Let (X_i) be an adapted cover of X; we prove by induction on p, that Φ_Z is an isomorphism when $Z=X_{i_1}\cup\cdots\cup X_{i_p}$, for any space Y. If $p=1$, Proposition 3.24 is simply a reformulation of the Bott periodicity theorem in the complex case: $K^*(\mathbb{R}^n)\otimes K^*(Y)\approx K^*(\mathbb{R}^n\times Y)$. To pass from stage $(p-1)$ to stage p, we let $Z'=X_{i_1}\cup\cdots\cup X_{i_{p-1}}$ and $Z''=X_{i_p}$. Then the subsets $X_{i_r}\cap X_{i_p}$, for $r=1,\ldots,p-1$, form an adapted cover of $Z'\cap Z''$. Therefore, by the induction hypothesis, $\Phi_{Z'}$, $\Phi_{Z''}$, and $\Phi_{Z'\cap Z''}$, are isomorphisms. Since $K^*(Y)$ is free, we have the following commutative diagram, where the vertical sequences are exact (II.4.18 and II.4.9).

$$\begin{array}{ccc}
[K^*(Z'\times\mathbb{R})\oplus K^*(Z''\times\mathbb{R})]\otimes K^*(Y) & \xrightarrow{\Phi_{Z'\times\mathbb{R}}\oplus\Phi_{Z''\times\mathbb{R}}} & K^*(Z'\times Y\times\mathbb{R})\oplus K^*(Z''\times Y\times\mathbb{R}) \\
\downarrow & & \downarrow \\
K^*((Z'\cap Z'')\times\mathbb{R})\otimes K^*(Y) & \xrightarrow{\Phi_{(Z'\cap Z'')\times\mathbb{R}}} & K^*((Z'\cap Z'')\times Y\times\mathbb{R}) \\
\downarrow & & \downarrow \\
K^*(Z'\cup Z'')\otimes K^*(Y) & \xrightarrow{\Phi_{Z'\cup Z''}} & K^*(Z'\cup Z'')\times Y) \\
\downarrow & & \downarrow \\
[K^*(Z')\oplus K^*(Z'')]\otimes K^*(Y) & \xrightarrow{\Phi_{Z'}\oplus\Phi_{Z''}} & K^*(Z'\times Y)\oplus K^*(Z''\times Y) \\
\downarrow & & \downarrow \\
K^*(Z'\cap Z'')\otimes K^*(Y) & \xrightarrow{\Phi_{Z'\cap Z''}} & K^*((Z'\cap Z'')\times Y)
\end{array}$$

By our induction hypothesis (stated for any space X), $\Phi_{Z'\times\mathbb{R}}$, $\Phi_{Z''\times\mathbb{R}}$, and $\Phi_{(Z'\cap Z'')\times\mathbb{R}}$, are also isomorphisms (since $(X_i\times\mathbb{R})$ is an adapted cover of $X\times\mathbb{R}$). Hence $\Phi_{Z'\cup Z''}$ is an isomorphism by the five lemma. □

3.25. *Remark.* Let Y' and Y'' be compact spaces such that $K(Y', Y'')$ is a free abelian group. Then we also have $K^*(X\times Y', X\times Y'')\approx K^*(X)\otimes K^*(Y', Y'')$ by the proposition above applied to $Y=Y'-Y''$. Of course, this isomorphism is given by the cup-product.

3.26. Lemma. *Let Z be a compact space such that $K^*(Z)$ is a finitely generated abelian group. Then there exists a compact space Y_1 and a continuous map $f\colon S'(Z)\to Y_1$, such that*

a) *$K^*(Y_1)$ is a finitely generated free abelian group, and*

b) *the induced homomorphism $K^*(Y_1)\to K^*(S'(Z))$ is surjective* (the suspension $S'(Z)$ is defined as in I.3.14; note that $K^*(S'(Z))\approx K^*(S(Z))$; cf. II.3.27).

Proof. Let $G_{p,n}=\{-n,n\}\times G_p(\mathbb{C}^n)$, and let $\alpha_{p,n}\in K^0(G_{p,n})$ be the element represented over $\{i\}\times G_p(\mathbb{C}^n)$ by $[\xi]-p+i$, where ξ is the canonical p-plane bundle over $G_p(\mathbb{C}^n)$. According to II.1.33, each element of $K^0(Z)$ may be written as $f^*(\alpha_{p,n})$, for suitable integers p and n, and a suitable continuous map $f\colon Z\to G_{p,n}$. Now let $a_1,\ldots,a_r$ be generators of $K^0(Z)$, and let $b_1,\ldots,b_s$ be generators of $K^{-1}(Z)=\tilde{K}^0(S'(Z))$. Let $f_i\colon Z\to G_{p_i,n_i}$ (resp. $g_j\colon S'(Z)\to G_{p_j,n_j}$) be continuous maps such that $f_i^*(\alpha_{p_i,n_i})=a_i$ (resp. $g_j^*(\alpha_{p_j,n_j})=b_j$). If we consider the space $Y_1=\prod_{j=1}^{s} G_{p_j,n_j}\times S'\left(\prod_{i=1}^{r} G_{p_i,n_i}\right)$, we have

$$K^*(Y_1)\approx[\otimes K^*(G_{p_j,n_j})]^2\otimes K^{*+1}(G_{p_i,n_i}),$$

which is free by 3.5 applied $r+s$ times. Moreover, if $f\colon S'(Z)\to Y_1$ is the obvious map, then the induced homomorphism $K^*(Y_1)\to K^*(S'(Z))$ is surjective by our construction. □

3.27. Theorem (compare with Atiyah [2]). *Let Y be a compact space such that $K^*(Y)$ is a finitely generated abelian group, and let X be a compact space of finite type. Then we have a natural exact sequence*

$$0 \to \sum_{i+j=n} K^i(X) \otimes K^j(Y) \to K^n(X \times Y) \to \sum_{i+j=n+1} \operatorname{Tor}(K^i(X), K^j(Y)) \to 0,$$

where i and j are integers mod 2.

Proof. Let us first assume that Y may be written as $S'(Z)$, and let Y_1 be the space constructed in 3.26. Let Y_2 be the mapping cylinder of f, i.e. the quotient of $Y \times [0, 1] \sqcup Y_1$ by the relation identifying $(y, 1)$ with $f(y)$. Thus we have the commutative diagram (up to homotopy)

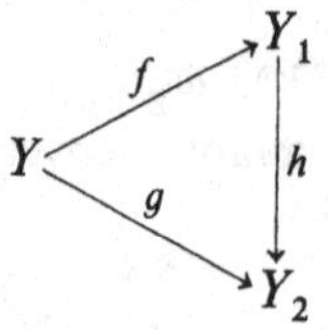

where $g(y)$ is the class of $(y, 0) \in Y \times [0, 1]$, and $h(y_1)$ is the class of $y_1 \in Y_1$. Since h is a homotopy equivalence, we may replace the pair (Y_1, f) by (Y_2, g), where g is an inclusion.

The exact sequence II.4.13, associated with the pair (Y_2, Y). may be written (using III.1.3) as

$$K^{*-1}(Y) \longrightarrow K^*(Y_2/Y, P) \longrightarrow K^*(Y_2) \longrightarrow K^*(Y) \longrightarrow K^{*+1}(Y_2/Y, P),$$

where P is a point. Since the homomorphism $K^*(Y_2) \to K^*(Y)$ is surjective, this exact sequence may be reduced to

$$0 \longrightarrow K^*(Y_2/Y, P) \longrightarrow K^*(Y_2) \longrightarrow K^*(Y) \longrightarrow 0.$$

In particular, $K^*(Y_2/Y, P)$ is free and finitely generated. Hence, the exact sequence above, defines a free resolution of the abelian group $K^*(Y)$, with two terms.

Similarly, the exact sequence II.4.13, associated with the pair $(X \times Y_2, X \times Y)$, may be written as

$$K^{*-1}(X \times Y) \longrightarrow K^*(X \times Y_2/Y, X \times P) \xrightarrow{\alpha} K^*(X \times Y_2)$$
$$\longrightarrow K^*(X \times Y) \xrightarrow{\beta} K^{*+1}(X \times Y_2/Y, X \times P).$$

Since $K^*(X \times Y_2) \approx K^*(X) \otimes K^*(Y_2)$ and

$$K^*(X \times Y_2/Y, X \times P) \approx K^*(X) \otimes K^*(Y_2/Y, P)$$

by 3.24 and 3.25, Coker α may be identified with $K^*(X) \otimes K^*(Y)$. On the other hand, Im $\beta = \operatorname{Ker}(\alpha^{*+1}) = \operatorname{Tor}(K^*(X), K^{*+1}(Y))$ by the definition of the *Tor* functor. Hence the exact sequence may be written as

$$0 \longrightarrow K^*(X) \otimes K^*(Y) \longrightarrow K^*(X \times Y) \longrightarrow \operatorname{Tor}(K^*(X), K^{*+1}(Y)) \longrightarrow 0,$$

and it is easy to show (using the natural properties of the Tor functor) that it is independent of the choice of (Y_2, g), and is natural in X and $Y = S'(Z)$. This proves 3.27 for this case.

Now if Y is an arbitrary space, we may write two exact sequences

$$\begin{array}{ccccccc}
0 \to \sum\limits_{i+j=n} K^i(X) \otimes K^j(S'^2(Y)) & \to & K^n(X \times S'^2(Y)) & \to & \sum\limits_{i+j=n+1} \operatorname{Tor}(K^i(X), K^j(S'^2(Y)) \to (\\
\downarrow & & \downarrow & & \downarrow \\
0 \to \sum\limits_{i+j=n} K^i(X) \otimes K^j(P) & \longrightarrow & K^n(X \times P) & \longrightarrow & \sum\limits_{i+j=n+1} \operatorname{Tor}(K^i(X), K^j(P) \to 0. \\
 & & & & \| \\
 & & & & 0
\end{array}$$

Since $K^n(X \times Y) \approx \operatorname{Ker}[K^n(X \times S'^2(Y)) \to K^n(X \times P)]$ by Bott periodicity, we obtain the general theorem. □

In 3.6 we used the following theorem, which we now prove:

3.28. Theorem. *Let A be any arbitrary ring with unit, and let $A[x_1, \ldots, x_n]$ be the ring of polynomials of n variables with coefficients in A. Let $\sigma_1, \ldots, \sigma_n$ be the elementary symmetric functions of the x_i. Then, with respect to the obvious imbedding $A[\sigma_1, \ldots, \sigma_n] \to A[x_1, \ldots, x_n]$ (cf. Lang [1]), the second ring is a free module over the first with basis the monomials $x_1^{h_1} \cdots x_{n-1}^{h_{n-1}}$, for $0 \leqslant h_i \leqslant n-i$.*

Proof. We prove this theorem by induction on n. For $n=1$, the theorem is clear. Let us now assume $n>1$, and let $\sigma'_1, \ldots, \sigma'_{n-1}$ be the elementary symmetric functions of $x_2, \ldots, x_n$. Then we have the imbeddings

$$A[\sigma_1, \ldots, \sigma_n] \subset A[\sigma'_1, \ldots, \sigma'_{n-1}][x_1] \subset A[x_1, \ldots, x_n] = A[x_1, \ldots, x_{n-1}][x_1].$$

By the induction hypothesis, we know that the products $x_2^{h_2} \cdots x_{n-1}^{h_{n-1}}$ are a basis for the third ring regarded as a module over the second. Therefore, it suffices to prove that the monomials $1, x_1, \ldots, x_1^{n-1}$ are a basis for the second ring regarded as a module over the first.

The relations

$$\sigma_1 = \sigma'_1 + x_1; \qquad \sigma_2 = \sigma'_2 + \sigma'_1 x_1; \qquad \ldots; \qquad \sigma_{n-1} = \sigma'_{n-1} + \sigma'_{n-2} x_1$$

obviously show that $1, x_1, \ldots, x_1^{n-1}$ generate the second ring as a module over the first. Now if we have a relation

$$P_0+P_1x_1+\cdots+P_{n-1}x_1^{n-1}=0,$$

where the P_i are polynomials of the σ_j, we also have the relation

$$P_0+P_1x_r+\cdots+P_{n-1}x_r^{n-1}=0,$$

by the action of the symmetric group, when we consider $A[\sigma_1,\ldots,\sigma_n]$ as a subring of $A[x_1,\ldots,x_n]$.

Therefore, we obtain n equations involving the P_i, which only have 0 as a solution, since the determinant

$$\begin{vmatrix} 1 & x_1 & \cdots & x_1^{n-1} \\ \vdots & & & \vdots \\ 1 & x_n & \cdots & x_n^{n-1} \end{vmatrix} = \prod_{i>j}(x_i-x_j)$$

is not zero. □

Exercises (Section IV.8) 4, 12.

4. Complements in Clifford Algebras

4.1. Let V be a finite dimensional real vector space provided with a nondegenerate quadratic form Q. By III.3.12, the canonical map $V\to C(V)$ is injective, and thus we identify V with its image in $C(V)$. On the other hand, the endomorphism $v\mapsto -v$ of V induces an involution on $C(V)$, which we denote by $x\mapsto \bar{x}$. Therefore $\bar{x}=x$ (resp. $\bar{x}=-x$) if $x\in C^{(0)}(V)$ (resp. $x\in C^{(1)}(V)$) (III.3.6).

4.2. Definition (Atiyah-Bott-Shapiro [1]). The *twisted Clifford group* $\tilde{\Gamma}(V)$ is the set of elements x of $C(V)^*$, such that $\bar{x}Vx^{-1}\subset V$.

It is clear that $\tilde{\Gamma}(V)$ is also the set of elements x of $C(V)$ such that $\bar{x}Vx^{-1}=V$, and in this form, $\tilde{\Gamma}(V)$ appears as a subgroup of $C(V)^*$; this justifies our terminology. Let $\tilde{\rho}:\tilde{\Gamma}(V)\to \mathrm{GL}(V)$ be the homomorphism $x\mapsto\tilde{\rho}_x$, where $\tilde{\rho}_x(v)=\bar{x}vx^{-1}$.

4.3. Proposition. *The kernel of the homomorphism* $\tilde{\rho}:\tilde{\Gamma}(V)\to\mathrm{GL}(V)$ *is* $\mathbb{R}^*\subset\tilde{\Gamma}(V)$.

Proof. Let $\{e_i\}$ be an orthogonal basis of V such that $Q(e_i)\neq 0$. Let $x\in\mathrm{Ker}(\tilde{\rho})$, and let $x=x^0+x^1$ be the decomposition of x into homogeneous elements, with respect to the $\mathbb{Z}/2$-grading of the Clifford algebra. We may write

$x^0 = a_i^0 + e_i b_i^1$, where a_i^0 and b_i^1 do not contain e_i, and are of degree 0 and 1, respectively. The identity $x^0 e_i = e_i x^0$ therefore implies

$$a_i^0 + e_i b_i^1 = e_i(a_i^0 + e_i b_i^1)e_i^{-1} = a_i^0 - e_i b_i^1.$$

Hence, $b_i^1 = 0$ for each i, and x^0 must be an element of $\mathbb{R}^* \subset \tilde{\Gamma}(V)$.

Similarly, the identity $x^1 e_i = -e_i x^1$ may be written as

$$a_i^1 + e_i b_i^0 = -e_i(a_i^1 + e_i b_i^0)e_i^{-1} = a_i^1 - e_i b_i^0.$$

Hence, $b_i^0 = 0$ for each i, and x^1 must be 0. Therefore $x = x^0 + x^1 = x^0 \in \mathbb{R}^*$. □

4.4. Let $C(V)^0$ be the opposite algebra of $C(V)$ ($C(V)^0$ has the same underlying group as $C(V)$, but the product xy in $C(V)^0$ is defined as the product yx in $C(V)$). The canonical map $j\colon V \to C(V)$ may be interpreted as a linear map $V \to C(V)^0$, which satisfies the universal property of Clifford algebras (III.3.1; take $A = C(V)^0$). From this we obtain an algebra homomorphism $C(V) \to C(V)^0$, which we may interpret as an anti-involution of $C(V)$, denoted by $x \mapsto {}^t x$. More precisely, if y is an element of $C(V)$, written as $y = v_1 \cdots v_n$ where $v_i \in V$, we have ${}^t y = v_n \cdots v_1$. It is easy to check that $\overline{({}^t y)} = {}^t\bar{y} = (-1)^n v_n \cdots v_1$. Since every element x of $C(V)$ is the sum of elements of this form, the formula $\overline{{}^t x} = {}^t\bar{x}$ is valid in general. We define the *spinorial norm* of any element x of $C(V)$, as $N(x) = {}^t\bar{x} \cdot x \in C(V)$.

4.5. Proposition. *If x is an element of $\tilde{\Gamma}(V)$, we have $N(x) \in \mathbb{R}^*$. Moreover, the map $x \mapsto N(x)$ induces a homomorphism from $\tilde{\Gamma}(V)$ to $\mathbb{R}^*$.*

Proof. By the definition of $\tilde{\Gamma}(V)$, we have $\bar{x}$ and ${}^t x \in \tilde{\Gamma}(V)$. Therefore, $N(x) \in \tilde{\Gamma}(V)$ if $x \in \tilde{\Gamma}(V)$. To prove $N(x) \in \mathbb{R}^*$ by 4.3, it suffices to prove that $N(x) \in \operatorname{Ker}(\tilde{\rho})$. For $v \in V$, we have ${}^t(\tilde{\rho}_x(v)) = \rho_x(v)$, i.e. ${}^t(\bar{x}vx^{-1}) = \bar{x}vx^{-1}$. Hence ${}^t x^{-1} v {}^t\bar{x} = \bar{x}vx^{-1}$, and so $\overline{N(x)} v N(x)^{-1} = v$, which proves the first assertion.

Now, if x and x' are elements of $\tilde{\Gamma}(V)$, we have

$$N(xx') = ({}^t\bar{x}'\,{}^t\bar{x})(xx') = {}^t\bar{x}'({}^t\bar{x}x)x' = ({}^t\bar{x}x)({}^t\bar{x}'x') = N(x)N(x').$$

Therefore N defines a homomorphism from $\tilde{\Gamma}(V)$ to $\mathbb{R}^*$, since $N(1) = 1$. □

4.6. *Remark.* If $v \in V$, we have $N(v) = -Q(v)$, and the same proof as above shows that $N(xv) = N(x)N(v) = N(vx)$ for any $x \in \tilde{\Gamma}(V)$. In the same way, $N(xvy) = N(x)N(v)N(y)$ for $x \in \tilde{\Gamma}(V)$ and $y \in \tilde{\Gamma}(V)$.

4.7. Theorem. *For any element x of $\tilde{\Gamma}(V)$, $\tilde{\rho}(x)$ belongs to the orthogonal group $\mathrm{O}(V) \subset \mathrm{GL}(V)$, where V is provided with the quadratic form Q. Conversely, any element of $\mathrm{O}(V)$ is of the form $\tilde{\rho}(x)$, where x is determined up to multiplication by a scalar. Hence we have the exact sequence of groups*

$$1 \longrightarrow \mathbb{R}^* \longrightarrow \tilde{\Gamma}(V) \xrightarrow{\tilde{\rho}} \mathrm{O}(V) \longrightarrow 1.$$

Finally, any element of $\tilde{\Gamma}(V)$ *may be written as* $v_1 \cdots v_n$, *where* $v_i \in V$ *and* $Q(v_i) \neq 0$. *In particular, every element of* $\tilde{\Gamma}(V)$ *is homogeneous with respect to the* $\mathbb{Z}/2$*-grading of the Clifford algebra* $C(V)$.

Proof. If $x \in \tilde{\Gamma}(V)$ and $v \in V$, we have $N(\bar{x}vx^{-1}) = N(\bar{x})N(v)N(x)^{-1}$ by 4.6. On the other hand, $N(\bar{x}) = \overline{N(x)} = N(x)$ since $N(x) \in \mathbb{R}^*$. Hence $Q(\tilde{\rho}_x(v)) = -N(\bar{x}vx^{-1}) = -N(\bar{x})N(v)N(x)^{-1} = -N(v) = Q(v)$, and $\tilde{\rho}(x) \in \mathrm{O}(V)$.

It is well known that any element of $\mathrm{O}(V)$ is the product of orthogonal symmetries with respect to hyperplanes. Therefore, to prove that $\tilde{\rho}: \tilde{\Gamma}(V) \to \mathrm{O}(V)$ is surjective, it suffices to show that any such symmetry belongs to the image of $\tilde{\rho}$. If H is the hyperplane orthogonal to the vector v with $Q(v) \neq 0$, we may write any vector w of V in the form $w = \lambda v + v'$, where $v' \in H$ and $\lambda \in \mathbb{R}$. Therefore $\tilde{\rho}(v)(w) = -v(\lambda v + w')v^{-1} = -\lambda v + w$, since v and w' anticommute in $C(V)$ (III.3.8). Hence, $\tilde{\rho}(v)$ is the symmetry required, and the exact sequence is proved.

For any vector v of V such that $Q(v) \neq 0$, let us denote the orthogonal symmetry with respect to the hyperplane orthogonal to v by S_v. If $x \in \tilde{\Gamma}(V)$, we have $\tilde{\rho}(x) = S_{v_1} \cdot S_{v_2} \cdots S_{v_n}$ for some vectors v_i. Hence, $\tilde{\rho}(x) = \tilde{\rho}(v_1 \cdot v_2 \cdots v_n)$, and $x = \lambda v_1 \cdots v_n$ for $\lambda \in \mathbb{R}^*$. □

4.8. Corollary. *Let* $\Gamma^0(V) = \tilde{\Gamma}(V) \cap C^{(0)}(V)$, *and let* $\mathrm{SO}(V) = \{u \in \mathrm{O}(V) \mid \mathrm{Det}(u) = 1\}$. *Then we have the exact sequence (where* $\rho^0 = \tilde{\rho} \mid \Gamma^0(V)$*)*

$$1 \longrightarrow \mathbb{R}^* \longrightarrow \Gamma^0(V) \xrightarrow{\rho^0} \mathrm{SO}(V) \longrightarrow 1.$$

Proof. Any element of $\mathrm{SO}(V)$ may be written as $\tilde{\rho}(v_1) \cdot \tilde{\rho}(v_2) \cdots \tilde{\rho}(v_n)$, where n is even. Hence ρ^0 is surjective; moreover, $\mathrm{Ker}\,\rho^0 = \mathbb{R}^*$ since $\mathbb{R}^* \subset C^0(V)$. The group $\Gamma^0(V)$ is called the *special Clifford group*. □

4.9. Corollary. *Let* $\mathrm{Pin}(V) = \{x \in \tilde{\Gamma}(V) \mid |N(x)| = 1\}$ *and* $\mathrm{Spin}(V) = \mathrm{Pin}(V) \cap C^{(0)}(V)$. *Then we have the exact sequences*

$$1 \longrightarrow \mathbb{Z}/2 \longrightarrow \mathrm{Pin}(V) \longrightarrow \mathrm{O}(V) \longrightarrow 1,$$

and

$$1 \longrightarrow \mathbb{Z}/2 \longrightarrow \mathrm{Spin}(V) \longrightarrow \mathrm{SO}(V) \longrightarrow 1.$$

Therefore, if we let $\mathrm{Spin}(n)$ *denote the group* $\mathrm{Spin}(V)$ *when* $V = \mathbb{R}^n$, *provided with the quadratic form* $\sum_{i=1}^{n} x_i^2$, *we have the exact sequence*

$$1 \longrightarrow \mathbb{Z}/2 \longrightarrow \mathrm{Spin}(n) \longrightarrow \mathrm{SO}(n) \longrightarrow 1.$$

Proof. If $x \in \tilde{\Gamma}(V)$ (resp. $\Gamma^0(V)$), then $\lambda x \in \mathrm{Pin}(V)$ (resp. $\mathrm{Spin}(V)$) for $\lambda = 1/\sqrt{|N(x)|}$. This shows the surjectivity of the homomorphisms $\mathrm{Pin}(V) \to \mathrm{O}(V)$ and $\mathrm{Spin}(V) \to \mathrm{SO}(V)$. The kernel of these maps is the set of elements $\lambda \in \mathbb{R}^*$ such that $N(\lambda) = \lambda^2 = 1$, i.e. the kernel is isomorphic to $\mathbb{Z}/2$. □

4.10. The choice of an orthogonal basis (e_i) of V such that $Q(e_i)=\pm 1$, enables us to identify $C(V)$ with $\mathbb{R}^n$ (III.3.11), and to provide $C(V)$ with a topology which is independent of the choice of basis. More precisely, a change of the above basis is given by an orthogonal transformation, which induces a linear transformation in $C(V)$, depending continuously on the basis. Hence, the functor $V \mapsto C(V)$ may be regarded as a continuous functor, in the sense of I.4.1, from the category of finite dimensional vector spaces provided with a quadratic form, to the category of finite dimensional algebras.

From these observations, it follows that the groups $\tilde{\Gamma}(V)$, $\mathrm{Pin}(V)$, $\Gamma^0(V)$, and $\mathrm{Spin}(V)$ are naturally topological groups, and that the maps $\tilde{\rho}$ and ρ^0 are continuous maps. In fact, the maps

$$\tilde{\Gamma}(V) \to \mathrm{O}(V),\ \Gamma^0(V) \to \mathrm{SO}(V),\ \mathrm{Pin}(V) \to \mathrm{O}(V), \text{ and } \mathrm{Spin}(V) \to \mathrm{SO}(V),$$

define locally trivial fibrations. Let us show this for the map $\tilde{\Gamma}(V) \to \mathrm{O}(V)$, for instance, (the proof for the three other cases is similar). Let $e_1, \ldots, e_n$ be an orthogonal basis such that $Q(e_i)=\pm 1$. By induction on p, where $0 \leqslant p \leqslant n$, we define a neighbourhood V_p of 1 in $\mathrm{O}(V)$, and a continuous map

$$s_p : V_p \longrightarrow \tilde{\Gamma}(V)$$

such that

(i) $V_0 = \mathrm{O}(V)$, $s_0(\alpha)=1$,

(ii) $s_{p+1}(\alpha)=(1+tw)s_p(\alpha)$ with $t=\alpha(e_{p+1})$ and $w=\tilde{\rho}(s_p(\alpha))(e_{p+1})$ if $Q(e_{p+1})=1$, $=(1-tw)s_p(\alpha)$ if $Q(e_{p+1})=-1$, and

(iii) V_{p+1} is the subset of V_p defined by the condition $Q(t+w)\neq 0$

Then by induction on p, we can prove that $\alpha^{-1}\tilde{\rho}(s_p(\alpha))$ leaves $e_1, \ldots, e_p$ fixed. Hence $\tilde{\rho}(s_n(\alpha))=\alpha$, and the map $\alpha \mapsto s_n(\alpha)$ is a section of $\tilde{\rho}$ defined on the neighbourhood V_n. It follows that the map $V_n \times \mathbb{R}^* \to \tilde{\rho}^{-1}(V_n)$, defined by $(\alpha, \lambda) \mapsto \lambda s_n(\alpha)$, is a homeomorphism, and by translation, we have $\tilde{\rho}^{-1}(aV_n) \approx aV_n \times \mathbb{R}^*$ for any element a of $\mathrm{O}(V)$.

4.11. Since the functor $V \mapsto C(V)$ is continuous, we may extend this functor to the category of vector bundles. If V is now a vector bundle, we define a bundle of algebras, again denoted by $C(V)$, such that $C(V)_x = C(V_x)$. The multiplication in each fiber defines a continuous map $(C(V) \times_X C(V) \to C(V)$. If E is a k-vector bundle ($k=\mathbb{R}$ or $\mathbb{C}$), a $C(V)$-module structure on E is given by a continuous map $C(V) \times_X E \to E$, such that each fiber E_x is provided with a $C(V_x)$-module structure compatible with the k-module structure. Since $V \subset C(V)$, we obtain a fiberwise map $V \times_X E$ denoted by $(v, e) \mapsto v \cdot e$ such that $v \cdot (v \cdot e) = Q(v)e$, and such that the induced map on each fiber $V_x \times E_x \to E_x$ is $\mathbb{R}$-linear (resp. k-linear) with respect to the first (resp. second) factor. Conversely, every such bilinear map defines a $C(V)$-module structure on E by the universal property of Clifford algebras (III.3.1). Equivalently, the above bilinear map defines a morphism on the underlying real vector bundles $m: V \to \mathrm{HOM}(E, E)$, such that $(m(v))^2 = Q(v)$ over each point of the base (I.4.8).

The vector bundles provided with $C(V)$-module structures are the objects of a category, whose morphisms are vector bundle morphisms which induce $C(V_x)$-module morphisms over each point x of the base. We denote this category by $\mathscr{E}^V(X)$, and notice that it is a Banach category contained in $\mathscr{E}(X)$.

4.12. *Example.* Assume that $V = X \times \mathbb{R}^{p+q}$, provided with the "trivial quadratic form" $-x_1^2 - \cdots - x_p^2 + x_{p+1}^2 + \cdots + x_{p+q}^2$. Then a $C(V)$-module structure on the vector bundle E is given by $p+q$ automorphisms e_i, where $e_i e_j + e_j e_i = 0$ for $i \neq j$, $e_i^2 = -1$ if $1 \leqslant i \leqslant p$, and $e_i^2 = 1$ if $p+1 \leqslant i \leqslant p+q$. Hence, the category $\mathscr{E}^V(X)$ is isomorphic to the category $\mathscr{E}^{p,q}(X)$ considered in III.4.

4.13. Definition. Let V be a real vector bundle with compact base X. An *orientation* on V is an element α of $H^1(X; \mathrm{SL}_n(\mathbb{R}))$, whose image under the map $H^1(X, \mathrm{SL}_n(\mathbb{R})) \to H^1(X; \mathrm{GL}_n(\mathbb{R}))$ is the class of the bundle V (I.3.5). A *spinorial structure* on V is an element β of $H^1(X; \mathrm{Spin}(n))$, whose image under the composition $H^1(X; \mathrm{Spin}(n)) \to H^1(X; \mathrm{SO}(n)) \to H^1(X; \mathrm{GL}_n(\mathbb{R}))$ is the class of V.

4.14. This definition could be made in terms of "principal bundles". If G is a topological group, and if (g_{ji}) is a G-cocycle in the sense of I.3.5, we consider the space P which is the quotient of the disjoint union $\bigsqcup U_i \times G$, by the equivalence relation $(x_i, g_i) \sim (x_j, g_j)$ if $x_i = x_j \in U_i \cap U_j$, and $g_j = g_{ji}(x_i) g_i$. The group G acts on the right on P, by the formula $(x_i, g_i) \cdot g = (x_i, g_i g)$. Moreover, this action is free, and we have $X \approx P/G$. If G acts on the left on a k-vector space F of dimension n, we associate P with the quasi-vector bundle $E = P \times_G F$, the quotient of $P \times F$ by the equivalence relation $(p, f) \sim (pg, g^{-1} f)$ for $g \in G$.

4.15. Proposition. *The quasi-vector bundle $E = P \times_G F$ above is actually a vector bundle. It is the k-vector bundle associated with the cocycle g_{ji}.*

Proof. The space $P \times_G F$ may be identified with the quotient of $\bigsqcup U_i \times G \times F$, by the equivalence relation generated by $(x_i, g, f) \sim (x_j, g_{ji}(x_i) g, f) \sim (x_j, g_{ji}(x_i), gf)$. In particular, each element of $P \times_G F = \bigsqcup U_i \times G \times F/\sim$ is the class of a triple $(x_i, 1, f) = [x_i, f]$. Therefore, $P \times_G F$ may be identified with the quotient of $\bigsqcup U_i \times F$, by the equivalence relation $[x_i, f_i] \sim [x_j, f_j]$ if $f_j = g_{ji}(x_i)(f_i)$; this is the bundle associated with the cocycle (g_{ji}) in the sense of I.3.6. □

4.16. Corollary. *Let V be a real vector bundle of rank n with compact base. Then V may be provided with an orientation (resp. a spinorial structure) if there exists a principal bundle P with structural group* $\mathrm{SO}(n)$ *(resp.* $\mathrm{Spin}(n)$*), such that $V \approx P \times_G \mathbb{R}^n$ where $G = \mathrm{SO}(n)$ (resp. $\mathrm{Spin}(n)$). This bundle P is associated with $\alpha \in H^1(X; \mathrm{SO}(n))$ (resp. $\beta \in H^1(X; \mathrm{Spin}(n))$ defined in* 4.13.

4.17. *Example.* Let W be a real vector bundle, and let P be a principal bundle with group $O(n)$, such that $W \approx P \times_{O(n)} \mathbb{R}^n$ (I.8.7; choose trivializations $W|_{U_i} \xrightarrow{\approx} U_i \times \mathbb{R}^n$ compatible with the metric, and apply the construction of I.3.6; see also Remark 4.21 below). The principal bundle P is associated with the cocycle (g_{ji}), where $g_{ji}(x) \in O(n)$. The bundle $W \oplus W$ may be written as $P' \times_{O(2n)} \mathbb{R}^{2n}$, where P' is associated with the cocycle

$$h_{ji}(x) = \begin{pmatrix} g_{ji}(x) & 0 \\ 0 & g_{ji}(x) \end{pmatrix}.$$

Since $h_{ji}(x) \in SO(2n)$, we see that V is canonically an oriented bundle.

4.18. *Example.* Let W be an oriented vector bundle, and let P be a principal bundle with structural group $SO(n)$, such that $W \approx P \times_{SO(n)} \mathbb{R}^n$. We will show that $W \oplus W$ may be provided with a spinorial structure. The argument used in 4.17 shows that it suffices to prove that the composite homomorphism

$$SO(n) \xrightarrow{\Delta} SO(n) \times SO(n) \longrightarrow SO(2n),$$

where Δ is the diagonal homomorphism, may be lifted to a homomorphism j from $SO(n)$ to $Spin(2n)$, making the diagram

$$\begin{array}{ccc} & & Spin(2n) \\ & \overset{j}{\nearrow} & \downarrow \\ SO(n) & \longrightarrow & SO(2n) \end{array}$$

commutative.

Let $D\colon Spin(n) \times Spin(n) \to Spin(2n)$ be the homomorphism defined by $D(\alpha_1, \alpha_2) = i_1(\alpha_1) i_2(\alpha_2)$, where i_1 (resp. i_2) is induced by the canonical inclusion of $\mathbb{R}^n \oplus 0$ (resp. $0 \oplus \mathbb{R}^n$) in $\mathbb{R}^n \oplus \mathbb{R}^n = \mathbb{R}^{2n}$ (cf. III.3.5). We clearly have $D(\varepsilon\alpha_1, \eta\alpha_2) = \varepsilon\eta D(\alpha_1, \alpha_2)$, where $\varepsilon = \pm 1$ and $\eta = \pm 1$. Now if u is an element of $SO(n)$, and if $\tilde{u}$ is some element of $Spin(n)$ such that $\rho^0(\tilde{u}) = u$, then $v = D(\tilde{u}, \tilde{u})$ is a well-defined element of $Spin(2n)$, independent of the choice of $\tilde{u}$. The correspondence $u \mapsto v$ (which is clearly continuous by 4.10) defines the required homomorphism.

4.19. *Remark.* From 4.17 and 4.18, it follows that for every real bundle W, the bundle $W \oplus W \oplus W \oplus W$ may be canonically provided with a spinorial structure.

* **4.20.** For the reader who is familiar with algebraic topology, we remark that a real vector bundle is orientable if and only if its first *Stiefel-Whitney class*, $w_1(V) \in H^1(X; \mathbb{Z}/2)$, is 0. This may be shown by associating each bundle which has a certain class $\alpha \in H^1(X; O(n))$, with the element in $H^1(X; \mathbb{Z}/2)$ obtained by the determinant homomorphism $O(n) \to \mathbb{Z}/2$. Since this correspondence in "natural", it is determined by a well-defined element $w_1(V) \in H^1(BO(n); \mathbb{Z}/2) \approx \mathbb{Z}/2$. This element is nontrivial: consider $X = RP_1$ and $V = \xi \oplus \eta$, where ξ is the canonical bundle and η is the trivial bundle of rank $n-1$.

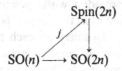

Now let V be an oriented vector bundle defined by a cocycle (g_{ji}), where $g_{ji}(x) \in \mathrm{SO}(n)$. If the cover (U_i) is fine enough, by using 4.10 we obtain continuous maps $\tilde{g}_{ji}\colon U_i \cap U_j \to \mathrm{Spin}(n)$, such that $\rho^0(\tilde{g}_{ji}(x)) = g_{ji}(x)$, $\tilde{g}_{ii}(x) = 1$. Then for $x \in U_i \cap U_j \cap U_k$, we consider $h_{ijk}(x) = g_{ij}(x) g_{jk}(x) g_{ki}(x)$; this defines an element of $Z^2(X; \mathbb{Z}/2)$, hence a well-defined element $w_2(V)$ of $H^2(X, \mathbb{Z}/2)$ (second Čech cohomology group with coefficients in $\mathbb{Z}/2$). By naturality, since $H^2(BSO(n); \mathbb{Z}/2) = \mathbb{Z}/2$, we see that the correspondence $V \mapsto w_2(V)$ is either trivial, or the second Stiefel-Whitney class. Moreover, $w_2(V) = 0$ if and only if the bundle may be provided with a spinorial structure.

To see that $w_2(V)$ is nontrivial, we consider bundles over S^2. Then the argument in I.7.6 shows that $w_2(V)$ trivial implies that the map $\pi_1(\mathrm{Spin}(n)) \to \pi_1(SO(n))$ is surjective. For $n=2$, $\mathrm{Spin}(n) \approx \mathrm{SO}(2) \approx S^1$, and the map ρ^0 is essentially the map $z \mapsto z^2$ on the circle. Hence, $\pi_1(\mathrm{Spin}(n)) \approx \pi_1(\mathrm{SO}(n)) \approx \mathbb{Z}$, and the homomorphism $\mathbb{Z} \approx \pi_1(\mathrm{Spin}(n)) \to \mathbb{Z} \approx \pi_1(\mathrm{SO}(n))$ is multiplication by 2, which is not surjective. If $n > 2$, then $\pi_1(\mathrm{Spin}(n)) = 0$ and $\pi_1(SO(n)) \approx \mathbb{Z}/2$ by elementary topological observations (note that $\mathrm{Spin}(3) \approx S^3$, and that $\mathrm{Spin}(3) \to \mathrm{SO}(3)$ is the nontrivial covering). *

4.21. *Remark.* If V is a real vector bundle with paracompact base X, we can always find a principal bundle P with group $\mathrm{O}(n)$ such that $V \approx P \times_{\mathrm{O}(n)} \mathbb{R}^n$. Essentially this follows from I.8.7, since we can always provide V with a positive definite quadratic form. If $\varphi\colon U \times \mathbb{R}^n \to E_U$ is a trivialization of E, then φ may be uniquely written as $\varphi = \psi \cdot \theta$, where $\theta^* = \theta > 0$, and ψ is an isometry on each fiber (take $\theta = \sqrt{\varphi^* \varphi}$ and $\psi = \varphi \theta^{-1}$). If we replace φ by ψ, we may assume that φ is an isometry on each fiber. Hence, the vector bundle V may be constructed from a $\mathrm{O}(n)$-cocycle. The argument in 4.20 shows that in general V cannot always be constructed from a $\mathrm{SO}(n)$-cocycle, nor a $\mathrm{Spin}(n)$-cocycle.

4.22. Theorem. *Let V be a real vector bundle of rank n, provided with a spinorial structure. Then, with respect to a positive definite quadratic form on V, the categories $\mathscr{E}^{0,n}(X)$ and $\mathscr{E}^V(X)$ are equivalent* (cf. 4.11).

Proof. Let T denote the trivial vector bundle of rank n, provided with the "trivial" quadratic form $\sum_{i=1}^{n} (x_i)^2$. By 4.12, the categories $\mathscr{E}^{0,n}(X)$ and $\mathscr{E}^T(X)$ are equivalent. On the other hand, the category $\mathscr{E}^V(X)$ does not depend on the metric chosen, since any two metrics are isomorphic (I.8.8). If we write $V = P \times_{\mathrm{Spin}(n)} \mathbb{R}^n$, we may choose the metric to be the one induced by the canonical quadratic form on $\mathbb{R}^n$.

Using the spinorial structure on V, we now define two category equivalences

$$\theta\colon \mathscr{E}^T(X) \longrightarrow \mathscr{E}^V(X) \quad \text{and} \quad \varphi\colon \mathscr{E}^V(X) \longrightarrow \mathscr{E}^T(X),$$

inverse to each other (up to isomorphism). To define θ, we write $V = P \times_{\mathrm{Spin}(n)} \mathbb{R}^n$. For $E \in Ob\mathscr{E}^T(X)$, we let $\theta(E) = P\,\Delta_{\mathrm{Spin}(n)} E$, where $P\,\Delta_{\mathrm{Spin}(n)} E$ denotes the quotient of the fiber product $P \times_X E$, by the equivalence relation $(p, e) \sim (pg, g^{-1}e)$, where

$g \in \mathrm{Spin}(n) \subset C^{0,n}$ acts naturally on E. If $f: E \to E'$ is a morphism, then $\theta(f) = (\mathrm{Id}_P, f)$ defines a morphism from $\theta(E)$ to $\theta(E')$.

To define φ, we notice that $C(V) = P \times_{\mathrm{Spin}(n)} C^{0,n}$, where $\mathrm{Spin}(n)$ acts on $C^{0,n}$ by inner automorphisms. More precisely, $C(V)$ may be identified with the quotient of the product $P \times C^{0,n}$, by the equivalence relation $(p, \lambda) \sim (pg, g^{-1}\lambda g)$, where $g \in \mathrm{Spin}(n) \subset C^{0,n}$. Hence, the "bundle of groups", $\mathrm{Spin}(V) = Q$, may be identified with the quotient of $P \times G$, where $G = \mathrm{Spin}(n)$, by the equivalence relation $(p, g) \sim (ph, h^{-1}gh)$. Now if F is a $C(V)$-module, we define a "*right action*" of Q on P, by the formula $p \cdot (p, \alpha) = p\alpha^{-1}$, and we define $\varphi(F)$ as $P \Delta_Q F$, the quotient of $P \times_X F$ by the equivalence relation $(p, f) \sim (p\lambda, \lambda^{-1} f)$, where $\lambda = (p, g) \in Q \subset C(V)$.

The $C(T)$-module $(\varphi\theta)(E)$ is the quotient of $P \times_X P \times_X E$ by the equivalence relation generated by $(p', p, e) \sim (p, p'g, g^{-1}e)$ and $(p, p, e) \sim (pg^{-1}, p, g^{-1}e)$. Therefore, the map $e \mapsto (p, p, e) \sim (pg, pg, e)$, where p is in P_x for $e \in E_x$, defines a natural isomorphism between E and $(\varphi\theta)(E)$.

Conversely, $(\theta\varphi)(F)$ is the quotient of $P \times_X P \times_X F$ by the equivalence relation generated by $(p', p, f) \sim (p', p\lambda, \lambda^{-1} f)$ for $\lambda = (p, g) \in Q$, and $(p, p, f) \sim (p\alpha, p, \alpha^{-1} f)$ for $\varphi = (p, \alpha) \in Q$. Therefore, the map $f \mapsto (p, p, f) \sim (p\alpha, p\alpha, f)$, where $p \in P_x$ for $f \in F_x$, defines a natural isomorphism between F and $(\theta\varphi)(F)$. □

4.23. *Remark.* Let W be an arbitrary vector bundle, provided with a non degenerate quadratic form. Then the same type of proof shows that $\mathscr{E}^{T \oplus W}(X) \sim \mathscr{E}^{V \oplus W}(X)$.

4.24. *Remark.* If V and V' are spinorial vector bundles of rank n and n', respectively, then $V \oplus V'$ is a spinorial vector bundle of rank $n+n'$. More precisely, the inclusions of $C^{0,n}$ and $C^{0,n'}$ in $C^{0,n+n'}$ induce a homomorphism from $\mathrm{Spin}(n) \times \mathrm{Spin}(n')$ to $\mathrm{Spin}(n+n')$, which makes the diagram

$$\begin{array}{ccc} \mathrm{Spin}(n) \times \mathrm{Spin}(n') & \xrightarrow{s} & \mathrm{Spin}(n+n') \\ \downarrow & & \downarrow \\ \mathrm{O}(n) \times \mathrm{O}(n) & \longrightarrow & \mathrm{O}(n+n') \end{array}$$

commutative. Hence $\mathscr{E}^{V \oplus V'}(X) \sim \mathscr{E}^{0, n+n'}(X)$.

4.25. Assume now that the basic field k is the field of complex numbers $\mathbb{C}$. We slightly modify the previous arguments by considering the group $\mathrm{Spin}^c(n)$ instead of $\mathrm{Spin}(n)$ (where $\mathrm{Spin}^c(n) = \mathrm{Spin}(n) \times_{\mathbf{Z}/2} \mathrm{U}(1)$, the quotient of $\mathrm{Spin}(n) \times \mathrm{U}(1)$ by the equivalence relation $(p, z) \sim (-p, -z)$). Then we have the exact sequence

$$1 \longrightarrow \mathrm{U}(1) \longrightarrow \mathrm{Spin}^c(n) \xrightarrow{\rho^c} \mathrm{SO}(n) \longrightarrow 1,$$

where $\rho^c(p, z) = \rho^0(p)$. If V is an oriented real vector bundle of rank n, a *^{c}spinorial structure* on V is given by a principle bundle P with structural group $\mathrm{Spin}^c(n)$, and an isomorphism $V \approx P \times_{\mathrm{Spin}^c(n)} \mathbb{R}^n$. For such bundles V, we define a category equivalence

$$\theta^c : \mathscr{E}_{\mathbb{C}}^T(X) \longrightarrow \mathscr{E}_{\mathbb{C}}^V(X),$$

by the formula $\theta^c(E) = P \Delta_{\mathrm{Spin}^c(n)} E$, the quotient of $P \times_X E$ by the equivalence relation $(p, e) \sim (pg^{-1}, ge)$, where $g \in \mathrm{Spin}^c(n)$ acts on E via the imbedding of $\mathrm{Spin}(n)$ in $C^{0,n}$, and via the natural action of $\mathrm{U}(1) \subset \mathbb{C}$ on the *complex* vector bundle E. The argument used in the proof of 4.22 also shows that θ^c is a category equivalence.

4.26. It is useful to consider ^{c}spinorial vector bundles, since it is possible for a real vector bundle to be ^{c}spinorial without being spinorial. This is the case for example when V underlies a complex vector bundle of rank n (again denoted V). Since any complex vector bundle may be provided with a metric (I.8.7), we can always find a principal bundle P, with structural group $\mathrm{U}(n)$, such that $V \approx P \times_{\mathrm{U}(n)} \mathbb{R}^{2n}$. Thus to prove the assertion above, it suffices to show the existence of a homomorphism $\sigma : \mathrm{U}(n) \to \mathrm{Spin}^c(2n)$ which makes the following diagram commutative:

$$\begin{array}{ccc} & & \mathrm{Spin}^c(2n) \\ & \overset{\sigma}{\nearrow} & \big\downarrow \rho^c \\ \mathrm{U}(n) & \longrightarrow & \mathrm{SO}(2n). \end{array}$$

Let $\alpha \in \mathrm{U}(n)$, and let $\tau(t)$ be a path in $\mathrm{U}(n)$ such that $\tau(0) = 1$ and $\tau(1) = \alpha$ ($\mathrm{U}(n)$ is arcwise connected). Since the map $\mathrm{Spin}(2n) \to \mathrm{SO}(2n)$ is a covering (4.10), there exists a unique path $\tilde{\tau}$ in $\mathrm{Spin}(2n)$ such that $\tilde{\tau}(0) = 1$ and $\rho^c(\tilde{\tau}(t)) = \tau(t)$ (Godbillon [2]). Similarly, the path $t \mapsto \gamma(t) = Det(\tau(t)) \in \mathrm{U}(1) = \mathrm{SO}(2)$ may be lifted to $\bar{\gamma}(t) \in \mathrm{Spin}(2) \approx \mathrm{U}(1)$. Then the pair $(\tilde{\tau}(1), \bar{\gamma}(1))$ defines an element of $\mathrm{Spin}^c(2n)$, which is independent of the path $\tau(t)$ connecting α to 1. To prove this, we consider another path τ', connecting α to 1. Then τ and τ' differ by a loop in $\mathrm{U}(n)$, and $\tilde{\tau}(1)$ and $\tilde{\tau}'(1)$ differ by a sign obtained by applying the homomorphism

$$\pi_1(\mathrm{U}(n)) \longrightarrow \mathbb{Z}/2,$$

induced by the covering $\mathrm{Spin}(2n)|_{\mathrm{U}(n)}$, to the class of the loop above (cf. Godbillon [2]). Since the determinant map induces an isomorphism from $\pi_1(\mathrm{U}(n))$ to $\pi_1(\mathrm{U}(1))$, inverse to the isomorphism $\pi_1(\mathrm{U}(1)) \to \pi_1(\mathrm{U}(n))$ induced by the inclusion of $\mathrm{U}(1)$ in $\mathrm{U}(n)$, we must have $(\tilde{\tau}(1), \bar{\gamma}(1)) = (\tilde{\tau}'(1), \bar{\gamma}'(1)$, or $(\tilde{\tau}(1), \bar{\gamma}(1)) = (-\tilde{\tau}'(1), -\bar{\gamma}'(1))$. Hence σ is well-defined. Moreover, σ is a homomorphism since $\mathrm{U}(n)$ is connected, and since $\sigma(xy) = \sigma(x)\sigma(y)$ for x and y close enough to 1.

4.27. (Atiyah-Bott-Shapiro [1]). The homomorphism σ may be examined more closely in the following way. Let $f_1, \ldots, f_n$ be an orthonormal basis of $\mathbb{C}^n$ such that $\alpha(f_r) = \exp(i\theta_r) \cdot f_r$. Let $e_{2r-1} = f_r$ and $e_{2r} = if_r$, be the corresponding orthonormal basis of $\mathbb{R}^{2n}$, and let S be the element of $\mathrm{Spin}^c(2n)$, defined by the product

$$S=\prod_{r=1}^{n}(\cos\theta_r/2-e_{2r-1}e_{2r}\sin\theta_r/2)\exp(i\theta_r/2).$$

Then $\rho^c(S)=\alpha\in \mathrm{U}(n)\subset \mathrm{SO}(2n)$, and

$$\tilde{\tau}(t)=\prod_{r=1}^{n}(\cos t\theta_r/2-e_{2r-1}e_{2r}\sin t\theta_r/2)\exp(it\theta_r/2)$$

is a lifting of a path from 1 to S, such that $\tilde{\tau}(0)=1$. Hence $S=\sigma(\alpha)$.

4.28. By 1.5, $\Lambda(\mathbb{C}^n)$ is a $C^{0,2n}\otimes_{\mathbb{R}}\mathbb{C}$-module. Since $C^{0,2n}\otimes_{\mathbb{R}}\mathbb{C}\approx M_{2^n}(\mathbb{C})$, and since $\Lambda(\mathbb{C}^n)$ is of dimension 2^n, $\Lambda(\mathbb{C}^n)$ is a generator of $K(\mathscr{E}_{\mathbb{C}}^{0,2n})\approx\mathbb{Z}$ by III.4.4. Therefore, the functor

$$\mathscr{E}_{\mathbb{C}}(X)\longrightarrow\mathscr{E}_{\mathbb{C}}^{0,2n}(X),$$

defined by $E\mapsto\Lambda(\mathbb{C}^n)\otimes E$, is the category equivalence described in III.4.6.

4.29. If V is a complex bundle of rank n, it may be provided with a ^{c}spinorial structure by 4.26. Therefore, we have a category equivalence

$$\mathscr{E}_{\mathbb{C}}^{0,2n}(X)\longrightarrow\mathscr{E}_{\mathbb{C}}^{V}(X)$$

by 4.25. Now we claim that the composition $\mathscr{E}_{\mathbb{C}}(X)\sim\mathscr{E}_{\mathbb{C}}^{0,2n}(X)\sim\mathscr{E}_{\mathbb{C}}^{V}(X)$ is simply the functor $E\mapsto\Lambda(V)\otimes E$, which is well-defined by 1.5 and 4.11. Essentially, if we write V as $P\times_{\mathrm{U}(n)}\mathbb{C}^n$, the composition of the two functors is defined by $E\mapsto(P'\times_{\mathrm{Spin}^c(2n)}\Lambda\mathbb{C}^n)\otimes E$, where P' is the principal bundle with structural group $\mathrm{Spin}^c(2n)$, associated with P, i.e. $P'=P\times_{\mathrm{U}(n)}\mathrm{Spin}^c(2n)$. On the other hand, the functor $E\mapsto\Lambda(V)\otimes E$ is $E\mapsto(P\times_{\mathrm{U}(n)}\Lambda\mathbb{C}^n)\otimes E$. Thus the fact that the two functors are isomorphic, will follow directly from the commutativity of the diagram

$$\begin{array}{ccc}
\mathrm{U}(n) & \xrightarrow{\ \sigma\ } & \mathrm{Spin}^c(2n)\\
\big\downarrow & & \big\downarrow{\scriptstyle\gamma}\\
\mathrm{End}(\mathbb{C}^n) & \xrightarrow{\ \Lambda\ } & \mathrm{End}(\Lambda(\mathbb{C}^n)),
\end{array}$$

where σ is the homomorphism described in 4.26, γ is the homomorphism describing the action of $\mathrm{Spin}^c(2n)\subset C^{0,2n}\otimes\mathbb{C}$ on $\Lambda(\mathbb{C}^n)$ (1.5), and Λ is induced by the functoriality of exterior powers. If $(f_1,\ldots,f_n)$ is an orthonormal basis of $\mathbb{C}^n$ such that $\alpha(f_r)=\exp(i\theta_r)f_r$, for $\alpha\in\mathrm{U}(n)$, we have

$$\sigma(\alpha)=\prod_{r=1}^{n}(\cos\theta_r/2-e_{2r-1}e_{2r}\sin\theta_r/2)\exp(i\theta_r/2)$$

by 4.27. Now let $x=a+f_r \wedge b$ be any element of $\Lambda\mathbf{C}^n$ such that a and b do not contain f_r. Then we have

$$\gamma(\sigma(\alpha))(x)=(\cos\theta_r/2-e_{2r-1}e_{2r}\sin\theta_r/2)\exp(i\theta_r/2)(a+f_r\wedge b)=a+\exp(i\theta_r)f_r\wedge b.$$

Therefore $\gamma(\sigma(\alpha))(f_{r_1}\wedge\cdots\wedge f_{r_s})=\alpha(f_{r_1})\wedge\cdots\wedge\alpha(f_{r_s})$.

4.30. Let us return to the general case where V is a ^{c}spinorial bundle of rank n. Let

$$l\colon \mathrm{Spin}^c(n)\times_{\mathbf{Z}/2}\mathrm{U}(1)\longrightarrow \mathrm{U}(1)$$

be the homomorphism defined by $l(\beta, z)=z^2$. If P is a principal bundle of group $\mathrm{Spin}^c(n)$, we associate P with the complex line bundle $L(V)=P\times_{\mathrm{Spin}^c(n)}\mathbf{C}$, the quotient of $P\times\mathbf{C}$ by the equivalence relation $(pg,\lambda)\sim(p, l(g)^{-1}\lambda)$, where $g\in\mathrm{Spin}^c(n)$. The line bundle $L(V)$ is said to be associated with the ^{c}spinorial structure of V.

If $n=2p$, and if P is associated with a complex structure on V as in 4.26, the explicit formula given in 4.27 shows that $L=P'\times_{U(p)}\mathbf{C}$. In this notation, P' is the principal bundle, with group $\mathrm{U}(p)$, which defines the complex structure on V, and $P'\times_{\mathrm{U}(p)}\mathbf{C}$ is the quotient of $P'\times\mathbf{C}$ by the equivalence relation $(p'g,\lambda)\sim(p', d(g)^{-1}\lambda)$, where $d\colon \mathrm{U}(p)\to\mathrm{U}(1)$ is the determinant. Hence, $L(V)$ is isomorphic to $\lambda^p(V)$ (p^{th} complex exterior power of V).

If V and V' are ^{c}spinorial bundles, then $V\oplus V'$ is naturally a ^{c}spinorial bundle. This follows from the homomorphism

$$s^c\colon \mathrm{Spin}^c(n)\times\mathrm{Spin}^c(n')\longrightarrow\mathrm{Spin}^c(n+n'),$$

defined by $[(\beta,z),(\beta',z')]\mapsto[s(\beta,\beta'), zz']$, where s is defined in 4.24, and where $z, z'\in\mathrm{U}(1)$. Note that $L(V\oplus V')\approx L(V)\otimes L(V')$.

4.31. Theorem. *Let V be a ^{c}spinorial bundle of rank n, and let W be the real bundle $V\oplus V$ provided with the complex structure defined in I.4.8: multiplication by i is represented by the matrix*

$$\begin{pmatrix}0 & -1\\ 1 & 0\end{pmatrix}.$$

Up to isomorphism, we then have the commutative diagram

$$\begin{array}{ccc} & & \mathscr{E}_{\mathbf{C}}^{V\oplus V}(X) \\ & \overset{\theta^c}{\nearrow} & \uparrow \mathscr{L} \\ \mathscr{E}_{\mathbf{C}}^{0,2n}(X) & & \\ & \underset{\theta'^c}{\searrow} & \\ & & \mathscr{E}_{\mathbf{C}}^{W}(X) \end{array}$$

where θ^c and θ'^c are defined as in 4.25 *(using the* ^{c}*spinorial structures presented in* 4.30 *and* 4.26 *respectively), and where $\mathcal{L}$ is the functor $E \mapsto E \otimes L(V)$.*

Proof. Let us compute the composition j'^c of the homomorphisms

$$\mathrm{Spin}^c(n) \xrightarrow{\pi} \mathrm{SO}(n) \xrightarrow{\iota} \mathrm{U}(n) \xrightarrow{\sigma} \mathrm{Spin}^c(2n) = \mathrm{Spin}(2n) \times_{\mathbf{Z}/2} \mathrm{U}(1).$$

If $\delta = (\beta, z) \in \mathrm{Spin}^c(n)$, and if $\alpha = \pi(\delta) \in \mathrm{SO}(n) \subset \mathrm{U}(n)$, there is an orthonormal basis $f_1, \ldots, f_p, \bar{f}_1, \ldots, \bar{f}_p, f_{2p+1}, \ldots, f_n$ of $\mathbf{C}^n$ such that $\alpha(f_r) = \exp(i\theta_r)(f_r)$, $\alpha(\bar{f}_r) = \exp(-i\theta_r) f_r$ for $r \leqslant p$, and $\alpha(f_r) = f_r$ for $r > 2p$. By the formula given in 4.27, it follows that $(\sigma\iota\pi)(\delta) \in \mathrm{Spin}(2n) \subset \mathrm{Spin}^c(2n)$. Hence, $\sigma\iota$ defines a homomorphism $j\colon \mathrm{SO}(n) \to \mathrm{Spin}(2n)$ which makes the diagram

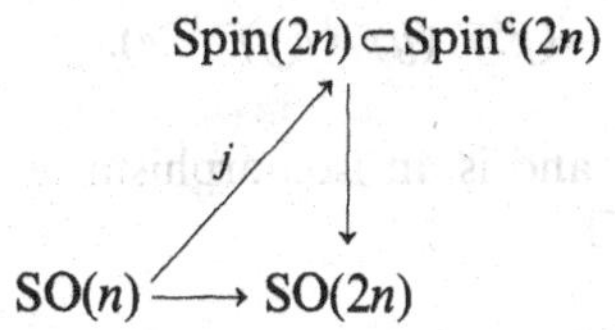

commutative. Since such a homomorphism is unique, it must coincide with the homomorphism (also denoted by j) defined in 4.18, and we have the formula $(j\pi)(\beta, z) = (s(\beta, \beta), 1) \in \mathrm{Spin}(2n) \times_{\mathbf{Z}/2} \mathrm{U}(1)$, where s is defined as in 4.24. If we let j^c denote the composition

$$\mathrm{Spin}^c(n) \xrightarrow{\Delta} \mathrm{Spin}^c(n) \times \mathrm{Spin}^c(n) \xrightarrow{s^c} \mathrm{Spin}^c(2n),$$

where s^c is defined as in 4.29, and where Δ is the diagonal homomorphism, then we have the formula

$$j^c(g) = j'^c(g) l(g).$$

We are now approaching the proof of the theorem. We consider a principal $\mathrm{Spin}^c(n)$-bundle P, such that $V \approx P \times_{\mathrm{Spin}^c(n)} \mathbf{R}^n$. By definition, the principal $\mathrm{Spin}^c(2n)$-bundle, associated with $V \oplus V$, is $P \times_{\mathrm{Spin}^c(n)} \mathrm{Spin}^c(2n)$, the quotient of $P \times \mathrm{Spin}^c(2n)$ by the equivalence relation $(pg, h) \sim (p, j^c(g)^{-1} h)$. Therefore $\theta^c(E)$ is $P \times_{\mathrm{Spin}^c(n)} \mathrm{Spin}^c(2n)\, \Delta_{\mathrm{Spin}^c(2n)}\, E$, using the notation of 4.22 and 4.25. More precisely, $\theta^c(E)$ is the quotient of $P \times \mathrm{Spin}^c(2n) \times_X E$ by the equivalence relation generated by $(pg, h, e) \sim (p, j^c(g)^{-1} h, e)$ and $(p, h_1 h_2, e) \sim (p, h_1, h_2^{-1} e)$ (where the projections of p and e on the base X are equal). Since any element of this space is the class of a triple $(p, 1, e)$, we could also write this space as $P\, \Delta_{\mathrm{Spin}^c(n)}\, E$, i.e. the quotient of $P \times_X E$ by the equivalence relation $(pg, e) \sim (p, j^c(g)^{-1} e)$, where $g \in \mathrm{Spin}^c(n)$.

On the other hand, the principal $\mathrm{Spin}^c(2n)$-bundle associated with the complex structure of $V \oplus V$, is $P \times_{\mathrm{Spin}^c(n)} \mathrm{Spin}^c(2n)$, the quotient of $P \times \mathrm{Spin}^c(2n)$ by the equivalence relation $(pg, h) \sim (p, j'^c(g)^{-1} h)$. Therefore, as above, we may write $\theta'^c(E)$ as $P\, \Delta_{\mathrm{Spin}^c(n)}\, E$, i.e. the quotient of $P\, \Delta\, E$ by the equivalence relation $(pg, e) \sim (p, j'^c(g)^{-1} e)$, where $g \in \mathrm{Spin}^c(n)$.

Finally, we write $L(V)$ as the quotient of $P \times_{\mathrm{Spin}^c(n)} \mathbb{C}$ by the equivalence relation $(pg, \lambda) \sim (p, l(g)^{-1}\lambda)$, and we define a morphism

$$\theta'^c(E) \otimes L(V) \longrightarrow \theta^c(E)$$

by the formula

$$(p, e) \otimes (p, \lambda) \longmapsto (p, \lambda e).$$

We notice that

$$(pg, e) \otimes (pg, \lambda) \longmapsto (pg, \lambda e) = (p, j^c(g)^{-1}\lambda e)$$

$$(p, j'^c(g)^{-1}e) \otimes (p, l(g)^{-1}e) \longmapsto (p, j'^c(g)^{-1} l(g)^{-1}\lambda e).$$

Therefore, this morphism is well-defined, and is an isomorphism on each fiber. Thus we have an isomorphism by I.2.7. □

4.32. Corollary. *Let V be a ^{c}spinorial bundle of rank n, and let W be the real bundle $V \oplus V$, provided with the complex structure defined in* I.4.8. *Then we have a commutative diagram (up to isomorphism),*

$$\begin{array}{ccc} & & \mathscr{E}_{\mathbb{C}}^{V \oplus V}(X) \\ & \overset{\psi}{\nearrow} & \uparrow \mathscr{L} \\ \mathscr{E}_{\mathbb{C}}(X) & & \\ & \underset{\varphi}{\searrow} & \\ & & \mathscr{E}_{\mathbb{C}}^{W}(X). \end{array}$$

In this diagram, φ is the functor $E \mapsto \Lambda(W) \otimes E$ described in 4.29; *ψ is the composition $\mathscr{E}_{\mathbb{C}}(X) \sim \mathscr{E}_{\mathbb{C}}^{0,2n}(X) \xrightarrow{\theta^c} \mathscr{E}_{\mathbb{C}}^{V \oplus V}(X)$, where θ^c is the category equivalence induced by the ^{c}spinorial structure of $V \oplus V$* (cf. 4.25 *and* 4.29). *Finally $\mathscr{L}$ is the functor $E \mapsto E \otimes L(V)$, where $L(V)$ is the line bundle associated with the ^{c}spinorial structure of V* (cf. 4.30).

Proof. This is an immediate consequence of the theorem above, and the commutative diagram

$$\begin{array}{ccccc} & & & & \mathscr{E}_{\mathbb{C}}^{V \oplus V}(X) \\ & & & \overset{\theta^c}{\nearrow} & \uparrow \mathscr{L} \\ \mathscr{E}_{\mathbb{C}}(X) & \xrightarrow{\sim} & \mathscr{E}_{\mathbb{C}}^{0,2n}(X) & & \\ & & & \underset{\theta'^c}{\searrow} & \\ & & & & \mathscr{E}_{\mathbb{C}}^{W}(X), \end{array}$$

where the composition $\mathscr{E}_{\mathbb{C}}(X) \xrightarrow{\sim} \mathscr{E}_{\mathbb{C}}^{0,2n}(X) \xrightarrow{\theta^c} \mathscr{E}_{\mathbb{C}}^{W}(X)$ is identified with φ (cf. 4.29). □

Exercise (IV.8.1).

5. The Thom Isomorphism in Real and Complex K-Theory for Real Vector Bundles

5.1. Let V be a real vector bundle provided with a nondegenerate symmetric bilinear form (or quadratic form) (I.8.4). We let 1 denote the trivial bundle of rank one provided with the form $\lambda \mapsto \lambda^2$. Let φ^V denote the functor from $\mathscr{E}^{V \oplus 1}(X)$ to $\mathscr{E}^V(X)$ which associates each $C(V \oplus 1)$-module with its underlying $C(V)$-module (note that $C(V) \subset C(V \oplus 1)$). Now we define the group $K^V(X)$ as the Grothendieck group of the Banach functor φ^V (cf. II.2.13). If $V = X \times \mathbb{R}^{p+q}$ provided with the quadratic form

$$(x, \lambda_1, \ldots, \lambda_{p+q}) \longmapsto -(\lambda_1)^2 - \cdots - (\lambda_p)^2 + (\lambda_{p+1})^2 + \cdots + (\lambda_{p+2})^2,$$

then Example 4.12 shows that $K^V(X)$ is isomorphic to the group $K^{p,q}(X)$ introduced in III.4.11. An equivalent description of the group $K^V(X)$ may be given along the lines of III.4.14, ..., III.4.23 in terms of gradations. More precisely, a gradation of a bundle E provided with a $C(V)$-module structure (i.e. with a morphism $m: V \to \mathrm{HOM}(E, E)$ such that $(m(v))^2 = Q(v)$; cf. 4.11) is a morphism $\eta: E \to E$, such that $n^2 = 1$ and $\eta m(v) = -m(v)\eta$ for each $v \in V$. Then, as in III.4, we consider the set of triples (E, η_1, η_2), where E is a $C(V)$-module and η_1 and η_2 are gradations. Now $K^V(X)$ is the quotient of the free group generated by these triples, by the subgroup generated by the relations

(*i*) $\quad (E, \eta_1, \eta_2) + (F, \zeta_1, \zeta_2) = (E \oplus F, \eta_1 \oplus \zeta_1, \eta_2 \oplus \zeta_2)$, and

(*ii*) $\quad (E, \eta_1, \eta_2) = 0$

if η_1 is homotopic to η_2 within the gradations of E.

5.2. Theorem. *Assume that V is provided with a nondegenerate positive quadratic form and also with a spinorial structure (resp. a ^{c}spinorial structure). Then the category equivalence θ (resp. θ^c) defined in* 4.22 *(resp.* 4.25*) induces an isomorphism*

$$K^{0,n}(X) \approx K^V(X) \quad (\textit{resp. } K_{\mathbb{C}}^{0,n}(X) \approx K_{\mathbb{C}}^V(X)),$$

where $n = \mathrm{rank}(V)$.

Proof. In general, we let T_p denote the trivial bundle of rank p. If V is a Spin(n)-bundle (resp. a Spin^c(n)-bundle), then $V \oplus T_1$ is naturally a Spin($n+1$)-bundle (resp. a Spin^c($n+1$)-bundle) by 4.24 (resp. 4.29). Hence, we have the commutative category diagram (up to isomorphism)

$$\begin{array}{ccc} \mathscr{E}^{T_{n+1}}(X) & \xrightarrow{\theta} & \mathscr{E}^{V \oplus 1}(X) \\ \downarrow{\scriptstyle \varphi^{T_n}} & & \downarrow{\scriptstyle \varphi^V} \\ \mathscr{E}^{T_n}(X) & \xrightarrow{\theta} & \mathscr{E}^V(X) \end{array}, \text{ resp. } \begin{array}{ccc} \mathscr{E}_{\mathbb{C}}^{T_{n+1}}(X) & \xrightarrow{\theta^c} & \mathscr{E}_{\mathbb{C}}^{V \oplus 1}(X) \\ \downarrow{\scriptstyle \varphi_{\mathbb{C}}^{T_n}} & & \downarrow{\scriptstyle \varphi_{\mathbb{C}}^V} \\ \mathscr{E}_{\mathbb{C}}^{T_n}(X) & \xrightarrow{\theta^c} & \mathscr{E}_{\mathbb{C}}^V(X). \end{array}$$

Since the horizontal arrows are Banach category equivalences, they induce isomorphisms $K^{0,n}(X)\approx K(\varphi^{T_n})\approx K(\varphi^V)$ (resp. $K_{\mathbb{C}}^{0,n}(X)\approx K(\varphi_{\mathbb{C}}^{T_n})\approx K(\varphi_{\mathbb{C}}^V)$). Note that $K^{0,n}(X)\approx K^{-n}(X)$ and $K_{\mathbb{C}}^{0,n}(X)\approx K_{\mathbb{C}}^{-n}(X)$ in general (cf. III.5.12). □

5.3. If V is provided with a complex structure, it may also be provided with a ^{c}spinorial structure (4.26). Hence, if we write $\operatorname{rank}(V)=2p$, then we have $K_{\mathbb{C}}(X)\approx K_{\mathbb{C}}^{0,2p}(X)\approx K_{\mathbb{C}}^V(X)$. We will make the composition of these isomorphisms more explicit. This will be the pattern followed in the generalization, Theorem 5.8.

First of all, the isomorphism $K_{\mathbb{C}}(X)\approx K_{\mathbb{C}}^{0,2p}(X)$ is obtained from the diagram

$$\begin{array}{ccc} \mathscr{E}_{\mathbb{C}}^{0,1}(X) & \longrightarrow & \mathscr{E}_{\mathbb{C}}^{0,2p+1}(X) \\ \downarrow & & \downarrow \\ \mathscr{E}_{\mathbb{C}}(X) & \longrightarrow & \mathscr{E}_{\mathbb{C}}^{0,2p}(X), \end{array}$$

where the horizontal arrows are category equivalences (III.4.6). More precisely, if we write $D^{0,q}$ for $C^{0,q}\otimes_{\mathbb{R}}\mathbb{C}$ in general, then we have an isomorphism $D^{0,2p+1}\to D^{0,2p}\otimes D^{0,1}$. It is induced by the $\mathbb{R}$-linear map $\mathbb{R}^{2p}\oplus\mathbb{R}\to D^{0,2p}\otimes D^{0,1}$, defined by $(v,\lambda)\mapsto v\otimes 1+\varepsilon\otimes\lambda$, with $\varepsilon=e_1(ie_2)e_3(ie_4)\cdots(ie_{2p})$ where $e_1,\ldots,e_{2p}$ is the canonical basis of $\mathbb{R}^{2p}$ imbedded in $C^{0,2p}\subset D^{0,2p}$ (compare with III.3.16; note that $(\varepsilon)^2=1$). If M is an irreducible $D^{0,2p}$-module, then the category equivalence $\mathscr{E}_{\mathbb{C}}^{0,1}(X)\xrightarrow{\sim}\mathscr{E}_{\mathbb{C}}^{0,2p+1}(X)$ is thus defined by $E\mapsto M\otimes E$, where $\mathbb{R}^{2p}\oplus\mathbb{R}=\mathbb{R}^{2p+1}\subset D^{0,2p+1}$ acts on $M\otimes E$ by the map $[(v,\lambda),(u\otimes e)]\mapsto vu\otimes e+\varepsilon u\otimes\lambda e$ (cf. III.4.6). By 1.5 we may choose $M=\Lambda(\mathbb{C}^p)$. To proceed any further, we require the following lemma.

5.4. Lemma. *Let us identify $\mathbb{R}^{2p}$ with $\mathbb{C}^p$ by the map*

$$(x_1,\ldots,x_{2p})\longmapsto(x_1+ix_2,\ldots,x_{2p-1}+ix_{2p}).$$

Then $\varepsilon=e_1(ie_2)\cdots(ie_{2p})$ is equal to $+1$ on $\Lambda^{(0)}(\mathbb{C}^p)$, and to -1 on $\Lambda^{(1)}(\mathbb{C}^p)$.

Proof. If V and W are complex vector spaces provided with positive nondegenerate Hermitian forms, then the canonical isomorphism

$$f\colon \Lambda(V)\otimes\Lambda(W)\approx\Lambda(V\oplus W)$$

(cf. III.3.10) is an isomorphism of $C(V)\hat{\otimes}C(W)\approx C(V\oplus W)$-modules. Essentially, f is induced by $w_1\otimes w_2\mapsto w_1\wedge w_2$, and we have the identities

$$d_v(w_1\wedge w_2)=d_v(w_1)\wedge w_2 \quad \text{if } v\in V,$$
$$d_w(w_1\wedge w_2)=(-1)^{\omega(w)}w_1\wedge d_w(w_2) \quad \text{if } w\in W \quad \text{(where } \omega \text{ denotes the degree)},$$

$$\partial_v(w_1\wedge w_2)=\partial_v(w_1)\wedge w_2 \quad \text{if } v\in V,$$

and $$\partial_w(w_1\wedge w_2)=(-1)^{\omega(w)}w_1\wedge\partial_w(w_2) \quad \text{if } w\in W.$$

Hence $$d_{v+w}+\partial_{v+w}=(d_v+\partial_v)\hat{\otimes}1+1\hat{\otimes}(d_w+\partial_w)$$

on $\Lambda(V) \otimes \Lambda(W)$ (compare with III.3.10). Therefore we need only verify the lemma for the case where $V=\mathbb{C}$ with basis f_1. In this case, we have $\Lambda(V)=\mathbb{C}\oplus\mathbb{C}$ with basis 1 and f_1. Moreover

$$d_{f_1}(1)=f_1, \qquad \partial_{f_1}(1)=0,$$
$$d_{f_1}(f_1)=0, \quad \text{and} \quad \partial_{f_1}(f_1)=1.$$

If $\rho(v)$ denotes the action of $v\in\mathbb{C}\approx\mathbb{R}^2$ on $\Lambda(V)$, we have

$$\rho(e_1)=d_{f_1}+\partial_{f_1}=\begin{pmatrix}0 & 1\\ 1 & 0\end{pmatrix}; \ \rho(e_2)=d_{if_1}+\partial_{if_1}=i(d_{f_1}-\partial_{f_1})=\begin{pmatrix}0 & i\\ -i & 0\end{pmatrix}.$$

Hence $\quad i\rho(e_1)\rho(e_2)=\begin{pmatrix}1 & 0\\ 0 & -1\end{pmatrix}.$ □

5.5. Proposition. *The category equivalence*

$$\mathscr{E}_{\mathbb{C}}^{0,1}(X)\longrightarrow \mathscr{E}_{\mathbb{C}}^{0,2p+1}(X)$$

defined in III.4 *is* $E\mapsto\Lambda(\mathbb{C}^p)\hat{\otimes}E$ *(graded tensor product), where* $\Lambda(\mathbb{C}^p)\hat{\otimes}E$ *is regarded as a module over* $D^{0,2p}\hat{\otimes}D^{0,1}\approx D^{0,2p+1}$

Proof. This follows directly from Lemma 5.4 and the preceding observations (with $M=\Lambda(\mathbb{C}^p)$). □

5.6. Proposition. *Let V be a complex vector bundle of rank p (real rank n=2p). Then the composition of the category equivalences*

$$\mathscr{E}_{\mathbb{C}}^{0,1}(X)\longrightarrow \mathscr{E}_{\mathbb{C}}^{0,2p+1}(X)\xrightarrow{\theta^c}\mathscr{E}_{\mathbb{C}}^{V\oplus 1}(X)$$

is defined by $E\mapsto\Lambda(V)\hat{\otimes}E$, *where* $\Lambda(V)\hat{\otimes}E$ *is regarded as a module over* $C(V)\hat{\otimes}C(T_1)\approx C(V\oplus 1)$.

Proof. The explicit computation in 4.29 can be repeated almost verbatim by placing gradations on all the vector bundles involved. □

5.7. Theorem. *Let V be a complex vector bundle of rank p (real rank n=2p). Then the composition of the isomorphisms*

$$K_{\mathbb{C}}(X)\approx K_{\mathbb{C}}^{0,0}(X)\longrightarrow K_{\mathbb{C}}^{0,n}(X)\longrightarrow K_{\mathbb{C}}^{V}(X)$$

defined in 4.22 *and* III.4 *is induced by*

$$[E]\longmapsto d(\Lambda(V)\otimes E, \eta\otimes 1, -\eta\otimes 1),$$

where η *is the canonical* $\mathbb{Z}/2$*-gradation of the exterior algebra* $\Lambda(V)$.

Proof. Let E be an object of $\mathscr{E}_{\mathbb{C}}(X)$. Its associated element of $K^{0,0}(X)$ is $d(E^{(0)}, E^{(1)}, \alpha)$, where $E^{(i)}$ is the concentration of E in degree i, and α is the identity on the underlying nongraded modules (III.4.12). Hence, the image of $[E]$ under the composition of the isomorphisms is the element

$$d(\Lambda(V)\,\hat{\otimes}\,E^{(0)}, \Lambda(V)\,\hat{\otimes}\,E^{(1)}, 1\,\hat{\otimes}\,\alpha)$$

by 5.6. With respect to the definition of $K^V(X)$ in terms of gradations, this is also $d(\Lambda(V)\otimes E, \eta\otimes 1, -\eta\otimes 1)$. □

5.8. Theorem. *Let V be a* Spin(8*n*)*-bundle* (*resp. a* Spin^c(2*n*)*-bundle*) *such that $V\approx P\times_{\mathrm{Spin}(8n)}\mathbb{R}^{8n}$* (*resp. $V\approx P\times_{\mathrm{Spin}^c(2n)}\mathbb{R}^{2n}$*). *Let M be an irreducible $C^{0,8n}$-module* (*resp. an irreducible $D^{0,2n}=C^{0,2n}\otimes_{\mathbb{R}}\mathbb{C}$-module*), *and let η_M be the gradation defined by $\eta_M=e_1\cdots e_{8n}$* (*resp. $\eta_M=(-1)^n e_1(ie_2)\cdots e_{2n}(ie_{2n})$*). *Then the composition of the isomorphisms*

$$K_{\mathbb{R}}(X)\approx K_{\mathbb{R}}^{0,0}(X)\approx K_{\mathbb{R}}^{0,8n}(X)\approx K_{\mathbb{R}}^{V}(X)$$

(*resp.* $\quad K_{\mathbb{C}}(X)\approx K_{\mathbb{C}}^{0,0}(X)\approx K_{\mathbb{C}}^{0,2n}(X)\approx K_{\mathbb{C}}^{V}(X)$)

is induced by

$$E\mapsto d(M(V)\otimes E, \eta\otimes 1, -\eta\otimes 1),$$

where $\quad M(V)=P\times_{\mathrm{Spin}(8n)}M$ (*resp.* $M(V)=P\times_{\mathrm{Spin}^c(2n)}M$) *and* $\eta=(Id_p, \eta_M)$.

Proof. The proof of this theorem is analogous to the proof of Theorem 5.7, using the explicit category equivalences θ and θ^c (cf. 4.22 and 4.25). The only point needing clarification is the gradation η_M, which is used to identify $C^{0,8n+1}$ with $C^{0,8n}\otimes C^{0,1}$ (resp. $D^{0,2n+1}$ with $D^{0,2n}\otimes D^{0,1}$). In fact, there are only two gradations on M, η_M and $-\eta_M$: they correspond to the two irreducible $C^{0,8n+1}$-modules (resp. $D^{0,2n+1}$-modules) of real dimension 16^n (resp. complex dimension 2^n): cf. III.3. In order to avoid ambiguity of sign, we must remark that η_M acts as $(-1)^n$ on $\Lambda^0\mathbb{C}^n\subset\Lambda\mathbb{C}^n$, regarded as an irreducible $D^{0,2n}$-module (cf. 5.4). In general, changing the sign of η_M results in a change of sign for the isomorphism $K_{\mathbb{R}}(X)\approx K_{\mathbb{R}}^V(X)$ (resp. $K_{\mathbb{C}}(X)\approx K_{\mathbb{C}}^V(X)$). □

5.9. Proposition. *Let V be a* Spin^c(*n*)*-bundle, and let W* (*resp. W'*) *be the real bundle $V\oplus V$ provided with the* Spin^c(2*n*)*-bundle structure explicitly described in* 4.30 (*resp. provided with the complex structure described in* I.4.8 *and* 4.31). *Then we have the commutative diagram*

$$\begin{array}{ccc} & & K_{\mathbb{C}}^{W}(X)=K_{\mathbb{C}}^{V\oplus V}(X) \\ & \overset{\psi^*}{\nearrow} & \uparrow \mathscr{L}^* \\ K_{\mathbb{C}}(X) & \underset{\varphi^*}{\searrow} & \\ & & K_{\mathbb{C}}^{W'}(X)=K_{\mathbb{C}}^{V\oplus V}(X) \end{array}$$

where ψ^* *and* φ^* *are defined as in* 4.32, *and where* $\mathscr{L}^*$ *is defined by the tensor product with the line bundle* $L(V)$ (cf. 4.30).

Proof. This proposition follows directly from 4.32. □

5.10. We now consider an arbitrary real vector bundle V with compact base provided with a positive quadratic form. Let $B(V)$ (resp. $S(V)$) be the ball bundle (resp. the sphere bundle) of V. Recall that $K(B(V), S(V)) \approx K(V) \approx \tilde{K}(\dot{V})$, where $\dot{V}$ is the one point compactification of V, i.e. the Thom space of V. We wish to define a fundamental homomorphism

$$t: K^V(X) \longrightarrow K(B(V), S(V))$$

which generalizes the homomorphism t defined in III.5.9 in some sense (see 6.21 for a precise analogy).

More explicitly, we identify the ball bundle $B(V)$ with the upper hemisphere $S^+(V \oplus 1)$ of the sphere bundle $S(V \oplus 1)$:

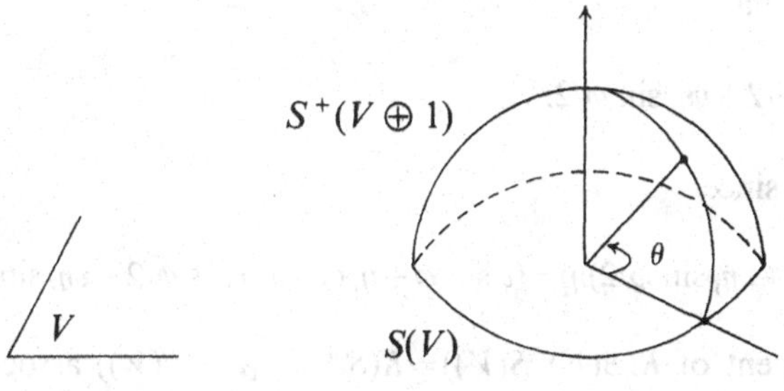

Fig. 21

this is the set of points (v, λ), where $v \in V$ and $\lambda \in \mathbb{R}^+$ such that $Q(v) + \lambda^2 = 1$. We let $\pi: S^+(V \oplus 1) \approx B(V) \longrightarrow X$ denote the canonical projection.

Let $x = d(E, \eta_1, \eta_2) \in K^V(X)$, and let $E' = \pi^* E$. Over any point (v, λ) of $S^+(V \oplus 1)$, we may consider the two gradations of $\pi^* E$ defined by $\rho(v) + \lambda\eta_1$ and $\rho(v) + \lambda\eta_2$ (where $\rho: V \to \mathrm{HOM}(E, E)$ denotes the action of V on E arising from the $C(V)$-module structure cf. 4.11). Using polar coordinates and writing v instead of $\rho(v)$ for simplicity, the two gradations above may also be written as $v \cos\theta + \eta_i \sin\theta$, $i = 1, 2$, over the point with polar coordinates (v, θ), where $v \in S(V)$ and $0 \leqslant \theta \leqslant \pi/2$. We define the homomorphism t by the formula

$$t(d(E, \eta_1, \eta_2)) = d(\pi^* E, v \cos\theta + \eta_1 \sin\theta, v \cos\theta + \eta_2 \sin\theta),$$

where the right side must be interpreted as an element of $K^{0,0}(B(V), S(V)) \approx K(B(V), S(V))$ (cf. III.5.7). This is well-defined since the two gradations $\zeta_i(\theta) = v \cos\theta + \eta_i \sin\theta$, for $i = 1, 2$, agree on $S(V) \subset S^+(V \oplus 1)$. The following theorem will be proved later (6.21) in a more general form.

5.11. Theorem. *The homomorphism defined above,*

$$t: K^V(X) \longrightarrow K(B(V), S(V))$$

is an isomorphism.

In this section we only prove this theorem for the case where $\mathrm{Rank}(V) \equiv 0 \bmod 8$ (resp. $\equiv 0 \bmod 2$), where V is provided with a spinorial structure (resp. ^{c}spinorial structure), and where the K-theory considered is real (resp. complex) K-theory. Before doing this, we explicitly compute the composition

$$K^V(X) \xrightarrow{t} K^{0,0}(B(V)), S(V)) \xrightarrow{g} K(B(V), S(V))$$

where g is defined as in III.5.7.

Let $d(E, \eta_1, \eta_2)$ be an element of $K^V(X)$. The bundle π^*E, provided with the gradation η_i, is isomorphic to the bundle π^*E provided with the gradation $\zeta_i(\theta) = v \cos\theta + \eta_i \sin\theta$ due to the isomorphism

$$f_i: (E, \eta_i) \longrightarrow (E, v \sin\varphi + \eta_i \cos\varphi), \quad \text{for} \quad \varphi = \pi/2 - \theta,$$

defined by the formula

$$f_i = \cos\varphi/2 + v\eta_i \sin\varphi/2.$$

This is well-defined since

$$(\cos\varphi/2 + v\eta_i \sin\varphi/2)\eta_i = (v \sin\varphi + \eta_i \cos\varphi)(\cos\varphi/2 + v\eta_i \sin\varphi/2).$$

By III.5.7, the element of $K(B(V), S(V)) = K(S^+(V \oplus 1), S(V))$ associated with $d(\pi^*E, \zeta_1(\theta), \zeta_2(\theta))$ is $d(E_{1,\theta}, E_{2,\theta}, \alpha)$, where $E_{i,\theta}$ denotes the bundle $\mathrm{Ker}\left(\frac{1-\zeta_i(\theta)}{2}\right)$ on $S^+(V \oplus 1)$, and α is the identification of these two bundles over $S(V)$ (this is possible since $\zeta_1(\theta) = \zeta_2(\theta)$ over $S(V)$). Using the isomorphisms f_i, we may also write $d(E_{1,\theta}, E_{2,\theta}, \alpha)$ as $d(\pi^*E_1, \pi^*E_2, \beta)$, where $E_i = \mathrm{Ker}\left(\frac{1-\eta_i}{2}\right)$, and β is $f_2^{-1}f_1$ restricted to $S(V)$.

$$\begin{array}{ccc} (E, \eta_1) & \xrightarrow{\alpha} & (E, \eta_2) \\ \downarrow f_1 & & \downarrow f_2 \\ (E, v\sin\varphi + \eta_1\cos\varphi) & \xrightarrow{\beta} & (E, v\sin\varphi + \eta_2\cos\varphi) \end{array}$$

If $\varphi = \pi/2$, we see that $\beta = 1/2(1 - v\eta_2)(1 + v\eta_1)$. In particular, taking $\eta_2 = -\eta_1$, we obtain the following proposition (cf. III.4.12).

5.12. Proposition. *Let V be a* Spin(8n)*-bundle* (*resp. a* Spin^c(2n)*-bundle*). *Then the composition of the homomorphisms*

$$K_{\mathbb{R}}(X) \xrightarrow{\approx} K_{\mathbb{R}}^{V}(X) \xrightarrow{t} K_{\mathbb{R}}(B(V), S(V))$$

$$(\text{resp.}\quad K_{\mathbb{C}}(X) \xrightarrow{\approx} K_{\mathbb{C}}^{V}(X) \xrightarrow{t} K_{\mathbb{C}}(B(V), S(V))),$$

is induced by

$$[E] \longmapsto d(\pi^*M(V)^0 \otimes E, \pi^*M(V)^1 \otimes E, \alpha \otimes 1),$$

where $M(V) = M(V)^0 \oplus M(V)^1$ *is the graded* $C(V)$*-module defined in* 5.8, *and* $\alpha : \pi^*M(V)^0|_{S(V)} \to \pi^*M(V)^1|_{S(V)}$ *is the isomorphism defined over each point* $v \in S(V)$ *by multiplication with the Clifford number* $v \in V \subset C(V)$.

5.13. The element $d(\pi^*M(V)^0, \pi^*M(V)^1, \alpha)$ is called the *Thom class* T_V of the Spin(8n)-bundle V (resp. the Spin^c(2n)-bundle V). It belongs to the group $K(B(V), S(V))$ (resp. $K_{\mathbb{C}}(B(V), S(V))$). The explicit formulas given in II.5.21, show that the composition of the homomorphisms $K_{\mathbb{R}}(X) \to K_{\mathbb{R}}^V(X) \to K_{\mathbb{R}}(B(V), S(V))$ (resp. $K_{\mathbb{C}}(X) \to K_{\mathbb{C}}^V(X) \to K_{\mathbb{C}}(B(V), S(V))$) is defined by the product with the Thom class. When V is provided with a complex structure (hence with a Spin^c(2n)-structure), it follows from 1.6, 4.28, 5.8, and 5.12, that up to the sign $(-1)^n$ the Thom class T_V is the same as the Thom class U_V, introduced in 1.6 (see the convention of signs made in V.4.8). From Theorem 1.9, we see that the composition $K_{\mathbb{C}}(X) \to K_{\mathbb{C}}^V(X) \to K_{\mathbb{C}}(B(V), S(V))$ is an isomorphism in this case.

5.14. Theorem (Thom isomorphism). *Let V be a* Spin(8n)*-bundle* (*resp. a* Spin^c(2n)*-bundle*) *with compact base X. Then the product with* $T_V \in K_{\mathbb{R}}(B(V), S(V))$ (*resp.* $T_V \in K_{\mathbb{C}}(B(V), S(V))$) *induces an isomorphism* $K_{\mathbb{C}}(X) \approx K_{\mathbb{C}}(B(V), S(V))$. *More generally,* $K_{\mathbb{R}}^*(V) \approx K_{\mathbb{R}}^*(B(V), S(V))$ (*resp.* $K_{\mathbb{C}}^*(V) \approx K_{\mathbb{C}}^*(B(V), S(V))$) *is a free* $K_{\mathbb{R}}^*(X)$*-module* (*resp.* $K_{\mathbb{C}}^*(X)$*-module*) *generated by the Thom class* T_V.

Proof. By Theorem 1.3, it suffices to prove the theorem for V trivial. In the complex case ($K = K_{\mathbb{C}}$), the trivial bundle V of rank $2n$ may be provided with a complex structure, and $T_V = (-1)^n U_V$ (5.13). Therefore, the theorem is true in this case. For the real case ($K = K_{\mathbb{R}}$), the Thom class T_V is $p^* u_{8n}$, where u_{8n} is the Thom class of $\mathbb{R}^{8n}$ regarded as bundle over the point $P (p : X \to P)$. If M is an irreducible $C^{0,8n}$-module, then $M \otimes_{\mathbb{R}} \mathbb{C}$ is an irreducible $C^{0,8n} \otimes_{\mathbb{R}} \mathbb{C}$-module (III.3.22). Therefore, the complexification of T_V in $K_{\mathbb{C}}(\mathbb{R}^{8n})$ is the generator. Since the complexification

$$K_{\mathbb{R}}(\mathbb{R}^{8n}) \longrightarrow K_{\mathbb{C}}(\mathbb{R}^{8n})$$

is an isomorphism (III.5.19), u_{8n} is the generator of $K_{\mathbb{R}}(\mathbb{R}^{8n}) \approx \mathbb{Z}$. By 1.3, it follows that $K_{\mathbb{R}}^*(V)$ is a free $K_{\mathbb{R}}^*(X)$-module generated by T_V. □

5.15. As an application of the preceding theorem, we consider $V=\underbrace{\xi^* \oplus \cdots \oplus \xi^*}_{m}$, where ξ is the canonical line bundle over RP_n. Then V may be identified with the quotient $Q/\mathbb{R}^*$, where Q is the set of points $(x_0, \ldots, x_{n+m}) \in \mathbb{R}^{n+m+1}$, with $(x_0, \ldots, x_n) \neq (0, \ldots, 0)$. Hence, the Thom space $\dot{V}$ is homeomorphic to the one point compactification of $RP_{n+m} - RP_{m-1}$, i.e. RP_{n+m}/RP_{m-1}. In particular, if $m \equiv 0 \bmod 8$ (resp. $m \equiv 0 \bmod 2$), we have $\tilde{K}_{\mathbb{R}}(RP_{n+m}/RP_{m-1}) \approx K_{\mathbb{R}}(RP_n)$ (resp. $\tilde{K}_{\mathbb{C}}(RP_{n+m}/RP_{m-1}) \approx K_{\mathbb{C}}(RP_n)$) since V is spinorial (resp. ^{c}spinorial) by 4.19 (resp. 4.26).

5.16. As in 1.10, we may define a "Thom homomorphism"

$$K_{\mathbb{R}}(X) \longrightarrow K_{\mathbb{R}}(V) \qquad (\text{resp. } K_{\mathbb{C}}(X) \longrightarrow K_{\mathbb{C}}(V))$$

when the base X is locally compact, but not necessarily compact. In fact, $\sigma(M(V), \Delta)$, where Δ_v is multiplication by $v \in V$, is naturally an element of $K(V)$ if X is compact, and coincides with the Thom class T_V, modulo the isomorphism between the groups K and K_0 described in II.5.20. Therefore, the Thom homomorphism

$$K_{\mathbb{R}}(X) \longrightarrow K_{\mathbb{R}}(V) \qquad (\text{resp. } K_{\mathbb{C}}(X) \longrightarrow K_{\mathbb{C}}(V))$$

is defined by $\sigma(E, D) \mapsto \sigma(\pi^* E \otimes M(V), \pi^* D \hat{\otimes} 1 + 1 \hat{\otimes} \Delta)$ as in 1.10. However, it can be seen that this formula is still well-defined when X is locally compact.

5.17. Theorem. *Let X be a locally compact space, and let V be a* Spin(8*n*)*-bundle (resp. a* Spin^c(2*n*)*-bundle). Then the Thom homomorphism defined above,*

$$K_{\mathbb{R}}(X) \longrightarrow K_{\mathbb{R}}(V) \qquad (\textit{resp. } K_{\mathbb{C}}(X) \longrightarrow K_{\mathbb{C}}(V)),$$

is an isomorphism.

Proof. The proof of this theorem is analogous to the proof described in 1.11. □

5.18. Theorem. *Let X be a compact space, and let V and V' be spinorial bundles (resp. ^{c}spinorial bundles) of rank $\equiv 0 \bmod 8$ (resp. $\equiv 0 \bmod 2$). Then*

$$T_{V \boxplus V'} = T_V \cup T_{V'}$$

in the group $K(V \boxplus V') = K(V \times V')$. In particular, $T_{V \oplus V'} = i^*(T_V \cup T_{V'})$, where $i\colon V \oplus V' \to V \times V'$ is the canonical inclusion (I.6.1).

Proof. We only consider the real case, since the complex case is completely analogous. If $\operatorname{rank}(V) = 8n$ and $\operatorname{rank}(V') = 8n'$, then we have $M(V) = P \times_{\mathrm{Spin}(8n)} M_{8n}$, where P is the principal bundle with group Spin(8*n*) which defines the spinorial structure on V. In fact, $T_V = \sigma(M(V), \Delta)$ where Δ is defined as in 5.16. In the same way, $T_{V'}=\sigma(M(V'),\Delta')$ where $M(V')=P'\times_{\mathrm{Spin}(8n')}M_{8n'}$. As graded modules, it is clear that $M(V)\hat{\boxtimes}M(V')\approx P\times P'\times_{G_n\times G_{n'}}M_{8n}\hat{\otimes}M_{8n'}$, where G_q denotes Spin$(8q)$ in general. The principal bundle associated with $V\boxplus V'$ is $P\times P'\times_{G_n\times G_{n'}}G_{n+n'}$, and $M_{8n+8n'}=M_{8n}\hat{\otimes}M_{8n'}$. Hence $M(V\boxplus V')\approx M(V)\hat{\boxtimes}M(V')$. Using these identifications, we therefore have (cf. II.5.21)

$$T_V\cup T_{V'}=\sigma(M(V)\hat{\boxtimes}M(V'),\Delta\hat{\boxtimes}1+1\hat{\boxtimes}\Delta')=\sigma(M(V\boxplus V'),\Delta''),$$

where Δ'' is defined over the point $(v,v')\in V\times V'=V\boxplus V'$ by $\Delta_v\hat{\otimes}1+1\hat{\otimes}\Delta'_{v'}=\Delta_{v,v'}$ (compare with III.3.10; note that $C(V\boxplus V')\approx C(V)\hat{\boxtimes}C(V')$). □

5.19. Theorem. *Let V and V' be spinorial bundles (resp. ^{c}spinorial bundles) of rank $\equiv 0 \bmod 8$ (resp. $\equiv 0 \bmod 2$) with a locally compact base X. Then the composition of the Thom homomorphisms*

$$K(X)\xrightarrow{\beta}K(V)\xrightarrow{\beta'}K(V\oplus V'),$$

where $V\oplus V'$ is regarded as a bundle over V, coincides with the Thom homomorphism

$$K(X)\xrightarrow{\beta''}K(V\oplus V')$$

associated with $V''=V\oplus V'$ (transitivity of Thom homomorphisms).

Proof. As in 5.17, we identify $M(V)\hat{\otimes}M(V')$ with $M(V\oplus V')$, and $C(V\oplus V')$ with $C(V)\hat{\otimes}C(V')$. If $\pi\colon V\to X$, $\pi'\colon V'\to X$, and $\pi''\colon V\oplus V'\to X$; then we have

$\beta(\sigma(E,D))=\sigma(\pi^*E\hat{\otimes}M(V),\pi^*D\hat{\otimes}1+1\hat{\otimes}\Delta)$, and

$\beta'(\beta(\sigma(E,D)))=\sigma(\pi''^*E\hat{\otimes}M(V)\hat{\otimes}M(V'),D\hat{\otimes}1\hat{\otimes}1+1\hat{\otimes}\Delta\hat{\otimes}1+1\hat{\otimes}1\hat{\otimes}\Delta')$.

By making the identifications above, we see that

$$D''=\pi''^*D\hat{\otimes}1\hat{\otimes}1+1\hat{\otimes}\Delta\hat{\otimes}1+1\hat{\otimes}1\hat{\otimes}\Delta'$$

is the admissible endomorphism associated with $\sigma(E,D)$ by the Thom isomorphism $K(X)\to K(V\oplus V')$ which is defined by

$$\sigma(E,D)\longmapsto\sigma(\pi''^*E\hat{\otimes}M(V\oplus V'),\pi''^*D\hat{\otimes}1+1\hat{\otimes}\Delta'')$$

(note that $\Delta''_{v,v'}=\Delta_v\hat{\otimes}1+1\hat{\otimes}\Delta_{v'}$). □

5.20. Definition. Let X be a compact space, and let V be a Spin$(8n)$-bundle (resp. a Spin$^c(2n)$-bundle). We define the *Euler class* $\chi(V)$ of V to be the restriction of T_V to the zero section $i\colon X\to V$.

If V is provided with a complex structure, hence with the associated canonical Spin$^c(2n)$-bundle structure, then the Euler class $\chi(V)$ is the same as the one defined in complex K-theory (1.13). Therefore $\chi(V)=\sum(-1)^i\lambda^i(V)$ (complex exterior powers). Moreover, if $f\colon Y\to X$ is a continuous map, then $\chi(f^*(V))=f^*(\chi(V))$ (naturality).

If W is another Spin$(8p)$ (resp. Spin$^c(2p)$ bundle), then $V\oplus W$ is naturally a Spin$(8n+8p)$ (resp. Spin$^c(2n+2p)$) bundle. By 5.18 we have $T_{V\boxplus W}=T_V\cup T_W$. Hence $\chi(V\oplus W)=\chi(V)\chi(W)$.

Finally, if V is a Spin$^c(2n)$-bundle, and if $L(V)$ is its associated line bundle (4.30), then the vector bundle $W=V\oplus V$ may be provided with a complex structure (I.4.8e)). By 5.9, we have $T_{V\oplus V}=L(V)T_W$; hence,

$$(\chi(V))^2=[L(V)]\left(\sum_{i=0}^{\infty}(-1)^i\lambda^i(W)\right).$$

5.21. We now consider two locally compact manifolds X and Y, such that $\mathrm{Dim}(Y)-\mathrm{Dim}(X)\equiv 0 \bmod 8$ (resp. $\mathrm{Dim}(Y)-\mathrm{Dim}(X)\equiv 0 \bmod 2$ for $K=K_{\mathbb{C}}$). Let $f\colon X\to Y$ be a *proper* imbedding such that the normal bundle of X in Y is provided with a spinorial structure (resp. a ^{c}spinorial structure if $K=K_{\mathbb{C}}$), assuming we have a Riemannian metric on X (cf. Lang [2]). We identify this normal bundle with a tubular neighbourhood N of X in Y, and with $f^*(TY)/TX$.

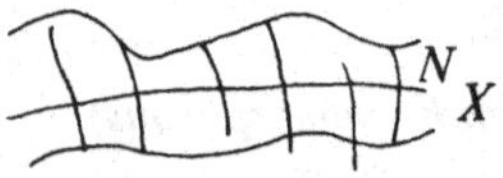

Fig. 22

For example, if X and Y are complex manifolds, then the normal bundle N is canonically ^{c}spinorial (4.26). However, in order to have our notation agree with that in Hirzebruch's book [2], we shall adopt the convention of providing N with the ^{c}spinorial structure given by the *conjugate* of the complex structure of N.

In general, we are now able to define a "*Gysin homomorphism*"

$$f_*\colon K(X)\longrightarrow K(Y)$$

as the composition of the Thom isomorphism $K(X)\to K(N)$ (5.16) with the homomorphism $K(N)\to K(Y)$, induced by the morphism $N\dashrightarrow X$ of locally compact spaces (i.e. $\dot{Y}\to\dot{N}$, cf. II.4.1). This Gysin homomorphism does not depend on the tubular neighbourhood. More precisely, if N' is another neighbourhood associated with the same normal bundle (with the given spinorial or ^{c}spinorial structure), we can find a tubular neighbourhood N'' of $X\times\mathbb{R}$ in $Y\times\mathbb{R}$ such that $N''|_{X\times\{0\}}=N$ and $N''|_{X\times\{1\}}=N'$ (cf. Lang [2]). Since the K functor is a homotopy invariant (II.1.25), the Gysin homomorphisms associated with N and N' coincide. For the same reason, f_* only depends on the homotopy class of f (within the proper imbeddings of X in Y provided with a normal spinorial or ^{c}spinorial structure).

For the next proposition we need the following definition. Let Z be any locally compact space, and let $\mathscr{E}'(Z)$ denote the category considered in I.6.24. Two objects E_0 and E_1 are called homotopic if there exists an object E of $\mathscr{E}'(X\times I)$, whose restrictions to $X \times \{0\}$ and $X \times \{1\}$ are isomorphic to E_0 and E_1, respectively. The set of homotopy classes is obviously an abelian monoid. We let $\overline{K}(Z)$ denote the symmetrized group associated with it. The tensor product of bundles provides $\overline{K}(Z)$ with a ring structure. The group $K(Z) \approx K_0(Z)$ is a module over this ring, and is free of rank one if Z is compact.

5.22. Proposition. *The Gysin homomorphism*

$$f_*: K(X) \to K(Y),$$

defined for any proper imbedding $f: X \to Y$ which satisfies the hypothesis above, has the following properties:

a) *If $f: X \to Y$ and $g: Y \to Z$ are two such proper imbeddings then $(g \cdot f)_* = g_* \cdot f_*$.*

b) *We have the formula*

$$f_*(x \cdot f^*(y)) = f_*(x) \cdot y, \quad \text{for} \quad x \in K(X) \quad \text{and} \quad y \in \overline{K}(Y).$$

c) *For X compact, $f^*(f_*(x)) = x \cdot \chi(N)$, where N is the normal bundle of the imbedding, and $\chi(N)$ its Euler class* (5.20).

Proof. Let N' be the normal bundle of the imbedding $Y \to Z$, and let $N_1' = N'|_X$. Then $(g \cdot f)_*$ is the composition

$$K(X) \longrightarrow K(N \oplus N_1') \longrightarrow K(Z).$$

On the other hand, $g_* \cdot f_*$ is the composition

$$K(X) \longrightarrow K(N) \longrightarrow K(Y) \longrightarrow K(N') \longrightarrow K(Z).$$

Since the Thom isomorphism is "natural" in an obvious sense, we have the commutative diagram

$$\begin{array}{ccccccc}
K(X) & \longrightarrow & K(N) & \longrightarrow & K(N \oplus N_1') & \longrightarrow & K(Z) \\
 & & \downarrow & & \downarrow & & \\
 & & K(Y) & \longrightarrow & K(N'). & &
\end{array}$$

Therefore, $g_* \cdot f_*$ coincides with the composition

$$K(X) \longrightarrow K(N) \longrightarrow K(N \oplus N_1') \longrightarrow K(Z),$$

hence with $(g \cdot f)_*$ by 5.19.

For the proof of b) we show that the homomorphism $K(X) \to K(N)$ is a $\overline{K}(N)$-module homomorphism and that the homomorphism $K(N) \xrightarrow{\alpha} K(Y)$ is a $\overline{K}(Y)$-module homomorphism. Since X is a deformation retract of N, any vector bundle E over N has the homotopy type of π^*F, where $F=E|_X$ and $\pi: N \to X$. Therefore $K(X) \to K(N)$ is a $\overline{K}(N)$-module map if and only if it is a $\overline{K}(X)$-module map, which is clear from the definition. Now let $[G]$ be an element of $\overline{K}(Y)$, and $\sigma(H, D)$ be an element of $K(N)$ (II.5.20). Without loss of generality, we may assume that $H=H'|_N$, and that $D=D'|_N$, where D' is acyclic outside a compact set in N. The element of $K(Y)$ associated with $\sigma(H, D)$ by α, is $\sigma(H', D')$ by II.5.19. Therefore $\alpha(\sigma(H, D)\cdot[G|_N])=\alpha(\sigma(H\otimes G_N, D\otimes 1))=(\sigma(H'\otimes G, D\otimes 1)=\alpha(\sigma(H, D))\cdot[G]$.

To prove c), we notice that for any locally compact space T, there is a well-defined homomorphism

$$K(T) \longrightarrow \overline{K}(T),$$

obtained from the correspondence $\sigma(E, D) \mapsto [E_0]-[E_1]$, using the notation of II.5.15. Moreover, we have the commutative diagram

$$\begin{array}{ccc} K(N) & \longrightarrow & K(Y) \\ \downarrow & & \downarrow \\ \overline{K}(N) & \longleftarrow & \overline{K}(Y) \end{array}$$

by II.5.15. Therefore, $f^* \cdot f_*$ coincides with the composition

$$K(X) \longrightarrow K(N) \longrightarrow \overline{K}(N) \longrightarrow \overline{K}(X)=K(X).$$

Since this composition is a $K(X)$-module map, and since the image of 1 is $\chi(N)$ by the definition of the Euler class, we have $f^*(f_*(x))=\chi(N)\cdot x$. □

5.23. Now let $f: X \longrightarrow Y$ be any differentiable *proper* map (not necessarily an imbedding) such that $\mathrm{Dim}(Y)-\mathrm{Dim}(X)\equiv 0 \bmod 8$ (resp. $\mathrm{Dim}(Y)-\mathrm{Dim}(X)\equiv 0 \bmod 2$ for $K=K_{\mathbb{C}}$). Under some appropriate hypothesis, we would like to show the existence of a Gysin map (again denoted by f_*) between $K(X)$ and $K(Y)$. More precisely, let $v_f=[f^*(TY)]-[TX]$. Then $v_f+[T_n]$, where $T_n=X\times\mathbb{R}^n$, is the class of a bundle E_n for n large enough (since X and Y can be imbedded in a Euclidean space, TX and $f^*(TY)$ are objects of the category $\mathscr{E}'(X)$ considered in I.6.24). This bundle E_n is uniquely determined up to stable equivalence (i.e. modulo addition with a trivial bundle). By definition, a spinorial structure on v_f is a spinorial structure on E_n, for n large enough. We use the convention of identifying two spinorial structures on E_n, if they are homotopic on $E_n \oplus T_p$ for some p large enough. In an analogous way, we define a ^{c}spinorial structure on v_f.

According to a well known theorem already used above, there exists a proper imbedding $i: X \to \mathbb{R}^m$, with $m\equiv 0 \bmod 8$ (mod 2 when we are dealing with complex K-theory). The map f may be factored into an imbedding $j: X \to Y\times\mathbb{R}^m$, followed by a projection $Y\times\mathbb{R}^m \to Y$. Then $v_f+[T_m]$ is the class of the bundle $E_m= j^*(T(Y\times\mathbb{R}^m))/TX$, and a spinorial (resp. ^{c}spinorial) structure on v_f is essentially a spinorial (resp. ^{c}spinorial) structure on E_m for some m large enough. We now define the Gysin homomorphism

$$f_*: K(X) \longrightarrow K(Y)$$

as the composition of $j_*: K(X) \to K(Y \times \mathbb{R}^m)$ as defined in 5.21, with the periodicity isomorphism $K(Y \times \mathbb{R}^m) \approx K(Y)$ (III.5.17).

5.24. Proposition. *The Gysin homomorphism*

$$f_*: K(X) \longrightarrow K(Y)$$

does not depend on the choice of the proper imbedding j, but only on the homotopy class of f (within the proper differentiable maps provided with spinorial or ^{c}spinorial structures). Moreover, it has the following properties:

a) *If $f: X \to Y$ and $g: Y \to Z$ are proper differentiable maps, we have $(g \cdot f)_* = g_* \cdot f_*$.*

b) *We have the formula $f_*(x \cdot f^*(y)) = f_*(x) \cdot y$ for $x \in K(X)$, $y \in \overline{K}(Y)$.*

Proof. We first show that f does not depend on the choice of the imbedding $X \to \mathbb{R}^m$. If $X \to \mathbb{R}^p$ is another imbedding, because of 5.22 a), we have the commutative diagram

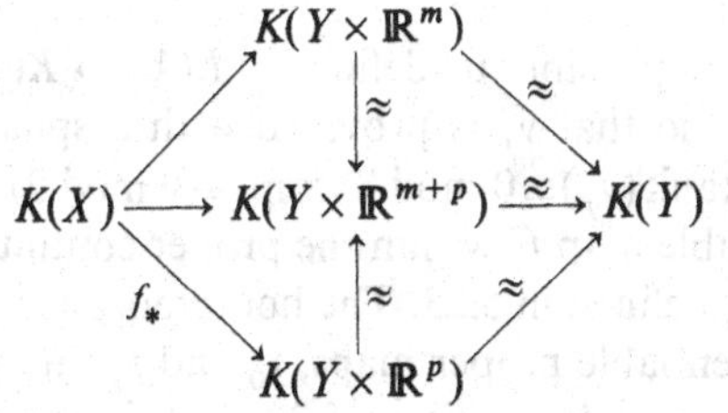
K(Y × ℝ^m)
K(X)
K(Y × ℝ^{m+p})
K(Y)
f_*
K(Y × ℝ^p)

(note that the periodicity isomorphism $K(Z) \approx K(Z \times \mathbb{R}^q)$ is also r_*, where $r: Z \to Z \times \mathbb{R}^q$ is the canonical imbedding, and $q \equiv 0 \bmod 8$ or 2, according to which K-theory is used).

If f_0 and f_1 are homotopic, they both may be factored by homotopic proper imbeddings from X into $Y \times \mathbb{R}^m$, for m large enough. Therefore $(f_0)_* = (f_1)_*$ by 5.21.

If $f: X \to Y$ and $g: Y \to Z$ are arbitrary proper differentiable maps, and if $X \to \mathbb{R}^m$ and $Y \to \mathbb{R}^p$ are proper imbeddings, then we have the commutative diagram

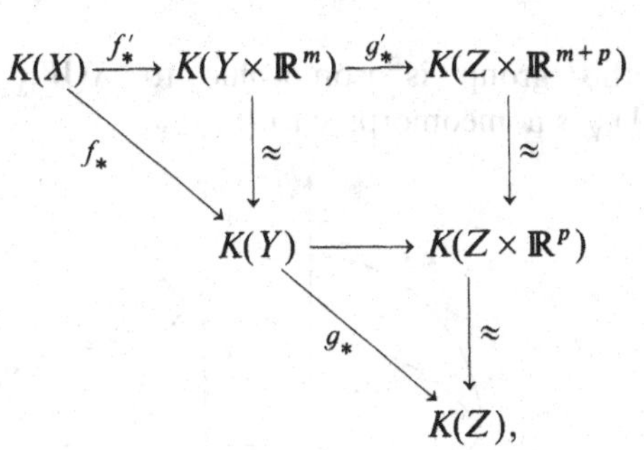
K(X)
f'_*
K(Y × ℝ^m)
g'_*
K(Z × ℝ^{m+p})
f_*
K(Y)
K(Z × ℝ^p)
g_*
K(Z),

which gives the identity $(g \cdot f)_* = g_* \cdot f_*$ since $(g' \cdot f')_* = g'_* \cdot f'_*$.

Finally, the formula $f_*(x \cdot f^*(y)) = f_*(x) \cdot y$ may be proved by the same method as in 5.22. We need only notice that the isomorphism $K(X) \approx K(X \times \mathbb{R}^m)$ is a $\overline{K}(X)$-module isomorphism. □

5.25. *Example.* Let X be a compact differentiable manifold of dimension $\equiv 0 \bmod 8$ (resp. $\equiv 0 \bmod 2$). We assume that the normal bundle of X in some imbedding, is provided with a spinorial structure (resp. a ^{c}spinorial structure). Then, if Y is chosen to be a point P, we have a homomorphism

$$K(X) \longrightarrow K(Y) \approx \mathbb{Z}.$$

The image of $[TX]$ (or $[TX \otimes \mathbb{C}]$ for $K = K_{\mathbb{C}}$) is an important invariant of the manifold, which is computed in the next chapter (V.4).

5.26. *Remark.* It is possible to prove that there exists a bijective correspondence between spinorial (resp. ^{c}spinorial) structures on ν_f, where $f: X \to$ *Point*, and stable spinorial (resp. ^{c}spinorial) structures on TX.

5.27. Finally, we remark that it is possible to define $f_*: K(X) \to K(Y)$ for any proper *continuous* map $f: X \to Y$, so that ν_f is provided with a spinorial (resp. ^{c}spinorial) structure, and so that $\operatorname{Rank}(\nu_f) \equiv 0 \bmod 8$ (resp. $\equiv 0 \bmod 2$). In fact, f is homotopic to a proper differentiable map f' within the proper continuous maps, and thus we define f_* to be the f'_* defined in 5.23. The homomorphism f_* is well-defined since any two such differentiable proper maps, f'_0 and f'_1, are homotopic within the proper differentiable maps. Proposition 5.24 is still true in the continuous case.

6. Real and Complex K-Theory for Real Projective Spaces and Real Projective Bundles

6.1. In this section, we would like to explicitly compute the K-theory of $P(V)$, when V is a real vector bundle with compact base X. More generally, if W is a sub-bundle of V, and if Y is a closed subspace in X, we wish to compute the group $K(P(V), P(W) \cup P(V)|_Y)$.

If $V = W \oplus 1$, then this group is isomorphic to $K(W|_{X-Y})$, since $P(W \oplus 1) - P(W) \cup P(W \oplus 1)|_Y$ is homeomorphic to $W|_{X-Y}$.

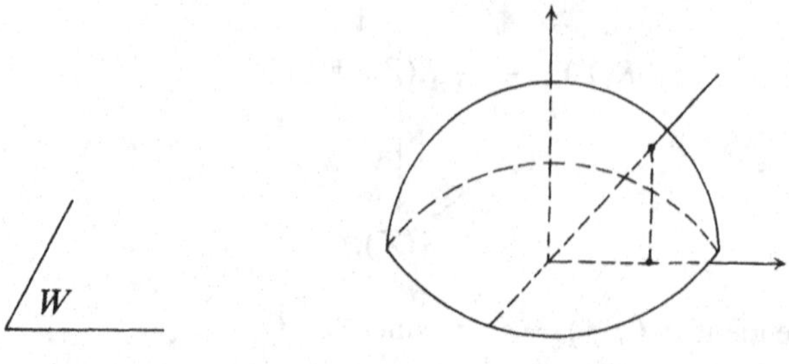

Fig. 23

For example, if W is a Spin(8n)-bundle (or a Spin^c(2n)-bundle when $K = K_{\mathbb{C}}$), we have $K(P(W \oplus 1), P(W) \cup P(W+1)_Y) \approx K(W|_{X-Y}) \approx K(X-Y)$ (5.14). This shows that the computations of this section generalize, in some sense, the computations of Section IV.5. Before beginning these computations (cf. 6.34), we require many technical definitions and lemmas about the Grothendieck group of the "restriction of scalars" functor:

$$\mathscr{E}^V(X) \longrightarrow \mathscr{E}^W(X),$$

where V is provided with a nondegenerate quadratic form (4.11), and W is provided with the induced quadratic form which is assumed to be nondegenerate.

6.2. Lemma. *Let E and F be objects of $\mathscr{E}^V(X)$, and let $\alpha: E|_Y \to F|_Y$ be a $C(V|_Y)$-module morphism, where Y is a closed subspace of X. Then there exists a morphism $\beta: E \to F$ such that $\beta|_Y = \alpha$. In particular, if α is an isomorphism, then β is an isomorphism over a neighbourhood of Y.*

Proof. By a classical argument involving partitions of unity (cf. I.5.7), we may assume that E, F and V are trivial bundles; hence, $E = X \times k^n$, $F = X \times k^p$, and $V = X \times \mathbb{R}^r$ (for $k = \mathbb{R}$ or $\mathbb{C}$ according to which K-theory is considered). Let e_i, for $i = 1, \ldots, r$, be the generators of the Clifford algebra of $\mathbb{R}^r$ (III.3.13), and let G be the finite group of order 2^{r+1}, generated multiplicatively by the $\pm e_i$: every element of this group may be written in the form $\pm e_{i_1} \cdots e_{i_s}$ where $i_1 < i_2 < \cdots < i_s$. Let $\beta': E \to F$ be a vector bundle morphism (not necessarily compatible with the module structure) such that $\beta'|_Y = \alpha$ (I.5.7), and let $\beta: E \to F$ be defined by the formula

$$\beta(e) = \frac{1}{2^{r+1}} \sum_{g \in G} g^{-1} \beta'(ge).$$

Then $\beta(he) = h\beta(e)$ for each element h of G. By linearity, it follows that $\beta(\lambda e) = \lambda\beta(e)$ for each element λ of the Clifford algebra (III.3.13). □

6.3. Lemma. *Let $\pi: X \times I \to X$, and let E be a vector bundle on $X \times I$ provided with a $C(\pi^* V)$-module structure. Then E is isomorphic to $\pi^* F$, where $F = E|_{X \times \{0\}}$.*

Proof. The proof of this lemma is analogous to the proof of Theorem I.7.3 using Lemma 6.2. □

6.4. Lemma. *Let E be a vector bundle on a closed subspace Y of X, which is provided with a $C(V_Y)$-module structure. Then there is a closed neighbourhood Z of Y, and a $C(V_Z)$-module F, such that $F|_Y = E$.*

Proof. For each point y of Y, there exists a neighbourhood U_y of y in X, and a vector bundle F_{U_y} on U_y, such that $F_{U_y}|_{U_y \cap Y} = E|_{U_y \cap Y}$. Since Y is compact, there exist closed subsets U_i, for $i = 1, \ldots, n$, such that $Y \subset \cup \mathring{U}_i$, and vector bundles E_i on U_i such that $E_i|_{U_i \cap Y} = E|_{U_i \cap Y}$. Moreover, E_i is a $C(V_i)$-module where $V_i = V|_{U_i}$.

Since $E_1|_{U_1 \cap U_2 \cap Y} = E_2|_{U_1 \cap U_2 \cap Y}$, there exists a closed subset U_1' of U_1, which is a neighbourhood of $U_1 \cap Y$ in U_1, and an isomorphism $g_{21}: E_1|_{U_1' \cap U_2} = E_2|_{U_1' \cap U_2}$. Let E_2' be the vector bundle obtained by clutching $E_1|_{U_1'}$ and E_2 via this isomorphism (I.3.2). Then E_2' is a $C(V_2')$-module, where $V_2' = V|_{U_1' \cup U_2}$ and $E'_2|_{(U_1' \cup U_2) \cap Y} \approx E|_{(U_1' \cup U_2) \cap Y}$. If we put $U'_2 = U'_1 \cup U_2$, $U'_3 = U_3, \ldots, U'_n = U_n$, we may repeat the above argument with U_2' and U_3' etc. Therefore, by induction on n, we construct the desired neighbourhood Z, and the required vector bundle F. □

6.5. Lemma (*Double extension of structures*). *Let E be a vector bundle over X, provided with a $C(W)$-module structure, and let v be a $C(V_Y)$-module structure on E_Y, which is compatible with the $C(W_Y)$-module structure. Then there exists a closed neighbourhood Z of Y and a $C(V_Z)$-module structure $\tilde{v}$ on E_Z, which is compatible with the $C(W_Z)$-module structure such that $\tilde{v}|_Y = v$.*

Proof. By Lemma 6.4, there exists a closed neighbourhood Z of Y, and a vector bundle F over Z, which is a $C(V_Z)$-module such that $F|_Y = E$. If Z is sufficiently small, then $F|_Z$ is isomorphic to $E|_Z$ as $C(W)$-modules by 6.2. □

6.6. Lemma. *Let E be a $C(V)$-module. Then E is a direct factor of a $C(V)$-module of the form $C(V) \otimes F$, where F is a trivial vector bundle (the $C(V)$-module structure on $C(V) \otimes F$ is induced from the factor $C(V)$).*

Proof. This lemma is true locally, since any $C^{p,q}$-module is a direct factor of some "trivial" $C^{p,q}$-module (III.4.8). Using a partition of unity as in I.6.5, we thus may prove that E is a direct factor of some $C(V) \otimes F$. □

6.7. Lemma. *Let E be a vector bundle over X, provided with a $C(V)$-module structure, and let $\alpha(t): E|_Y \to E|_Y$, for $t \in [0, 1]$, be a continuous family of automorphisms of $E|_Y$ such that $\alpha(0) = 1$. Then there exists a continuous family of automorphisms $\beta(t)$ of E, such that $\beta(t)|_Y = \alpha(t)$.*

Proof. Using Lemma 6.2, the proof of this lemma is completely analogous to the proof of Proposition II.2.24. □

6.8. Lemma. *Let E be a vector bundle over X, provided with a $C(V)$-module structure denoted by v (hence v represents a morphism $V \to \mathrm{END}(E)$ with the properties described in 4.11). Let Y be a closed subspace of X, and let $v'(t)$ be a continuous family of $C(V')$-module structures on $E' = E|_Y$, where $V' = V|_Y$ and $v'(0) = v|_Y$. Then there exists a continuous family $v(t)$ of $C(V)$-module structures on E, such that $v(t)|_Y = v'(t)$ and $v(0) = v$. Moreover, if $v'(t)|_W = v'(0)|_W$ where W is a sub-bundle of V, then we may choose $v(t)$ so that $v(t)|_W = v(0)|_W$.*

Proof. For simplicity, we use V, W, etc. to denote the restrictions of V, W, etc. to a subspace, or more generally, the inverse image of these bundles with respect to continuous maps $T \to X$, a convention which we systematically adopt for the rest of this section.

The existence of v and $v'(t)$ is equivalent to a $C(V)$-module structure on $\pi^*E|_{X\times\{0\}\cup Y\times I}$, where $\pi: X\times I\to X$. By lemma 6.5, there exists a neighbourhood U of Y in X, and a $C(V)$-module structure on $\pi^*E|_{X\times\{0\}\cup U\times I}$, which extends the $C(V)$-module structure on $\pi^*E|_{X\times\{0\}\cup Y\times I}$, and such that $\tilde{v}|_W=\pi^*v|_W$. Now let $\eta: X\to[0,1]$ be a continuous function which is 1 on Y, and 0 outside U. Let $\theta: X\times I\to X\times\{0\}\cup U\times I$ be the continuous function defined by $(x,t)\mapsto(x,\eta(x)t)$. Then $\theta(x,t)=(x,t)$ if $(x,t)\in X\times\{0\}\cup Y\times I$. Therefore $\theta^*\tilde{v}$ is the extension required. $\square$

6.9. Definition. Let V be a real vector bundle provided with a non degenerate quadratic form, and let W be a sub-bundle provided with the induced quadratic form which we assume non degenerate. Then the group $K^{V,W}(X)$ is the *Grothendieck group* of the "restriction of scalars" functor (cf. II.2.13):

$$\mathscr{E}^V(X)\longrightarrow\mathscr{E}^W(X).$$

If $V=W\oplus 1$, then the group $K^{V,W}(X)$ is simply the group $K^V(X)$, defined in 5.1. As in 5.1, the group $K^{V,W}(X)$ may be described in the following way. Let $\mathscr{T}^{V,W}(X)$ be the set of triples (E,v_1,v_2), where E is a $C(W)$-module, and where the v_i represent two $C(V)$-module structures which extend the $C(W)$-module structure (again note that $C(W)\subset C(V)$). Such a triple is said to be *elementary* if v_1 and v_2 are homotopic (where the restriction of this homotopy to W is constant). Then the group $K^{V,W}(X)$ is the quotient of $\mathscr{T}^{V,W}(X)$, by the equivalence relation generated by the sum of elementary triples (cf. III.4.16, ..., III.4.22, together with 6.3). Using the same method, we may also define a relative group $K^{V,W}(X,Y)$, for Y closed in X (cf. III.5.1). Note that $K^W(X,Y)\approx K^{W\oplus 1,W}(X,Y)$, and that $K^{V,W}(X,Y)$ has the same formal properties as those described in III.4 and III.5 for the group $K^{p,q}(X,Y)$.

6.10. Proposition. *Let X_1 and X_2 be closed subspaces of X such that $X_1\cup X_2=X$. Then the natural homomorphism*

$$g: K^{V,W}(X_1\cup X_2, X_1)\longrightarrow K^{V,W}(X_2, X_1\cap X_2)$$

is an isomorphism.

Proof. a) *g is surjective*. Let $d(E,v_1,v_2)$ be an element of $K^{V,W}(X_2,X_1\cap X_2)$. By Lemma 6.6, we may assume that (E,v_1) is a $C(V)$-module of the form $C(V)\otimes F$, where F is a trivial bundle. Therefore, (E,v_1) is the restriction to X_2 of a $C(V)$-module on X (denoted by $(E,\tilde{v}_1)$). Let $\tilde{v}_2$ be the $C(V)$-module structure on E over X, which coincides with $\tilde{v}_1$ on X_1 and with v_2 on X_2 (I.3.1). Then $g(d(E,\tilde{v}_1,\tilde{v}_2))=d(E,v_1,v_2)$.

b) *g is injective*. Let $d(E,v_1,v_2)$ be an element of $K^{V,W}(X_1\cup X_2,X_1)$ whose restriction to $(X_2,X_1\cap X_2)$ is zero. By adding an elementary triple, we may assume that $v_1|_{X_2}$ and $v_2|_{X_2}$ are homotopic, with the homotopy constant over $X_1\cap X_2$ (cf. III.4.20). Let $v'(t)$ be such a homotopy, and let $v(t)$ be a continuous family of $C(V)$-module structures on E over X, such that $v(t)|_{X_2}=v'(t)$, $v(t)|_{X_1}=v_1|_{X_1}$, and $v(t)|_W=v_1|_W$ (Lemma 6.8). Then $d(E, v_1, v_2)=d(E, v(1), v_2)=0$, since $v(1)|_{X_2}=v_2|_{X_2}$ and $v(1)|_{X_1}=v_2|_{X_1}$. □

6.11. Proposition. *Let Y be a closed subspace of X, and let us define, in general,*

$$K_n^{V,W}(X, Y)=K^{V,W}(X\times B^n, X\times S^{n-1}\cup Y\times B^n)$$

Then, with the notation of 6.10, the natural homomorphism

$$K_n^{V,W}(X_1\cup X_2, X_1)\longrightarrow K_n^{V,W}(X_2, X_1\cap X_2)$$

is an isomorphism.

Proof. This follows immediately from proposition 6.10 applied to $X\times B^n=Y_1\cup Y_2$, where $Y_2=X_2\times B^n\cup(X_1\cup X_2)\times S^{n-1}$ and $Y_1=X_1\times B^n$. □

6.12. For any triple (X, Y, Z) where Z is closed in Y and Y is closed in the compact space X, we define a "connecting homomorphism"

$$\partial: K_1^{V,W}(Y, Z)\longrightarrow K^{V,W}(X, Y).$$

To do this, we consider a triple (E, v_1, v_2), where E is a $C(W)$-module over $Y\times B^1$, where v_1 and v_2 are two $C(V)$-module structures on E, extending the original $C(W)$-module structure, and such that $v_1|_{Y\times S^0\cup Z\times B^1}=v_2|_{Y\times S^0\cup Z\times B^1}$. Such a triple is said to be normalized if E is of the form $C(V)\otimes F$, where F is a trivial vector bundle, and v_1 is the $C(V)$-module structure on $C(V)\otimes F$ induced from the factor $C(V)$. We define $\delta(F, v_2)$ or $\delta(F, v_2(t))$, to be the class of the triple (E, v_1, v_2) in the group $K_1^{V,W}(Y, Z)$.

6.13. Lemma. *Each element of $K_1^{V,W}(Y, Z)$ may be written in the form $\delta(F, v_2)$. We have $\delta(F, v_2)=\delta(F, v_2')$ if and only if there exists a trivial bundle G, such that $v_2\oplus\bar{v}_1$ is homotopic to $v_2'\oplus\bar{v}_1$ (where $\bar{v}_1$ denotes the $C(V)$-module structure on $C(V)\otimes G$ induced from the factor $C(V)$), with the homotopy constant over $Y\times S^0\cup Z\times B^1$.*

Proof. This lemma is an immediate consequence of Lemma 6.6. □

6.14. Let $\delta(F, v_2(t))$ be an element of $K_1^{V,W}(Y, Z)$. By 6.8, there exists a continuous family $\tilde{v}_2(t)$ of $C(V)$-module structures on $C(V)\otimes F$ over X, such that $\tilde{v}_2(t)|_Y=v_2(t)$, $\tilde{v}_2(0)=v_1$, and $\tilde{v}_2(t)|_W=\tilde{v}_2(0)|_W=v_1|_W$. Now we define

$$\partial(\delta(F, v_2(t)))=d(C(V)\otimes F, v_1, \tilde{v}_2(1))\in K^{V,W}(X, Y).$$

This definition is independent of the choice of $v_2(t)$ and its extension to X. Suppose $v_2'(t)$ is another choice and $\tilde{v}_2'(t)$ is another extension. We must prove the equality $d(C(V)\otimes F, v_1, \tilde{v}_2(1))=d(C(V)\otimes F, v_1, \tilde{v}_2'(1))$ in the group $K^{V,W}(X, Y)$.

By Lemma 6.13 (after addition of an elementary triple), there exists a continuous family $v_2(t, u)$, for $t \in [0, 1]$ and $u \in [0, 1]$, of $C(V)$-module structures on $F = C(V) \otimes F$ over Y, such that $v_2(t, u) = v_1$ over $Y \times S^0 \cup Z \times B^1$, $v_2(t, u)|_W = v_1$, $v_2(t) = v_2(t, 0)$, and $v_2'(t) = v_2(t, 1)$. Let K be the square $[0, 1] \times [0, 1]$, and let L be the closed subset defined as the union of $[0, 1] \times \{0\}$, $[0, 1] \times \{1\}$, and $\{0\} \times [0, 1]$. Then K is homeomorphic to $L \times [0, 1]$, with $L \times \{0\}$ identified with L. Therefore, by Lemma 6.8, there exists a continuous family $\tilde{v}_2(t, u)$ of $C(V)$-module structures on E over X, such that $\tilde{v}_2(t, u)|_W = v_1$, $\tilde{v}_2(t, 0) = \tilde{v}_2(t)$, $\tilde{v}_2(t, 1) = \tilde{v}_2'(t)$, and $\tilde{v}_2(t, u)|_Y = v_2(t, u)$. Hence, $\tilde{v}_2(1)$ and $\tilde{v}_2'(1)$ are homotopic via the map $u \mapsto \tilde{v}_2(1, u)$ (which is constant over Y).

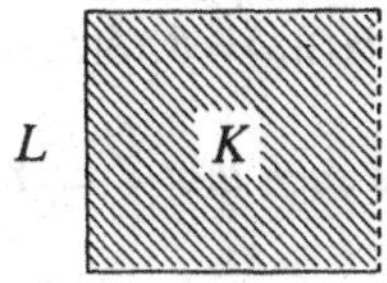

Fig. 24

6.15. Theorem. *Let X, Y, Z be compact spaces such that $Z \subset Y \subset X$. Then the sequence*

$$K_1^{V,W}(X, Z) \xrightarrow{\beta_1} K_1^{V,W}(Y, Z) \xrightarrow{\partial} K^{V,W}(X, Y) \xrightarrow{\alpha} K^{V,W}(X, Z) \xrightarrow{\beta} K^{V,W}(Y, \dot{Z})$$

is exact.

Proof. a) *Exactness at $K^{V,W}(X, Z)$.*
— Let $d(E, v_1, v_2)$ be an element of $K^{V,W}(X, Y)$. Then $(\beta\alpha)(d(E, v_1, v_2)) = d(E, v_1|_Y, v_2|_Y) = 0$, since $v_1|_Y = v_2|_Y$.
— Conversely, let $d(E, v_1, v_2)$ be an element of $K^{V,W}(X, Z)$ such that

$$d(E, v_1|_Y, v_2|_Y) = 0.$$

By adding an elementary triple of the form $(C(V) \otimes F, \zeta, \zeta)$, where ζ is the canonical $C(V)$-module structure on $C(V \otimes F$ and F is trivial, we may assume that $v_1|_Y$ and $v_2|_Y$ are homotopic. More precisely, let $v(t)$ be a continuous family of $C(V)$-module structures on $E|_Y$, such that $v(0) = v_1|_Y$, $v(1) = v_2|_Y$, $v(t)|_Z = v_1|_Z$, and $v(t)|_W = v_1|_W$. By lemma 6.8, there exists a continuous family $\tilde{v}(t)$ of $C(V)$-module structures on E over X, such that $\tilde{v}(t)|_Y = v(t)$ and $\tilde{v}(0) = v_1$. Therefore $d(E, v_1, v_2) = d(E, \tilde{v}(0), v_2) = d(E, \tilde{v}(1), v_2)$, which belongs to the image of α since $\tilde{v}(1)|_Y = v_2|_Y$.

b) *Exactness at $K^{V,W}(X, Y)$.*
— Let $x = \delta(E, v_2(t)) = d(C(V) \otimes F, v_1, v_2(t))$ be an element of $K_1^{V,W}(Y, Z)$. Using the notation of 6.14, we have $(\alpha\partial)(x) = d(E, v_1|_Y, \tilde{v}_2(1)|_Y) = 0$ since $v_1|_Y = \tilde{v}_2(1)|_Y$.
— Conversely, let $d(E, v_1, v_2)$ be an element of $K^{V,W}(X, Y)$ where $(E, v_1) = C(V) \otimes F$, such that $\alpha(d(E, v_1, v_2)) = 0$. By adding an elementary triple of the form $(C(V) \otimes G, \zeta, \zeta)$, we may assume that v_1 and v_2 are homotopic (with the homotopy constant over Z). If we denote this homotopy by $v_2(t)$, we have $d(E, v_1, v_2) = \partial(\delta(F, v_2(t)))$.

c) *Exactness at $K_1^{V,W}(Y, Z)$.*
— Let $x=\delta(F, v_2(t))$ be an element of $K_1^{V,W}(Y, Z)$. Then

$$(\alpha\partial)(x)=d(C(V)\otimes F, v_1, v_2(1))=0,$$

since $v_2(1)=v_1$.
— Conversely, let $\delta(F, v_2(t))$ be an element of $K_1^{V,W}(Y, Z)$ such that

$$d(C(V)\otimes F, v_1, \tilde{v}_2(1))=0.$$

Again by adding an elementary triple of the form $C(V)\otimes G$, we may assume that this homotopy is constant over Y. If we denote the "composition" of $v_2(t)$ with this last homotopy by $v(t)$, we have $\delta(F, v_2(t))=\delta(F, v(t)|_Y)=\beta_1(\delta(F, v(t)))$. □

6.16. Corollary. *Let X, Y, and Z, be compact spaces such that $Z\subset Y\subset X$. Then we have the exact sequence*

$$K_{n+1}^{V,W}(X, Z)\to K_{n+1}^{V,W}(Y, Z)\to K_n^{V,W}(X, Y)\to K_n^{V,W}(X, Z)\to K_n^{V,W}(Y, Z),$$

for $n\geqslant 0$ (with $K_0^{V,W}=K^{V,W}$).

Proof. Consider the triple $(X\times B^n, X\times S^{n-1}\cup Y\times B^n, X\times S^{n-1}\cup Z\times B^n)$. Then $K_i(X\times S^{n-1}\cup Y\times B^n, X\times S^{n-1}\cup Z\times B^n)\approx K_i(Y\times B^n, Y\times S^{n-1}\cup Z\times B^n)$ by 6.11 applied to $X_1=X\times S^{n-1}\cup Z\times B^n$ and $X_2=Y\times B^n$. □

6.17. Theorem. *Let (X, Y) be a pair of compact spaces, and let X_1 and X_2 be closed subspaces of X. Let $Y_i=Y\cap X_i$. Then we have the "Mayer-Vietoris exact sequence"*

$$\begin{aligned}&K_{n+1}^{V,W}(X_1, \tilde{Y}_1)\oplus K_{n+1}^{V,W}(X_2, \tilde{Y}_2)\longrightarrow K_{n+1}^{V,W}(X_1\cap X_2, \tilde{Y}_1\cap\tilde{Y}_2)\\&\quad\to K_n^{V,W}(X_1\cup X_2, \tilde{Y}_1\cup\tilde{Y}_2)\\&\quad\to K_n^{V,W}(X_1, \tilde{Y}_1)\oplus K_n^{V,W}(X_2, \tilde{Y}_2)\longrightarrow K_n^{V,W}(X_1\cap X_2, \tilde{Y}_1\cap\tilde{Y}_2).\end{aligned}$$

where
$$\begin{aligned}\tilde{Y}_1=Y_1\cup(X_1\cap Y_2),\ \tilde{Y}_2=Y_2\cup(X_2\cap Y_1),\ &\tilde{Y}_1\cup\tilde{Y}_2\\&=Y_1\cup Y_2,\ \tilde{Y}_1\cap\tilde{Y}_2=(Y_1\cap X_2)\cup(X_1\cap Y_2).\end{aligned}$$

Proof. Consider the triple $(X_1\cup X_2, X_1\cup Y_2, Y_1\cup Y_2)$. If we set

$$G_n=K_n^{V,W}(X_1\cup X_2, X_1\cup Y_2),\ H_n=K_n^{V,W}(X_1\cup X_2, Y_1\cup Y_2),$$

and $$L_n=K_n^{V,W}(X_1\cup Y_2, Y_1\cup Y_2),$$

then we have the exact sequence

$$H_{n+1}\longrightarrow L_{n+1}\longrightarrow G_n\longrightarrow H_n\longrightarrow L_n.$$

Moreover, $G_n\approx K_n^{V,W}(X_2, (X_1\cap X_2)\cup Y_2)$ and $L_n\approx K_n^{V,W}(X_1, \tilde{Y}_1)$ by 6.11.

In the same way we consider the triple $(X_2, (X_1\cap X_2)\cup Y_2, \tilde{Y}_2)$. From before

$K_n^{V,W}(X_2, (X_1 \cap X_2) \cup Y_2) \approx G_n$, so if we define $L'_n = K_n^{V,W}(X_2, \tilde{Y}_2)$ and $P_n = K_n^{V,W}((X_1 \cap X_2) \cup Y_2, \tilde{Y}_2) \approx K_n^{V,W}(X_1 \cap X_2, X_1 \cap Y_2 \cup X_2 \cap Y_1)$ we obtain the exact sequence

$$L'_{n+1} \longrightarrow P_{n+1} \longrightarrow G_n \longrightarrow L'_n \longrightarrow P_n,$$

and the commutative diagram

$$\begin{array}{ccccccccc} H_{n+1} & \longrightarrow & L_{n+1} & \longrightarrow & G_n & \longrightarrow & H_n & \longrightarrow & L_n \\ \downarrow & & \downarrow & & \| & & \downarrow & & \downarrow \\ L'_{n+1} & \longrightarrow & P_{n+1} & \longrightarrow & G_n & \longrightarrow & L'_n & \longrightarrow & P_n . \end{array}$$

By elementary diagram chasing, we obtain the exact sequence

$$L_{n+1} \oplus L'_{n+1} \longrightarrow P_{n+1} \longrightarrow H_n \longrightarrow L_n \oplus L'_n, \quad \text{for } n \geqslant 0.$$

All that remains is to verify the exactness of the sequence

$$H_0 \longrightarrow L_0 \oplus L'_0 \longrightarrow P_0,$$

i.e. the sequence

$$K^{V,W}(X_1 \cup X_2, \tilde{Y}_1 \cup \tilde{Y}_2) \xrightarrow{l} K^{V,W}(X_1, \tilde{Y}_1) \oplus K^{V,W}(X_2, \tilde{Y}_2) \xrightarrow{h} K^{V,W}(X_1 \cap X_2, \tilde{Y}_1 \cap \tilde{Y}_2).$$

Let $x_1 = d(E_1, v_1, v'_1)$ and $x_2 = d(E_2, v_2, v'_2)$ be elements of $K^{V,W}(X_1, \tilde{Y}_1)$ and $K^{V,W}(X_2, \tilde{Y}_2)$, respectively, such that $h(x_1, x_2) = 0$. Without loss of generality, we may assume that v_1 and v_2 are the restrictions to X_1 and X_2 of a $C(V)$-module structure v of the form $E = C(V) \otimes F$, where F is a trivial bundle, and v_1 and v_2 are the $C(V)$-module structures on E_1 and E_2, induced from the factor $C(V)$. We may also assume that $v'_1|_{X_1 \cap X_2}$ is homotopic to $v'_2|_{X_1 \cap X_2}$ by a homotopy $v(t)$, with $v(t)|_W = v_2|_W = v_1|_W$ and $v(t)|_{\tilde{Y}_1 \cap \tilde{Y}_2} = v(0)|_{\tilde{Y}_1 \cap \tilde{Y}_2}$. By Lemma 6.8, there exists a continuous family $\tilde{v}(t)$ of $C(V)$-module structures on $E|_{X_1}$, such that $\tilde{v}(t)|_W = v|_W$, $\tilde{v}(0) = v'_1$, $\tilde{v}(t)|_{X_1 \cap X_2} = v(t)$, and $v(t)|_{\tilde{Y}_1} = v|_{\tilde{Y}_1}$. Then $(x_1, x_2) = l(x)$, where $x = d(C(V) \otimes F, v, v')$ with $v'|_{X_1} = \tilde{v}(1)$ and $v'|_{X_2} = v'_2$. □

6.18. As an application of the Mayer-Vietoris exact sequence, we consider a map $f: (X, Y) \to (X', Y')$ between compact pairs, which induces a homeomorphism $X - Y \approx X' - Y'$. If V' is a vector bundle on X' provided with a nondegenerate quadratic form, and if W' is a subbundle with the induced nondegenerate quadratic form, then we have a homomorphism

$$K_n^{V',W'}(X', Y') \longrightarrow K_n^{f^*V', f^*W'}(X, Y)$$

which we claim is an isomorphism. Since $X'/Y' \approx X/Y$, the same proof as used in III.5.3, shows that this is an isomorphism when V' and W' are trivial. Now let (X'_i) be a finite closed cover of X' such that $V'|_{X'_i}$ and $W'|_{X'_i}$ are trivial. Let

$$X_i = f^{-1}(X'_i),\ Y'_i = X'_i \cap Y', \text{ and } Y_i = f^{-1}(Y'_i).$$

Finally, let

$$X''^i = X'_1 \cup \cdots \cup X'_i,\ Y''^i = X''^i \cap Y', X^i = f^{-1}(X''^i), \text{ and } Y^i = f^{-1}(Y''^i).$$

Then the argument used in the proof of 1.3, shows by induction on i, that the homomorphism

$$K_n^{V'|_{X''^i}, W'|_{X''^i}}(X''^i, Y''^i) \longrightarrow K_n^{f^*(V')|_{X^i}, f^*(W')|_{X^i}}(X^i, Y^i)$$

is an isomorphism.

6.19. Now assume that the bundle V, over the compact space X, is provided with a *positive* quadratic form, and let T be another bundle on X, provided with a nondegenerate quadratic form. We wish to define a homomorphism

$$t: K^{T \oplus V}(X, Y) \longrightarrow K^{\pi^* T}(B(V), S(V) \cup B(V)|_Y),$$

where $\pi: B(V) \to X$, which simultaneously generalizes the homomorphisms also denoted by t, defined in III.5.9 and in 5.10. As in 5.10, we identify the ball bundle $B(V)$ with the upper hemisphere $S^+(V \oplus 1)$ of the bundle $V \oplus 1$. The formula

$$t(d(E, \eta_1, \eta_2)) = d(\pi^* E, v \cos\theta + \eta_1 \sin\theta, v \cos\theta + \eta_2 \sin\theta),$$

where $\pi: S^+(V \oplus 1) \to X$, and $\pi^* E$ is provided with the underlying $C(T)$-module structure, defines the homomorphism t (with the conventions made in 5.10).

If $V = V_1 \oplus V_2, \pi_2: S^+(V_2 \oplus 1) \to X$, and $Z = S(V_2) \cup S^+(V_2 \oplus 1)|_Y$, we define a map

$$f: (S^+(\pi_2^* V_1 \oplus 1), S(\pi_2^* V_1) \cup S^+(\pi_2^* V_1 \oplus 1)|_Z)$$
$$\longrightarrow (S^+(V_1 \oplus V_2 \oplus 1), S(V_1 \oplus V_2) \cup S^+(V_1 \oplus V_2 \oplus 1)|_Y)$$

in the following way. Each point of $S^+(\pi_2^* V_1 \oplus 1) \approx S^+(V_1 \oplus 1) \times_X S^+(V_2 \oplus 1)$ may be written in the form $(v_1 \cos\theta_1 + e_1 \sin\theta_1, v_2 \cos\theta_2 + e_2 \sin\theta_2)$, where e_1 and e_2 denote the "unit vectors" of $S^+(V_1 \oplus 1)$ and $S^+(V_2 \oplus 1)$, respectively,

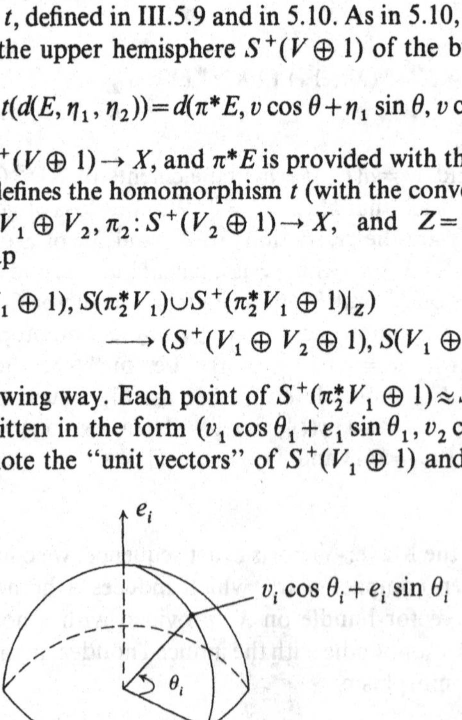

Fig. 25

and where $v_1 \in S(V_1)$ and $v_2 \in S(V_2)$ have the same projection on X. With this point, we associate the point of $S^+(V_1 \oplus V_2 \oplus 1)$, defined by $v_1 \cos\theta_1 + v_2 \sin\theta_1 \cos\theta_2 + e \sin\theta_1 \sin\theta_2$, where e is the "unit vector" of $S^+(V_1 \oplus V_2 \oplus 1)$. The map f thus defined induces a homeomorphism between

$$S^+(\pi_2^* V_1 \oplus 1) - S(\pi_2^* V_1) - S^+(\pi_2^* V_1 \oplus 1)|_Z$$

and $$S^+(V_1 \oplus V_2 \oplus 1) - S(V_1 \oplus V_2) \cup S^+(V_1 \oplus V_2 \oplus 1)|_Y.$$

6.20. Proposition. *The homomorphism t defined in 6.19 is "transitive" with respect to V. More precisely, we have the commutative diagram*

$$\begin{array}{ccc} K^{W \oplus V_1 \oplus V_2}(X, Y) & \xrightarrow{t_2} & K^{W \oplus V_1}(S^+(V_2 \oplus 1), S(V_2) \cup S^+(V_2 \oplus 1)|_Y) \\ \downarrow t & & \downarrow t_1 \\ K^W(S^+(V_1 \oplus V_2 \oplus 1), S(V_1 \oplus V_2) \cup S^+(V_1 \oplus V_2 \oplus 1)|_Y) & \xrightarrow{f^*} & K^W(S^+(\pi_2^* V_1 \oplus 1), S(\pi_2^* V_1) \cup S^+(\pi_2^* V_1 \oplus 1)|_Z), \end{array}$$

where f is the homermorphism defined in 6.19 (f^* is an isomorphism by 6.18).

Proof. Let $x = d(E, \eta, \eta')$ be an element of $K^{W \oplus V_1 \oplus V_2}(X, Y)$. Using the obvious notation, we have

$$t_2(x) = d(E, v_2 \cos\theta_2 + \eta \sin\theta_2, v_2 \cos\theta_2 + \eta' \sin\theta_2),$$

and
$$\begin{aligned} t_1(t_2(x)) &= d(E, v_1 \cos\theta_1 + (v_2 \cos\theta_2 + \eta \sin\theta_2) \sin\theta_1, v_1 \cos\theta_1 \\ &\quad + (v_2 \cos\theta_2 + \eta' \sin\theta_2) \sin\theta_1) \\ &= d(E, v_1 \cos\theta_1 + v_2 \cos\theta_2 \sin\theta_1 + \eta \sin\theta_2 \sin\theta_1, v_1 \cos\theta_1 \\ &\quad + v_2 \cos\theta_2 \sin\theta_1 + \eta' \sin\theta_1 \sin\theta_1). \end{aligned}$$

By definition of the homomorphism f^*, we therefore have

$$f^*(t(x)) = t_1(t_2(x)). \quad \square$$

6.21. Theorem. *The homomorphism*

$$t: K^{W \oplus V}(X, Y) \longrightarrow K^{\pi^* W}(B(V), S(V) \cup B(V)|_Y)$$

is an isomorphism. More generally, the homomorphism

$$t_n: K_n^{W \oplus V}(X, Y) \longrightarrow K_n^W(B(V), S(V) \cup B(V)|_Y),$$

obtained by the substitution $(X, Y) \mapsto (X \times B^n, X \times S^{n-1} \cup Y \times B^n)$, is also an isomorphism.

Proof. By a Mayer-Vietoris argument already used many times (6.17), we may assume that V and W are trivial. Moreover, using the transitivity of the homomorphism t (Proposition 6.20), we may assume that V is of rank one, provided with the quadratic form $+x^2$. Then Theorem 6.21 is simply a reformulation of the fundamental Theorem III.5.10, proved in III.6. □

6.22. *Remark.* This theorem generalizes Theorem 5.11 which we proved using additional spinorial or ^{c}spinorial structures. In fact it gives an independent (but more complicated) proof of Theorem 5.11.

6.23. Let us now consider vector bundles V, W, T, such that $V = W \oplus T$, where V is provided with a nondegenerate quadratic form whose restrictions to W and T are also nondegenerate and such that W is orthogonal to T. Our next objective is to compare the groups $K_n^{V,W}(X, Y)$, $K_n^{V,T}(X, Y)$, and $K_n^{W,T}(X, Y)$. More precisely, we will prove the exact sequence

$$K_{n+1}^{V,T}(X, Y) \to K_{n+1}^{W,T}(X, Y) \xrightarrow{\gamma} K_n^{V,W}(X, Y) \to K_n^{V,T}(X, Y) \to K_n^{W,T}(X, Y),$$

where all the homomorphisms are the obvious ones, except the "connecting homomorphism" γ, which we will define in 6.25.

6.24. Lemma. *Each element of $K^{W,T}(X, Y)$ may be written in the form $d(C(V) \otimes F, w_1, w_2)$, where F is a trivial bundle, and w_1 is the $C(W)$-module structure on $C(V) \otimes F$, induced from the factor $C(V)$ (note that $C(W) \subset C(V)$). Moreover, $d(C(V) \otimes F, w_1, w_2) = d(C(V) \otimes F, w_1, w_2')$ if and only if there exists a trivial bundle G, such that $w_2 \oplus w_3$ is homotopic to $w_2' \oplus w_3$, where w_3 is the canonical $C(W)$-module structure on $C(V) \otimes G \approx C(W) \otimes C(V/W) \otimes G$.*

Proof. The proof of this lemma is analogous to the proof of Proposition III.4.26 (note that every $C(W)$-module is a direct factor of some

$$C(V) \otimes F \approx C(W) \otimes C(V/W) \otimes F,$$

where F is a trivial bundle, by the argument used in the proof of Lemma 6.6). □

6.25. We denote the class of the triple $(C(V) \otimes F, w_1, w_2)$ in the group $K^{W,T}(X, Y)$ by $\delta(F, w_2)$. In particular, if we make the substitution $(X, Y) \mapsto (X \times B^1, X \times S^0 \cup Y \times B^1)$, we denote a typical element of $K_1^{W,T}(X, Y)$ by $\delta(F, w_2)$ or $\delta(F, w_2(t))$.

By 6.3 there exists a continuous family $\alpha(t)$ of automorphisms of $E = C(W) \otimes F$, regarded as an object of $\mathscr{E}^T(X)$, such that $w_2(t) = \alpha(t) w_2(0) \alpha(t)^{-1}$ with $\alpha(t)|_Y = 1$. Now we define

$$\gamma: K_1^{W,T}(X, Y) \longrightarrow K^{V,W}(X, Y)$$

by the formula $\gamma(\delta(F, w_2(t)) = d(C(V) \otimes F, v_1, \alpha(1) v_1 \alpha(1)^{-1})$. This definition is independent of the choice of $w_2(t)$ and $\alpha(t)$. Suppose $w_2'(t)$ and $\alpha'(t)$ are other choices. There exists a continuous family $w_2(t, u)$, for $(t, u) \in I \times I$, of $C(W)$-module structures on $E = C(V) \otimes F$, such that $w_2(t, 0) = w_2(t)$, $w_2(t, 1) = w_2'(t)$, $w_2(0, u) = w_2(1, u) = w_1$, $w_2(t, u)|_T = w_1|_T$, and $w_2(t, u)|_Y = w_1|_Y$. Let K be the unit square $I \times I$, and let L be the union $I \times \{0\} \cup I \times \{1\} \cup \{0\} \times I$. Since K is homeomorphic to $L \times I$, Lemma 6.7 implies the existence of a continuous family $\alpha(t, u)$ of automorphisms of the $C(T)$-module E, such that $\alpha(t, 0) = \alpha(t)$, $\alpha(t, 1) = \alpha'(t)$, $\alpha(0, u) = 1$, $\alpha(t, u)|_Y = 1$, and $w_2(t, u) = \alpha(t, u) w_1 \alpha(t, u)^{-1}$. Therefore, the continuous map $u \mapsto \alpha(1, u) v_1 \alpha(1, u)^{-1}$ provides a homotopy $v_1(u)$ between $\alpha(1) v_1 \alpha(1)^{-1}$ and $\alpha'(1) v_1 \alpha'(1)^{-1}$, such that $v_1(u)|_W = v_1|_W$ and $v_1(u)|_Y = v_1|_Y$.

6.26. Theorem. *The sequence*

$$K_1^{V,T}(X, Y) \xrightarrow{\chi_1} K_1^{W,T}(X, Y) \xrightarrow{\gamma} K^{V,W}(X, Y) \xrightarrow{\varphi} K^{V,T}(X, Y) \xrightarrow{\chi} K^{W,T}(X, Y)$$

is exact.

Proof. Direct inspection shows that the composition of each consecutive pair of homomorphisms is 0. Therefore, we need only verify that

$$\operatorname{Ker}(\chi) \subset \operatorname{Im}(\varphi),\ \operatorname{Ker}(\varphi) \subset \operatorname{Im}(\gamma), \quad \text{and} \quad \operatorname{Ker}(\gamma) \subset \operatorname{Im}(\chi_1).$$

a) <u>$\operatorname{Ker}(\chi) \subset \operatorname{Im}(\varphi)$</u>. Let $x = d(E, v_1, v_2)$ be an element of $K^{V,T}(X, Y)$ such that $\chi(x) = 0$. If we define $w_i = v_i|_W$, then there exists a homotopy $w(t)$ between w_1 and w_2 (up to addition of an elementary triple), such that $w(t)|_T = w_1|_T$ and $w(t)|_Y = w_1|_Y$. Now we define $w(t) = \alpha(t) w_1 \alpha(t)^{-1}$, where $\alpha(0) = 1$ and $\alpha(t)|_Y = 1$. From this we see that $d(E, v_1, v_2) = d(E, \alpha(1) v_1 \alpha(1)^{-1}, v_2) \in \operatorname{Im}(\varphi)$, since

$$\alpha(1) v_1 \alpha(1)^{-1}|_W = v_2|_W.$$

b) <u>$\operatorname{Ker}(\varphi) \subset \operatorname{Im}(\gamma)$</u>. Let $x = d(E, v_1, v_2)$ be an element of $K^{V,W}(X, Y)$ such that $\varphi(x) = 0$. Without loss of generality, we may assume that E is of the form $C(V) \otimes F$, where F is a trivial bundle, and that v_1 represents the canonical $C(V)$-module structure on $E = C(V) \otimes F$. Up to addition of an elementary triple, there exists a continuous family $v(t)$ of $C(V)$-module structures on E, such that $v(0) = v_1$, $v(1) = v_2$, $v(t)|_Y = v_1|_Y$, and $v(t)|_T = v_1|_T$. If we define $w(t) = v(t)|_W$, we thus have $d(E, v_1, v_2) = \gamma(\delta(E, w(t)))$.

c) <u>$\operatorname{Ker}(\gamma) \subset \operatorname{Im}(\chi_1)$</u>. Let $x = \delta(F, w_2(t))$ be an element of $K_1^{W,T}(X, Y)$ such that $\gamma(x) = 0$. If we define $v_2(t) = \alpha(t) v_1 \alpha(t)^{-1}$ using the notation of 6.24, we must have $v_2(1)$ homotopic to v_1 via a continuous family $v_2'(t)$ of $C(V)$-module structures. Moreover, we have $v_2'(0) = v_2(1)$, $v_2'(1) = v_1$, $v_2'(t)|_W = v_1|_W$, and $v_2'(t)|_Y = v_1|_Y$. Let $\bar{v}_2(t)$ denote the homotopy obtained by the composition of $v_2(t)$ and $v_2'(t)$, and let $\bar{w}_2(t) = \bar{v}_2(t)|_W$. Then $\delta(F, \bar{w}_2(t)) = \delta(F, w_2(t))$ since $v_2'(t)|_W = v_1|_W$, and $x = \chi_1(\delta(F, \bar{v}_2(t))$ since $\bar{v}_2(0) = \bar{v}_2(1) = v_1$, $\bar{v}_2(t)|_T = v_1|_T$, and $\bar{v}_2(t)|_Y = v_1|_Y$. □

6.27. Corollary. *For $n \geqslant 0$, we have the exact sequence*

$$K_{n+1}^{V,T}(X, Y) \to K_{n+1}^{W,T}(X, Y) \to K_n^{V,W}(X, Y) \to K_n^{V,T}(X, Y) \to K_n^{W,T}(X, Y).$$

6.28/29. The tensor product of vector bundles defines a bilinear map

$$K_n^{V,W}(X, Y) \times K(X') \longrightarrow K_n^{V,W}(X \times X', Y \times X').$$

By the argument used in II.5.6, there is a unique natural bilinear map

$$K_n^{V,W}(X, Y) \times K(X', Y') \longrightarrow K_n^{V,W}(X \times X', X \times Y' \cup Y \times X'),$$

which extends the map above when $Y' = \varnothing$. If we replace the pair (X', Y') by the pair $(X' \times B^p, X' \times S^{p-1} \cup Y' \times B^p)$, we may also define a bilinear map (also called the cup-product)

$$K_n^{V,W}(X, Y) \times K_p(X', Y') \longrightarrow K_{n+p}^{V,W}(X \times X', X \times Y' \cup Y \times X')$$

as in II.5.6.

6.30. Theorem. *Let α be a generator of $K_{\mathbb{R}}^{-8}(P)$ (resp. $K_{\mathbb{C}}^{-2}(P)$). Then the cup-product by α induces an isomorphism*

$$K_{\mathbb{R}}^{V,W}(X, Y) \approx K_{\mathbb{R}}^{V,W}(X \times B^8, X \times S^7 \cup Y \times B^8) \quad (\textit{resp. } K_{\mathbb{C}}^{V,W}(X, Y) \approx K_{\mathbb{C}}^{V,W}(X \times B^2, X \times S^1 \cup Y \times B^2)).$$

Proof. Due to the exact sequence proved in 6.15, we are reduced to proving $K_n^{V,W}(X) \approx K_{n+8}^{V,W}(X)$ (resp. $K_n^{V,W}(X) \approx K_{n+2}^{V,W}(X)$). By a Mayer-Vietoris argument (6.17), we are further reduced to the case where V and W are trivial, and $n=0$. Now we have the exact sequence

$$K^{-1}(\mathscr{E}^V(X)) \longrightarrow K^{-1}(\mathscr{E}^W(X)) \longrightarrow K^{V,W}(X) \longrightarrow K(\mathscr{E}^V(X)) \longrightarrow K(\mathscr{E}^W(X)),$$

associated with the functor $\varphi: \mathscr{E}^V(X) \to \mathscr{E}^W(X)$ (II.3.22). By III.4.8, the groups $K(\mathscr{E}^V(X))$, $K(\mathscr{E}^W(X))$, $K^{-1}(\mathscr{E}^V(X))$ and $K^{-1}(\mathscr{E}^W(X))$, are isomorphic to $K_F^n(X)$ for some $n=0$ or -1, and $F = \mathbb{R}, \mathbb{C}, \mathbb{H}, \mathbb{R} \times \mathbb{R}$, or $\mathbb{H} \times \mathbb{H}$. Therefore, the cup-product by α induces an isomorphism $K_F^n(X) \approx K_F^n(X \times B^8, X \times S^7)(K_F^n(X) \approx K_F^n(X \times B^2, X \times S^1)$ in the complex case). By the five lemma, it follows that the cup-product by α also induces an isomorphism

$$K_{\mathbb{R}}^{V,W}(X) \approx K_{\mathbb{R}}^{V,W}(X \times B^8, X \times S^7) \quad (\text{resp. } K_{\mathbb{C}}^{V,W}(X) \approx K_{\mathbb{C}}^{V,W}(X \times B^2, X \times S^1)). \quad \square$$

6.31. After this long series of lemmas, we are finally approaching our original problem, which is the computation of the K-theory of the pair $(P(V), P(W))$, and its relationship to the groups $K^{V,W}$, when V (hence W) is provided with a *positive* quadratic form. More precisely, if n denotes the trivial bundle $X \times \mathbb{R}^n$ provided with the "trivial" positive quadratic form, we define a fundamental homomorphism

$$p: K^{V \oplus n, W \oplus n}(X, Y) \longrightarrow K^{\xi \oplus n, n}(P(V), P(W) \cup P(V)|_Y)$$

(where ξ is the canonical line bundle over $P(V)$) in the following way. Let $d(E, u; v_1, v_2)$ be an element of $K^{V \oplus n, W \oplus n}(X, Y)$, where v_1 and v_2 represent two "actions" of the bundle V such that $v_1|_W = v_2|_W$ (6.9), and where u represents the action of the trivial bundle of rank n. Let $\pi' : S(V) \to X$, and let $\tilde{v}_i$ be the action on π'^*E of the trivial bundle of rank one, defined by the involution $v_i(v)$ over the point $v \in S(V)$. Since $v_i(-v) = -v_i(v)$, v_i induces an action of the canonical line bundle ξ on π^*E, where $\pi : P(V) \to X$. We again denote this action by $\tilde{v}_i$, and let u denote the action of the trivial bundle of rank n on π^*E induced by u. Then the homomorphism p above is defined by the formula

$$p(d(E, u, v_1, v_2)) = d(\pi^*E, \tilde{u}, \tilde{v}_1, \tilde{v}_2).$$

6.32. Theorem. *The homomorphism*

$$p : K^{V \oplus n, W \oplus n}(X, Y) \longrightarrow K^{\xi \oplus n, n}(P(V), P(W) \cup P(V)|_Y)$$

defined above, is an isomorphism.

Proof. If we replace the pair (X, Y) by the pair $(X \times B^r, X \times S^{r-1} \cup Y \times B^r)$, we obtain a slightly more general homomorphism,

$$p_r : K_r^{V \oplus n, W \oplus n}(X, Y) \longrightarrow K_r^{\xi \oplus n, n}(P(V), P(W) \cup P(V)|_Y),$$

and direct inspection shows that the homomorphism p_r is compatible with the various homomorphisms defined in the exact sequences 6.16 and 6.27. Therefore, using a Mayer-Vietoris argument (6.17), we may reduce the proof of the general isomorphism to the case where V and W are trivial. Moreover, Proposition 6.27 reduces this case to the case where $V = W \oplus 1$. The projection

$$\pi_1 : (S^+(W \oplus 1), S^+(W \oplus 1)|_Y \cup S(W)) \longrightarrow (P(W \oplus 1), P(W \oplus 1)|_Y \cup P(W))$$

induces an isomorphism between the groups

$$K_r^{\pi_1^*(\xi \oplus n), \pi_1^*(n)}(S^+(W \oplus 1), S^+(W \oplus 1)|_Y \cup S(W))$$

and $$K_r^{\xi \oplus n, n}(P(W \oplus 1), P(W \oplus 1)|_Y \cup P(W),$$

since the spaces

$$S^+(W \oplus 1) - S^+(W \oplus 1)|_Y \cup S(W) \quad \text{and} \quad P(W \oplus 1) - P(W \oplus 1)|_Y \cup P(W)$$

are homeomorphic (6.27). Moreover, by definition we have the commutative diagram

$$\begin{array}{ccc}
 & & K_r^{n \oplus 1, n}(S^+(W \oplus 1), S(W) \cup S^+(W \oplus 1)|_Y) \\
 & \overset{t}{\nearrow} & \uparrow \approx \pi_1^* \\
K_r^{n \oplus W \oplus 1, n \oplus W}(X, Y) & & \\
 & \underset{p_r}{\searrow} & \\
 & & K_r^{\xi \oplus n, n}(P(W \oplus 1), P(W) \cup P(W \oplus 1)|_Y),
\end{array}$$

where t is the isomorphism described in III.5.10 and 6.21. The fact that t is an isomorphism (III.5.10) implies that p_r is also an isomorphism. □

6.33. Consider the particular case $n=1$. Now each element x of the group $K^{\xi\oplus 1,1}(P(V), P(W)\cup P(V)|_Y)$ may be written in the form $x=d(E, \eta, \varepsilon_1, \varepsilon_2)$, where E is a vector bundle over $P(V)$ provided with an involution η arising from the action of the trivial bundle $X\times\mathbb{R}$. In this notation, ε_1 and ε_2 represent the two actions of ξ on E such that $\varepsilon_1|_{P(W)\cup P(V)|_Y}=\varepsilon_2|_{P(W)\cup P(V)|_Y}$. Since $\xi\otimes\xi=P(V)\times\mathbb{R}$, the product $\varepsilon_1\varepsilon_2$ defines an automorphism α of E which commutes with η. The restriction of α to $\operatorname{Ker}(\eta-1)$ defines an element of $K^{-1}(P(V), P(W)\cup P(V)|_Y)$ (II.3.25).

6.34. Proposition. *The correspondence above defines an isomorphism between the groups $K^{\xi\oplus 1,1}(P(V), P(W)\cup P(V)|_Y)$ and $K^{-1}(P(V), P(W)\cup P(V)|_Y)$.*

Proof. We define a homomorphism in the opposite direction. Let F be a vector bundle on $P(V)$, and let β be an automorphism of F such that $\beta|_{P(W)\cup P(V)|_Y}=1$. Let $E=F\oplus(\xi\otimes F)$, and let η be the involution defined by the matrix

$$\begin{pmatrix}1 & 0\\ 0 & -1\end{pmatrix}.$$

Finally, let $\varepsilon_1:\xi\times E\to E$ be the (bilinear) homomorphism induced by the isomorphism $\xi\otimes E\approx E$, and let $\varepsilon_2=\beta\varepsilon_1$. It is clear that the two correspondences are inverse to each other. □

6.35. Corollary. *The groups*

$$K_r^{\xi\oplus 1,1}(P(V), P(W)\cup P(V)|_Y) \quad \text{and} \quad K^{-r-1}(P(V), P(W)\cup P(V)|_Y)$$

are isomorphic.

6.36. Theorem. *Let r be an integer modulo 8 (modulo 2 in the complex case). Then the groups $K^{-r}(P(V), P(W)\cup P(V)|_Y)$ and $K_{r-1}^{V\oplus 1, W\oplus 1}(X, Y)$ are naturally isomorphic (as $K(X)$-modules).*

Proof. This follows directly from 6.34, 6.31 and 6.28/29. □

6.37. In order to use Theorem 6.36, we must compute the groups $K_r^{V,W}(X, Y)$ in terms of more classical invariants. For simplicity, we restrict ourselves to the case where $Y=\varnothing$. In general, we denote the Grothendieck group of the functor

$$\mathscr{E}^V(X\times B^r)\longrightarrow\mathscr{E}^V(X\times S^{r-1})$$

by $K_r^{(V)}(X)$. By excision, this is also the Grothendieck group of the functor

$$\mathscr{E}^V(X\times S^r)\longrightarrow\mathscr{E}^V(X\times P).$$

where P is a point.

6.38. Proposition. *We have the exact sequence*

$$K_{r+1}^{(V)}(X) \longrightarrow K_{r+1}^{(W)}(X) \xrightarrow{\partial} K_r^{V,W}(X) \longrightarrow K_r^{(V)}(X) \longrightarrow K_r^{(W)}(X).$$

Proof. Let $\varphi: \mathscr{E}^V(X) \to \mathscr{E}^W(X)$ be the obvious Banach functor. Then the exact sequence associated with the functor φ (II.3.22) may be written as

$$K^{-1}(\mathscr{E}^V(X)) \longrightarrow K^{-1}(\mathscr{E}^W(X)) \longrightarrow K^{V,W}(X) \longrightarrow K(\mathscr{E}^V(X)) \longrightarrow K(\mathscr{E}^W(X)).$$

Moreover, $K^{-1}(\mathscr{E}^V(X)) \approx K_1^{(V)}(X)$ by Theorem II.3.22 applied to the functor $\mathscr{E}^V(X \times B^1) \to \mathscr{E}^V(X \times S^0)$. Similarly, $K^{-1}(\mathscr{E}^W(X)) \approx K_1^{(W)}(X)$. Therefore, we have the exact sequence of the theorem for $r=0$. For $r>0$, we have

$$K_r^{V,W}(X) \approx \mathrm{Ker}[K^{V,W}(X \times S^r) \longrightarrow K^{V,W}(X \times P)] \quad \text{by 6.18.}$$

Similarly, $K_r^{(V)}(X) \approx \mathrm{Ker}[K^{(V)}(X \times S^r) \longrightarrow K^{(V)}(X \times P)]$,

and $K_{r+1}^{(V)}(X) \approx \mathrm{Ker}[K_1^{(V)}(X \times S^r) \longrightarrow K_1^{(V)}(X \times P)]$.

The case $r>0$ therefore follows from the case $r=0$. □

6.39. Let us now consider the case where W and V/W are spinorial bundles of rank p and $n-p$, respectively. Then V is spinorial of rank n, and from 4.22, we have category equivalences

$$\mathscr{E}^{0,n}(X) \sim \mathscr{E}^V(X) \quad \text{and} \quad \mathscr{E}^{0,p}(X) \sim \mathscr{E}^W(X).$$

Moreover, the composition

$$\alpha: \mathscr{E}^{0,n}(X) \sim \mathscr{E}^V(X) \longrightarrow \mathscr{E}^W(X) \sim \mathscr{E}^{0,p}(X)$$

is defined by

$$E \longmapsto P \Delta_{\mathrm{Spin}(n-p)} E,$$

where P is the principal bundle which defines the spinorial structure of V/W (4.22).

6.40. Theorem. *Let us assume that the vector bundles W and V/W are spinorial of rank p and $n-p$, respectively. Then we have the exact sequence*

$$\begin{array}{ccccccc} K^i(\mathscr{E}_{\mathbb{R}}^{0,n+1}(X)) & \xrightarrow{\alpha^i} & K^i(\mathscr{E}_{\mathbb{R}}^{0,p+1}(X)) & \longrightarrow & K_{\mathbb{R}}^i(P(V), P(W)) & \longrightarrow & K^{i+1}(\mathscr{E}_{\mathbb{R}}^{0,n+1}(X)) \\ & & & & & & \downarrow \alpha^{i+1} \\ & & & & & & K^{i+1}(\mathscr{E}_{\mathbb{R}}^{0,p+1}(X)), \end{array}$$

where α^i is induced by the functor α, and where the groups $K^i(\mathscr{E}_{\mathbb{R}}^{0,r}(X))$ are determined by the following table.

$r \bmod 8$	0	1	2	3
$K^i(\mathscr{E}_{\mathbb{R}}^{0,r}(X))$	$K_{\mathbb{R}}^i(X)$	$K_{\mathbb{R}}^i(X) \oplus K_{\mathbb{R}}^i(X)$	$K_{\mathbb{R}}^i(X)$	$K_{\mathbb{C}}^i(X)$
$r \bmod 8$	4	5	6	7
$K^i(\mathscr{E}_{\mathbb{R}}^{0,r}(X))$	$K_{\mathbb{H}}^i(X)$	$K_{\mathbb{H}}^i(X) \oplus K_{\mathbb{H}}^i(X)$	$K_{\mathbb{H}}^i(X)$	$K_{\mathbb{C}}^i(X)$

Proof. This is simply a reformulation of Proposition 6.38 and Theorem 6.36, using the category equivalences described in 4.22. □

6.41. *Example.* Let us assume that V/W is spinorial of rank $8t$, and that W is spinorial of rank $8r+1$. Then we have the exact sequence

$$K_{\mathbb{R}}^i(X) \xrightarrow{\alpha^i} K_{\mathbb{R}}^i(X) \longrightarrow K_{\mathbb{R}}^i(P(V), P(W)) \longrightarrow K_{\mathbb{R}}^{i+1}(X) \xrightarrow{\alpha^{i+1}} K_{\mathbb{R}}^{i+1}(X),$$

where α^i is induced by the functor $\varphi\colon \mathscr{E}_{\mathbb{R}}(X) \to \mathscr{E}_{\mathbb{R}}(X)$ defined by $\varphi(E) = F \otimes E$, where F is the vector bundle $P \times_{\mathrm{Spin}(8t)} M$, such that P is the principal bundle defining the spinorial structure of V/W, and where M is an irreducible $C^{0,8t}$-module. If V/W is trivial, then the functor $\varphi\colon \mathscr{E}_{\mathbb{R}}(X) \to \mathscr{E}_{\mathbb{R}}(X)$ is simply $E \mapsto \underbrace{E \oplus E \oplus \cdots \oplus E}_{n}$ where $n = 16^t$.

If V/W is not trivial, it is difficult to compute the vector bundle F in general. However, if V/W may be written in the form $U \oplus U$, where U is oriented (4.18), then V/W is a spinorial bundle and the vector bundle F may be identified with $\Lambda(U)$.

The above results could easily be stated in the framework of complex K-theory. For example, Theorem 6.39 has the following complex analog:

6.42. Theorem. *Let us assume that the vector bundles W and V/W are ^{c}spinorial of rank p and $n-p$, respectively. Then we have the exact sequence*

$$\begin{array}{c} K^i(\mathscr{E}_{\mathbb{C}}^{0,n+1}(X)) \xrightarrow{\alpha^i} K^i(\mathscr{E}_{\mathbb{C}}^{0,p+1}(X)) \longrightarrow K_{\mathbb{C}}^i(P(V), P(W)) \longrightarrow K^{i+1}(\mathscr{E}_{\mathbb{C}}^{0,n+1}(X)) \\ \qquad\qquad\qquad\qquad\qquad\qquad\qquad\qquad\qquad\qquad\qquad\qquad \Big\downarrow \alpha^{i+1} \\ \qquad\qquad\qquad\qquad\qquad\qquad\qquad\qquad\qquad\qquad\qquad\qquad K^{i+1}(\mathscr{E}_{\mathbb{C}}^{0,p+1}(X)), \end{array}$$

where α^i is induced by the functor $E \mapsto P\Delta_{\mathrm{Spin}^c(n-p)} E$ (where P is the principal bundle defining the ^{c}spinorial structure of V/W), and where $K^i(\mathscr{E}_{\mathbb{C}}^{0,r}(X)) = K_{\mathbb{C}}^i(X)$ if r is even, and $K^i(\mathscr{E}_{\mathbb{C}}^{0,r}(X)) = K_{\mathbb{C}}^i(X) \oplus K_{\mathbb{C}}^i(X)$ if r is odd.

6.43. A particular case of interest arises when V/W is provided with a complex structure (and W is ^{c}spinorial). Then the homomorphism α^i is induced by the functor $E \mapsto \Lambda(V/W) \otimes E$, where $\Lambda(V/W)$ is the *complex* exterior algebra bundle (1.4).

6.44. In the real case as in the complex case, we wish to describe more precisely the homomorphism

$$K^i(X) \oplus K^i(X) \approx K^i(\mathscr{E}^{0,1}(X)) \longrightarrow K^i(P(V))$$

in the exact sequence obtained in 6.40 or 6.42, when W is 0, For $i=-1$, this homomorphism is the composition

$$K^{-1}(\mathscr{E}^{0,1}(X)) \xrightarrow{\partial} K^{V\oplus 1,1}(X) \underset{\approx}{\xrightarrow{p}} K^{\xi\oplus 1,1}(P(V)) \underset{\approx}{\xrightarrow{u}} K^{-1}(P(V)),$$

where p is defined as in 6.31, where ∂ is defined as in 6.38, and u is defined as in 6.34. More precisely, an element x of $K^{-1}(\mathscr{E}^{0,1}(X))$ may be represented by a $\mathbb{Z}/2$-graded vector bundle $E=E_0 \oplus E_1$, and an automorphism α of the form

$$\alpha = \begin{pmatrix} \alpha_0 & 0 \\ 0 & \alpha_1 \end{pmatrix}.$$

These two automorphisms, α_0 and α_1, define the decomposition $K^{-1}(\mathscr{E}^{0,1}(X)) \approx K^{-1}(X) \oplus K^{-1}(X)$.

Since the functor $\mathscr{E}^{V\oplus 1}(X) \to \mathscr{E}^{0,1}(X)$ is quasi-surjective, we may assume without loss of generality, that E is provided with a graded $C(V)$-module structure with respect to the decomposition $E=E_0 \oplus E_1$, represented by v. Therefore, $\partial(x)$ is $d(E, v, \alpha^{-1}v\alpha)$, and $(up\partial)(x)$ is the restriction of the automorphism $\tilde{v}\alpha^{-1}\tilde{v}\tilde{\alpha} = (\tilde{v}\alpha^{-1}\tilde{v})\tilde{\alpha})$ to E_0, with the notation of 6.30. In fact $\tilde{v}$ induces an isomorphism between π^*E^0 and π^*E^1, where $\pi: P(V) \to X$, and the class of the automorphism $\tilde{v}\alpha^{-1}\tilde{v}$ in the group $K^{-1}(P(V))$ is simply $\pi^*\alpha^{-1} \otimes \mathrm{Id}_\xi$. Hence $(up\partial)(x)$ is the sum $\pi^*(x_0) - \theta^*(x_1)$, where x_0 and x_1 are the two components of x, where π^* is induced by the projection $\pi: P(V) \to X$, and where θ^* is induced by the projection $P(V) \to X$ and the product with the line bundle ξ. Replacing X by $X \times S^r$, we summarize these facts in the following theorem:

6.45. Theorem. *We have the exact sequence*

$$K^i(\mathscr{E}^{V\oplus 1}(X)) \longrightarrow K^i(X) \oplus K^i(X) \xrightarrow{(\pi^*, -\theta^*)} K^i(P(V))$$
$$\longrightarrow K^{i+1}(\mathscr{E}^{V\oplus 1}(X)) \longrightarrow K^{i+1}(X) \oplus K^{i+1}(X)$$

where π^* is induced by the projection $\pi: P(V) \to X$, and θ^* is induced by the functor $E \mapsto \xi \otimes \pi^*E$.

6.46. Corollary (Adams [1]). *Let RP_{n-1} be the projective space of $\mathbb{R}^n$. Then $\tilde{K}_{\mathbb{R}}(RP_{n-1})$ is generated by $\lambda_{n-1} = [\xi] - 1$, with the relations $(\lambda_{n-1})^2 = -2\lambda_{n-1}$ and*

$2^f\lambda_{n-1}=0$, *where f is the number of integers i such that* $0<i<n$ *with* $i=0, 1, 2$ *or* 4 *mod* 8.

Proof. From Theorem 6.45, we see that $K_{\mathbb{R}}(RP_{n-1})$ can be identified with the cokernel of the homomorphism

$$K(C^{0,n+1})\longrightarrow K(C^{0,1})$$

with the notation of III.3. Therefore

$$\tilde{K}_{\mathbb{R}}(RP_{n-1})\approx \operatorname{Coker}(K(C^{0,n+1})\longrightarrow K(C^{0,2}))\approx \operatorname{Coker}(K(C^{n-1,0})\longrightarrow K(C^{0,0}))\approx \mathbb{Z}/2^f\mathbb{Z}$$

by III.4.9. Moreover, the generator is λ_{n-1} by 6.45. The relation $(\lambda_{n-1})^2=-2\lambda_{n-1}$ follows from the relation $\xi\otimes\xi=1$. □

6.47. Corollary. *Let* RP_{n-1} *be the projective space of* $\mathbb{R}^n$. *Then* $\tilde{K}_{\mathbb{C}}(RP_{n-1})$ *is generated by* $\lambda'_{n-1}=[\xi']-1$, *where* $\xi'=\xi\otimes\mathbb{C}$, *with the relations* $(\lambda'_{n-1})^2=-2\lambda'_{n-1}$ *and* $2^g\lambda'_{n-1}=0$, *where g is the number of integers i such that* $0<i<n$ *and* $i=0$ *mod* 2.

6.48. Corollary. *Let X be a space such that* $K^1(\mathscr{E}^{n,0}_{\mathbb{R}}(X))=0$. *Then*

$$K_{\mathbb{R}}(X\times RP_n, X)\approx \operatorname{Coker}[K(\mathscr{E}^{n,0}_{\mathbb{R}}(X))\rightarrow K(\mathscr{E}_{\mathbb{R}}(X))]$$

Exercise IV.8.8.

7. Operations in K-Theory

7.1. An operation in *K*-theory is a natural map (not necessarily a homomorphism) defined from $K(X)$ to $K(X)$ for every compact space X, which is natural in X. For simplicity we begin with operations in complex *K*-theory $K_{\mathbb{C}}(X)$, denoted simply by $K(X)$ until 7.24. Very often we will write E instead of $[E]$ for the class of the vector bundle E in $K(X)$.

7.2. Our first example of an operation in *K*-theory was introduced by Grothendieck [1]. For any vector bundle E, we denote its ith exterior power (I.4.8) by $\lambda^i(E)$. By abuse of notation, we again write $\lambda^i(E)$ for its class in the group $K(X)$, and

$$\lambda_t(E)=1+t\lambda^1(E)+t^2\lambda^2(E)+\cdots=\sum_{i=0}^{\infty} t^i\lambda^i(E)\in K(X)[[t]].$$

Since $\lambda^n(E\oplus F)=\bigoplus_{i+j=n}\lambda^i(E)\lambda^j(F)$ (III.3.10), we have $\lambda_t(E\oplus F)=\lambda_t(E)\lambda_t(F)$. Therefore, the correspondence $E\mapsto\lambda_t(E)$ defines a homomorphism between the monoid $\Phi(X)$ and the multiplicative group of formal power series, whose constant term equal 1. From the universal property of Grothendieck groups (II.1.1), we obtain a homomorphism from the additive group $K(X)$ to the multiplicative group $1+tK(X)[[t]]$, which again we denote by λ_t. Explicitly we have

$$\lambda_t(E-F)=\lambda_t(E)\lambda_t(F)^{-1};$$

hence,

$$\lambda_t(E-n)=\lambda_t(E)(1+t)^{-n}$$

where n is the class of the trivial bundle of rank n.

If $x \in K(X)$, we define

$$\lambda_t(x)=\sum_{i=0}^{\infty} t^i \lambda^i(x).$$

The notation $\lambda^i(x)$ is a generalization of the notation $\lambda^i(E)$, and the map $x \mapsto \lambda^i(x)$ is a well-defined operation in K-theory. Moreover, we have the identity

$$\lambda^n(x+y)=\sum_{i+j=n} \lambda^i(x)\lambda^j(y).$$

7.3. In a parallel way, we introduce operations $x \mapsto \gamma^i(x)$ by the following method. For $x \in K(X)$, we set $\gamma_t(x)=\lambda_{t/1-t}(x) \in K(X)[[t]]$, and $\gamma_t(x)=\sum_{i=0}^{\infty} t^i\gamma^i(x)$. We have $\gamma^n(x+y)=\sum_{i+j=n} \gamma^i(x)_{\gamma^j}(y)$, since $\gamma_t(x+y)=\gamma_t(x)\gamma_t(y)$.

7.4. Proposition. *Let E be a vector bundle of rank n, and let $c_i(E)$ be its ith characteristic class in the sense of* 2.17. *Then $c_i(E)=(-1)^i\gamma^i(E-n)$. In particular, $\gamma^i(E-n)=0$ for $i>n$.*

Proof. By 2.17 it suffices to verify the assertion for a line bundle $E=L$. In this case $c_1(L)=1-L$, and $c_i(L)=0$ for $i>1$. On the other hand $\gamma_t(L-1)=1+t(L-1)$. Hence $\gamma^1(L-1)=-c_1(L)$, and $\gamma^i(L-1)=0$ for $i>1$. □

7.5. Corollary. *Let $\xi_{n,m}$ be the canonical bundle over $G_n(\mathbb{C}^m)$, and let γ^i be the image of $\gamma^i(\xi_{n,m}-n)$ in* proj lim $K(G_n(\mathbb{C}^m))=\mathscr{K}_{\mathbb{C}}(BU(n))$ (3.22). *Then $\mathscr{K}_{\mathbb{C}}(BU(n))$ is isomorphic to the algebra of formal power series $\mathbb{Z}[[\gamma^1,\ldots,\gamma^n]]$.*

Proof. This follows directly from 7.4 and 3.22. □

7.6. Let α be an element of $\mathscr{K}_{\mathbb{C}}((BU(n))$, and let E be a vector bundle of rank n with compact base X. Then E is isomorphic to the inverse image of $\xi_{n,m}$ under a suitable continuous map

$$f\colon X \longrightarrow G_n(\mathbb{C}^m).$$

If α_m denotes the "restriction" of α to $K(G_n(\mathbf{C}^m))$, then the element $\alpha_*(E)=f^*(\alpha_m)$ of the group $K(X)$ depends only on the vector bundle E and the class α (I.7.2).

Let us write $\mathrm{Op}(\Phi_n, K)$ for the set of natural maps from $\Phi_n(X)$ to $K(X)$. From the ring structure of $K(X)$, the set $\mathrm{Op}(\Phi_n, K)$ is obviously a commutative ring, and the correspondence $\alpha \mapsto [E \to \alpha_*(E)]$ defines a homomorphism

$$\theta: \mathscr{K}_{\mathbf{C}}(BU(n)) \longrightarrow \mathrm{Op}(\Phi_n, K).$$

7.7. Theorem. *The map θ defined above is an isomorphism between $\mathscr{K}_{\mathbf{C}}(BU(n))$ and $\mathrm{Op}(\Phi_n, K)$. In particular, $\mathrm{Op}(\Phi_n, K) \approx \mathbf{Z}[[\gamma^1, \ldots, \gamma^n]]$.*

Proof. We define a homomorphism in the opposite direction. If $c \in \mathrm{Op}(\Phi_n, K)$, then the elements $c(\xi_{n,m}) \in K(G_n(\mathbf{C}^m))$ form a projective system, which defines an element α of $\mathscr{K}_{\mathbf{C}}(BU(n))$. The correspondence $c \mapsto \alpha$ defines the inverse homomorphism. □

7.8. Since $K(X) \approx H^0(X; \mathbf{Z}) \oplus K'(X)$ (II.1.29), we see that the interesting operations in K-theory arise from operations from $K'(X)$ to $K(X)$. Let $\mathrm{Op}(K', K)$ denote the subset of $\mathrm{Op}(K, K)$ thus defined. Since $K'(X) \approx \mathrm{inj\,lim}\, \Phi_n(X)$ (II.1.31), we have $\mathrm{Op}(K', K) \approx \mathrm{proj\,lim}\, \mathrm{Op}(\Phi_n, K)$, which is a ring in the obvious sense. From this discussion, we obtain the following general theorem which determines almost all the operations in (complex) K-theory:

7.9. Theorem. *The map, which associates the variable t_i with the nilpotent operation γ^i, induces an isomorphism*

$$\mathbf{Z}[[t_1, \ldots, t_n, \ldots]] \longrightarrow \mathrm{Op}(K', K).$$

Proof. If $x \in K'(X)$, then $\gamma^i(x)=0$ for some i large enough by 7.4. Moreover, $\gamma^j(x)$ is nilpotent for every positive integer i, since it belongs to $K'(X)$. □

7.10. Let us now examine the operations in K-theory with nice "algebraic" properties; for instance, the operations γ from K' to K such that $\gamma(x+y)=\gamma(x)+\gamma(y)$. The set of such operations form a subgroup of $\mathrm{Op}(K', K)$, which we denote by $\mathrm{Op}^{++}(K', K)$. Let $\varphi: \mathrm{Op}^{++}(K', K) \to \mathrm{Op}(\Phi_1, K)$ be the group homomorphism which associates each such operation with its "restriction" to Φ_1 by the canonical map $\Phi_1 \to K'$. According to 7.7, $\mathrm{Op}(\Phi_1, K) \approx \mathbf{Z}[[u]]$ where u is interpreted as $\xi - 1$, and ξ is the canonical line bundle over $BU(1)$, which is the infinite complex projective space.

7.11. Proposition. *The homomorphism*

$$\varphi: \mathrm{Op}^{++}(K', K) \to \mathrm{Op}(\Phi_1, K) \approx \mathbf{Z}[[u]]$$

is injective. Its image is the group of formal power series without constant term.

Proof. Let c be an "additive" operation (i.e. an element of $\mathrm{Op}^{++}(K', K)$) whose restriction to Φ_1 is zero, and let $x = V - n$ be an element of $K'(X)$. Let $F(V)$ be the flag bundle over X described in Section 3. Then the homomorphism $K(X) \to K(F(V))$ is injective, and $\pi^*(V)$, where $\pi: F(V) \to X$, splits as a direct sum $\bigoplus_{i=1}^{n} L_i$ of line bundles. Therefore $\pi^*(c(V-n)) = c(\pi^* V - n) = \sum_{i=1}^{n} c(L_i - 1) = 0$; hence, $c(V-n)=0$ and φ is injective.

Let us now consider the image of φ. Since $K'(X)=0$ if X is a point, $\varphi(c)$ must be a series without constant term

$$f(u) = a_1 u + a_2 u^2 + \cdots,$$

where $a_i \in \mathbb{Z}$. We now show that any series of this type gives rise to an additive operation.

We let Q_k, for $k \geqslant 1$, denote the "Newton polynomials": they express the symmetric functions $\sum_{i=1}^{k} u_i^k$ as unique polynomials of the elementary symmetric functions, $\sigma_r = \sum_{i_1 < i_2 < \cdots < i_r} u_{i_1} \cdots u_{i_r}$, where $1 \leqslant r \leqslant k$. For instance

$$Q_1(\sigma_1) = \sigma_1,$$
$$Q_2(\sigma_1, \sigma_2) = (\sigma_1)^2 - 2\sigma_2,$$
$$Q_3(\sigma_1, \sigma_2, \sigma_3) = (\sigma_1)^3 - 3\sigma_1\sigma_2 + 3\sigma_3, \text{ etc.}$$

Then the series

$$S(\gamma^1, \gamma^2, \ldots, \gamma^k, \ldots) = a_1 Q_1(\gamma^1) + a_2 Q_2(\gamma^1, \gamma^2) + \cdots + a_k Q_k(\gamma^1, \gamma^2, \ldots, \gamma^k) + \cdots$$

converges in the ring $\mathbb{Z}[[\gamma^1, \gamma^2, \ldots, \gamma^k, \ldots]]$, since $Q_k(\sigma_1, \sigma_2, \ldots, \sigma_k)$ is of weight k, and the desired operation is $x \mapsto S(\gamma^1(x), \gamma^2(x), \ldots \gamma^k(x), \ldots)$ (note again that $\gamma^k(x)=0$ for k large enough, and that each $\gamma^i(x)$ is nilpotent). If x is $L-1$ where L is a line bundle, then we obtain $a_1 u + a_2 u^2 + \cdots + a_k u^k + \cdots$, where $u = \gamma^1(x) = x$.

We must verify that the operation c, defined by the formula above, is additive, i.e. $c(x+y) = c(x) + c(y)$ for $x = V-n$ and $y = W-p$. By the splitting principle (2.15), we may assume that $V = \bigoplus_{i=1}^{n} L_i$ and $W = \bigoplus_{j=1}^{p} R_j$, where the L_i and R_j are line bundles. If we set $u_i = L_i - 1 = \gamma^1(L_i - 1)$ and $v_j = R_j - 1 = \gamma^1(R_j - 1)$, then we have $\gamma_t(V-n) = \prod_{i=1}^{n} \gamma_t(L_i - 1) = \prod_{i=1}^{n} (1 + tu_i)$ and $\gamma_t(W-p) = \prod_{j=1}^{p} \gamma_t(R_j - 1) = \prod_{j=1}^{p} (1 + tv_j)$. Therefore $\gamma^r(x) = \gamma^r(V-n) = \sigma_r(u_1, \ldots, u_n)$ and $\gamma^s(y) = \gamma^s(W-p) = \sigma_s(v_1, \ldots, v_p)$. Similarly $\gamma^k(x+y) = \sigma_k(u_1, \ldots, u_n, v_1, \ldots, v_p)$. It follows that

$$Q_k(\gamma^1(x+y), \ldots, \gamma^k(x+y)) = \sum_{i=1}^{n} (u_i)^k + \sum_{j=1}^{p} (v_j)^k = Q_k(\gamma^1(x), \ldots, \gamma^k(x)) + Q_k(\gamma^1(y), \ldots, \gamma^k(y)).$$

By taking linear combinations of these relations, we obtain $c(x+y)=c(x)+c(y)$ as required. □

7.12. Let us now consider operations c from $K(X)$ to $K(X)$ which are ring homomorphisms (with unit). Such an operation makes the diagram

$$\begin{array}{ccc} K(X) & \xrightarrow{\ c\ } & K(X) \\ & \searrow \quad \swarrow & \\ & H^0(X;\mathbb{Z}) & \end{array}$$

commutative. By 7.11, it follows that these operations form a subset of $\mathbb{Z}[[u]]$, which we will now determine. If ξ is the canonical line bundle over $BU(1)$, we define $c(\xi)=1+a_1u+\cdots+a_nu^n=f(u)$, where $u=\xi-1=\gamma^1(\xi-1)$.

Now let L_1 and L_2 be line bundles over a compact space X. Then $c(L_i)=f(u_i)$ where $u_i=L_i-1$, and $c(L_1\otimes L_2)=f(u_1+u_2+u_1u_2)$, since $L_1L_2-1=L_1-1+L_2-1+(L_1-1)(L_2-1)$ in the group $K(X)$. In particular, if $X=P(\mathbb{C}^m)\times P(\mathbb{C}^m)$ for m large enough, and if L_1 and L_2 are the two canonical line bundles over $P(\mathbb{C}^m)\times P(\mathbb{C}^m)$, we have

$$K(X)\approx\mathbb{Z}[u_1,u_2]/I_m,\text{ where } I_m=(u_1)^{m+1}(u_2)^{m+1},\text{ by 2.11.}$$

Therefore, the formal series f must satisfy the equation $f(u_1+u_2+u_1u_2)=f(u_1)f(u_2)$ in the ring $\mathbb{Z}[u_1,u_2]/I_m$ for every integer m, hence $f(u_1+u_2+u_1u_2)=f(u_1)f(u_2)$ in the ring $\mathbb{Z}[[u_1,u_2]]$. If we derive this equation with respect to u_1, and then set $u_1=0$, we find that the only solutions of this equation are the formal power series of the form $f(u)=(1+u)^k$, where $k\in\mathbb{Z}$. In other words, the characteristic class c is determined on the line bundle L by the formula $c(L)=L^k$ (note that $L^k=\bar{L}^{-k}$ if $k<0$). More precisely, we have the following theorem:

7.13. Theorem. *Let $k\in\mathbb{Z}$. Then there exists an operation*

$$\psi^k: K(X)\longrightarrow K(X),$$

called the Adams operation, which is characterized by the following properties:

1) $\psi^k(x+y)=\psi^k(x)+\psi^k(y)$,
2) $\psi^k(L)=L^k$ *if L is the class of a line bundle. Moreover, we have the relations*
3) $\psi^k(xy)=\psi^k(x)\psi^k(y)$ *and* $\psi^k(1)=1$.

The Adams operations ψ^k, for $k\in\mathbb{Z}$, are the only operations in complex K-theory which are ring homomorphisms (i.e. which satisfy 1) *and* 3)).

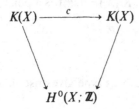

Proof. By 7.11, the operation ψ^k is determined and well-defined by Axioms 1) and 2). According to the splitting principle, it is enough to verify axiom 3) for x and y line bundles. This follows from the identity $(LR)^k=L^kR^k$, where L and R are line bundles. Finally, the last part of the theorem follows from the observations made in 7.12. □

7.14. *Remark*. Let $x \in K(X)$ and $x' \in K(X')$. Then we have $\psi^k(x \cup x') = \psi^k(x) \cup \psi^k(x')$ in the group $K(X \times X')$.

7.15. Theorem. *Let Q_k be the Newton polynomial defined in 7.11. Then $\psi^k(x) = Q_k(\lambda^1(x), \ldots, \lambda_k(x))$ for each element x of $K(X)$. Moreover, $\psi^k(\psi^l(x)) = \psi^{kl}(x)$ and $\psi^p(x) = x^p \bmod p$, if p is a prime number.*

Proof. In the ring $K(X)[[t]]$, let

$$\psi_{-t}(x) = -t\frac{\lambda_t'(x)}{\lambda_t(x)}.$$

Since $\lambda_t(x+y) = \lambda_t(x)\lambda_t(y)$, we have $\psi_t(x+y) = \psi_t(x) + \psi_t(y)$. If x is the class of a line bundle L, we have $\psi_{-t}(L) = -tL + t^2L^2 - t^3L^3 + \cdots$, or equivalently $\psi_t(L) = tL + t^2L^2 + t^3L^3 + \cdots$. According to the splitting principle, for each element x of $K(X)$, we have the identity $\psi_t(x) = t\psi^1(x) + t^2\psi^2(x) + t^3\psi^3(x) + \cdots$.

By the expression of $\psi_t(x)$ in terms of $\lambda_t(x)$ and $\lambda_t'(x)$, we see that $\psi_t(x)$ may be written as

$$\psi_t(x) = \sum_{m=1}^{\infty} t^m Q_m'(\lambda^1(x), \ldots, \lambda^m(x)),$$

where Q_m' is some polynomial in $\lambda^1(x), \ldots, \lambda^m(x)$. More precisely, let

$$\lambda_t = 1 + \sum_{m=1}^{k} t^m \lambda^m$$

in the quotient ring $\mathbf{Z}[\lambda^1, \ldots, \lambda^k][t]/(t^{k+1})$, and

$$\psi_{-t} = -t\,\lambda_t'/\lambda_t.$$

Then $\psi_t = \sum_{m=1}^{k} t^m Q_m'$, where Q_m' is some polynomial in $\lambda^1, \ldots, \lambda^m$ independent of k. To determine Q_m', we imbed $\mathbf{Z}[\lambda^1, \ldots, \lambda^k]$ in $\mathbf{Z}[u_1, \ldots, u_k]$ by sending λ^i to the ith elementary symmetric function of the u_α. Then we have the identities

$$\lambda_t = \prod_{i=1}^{k} (1 + tu_i)$$

and

$$\lambda_t'/\lambda_t = \sum_{i=1}^{k} u_i/1 + tu_i = S_1 - tS_2 + t^2S_3 - \cdots,$$

where $S_r = \sum_{r=1}^{k} (u_i)^r$. It follows that

$$\psi_t = \sum_{m=1}^{k} Q_m(\lambda^1, \ldots, \lambda^m)t^m.$$

Hence $\quad Q'_m(\lambda^1,\ldots,\lambda^m)=Q_m(\lambda^1,\ldots,\lambda^m)$, and $\psi^k(x)=Q_k(\lambda^1(x),\ldots,\lambda^k(x))$.

The relation $\psi^k(\psi^l(x))=\psi^{kl}(x)$ follows from the identity $\psi^k(\psi^l(L))=(L^l)^k=L^{lk}=\psi^{kl}(L)$, where L is a line bundle.

In the ring $\mathbb{Z}[u_1,\ldots,u_p]$, where p is a prime number, we have the relation $(u_1+\cdots+u_p)^p=(u_1)^p+\cdots(u_p)^p \bmod p$. Therefore $Q_p(\lambda^1,\ldots,\lambda^p)=(\lambda^1)^p \bmod p$, and $\psi^p(x)=x^p \bmod p$. □

7.16. Let $\dot{V}$ be the Thom space of a complex vector bundle V with compact base X. Let U_V be the Thom class of V (1.6), and let $\varphi_V\colon K(X)\to K(V)\approx\tilde{K}(\dot{V})$ be the Thom isomorphism defined by the product with U_V. We define $\rho^k(V)=\varphi_V^{-1}(\psi^k(U_V))\in K(X)$.

7.17. Proposition. *Let x be an element of $K(X)$. Then $\varphi_V^{-1}(\psi^k(\varphi_V(x)))=\psi^k(x)\rho^k(V)$. In other words, the following diagram is commutative*

$$\begin{array}{ccc} K(V) & \xrightarrow{\ \psi^k\ } & K(V) \\ {\scriptstyle\varphi_V}\big\uparrow & & \big\downarrow{\scriptstyle\varphi_V^{-1}} \\ K(X) & \xrightarrow{\ T_V^k\ } & K(X), \end{array}$$

where T_V^k is defined by $T_V^k(x)=\psi^k(x)\rho^k(V)$.

Proof. This follows from 7.13 and the fact that φ_V is a $K(X)$-module isomorphism. □

7.18. Theorem. *Let V and V' be complex vector bundles over X and X', respectively. Then $\rho^k(V\times V')=\rho^k(V)\rho^k(V')$. In particular, if $X=X'$, we have $\rho^k(V\oplus V')=\rho^k(V)\rho^k(V')$ in the ring $K(X)$.*

Proof. We have the commutative diagram

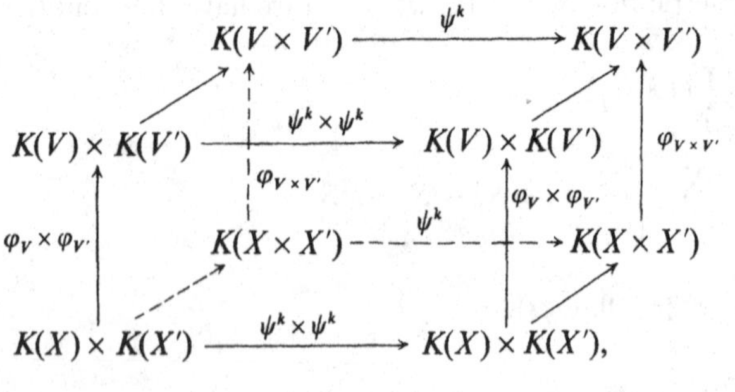

where the slanting arrows represent the cup-product. □

7.19. Corollary. *Let x be any element of $K(\mathbb{R}^{2n}) \approx \tilde{K}(S^{2n})$. Then $\psi^k(x) = k^n x$.*

Proof. We regard $\mathbb{C}^n$ as a vector bundle over a point. Then $\varphi_{\mathbb{C}^n}(1)$ is a generator of $K(\mathbb{C}^n) \approx \tilde{K}(S^{2n}) \approx \mathbb{Z}$, and we must verify that $\rho^k(\mathbb{C}^n) = k^n$. By the theorem above, it is enough to verify $\rho^k(\mathbb{C}^n) = k^n$ for $n=1$, i.e. to verify that $\rho^k(x) = kx$ for each element x of $\tilde{K}(S^2)$. But $\tilde{K}(S^2)$ is generated by $\xi - 1$, where ξ is the Hopf line bundle (III.1.1 and 2.5). Since $(\xi-1)^2 = 0$, we may write $\psi^k(\xi-1) = \xi^k - 1 = (1+(\xi-1))^k - 1 = k(\xi-1)$. □

7.20. Proposition. *The correspondence $V \mapsto \rho^k(V)$ from the set of isomorphism classes of complex vector bundles over X, to the group $K(X)$, is characterized by the following properties:*

1) *$\rho^k(f^*(V)) = f^*(\rho^k(V))$ for any continuous map $f: X' \to X$ (in other words ρ^k is "natural").*
2) *$\rho^k(V \oplus V') = \rho^k(V)\rho^k(V')$.*
3) *$\rho^k(L) = 1 + L + \cdots + L^{k-1}$ if L is a line bundle.*

Proof. By Theorem I.7, it is enough to verify Relation 3) when X is the projective space CP_n, and L is the canonical line bundle over CP_n. Moreover, the Thom space $\dot{L}$ may be identified with CP_{n+1} (2.3), and the Thom isomorphism

$$\begin{array}{ccc} K(CP^n) & \xrightarrow{\varphi_L} & \tilde{K}(CP_{n+1}) \\ \wr\wr & & \wr\wr \\ \mathbb{Z}[u]/u^{n+1} & & \mathbb{Z}[u]/u^{n+2} \end{array}$$

is simply defined by the product with u (cf. 2.4). Therefore

$$\rho^k(L) = \varphi_L^{-1}(\psi^k(1-L)) = (1-L^k)/(1-L) = 1 + L + \cdots + L^{k-1}. \quad \square$$

7.21. *Example.* If $k=2$, then the class $\rho^k(V)$ coincides with the class $\Lambda(V) = 1 + V + \lambda^2(V) + \lambda^3(V) \cdots$, because $\Lambda(V)$ satisfies Axioms 1), 2), and 3), of Proposition 7.20.

7.22. Let $\mathbb{Q}_k$ be the subring of $\mathbb{Q}$, consisting of fractions with denominator a power of k. We define an operation

$$\rho^k: K(X) \longrightarrow K(X) \otimes_{\mathbb{Z}} \mathbb{Q}_k$$

by the formula $\rho^k(x) = \rho^k(V)/k^n$, for each element x of $K(X)$, where x is written in the form $V - n$. Since $\rho^k(n) = k^n$ for each integer n (7.19), this operation is well-defined.

As an important example of a computation of the operation ρ^k, let us consider the *real* projective space RP_{n-1}. By 6.47 we see that $K(RP_{n-1}) = K_{\mathbb{C}}(RP_{n-1})$ is the quotient of the algebra $\mathbb{Z}[\lambda']$ by the relations $2^g\lambda' = 0$ and $(\lambda')^2 = -2\lambda'$, where g is the number of integers i such that $0<i<n$ and $i \equiv 0 \bmod 2$. In fact $\lambda' = \xi' - 1$, where ξ' is the complexification of the canonical line bundle.

7.23. Proposition. *For k odd, the operation*

$$\rho^k: K_{\mathbb{C}}(RP_{n-1}) \longrightarrow K_{\mathbb{C}}(RP_{n-1}) \otimes \mathbb{Q}_k$$

is determined by the relation

$$\rho^k(l\lambda') = 1 + \frac{k^l - 1}{2k^l}\lambda'.$$

Proof. For $l=1$, we have $\rho^k(\lambda') = \rho^k(\xi')/k = \frac{1}{k}(1+\xi'+\cdots+\xi'^{k-1})$. Since $\xi'^2 = 1$, we see that

$$\rho^k(\xi') = \frac{1}{k}\left(\frac{k-1}{2} + \frac{k-1}{2}\lambda'\right) = 1 + \frac{k-1}{2k}\lambda'.$$

By induction on l, we obtain the formula

$$\rho^k(l\lambda') = \left(1 + \frac{k-1}{2k}\lambda'\right)^l = 1 + \frac{k^l-1}{2k^l}\lambda'$$

as desired. □

7.24. Let us now examine operations in *real* K-theory. The operations λ^i, γ^i, λ_t, and γ_t, may be defined without difficulty. The Adams operations $\psi^k: K_{\mathbb{R}}(X) \to K_{\mathbb{R}}(X)$, for $k \in \mathbb{N}$, are defined by the formula $\psi^k(x) = Q_k(\lambda^1(x), \ldots, \lambda^k(x))$, where Q_k is the Newton polynomial. Then we have $\psi^k(x+y) = \psi^k(x) + \psi^k(y)$, and $\psi^k(L) = L^k$ if L is a real line bundle. *The only nontrivial results are the relations* $\psi^k(xy) = \psi^k(x) \cdot \psi^k(y)$ *and* $\psi^k(\psi^l(x)) = \psi^{kl}(x)$, *which will be proved at the end of this section with the aid of representation theory* (another proof is presented in Exercise 8.5).

7.25. Proposition. *We have the following commutative diagrams,*

$$\begin{array}{ccc} K_{\mathbb{R}}(X) & \xrightarrow{c} & K_{\mathbb{C}}(X) \\ \downarrow{\scriptstyle \lambda^k} & & \downarrow{\scriptstyle \lambda^k} \\ K_{\mathbb{R}}(X) & \xrightarrow{c} & K_{\mathbb{C}}(X) \end{array} \quad \text{and} \quad \begin{array}{ccc} K_{\mathbb{R}}(X) & \xrightarrow{c} & K_{\mathbb{C}}(X) \\ \downarrow{\scriptstyle \psi^k} & & \downarrow{\scriptstyle \psi^k} \\ K_{\mathbb{R}}(X) & \xrightarrow{c} & K_{\mathbb{C}}(X), \end{array}$$

where the horizontal arrows are the complexification homomorphisms.

Proof. The natural homomorphism $\lambda^i(E) \otimes_{\mathbb{R}} \mathbb{C} \to \lambda^i(E \otimes_{\mathbb{R}} \mathbb{C})$, where E is a real vector bundle, is an isomorphism. For each element x of $K_{\mathbb{R}}(X)$, we see that

$c(\lambda^i(x))=\lambda^i(c(x))$. Since ψ^k is a polynomial of the λ^i, for $i\leqslant k$, and since c is a ring homomorphism, we also have $c(\psi^k(x))=\psi^k(c(x))$. $\square$

7.26. Consider a real vector bundle V of rank $8n$, provided with a spinorial structure. By 5.12, we have a Thom isomorphism

$$\varphi_V\colon K_{\mathbb{R}}(X)\to K_{\mathbb{R}}(V).$$

Now we define $\rho_{\mathbb{R}}^k(V)$ as the class $\varphi_V^{-1}(\psi^k(U_V))$, where $U_V=\varphi_V(1)$ is the Thom class of V. By 7.24, we have $\rho_{\mathbb{R}}^k(V\oplus V')=\rho_{\mathbb{R}}^k(V)\rho_{\mathbb{R}}^k(V')$. We also have the commutative diagram

$$\begin{array}{ccc}
K_{\mathbb{R}}(V) & \xrightarrow{\psi^k} & K_{\mathbb{R}}(V)\\
\uparrow{\scriptstyle\varphi_V} & & \uparrow{\scriptstyle\varphi_V}\\
K_{\mathbb{R}}(X) & \xrightarrow{T_V^k} & K_{\mathbb{R}}(X),
\end{array}$$

where $T_V^k(x)=\psi^k(x)\rho_{\mathbb{R}}^k(V)$. To avoid confusion, we let $\rho_{\mathbb{C}}^k$ denote the class ρ^k, defined in complex K-theory (7.16).

7.27. Proposition. *Let V be an oriented real vector bundle of rank $4p$. Then $W=V\oplus V$ may be provided with a complex structure and a spinorial structure, and we have the relation*

$$c(\rho_{\mathbb{R}}^k(W_{\mathbb{R}}))=\rho_{\mathbb{C}}^k(W),$$

where $W_{\mathbb{R}}$ is the spinorial bundle underlying W.

Proof. By elementary observations, we have the commutative diagram

$$\begin{array}{ccc}
\mathrm{Spin}(8p) & \longrightarrow & \mathrm{Spin}^c(8p)\\
\uparrow & & \uparrow\\
\mathrm{SO}(4p) & \longrightarrow & \mathrm{U}(8p)
\end{array}$$

of group homomorphisms, where the vertical maps are defined as in 4.18 and 4.26. It follows from 5.12 that we have the commutative diagram of Thom isomorphisms

$$\begin{array}{ccc}
K_{\mathbb{R}}(X) & \xrightarrow{\varphi_{W_{\mathbb{R}}}} & K_{\mathbb{R}}(W_{\mathbb{R}})\\
\downarrow{\scriptstyle c} & & \downarrow{\scriptstyle c}\\
K_{\mathbb{C}}(X) & \xrightarrow{\varphi_W} & K_{\mathbb{C}}(W).
\end{array}$$

Therefore $c(U_{W_{\mathbb{R}}})=U_W$. Since the operation ψ^k commutes with complexification (7.25), the proposition follows. □

7.28. Corollary. *Let V be a spinorial bundle of rank $8r$. Then $c(\rho^k_{\mathbb{R}}(V)^2)=\rho^k_{\mathbb{C}}(W)$, where $W=V\otimes_{\mathbb{R}}\mathbb{C}$ is the complexification of V.*

∗ **7.29.** *Remark.* Let K spin(X) denote the subgroup of $K_{\mathbb{R}}(X)$, generated by spinorial bundles of rank$\equiv 0$ mod 8. By 4.20, it may be identified with the subgroup of $K_{\mathbb{R}}(X)$ consisting of elements x, such that rank$(x)\equiv 0$ mod 8, and $w_1(x)=w_2(x)=0$. If $x=[V]-[V']\in K\,\mathrm{spin}(X)$, then the spinorial structures of V and V' are well-defined up to multiplication by a line bundle. Therefore, if k is odd, the class $\rho^k(x)\in K_{\mathbb{R}}(X)\otimes_{\mathbb{Z}}\mathbb{Q}_k$ is well-defined. On the other hand, if k is even, then $\rho^k(x)$ is defined only up to multiplication by a line bundle. However, if $H^1(X;\mathbb{Z}/2)=0$, then all line bundles are trivial, and hence $\rho^k(x)$ is also well-defined in this case.∗

7.30. Proposition. *Let ξ be the canonical line bundle over RP_{n-1}, and let k be odd. Then*

$$\rho^k(4l\xi+4l)=k^{4l}\left(1+\frac{k^{2l}-1}{2k^{2l}}\lambda\right),$$

where $\lambda=\xi-1$.

Proof. Since the homomorphism $K_{\mathbb{R}}(RP_{m-1})\to K_{\mathbb{R}}(RP_{n-1})$ is surjective for $m\geqslant n$ (6.44), it is enough to consider the case $n-1\equiv 0$ mod 8. In this case, the complexification homomorphism

$$K_{\mathbb{R}}(RP_{n-1})\longrightarrow K_{\mathbb{C}}(RP_{n-1})$$

is an isomorphism (6.46). Since $2\xi=\xi\oplus\xi$ is oriented, Proposition 7.28 enables us to write $c(\rho^k_{\mathbb{R}}(4l\xi\oplus 4l))=\rho^k_{\mathbb{C}}(2l\xi'+2l)$, where $\xi'=\xi\otimes_{\mathbb{R}}\mathbb{C}$. By 7.23, we therefore have $c(\rho^k_{\mathbb{R}}(4l\xi+4l))=\rho^k_{\mathbb{C}}(2l\lambda'+4l)=k^{4l}\left(1+\frac{k^{2l}-1}{2k^{2l}}\lambda'\right)$. Since $c(\lambda)=\lambda'$, the relation follows. □

7.31. We now return to the relations which we described in 7.24. For this, we need some basic facts in the representation theory of compact Lie groups (the general reference for this material is Adams [3]).

Let G be a compact Lie group. We let $R_F(G)$, where $F=\mathbb{R}$ or $\mathbb{C}$, denote the Grothendieck group of the category of finite F-dimensional representations of G. In fact, $R_F(G)$ is the free group with basis, the set of irreducible representations of G. The composition $R_{\mathbb{R}}(G)\xrightarrow{c}R_{\mathbb{C}}(G)\xrightarrow{r}R_{\mathbb{R}}(G)$, where c is complexification, and r is realification, is multiplication by 2. Therefore $R_{\mathbb{R}}(G)$ may be regarded as a subgroup of $R_{\mathbb{C}}(G)$.

7.32. To each representation $\rho\colon G \to \mathrm{Aut}(V)$, we associate its character $\chi = \chi_\rho\colon G \to F$, defined by $\chi(g) = Tr(\rho(g))$. Note that $\chi(tgt^{-1}) = \chi(g)$. The map from $R_F(G)$ to $\mathscr{F}(G, F)$ (the space of continuous maps from G to F), which associates each representation ρ with its character, is injective (Adams [3]). Now let $\lambda^k(\rho)$ be the representation of G in $\lambda^k(V)$, defined by $\lambda^k(\rho)(g) = \lambda^k(\rho(g))\colon \lambda^k(V) \to \lambda^k(V)$. We define $\psi^k(\rho)$ as the element $Q_k(\lambda^1(\rho), \ldots, \lambda^k(\rho)) \in R_F(G)$, where Q_k is the Newton polynomial. Note that $\psi^k(\rho+\sigma) = \psi^k(\rho) + \psi^k(\sigma)$ by the same argument as in 7.15.

7.33. Proposition. *Let $\chi_\rho\colon G \to k$ be the character of the representation ρ. Then the character of $\psi^k(\rho)$ is $g \mapsto \chi_\rho(g^k)$* (cf. Adams [3]).

Proof. Since the map from $R_{\mathbb{R}}(G)$ to $R_{\mathbb{C}}(G)$ is injective, and since $\lambda^i(\rho \otimes \mathbb{C}) = \lambda^i(\rho) \otimes \mathbb{C}$, we may restrict ourselves to the case $F = \mathbb{C}$. Since G is compact, we may factor ρ into $G \to \mathrm{U}(n) \rightarrowtail \mathrm{Aut}(\mathbb{C}^n)$ with respect to a suitable isomorphism $\mathbb{C}^n \approx V$. If $t_1, \ldots, t_n$ are the eigenvalues of $\rho(g)$, we have $Tr(\rho(g)) = t_1 + \cdots + t_n$, $Tr(\lambda^2(\rho)(g)) = \sum_{i<j} t_i t_j$, and $Tr(\lambda^k(\rho)(g)) = \sigma_k$, where σ_k is the kth elementary symmetric function of the t_i. Hence $Tr(\psi^k(\rho)(g)) = Q_k(\sigma_1, \ldots, \sigma_k) = Tr(\rho(g)^k) = Tr(\rho(g^k)) = \chi_\rho(g^k)$. □

7.34. Proposition. *Let ρ and σ be representations of G. Then $\psi^k(\rho\sigma) = \psi^k(\rho)\psi^k(\sigma)$ in the ring $R_F(G)$. Hence $\psi^k\colon R_F(G) \to R_F(G)$ is a ring homomorphism.*

Proof. We need only verify the formula $\chi_{\psi^k(\rho\sigma)}(g) = \chi_{\psi^k(\rho)}(g)\chi_{\psi^k(\sigma)}(g)$. Since $\chi_{\psi^k(\rho)}(g) = \chi_\rho(g^k)$ and $\chi_{\psi^k(\sigma)} = \chi_\sigma(g^k)$, we have $\chi_{\psi^k(\rho\sigma)}(g) = \chi_{\rho\sigma}(g^k) = \chi_\rho(g^k)\chi_\sigma(g^k) = \chi_{\psi^k(\rho)}(g) \cdot \chi_{\psi^k(\sigma)}(g)$. □

7.35. *Remark.* If G_1 and G_2 are compact Lie groups, we have the bilinear pairing

$$R_F(G_1) \times R_F(G_2) \longrightarrow R_F(G_1 \times G_2),$$

denoted by $(\rho_1, \rho_2) \mapsto \rho_1 \cup \rho_2$ (this is the "external" tensor product of representations). If $\pi_i\colon G_1 \times G_2 \to G_i$ are the obvious projections, we have $\rho_1 \cup \rho_2 = \pi_1^*(\rho_1) \cdot \pi_2^*(\rho_2)$ in $R_F(G_1 \times G_2)$. From this, we arrive at the formula $\psi^k(\rho_1 \cup \rho_2) = \psi^k(\rho_1) \cup \psi^k(\rho_2)$.

7.36. Proposition. *We have the formula $\psi^k(\psi^l(\rho)) = \psi^{kl}(\rho)$ in $R_F(G)$.*

Proof. $\chi_{\psi^k(\psi^l(\rho))}(g) = \chi_{\psi^k}(g^l) = \chi((g^l)^k) = \chi(g^{lk}) = \chi_{\psi^{kl}(\rho)}(g)$. □

7.37. Theorem. *Let $c\colon R_{\mathbb{R}}(G) \to R_{\mathbb{C}}(G)$ (resp. $r\colon R_{\mathbb{C}}(G) \to R_{\mathbb{R}}(G)$) be the complexification homomorphism (resp. realification homomorphism). Then we have two* *commutative diagrams,*

$$\begin{array}{ccc} R_{\mathbb{R}}(G) & \xrightarrow{\psi^k} & R_{\mathbb{R}}(G) \\ \downarrow c & & \downarrow c \\ R_{\mathbb{C}}(G) & \xrightarrow{\psi^k} & R_{\mathbb{C}}(G) \end{array} \quad \text{and} \quad \begin{array}{ccc} R_{\mathbb{C}}(G) & \xrightarrow{\psi^k} & R_{\mathbb{C}}(G) \\ \downarrow r & & \downarrow r \\ R_{\mathbb{R}}(G) & \xrightarrow{\psi^k} & R_{\mathbb{R}}(G). \end{array}$$

Proof. Since c is a ring map, the commutativity of the first diagram follows from the identity $\lambda^i(\rho \otimes \mathbb{C}) = \lambda^i(\rho) \otimes \mathbb{C}$. Now let $\sigma: G \longrightarrow \operatorname{Aut}(V)$ be a complex representation, and let ρ be its underlying real representation. We have $\chi_\rho(g) = \chi_\sigma(g) + \overline{\chi_\sigma(g)}$, hence $\chi_{\psi^k(\rho)}(g) = \chi_\sigma(g^k) + \overline{\chi_\sigma(g^k)} = \chi_{r(\psi^k(\sigma))}(g)$. □

7.38. We now return to our original problem, which was the investigation of the properties of the Adams operations

$$\psi^k: K_{\mathbb{R}}(X) \longrightarrow K_{\mathbb{R}}(X).$$

In fact, our method also applies to the complex Adams operations $\psi^k: K_{\mathbb{C}}(X) \to K_{\mathbb{C}}(X)$, and provides another proof of Theorems 7.13 and 7.15, which is more "elementary" in style.

If E is a real vector bundle, we may write $E = P \times_G \mathbb{R}^n$, where P is a principal G bundle (4.14), and where G acts on $\mathbb{R}^n$ by a representation $\rho: G \to O(n)$. For example, we could choose $G = O(n)$ and P, the principal $O(n)$-bundle associated with E (4.14). Thus we may write $\lambda^k(E) = P \times_G \lambda^k(\mathbb{R}^n)$, where G acts on $\lambda^k(\mathbb{R}^n)$ in the natural way. More generally, if $V = V^+ - V^-$ is an element of $R_{\mathbb{R}}(G)$, where V^+ and V^- are orthogonal representations, we have a well-defined element $E_V = [P \times_G V^+] - [P \times_G V^-]$ of $K_{\mathbb{R}}(X)$; we simply write $E_V = P \times_G V$. In particular, $\psi^k(E)$ is E_V, where $V = \psi^k(\rho)$.

If $E_1 = P_1 \times_{O(n_1)} \mathbb{R}^{n_1}$ and $E_2 = P_2 \times_{O(n_2)} \mathbb{R}^{n_2}$, we have

$$E_1 \otimes E_2 = (P_1 \times_X P_2) \times_{O(n_1) \times O(n_2)} (\mathbb{R}^{n_1} \otimes \mathbb{R}^{n_2}),$$

where $\mathbb{R}^{n_1} \otimes \mathbb{R}^{n_2}$ is an $O(n_1) \times O(n_2)$-module in the natural way. Let us write $\psi^k(\mathbb{R}^{n_1}) = V_1^+ - V_1^-$ and $\psi^k(\mathbb{R}^{n_2}) = V_2^+ - V_2^-$. By 7.35, we have

$$\begin{aligned} \psi^k(\mathbb{R}^{n_1} \otimes \mathbb{R}^{n_2}) &= \psi^k(\mathbb{R}^{n_1}) \cup \psi^k(\mathbb{R}^{n_2}) \\ &= V_1^+ \otimes V_2^+ + V_1^- \otimes V_2^- - V_1^+ \otimes V_2^- - V_1^- \otimes V_2^+. \end{aligned}$$

Therefore, if we set $P = P_1 \times_X P_2$ and $G = O(n_1) \times O(n_2)$, we have

$$\begin{aligned} \psi^k(E_1 \otimes E_2) &= [P \times_G V_1^+ \otimes V_2^+] + [P \times_G V_1^- \otimes V_2^-] - [P \times_G V_1^+ \otimes V_2^-] \\ &\quad - [P \times_G V_1^- \otimes V_2^+] \\ &= (P_1 \times_{O(n_1)} V_1^+) \otimes (P_2 \times_{O(n_2)} V_2^+) + (P_1 \times_{O(n_1)} V_1^-) \otimes (P_2 \times_{O(n_2)} V_2^-) \\ &\quad - (P_1 \times_{O(n_1)} V_1^+) \otimes (P_2 \times_{O(n_2)} V_2^-) - (P_1 \times_{O(n_1)} V_1^-) \otimes (P_2 \times_{O(n_2)} V_2^+) \\ &= (P_1 \times_{O(n_1)} (V_1^+ - V_1^-)) \otimes (P_2 \times_{O(n_2)} (V_2^+ - V_2^-)) = \psi^k(E_1) \otimes \psi^k(E_2). \end{aligned}$$

Since $\psi^k: K_{\mathbb{R}}(X) \to K_{\mathbb{R}}(X)$ is a group homomorphism, we have the relation $\psi^k(xy) = \psi^k(x)\psi^k(y)$ for all $x, y \in K_{\mathbb{R}}(X)$.

7.39. If $E = P \times_{O(n)} \mathbb{R}^n$, we have $\psi^l(E) = P \times_{O(n)} V^+ - P \times_{O(n)} V^-$, where $\psi^l(\mathbb{R}^n) = V^+ - V^+$ in $R_{\mathbb{R}}(O(n))$. Similarly

$$\psi^k(\psi^l(E)) = P \times_{O(n)} W^+ - P \times_{O(n)} W^- - P \times_{O(n)} T^+ + P \times_{O(n)} T^-,$$

where $W^+ - W^- = \psi^k(V^+)$ and $T^+ - T^- = \psi^k(V^-)$ in $R_{\mathbb{R}}(O(n))$. Since $\psi^k(\psi^l(\mathbb{R}^n)) = \psi^{kl}(\mathbb{R}^n) = W^+ - W^- - T^+ + T^-$ (7.37), we have $\psi^k(\psi^l(E)) = \psi^{kl}(E)$. By additivity, we see that $\psi^k(\psi^l(x)) = \psi^{kl}(x)$ for any $x \in K_{\mathbb{R}}(X)$.

7.40. Proposition. *The following diagram*

$$\begin{array}{ccc} K_{\mathbb{C}}(X) & \xrightarrow{\psi^k} & K_{\mathbb{C}}(X) \\ \downarrow r & & \downarrow r \\ K_{\mathbb{R}}(X) & \xrightarrow{\psi^k} & K_{\mathbb{R}}(X) \end{array}$$

where r is the realification homomorphism, is commutative.

Proof. Let $E = P \times_{U(n)} \mathbb{C}^n$ be a complex vector bundle (4.14). Then $r(\psi^k(E)) = P \times_{U(n)} \psi^k(\mathbb{C}^n)$, where $\psi^k(\mathbb{C}^n)$ is regarded as an element of $R_{\mathbb{R}}(U(n))$. On the other hand, $\psi^k(r(E)) = P \times_{U(n)} \psi^k(\mathbb{C}^n)$, where $\mathbb{C}^n$ is regarded as the *real* $U(n)$-module $\mathbb{R}^{2n}$. Since the diagram

$$\begin{array}{ccc} R_{\mathbb{C}}(U(n)) & \xrightarrow{\psi^k} & R_{\mathbb{C}}(U(n)) \\ \downarrow r & & \downarrow r \\ R_{\mathbb{R}}(U(n)) & \xrightarrow{\psi^k} & R_{\mathbb{R}}(U(n)) \end{array}$$

is commutative (7.37), $\psi^k(\mathbb{R}^{2n}) = \psi^k(\mathbb{C}^n)$ as an element of $R_{\mathbb{R}}(U(n))$; hence $r(\psi^k(E)) = \psi^k(r(E))$, and by additivity, $r(\psi^k(x)) = \psi^k(r(x))$ for each element x of $K_{\mathbb{C}}(X)$. □

Exercises (Section IV.8) 5–7, 15, 16.

8. Exercises

***8.1.** Let E be an oriented bundle. Compute the number of spinorial structures (resp. ^{c}spinorial structures) which may be placed on E if $w_2(E) = 0$ (resp. if

$\beta(w_2(E))=0$, where $\beta: H^2(X;\mathbf{Z}/2)\to H^3(X;\mathbf{Z}/2)$ is the Bockstein function). Consider, in particular, the case where X is simply connected.*

8.2. Let $\rho: \mathrm{U}(n)\to \mathrm{U}(n)/\mathrm{U}(n-1)\approx S^{2n-1}$ be the canonical projection, and let γ be the canonical generator of $K_{\mathbf{C}}^{-1}(S^{2n-1})$. Now show that, up to sign, $\rho^*(\gamma)$ is the alternating sum $\sum_{i=0}^{n}(-1)^i\lambda^i$, where λ^i is the element of $K_{\mathbf{C}}^{-1}(\mathrm{U}(n))$ induced by the ith exterior power of the canonical representation $\mathrm{U}(n)\to \mathrm{Aut}(\mathbf{C}^n)$ (cf. II.3.17).

By induction on n, show that $K_{\mathbf{C}}^{*}(\mathrm{U}(n))$ is the exterior algebra generated by $\lambda^1,\ldots,\lambda^n$. (Apply Theorem 1.3.)

8.3. Let V be a Real vector bundle of rank n in the sense of III.7.13. Using the techniques developed in IV.2 and in Exercises III.7.13 and III.7.14, show that $KR(P(V))$ is a free $KR(X)$-module of rank $n=\mathrm{Rank}(V)$, with basis $1,h,h^2,\ldots,h^{n-1}$, where h is the class of the canonical line bundle over $P(V)$ (this bundle is a Real bundle). Moreover, prove the relation

$$h^n-\lambda^1(V)h^{n-1}+\lambda^2(V)h^{n-2}+\cdots+(-1)^n\lambda^n(V)=0 \quad \text{(compare with 2.16).}$$

8.4. (8.3. continued.) By the same method as in section 3, compute $KR(F(V))$ and $KR(G_k(V))$. In particular, prove that the canonical map $KR(X)\to KR(F(V))$ is injective, and that π^*V is the sum of Real line bundles (splitting principle for KR-theory).

8.5. Prove the existence and uniqueness of the Adams operations, $\psi^k: KR(X)\to KR(X)$, defined for any compact space X provided with an involution such that $\psi^k(x+y)=\psi^k(x)+\psi^k(y)$, and $\psi^k(L)=L^k$ when L is a Real line bundle. If X is provided with the trivial involution, and if $KR(X)$ is identified with $K_{\mathbf{R}}(X)$, prove that these Adams operations coincide with those defined in 7.24. Finally, show that ψ^k is a ring homomorphism (this gives another proof of the results in 7.38).

8.6. Let T^n be the n-dimensional torus regarded as a topological group. Now prove that $R_{\mathbf{C}}(T^n)$ is the algebra of Laurent polynomials $\mathbf{Z}[t_1,\ldots,t_n,t_1^{-1},\ldots,t_n^{-1}]$, where t_i is the class of the one-dimensional representation $T^n \xrightarrow{n^{th}\text{ projection}} T^1=\mathrm{U}(1)\to \mathrm{Aut}(\mathbf{C})$.

If $\mathscr{I}$ denotes the ideal of $R_{\mathbf{C}}(T^n)$ which is $\mathrm{Ker}[R_{\mathbf{C}}(T^n)\xrightarrow{\varepsilon}\mathbf{Z}]$ with $\varepsilon(t_i)=1$, show that the completion $\hat{R}_{\mathbf{C}}(T^n)$ of $R_{\mathbf{C}}(T^n)$ with respect to the $\mathscr{I}$-adic topology is $\mathbf{Z}[[x_1,\ldots,x_n]]$, where $x_i=t_i-1$ (by definition $\hat{R}_{\mathbf{C}}(T^n)=\mathrm{proj\,lim}\, R_{\mathbf{C}}(T^n)/\mathscr{I}^p$).

Finally, prove the existence of an isomorphism

$$\mathbf{Z}[[x_1,\ldots,x_n]]\approx\hat{R}_{\mathbf{C}}(T^n)\xrightarrow{\approx}\mathscr{K}_{\mathbf{C}}(\overbrace{BU(1)\times BU(1)\times\cdots\times BU(1)}^{n\text{-times}}).$$

8.7. Let $U(n)$ be the unitary group of rank n regarded as a topological group. Now prove that $R_{\mathbf{C}}(\mathrm{U}(n))$ is the algebra $\mathbf{Z}[\lambda^1,\ldots,\lambda^n,(\lambda^n)^{-1}]$, where λ^i is the ith exterior power of the natural representation $U(n) \to \mathrm{Aut}(\mathbb{C}^n)$. By the same method as in 8.6, prove the existence of an isomorphism

$$\mathbb{Z}[[y_1, \ldots, y_n]] \approx \hat{R}_{\mathbb{C}}(U(n)) \xrightarrow{\approx} \mathscr{K}_{\mathbb{C}}(BU(n)).$$

8.8. With the help of Theorem 6.40, completely compute the groups $K^i_{\mathbb{R}}(RP_n, RP_m)$ and $K^i_{\mathbb{C}}(RP_n, RP_m)$.

8.9. Let $S \xrightarrow{\pi} X$ be a spherical fibration between compact manifolds with $\mathrm{Dim}(S) - \mathrm{Dim}(X) \equiv 0 \bmod 8$, and let $i: X \to S$ be a section of π. If $TX - i^*(TS)$ is provided with a stable spinorial structure, prove that $K^*_{\mathbb{R}}(S)$ is a free $K^*_{\mathbb{R}}(X)$-module of rank 2. State and solve the analogous problem in complex K-theory.

8.10. Let HP_n be the projective space of $\mathbb{H}^{n+1}$ (where $\mathbb{H}$ is the field of quaternions). Show that $K_{\mathbb{C}}(HP_n) \approx \mathbb{Z}[\beta]/\beta^{n+1}$, where β is the class of the canonical bundle over HP_n, regarded as a complex bundle of rank 2. If $\pi: CP_{2n+1} \to HP_n$ is the obvious map, show that the homomorphism $K_{\mathbb{C}}(HP_n) \xrightarrow{\pi^*} K_{\mathbb{C}}(CP_{2n+1})$ is injective, and that $\pi^*(\beta) = h + \bar{h}$, where h is the class of the canonical line bundle over CP_{2n+1}. Compute $\psi^k(\beta)$. Make analogous computations for $K_{\mathbb{R}}(HP_n)$ and $K_{\mathbb{H}}(\mathrm{HP}_n)$.

8.11. Let G be a finite abelian group, and let V be a complex G-bundle over the G-space X (in the sense of I.9.29), which is the sum of line G-bundles. If K_G denotes complex equivariant K-theory (I.9.30), show that $K_G(V)$ is a free $K_G(X)$-module of rank one, generated by the Thom class described in 1.6, where G acts on all the bundles involved. If G acts freely on the sphere bundle $S(V)$, and if X is a point, prove the exact sequence

$$R_{\mathbb{C}}(G) \xrightarrow{\sigma} R_{\mathbb{C}}(G) \longrightarrow K_{\mathbb{C}}(S(V)/G) \longrightarrow 0,$$

where σ is multiplication by $\sum_{i=0}^{\mathrm{Dim}(V)} (-1)^i \lambda^i(V)$, with $\lambda^i(V)$ regarded as an element of $R_{\mathbb{C}}(G)$.

8.12. Let X be a space of finite type (3.23). Then we define the Euler-Poincaré characteristic $\chi(X)$ of the space X as $\mathrm{Dim}\, K^0_{\mathbb{C}}(X) \otimes \mathbb{Q} - \mathrm{Dim}\, K^{-1}_{\mathbb{C}}(X) \times \mathbb{Q}$ (by V.3.25, this is the usual Euler-Poincaré characteristic). Now prove the following well known properties of χ:

a) $\chi(X_1 \cup X_2) + \chi(X_1 \cap X_2) = \chi(X_1) + \chi(X_2)$.

b) $\chi(X \times Y) = \chi(X)\chi(Y)$.

c) More generally, if $E \to B$ is a fibration with fiber F, then $\chi(E) = \chi(B) \cdot \chi(F)$.

Conclude from the last fact that if a finite group G acts freely on a compact space X of finite type, and if X/G is of finite type, then the order of G divides the Euler-Poincaré characteristic of X.

8.13. Let V be an oriented vector bundle of rank n over a compact space X. Prove that $K(X) \otimes \mathbb{Z}[\frac{1}{2}] \approx K^n(V) \otimes \mathbb{Z}[\frac{1}{2}]$ as $K(X)$-modules (cf. the author [2]).

$*$ **8.14.** Let X be a locally compact space, and let Φ be a family of closed subsets of X such that:

1) Any finite union of elements of Φ belongs to Φ.
2) A closed subset of an element of Φ belongs to Φ.
3) Each element of Φ has a neighbourhood in the family Φ.

Then using the material developed in II.5 show how to define in a "reasonable way" complex K-theory with support in Φ (denoted $K_\Phi(X)$) and prove that $K_\Phi(X) \otimes \mathbb{Q} \approx H_\Phi^{\text{even}}(X; \mathbb{Q})$, where H_Φ denotes cohomology with supports in Φ.$*$

8.15. Let S^{2n} be the sphere of dimension $2n$

a) If $x \in K'_{\mathbb{C}}(S^{2n})$, prove that $\lambda_t(x)$ may be written as

$$1+\left[\frac{(-1)^{n-1}(n-1)!\,t^{n-1}}{(1+t)^n} f_n(t)-1\right]x,$$

where $f_n \in \mathbb{Z}[t]$ and $f_n(-1)=1$.

b) Deduce from a) that for any complex vector bundle of rank n on S^{2n}, we have $\lambda_{-1}(E)=(n-1)!\,x$, with $x=[E]-n$.

c) Let M be an irreducible complex $C^{0,2n}$-module, and let ε be a gradation of M (IV.5.2). The pair (M, ε) may be regarded as a $C^{0,2n+1}$-module, which we again denote by M. Over each point $w \in S^{2n} \subset \mathbb{R}^{2n+1}$, Clifford multiplication by vectors which are orthogonal to w, defines a $C(V)$-module structure on π^*M, where $\pi: S^{2n} \to$ Point and $V = TS^{2n}$ is the tangent bundle. On the other hand, Clifford multiplication by w defines a gradation of the $C(V)$-module π^*M, which we denote by η. Therefore, the triple $(\pi^*M, \eta, -\eta)$ defines an element of the group $K_{\mathbb{C}}^V(X)$ described in IV.5.1 where $X = S^{2n}$.

Now prove that the image of this element by the forgetful homomorphism

$$K_{\mathbb{C}}^V(X) \to K_{\mathbb{C}}(X) \to \widetilde{K}_{\mathbb{C}}(X),$$

is twice a generator of $\widetilde{K}_{\mathbb{C}}(X) \approx \mathbb{Z}$.

Also prove that this element is a generator of $K_{\mathbb{C}}^V(X) \approx K_{\mathbb{C}}(V) \approx \mathbb{Z}$ (IV.6.21).

d) Deduce from b) and c) that TS^{2n} may be provided with a complex structure only if $n=1$ or 3. (This exercise gives a purely K-theoretical proof of an old result of Borel-Serre [1]. Another proof will be given in V.3.)

8.16. Let X be a compact space such that $K(X)$ is generated by line bundles L_i, with $(L_i)^p = 1$ for p a fixed prime. Show that the torsion of $K(X)$ is p-primary.

9. Historical Note

The Thom isomorphism in complex K-theory is a key tool in the Atiyah-Hirzebruch Riemann-Roch theorem (for maps between manifolds provided with an almost complex structure). Thom isomorphism in real K-theory has analogous applications, and is due to Atiyah, Bott and Shapiro [1].

The computation of complex K-theory for complex projective bundles, flag bundles and Grassmann bundles, is based on the work of Atiyah [3] and Grothendieck [2]. The "Kunneth formula" in IV.3 is taken from Atiyah [2].

The computation of the K-theory of real projective space is essential to the solution of the vector field problem on the sphere (V.3). The result (due to Adams [1]) is generalized in IV.6 to real projective bundles.

Finally, the notion of operations in K-theory, which is important in applications, is due to Grothendieck [1], Adams [4] and Atiyah [5].

Chapter V

Some Applications of K-Theory

1. H-Space Structures on Spheres and the Hopf Invariant

1.1. Let $\alpha: S^{n-1} \to S^{n-1}$ be a continuous map. Then $\tilde{K}_{\mathbf{C}}^{n-1}(S^{n-1}) \approx \mathbf{Z}$, and α induces an endomorphism of $\mathbf{Z}$ of the form $x \mapsto \lambda x$ where $\lambda \in \mathbf{Z}$. The integer λ is called the *degree* of α, and is denoted by deg(α). In particular, if α is a homeomorphism, we have $\deg(\alpha) = \pm 1$. It is possible to prove (cf. Hu [1]) that $\pi_{n-1}(S^{n-1}) \approx \mathbf{Z}$, where the isomorphism is given by the degree. However, we do not require this result in this section.

1.2. Let $m: S^{n-1} \times S^{n-1} \to S^{n-1}$ be a continuous map. The map m is said to be of bidegree (p, q), if the maps $x \mapsto m(x, x_0)$ and $y \to m(x_0, y)$ are of degrees p and q, respectively (this definition does not depend on the choice of the point x_0). The sphere S^{n-1} is said to be an H-space (with respect to m) if the two maps above are homotopic to the identity of S^{n-1} (which implies $p = q = 1$). For example, S^1, S^3, and S^7, are H-spaces with respect to the multiplication of complex numbers, quaternions, and Cayley numbers, respectively. In fact, we will prove in this section that these spheres are the only one (other than S^0) which may be provided with an H-space structure. More precisely, we will prove that p odd and q odd can only occur when $n = 1, 2, 4$ or 8. This implies that the finite-dimensional real vector spaces which may be provided with an algebra structure (eventually non-associative and eventually without unit) without zero divisors, must have dimension 1, 2, 4 or 8. The essential tool used to prove these results is the notion of Hopf invariant.

1.3. *The Hopf invariant.* Let n be an even integer, and let $f: S^{2n-1} \to S^n$ be a continuous map which preserves base points. The Hopf invariant of f is an integer $h(f)$, depending only on the class of f in $\pi_{2n-1}(S^n)$, which is defined in the following way. We consider the Puppe sequence associated with f (II.3.29)

$$S^{2n-1} \longrightarrow S^n \xrightarrow{\ i\ } Cf \xrightarrow{\ j\ } S^{2n} \longrightarrow S^{n+1}$$

If we apply the functor $\tilde{K} = \tilde{K}_{\mathbf{C}}$ to it, we obtain the short exact sequence

$$0 \longrightarrow \tilde{K}(S^{2n}) \xrightarrow{\ j^*\ } \tilde{K}(Cf) \xrightarrow{\ i^*\ } \tilde{K}(S^n) \longrightarrow 0.$$

Therefore, $\tilde{K}(Cf) \approx \mathbf{Z} \oplus \mathbf{Z}$ with generators u and v, where $v = j^*(\beta_{2n})$ and $i^*(u) = \beta_n$ with β_n an arbitrary fixed generator of $\tilde{K}(S^n)$ and $\beta_{2n} = \beta_n \cup \beta_n$. In the ring $\tilde{K}(Cf)$, we thus have the relations $v^2 = uv = 0$ and $u^2 = \lambda v$ for some integer λ. This integer λ does not depend on the choice of u, since $(u+mv)^2 = u^2$ for any integer m. Now we define $\lambda = h(f)$, and it can be checked that the map $f \mapsto h(f)$ defines a homomorphism from $\pi_{2n-1}(S^n)$ to $\mathbb{Z}$. If $r: S^n \to S^n$ and $s: S^{2n-1} \to S^{2n-1}$ are continuous maps, we have the formula $h(rfs) = \deg(s)\deg(r)^2 h(f)$.

It is possible to give other definitions of the Hopf invariant; however, it can be shown that they are equivalent to this one.

1.4. We return now to our "multiplication" $m: S^{n-1} \times S^{n-1} \to S^{n-1}$ of bidegree (p, q), where we assume n even (the case n odd is easier and will be treated at the end of this section). To simplify matters, we may assume without loss of generality that $m(e, e) = e$, where $e = (1, 0, \dots, 0)$ is the base point of S^{n-1}. We now demonstrate how to obtain a map $f: S^{2n-1} \to S^n$ of Hopf invariant pq from m. To do this, we consider each of the factors S_1 and S_2 of the product $S^{n-1} \times S^{n-1}$ as the boundary of a ball B_1 (resp. B_2) of dimension n. Hence B_i is the quotient of $S_i \times [0,1]$ by the equivalence relation which identifies the subspace $S_i \times \{1\}$ to a point.

Let S^n_+ (resp. S^n_-) be the upper hemisphere (resp. the lower hemisphere) of S^n, defines by $x_{n+1} \geqslant 0$ (resp. $x_{n+1} \leqslant 0$). Then $S^n_+ \cup S^n_- = S^n$, $S^n_+ \cap S^n_- = S^{n-1}$, and both S^n_+ and S^n_- may be identified with the quotient of $S^{n-1} \times [0, 1]$, by the equivalence relation which identifies the space $S^{n-1} \times \{1\}$ to a single point.

From the map m, we obtain a map $f_1: S_1 \times B_2 \to S^n_+$ by the correspondence $(x, y, t) \mapsto (m(x, y), t)$, for $t \in [0, 1]$. By the same method, we also obtain a map $f_2: B_1 \times S_2 \to S^n_-$. In the space $B_1 \times B_2$, the subset $S_1 \times B_2 \cup B_1 \times S_2$ may be identified with S^{2n-1}. Now we define $f: S^{2n-1} \to S^n$ by the formula $f(x, y, t) = f_1(x, y, t)$ if $(x, y, t) \in S_1 \times B_2$, and $f(x, y, t) = f_2(x, y, t)$ if $(x, y, t) \in B_1 \times S_2$. The map f is well-defined since $f_1|_{S_1 \times S_2} = f_2|_{S_1 \times S_2}$. We now show that the Hopf invariant of f is equal to pq (with respect to *our* definition of the Hopf invariant).

1.5. By definition, the cone Cf of the map f, is the quotient of $Z = (B_1 \times B_2) \cup S^n$ by the equivalence relation which identifies x with $f(x)$, when $x \in S^{2n-1} = S_1 \times B_2 \cup B_1 \times S_2 \subset B_1 \times B_2$. We denote by $f_0: B_1 \times B_2 \to Cf$, the restriction to $B_1 \times B_2$ of the quotient map $\theta: Z \to Cf$. We notice that S^n (hence S^n_+ and S^n_-) are naturally subspaces of Cf. Let

$$g = (f_0, f_1, f_2): (B_1 \times B_2, S_1 \times B_2, B_1 \times S_2) \longrightarrow (Cf, S^n_+, S^n_-)$$

be the map of triples. Then g has the following properties:

a) g induces a relative homeomorphism between $(B_1 \times B_2) - (S_1 \times B_2 \cup B_1 \times S_2)$ and $Cf - S^n$. Hence g induces the isomorphism between

$$K(B_1 \times B_2, S_1 \times B_2 \cup B_1 \times S_2) \approx \tilde{K}(S^{2n}) \quad \text{and} \quad K(Cf, S^n) \approx \tilde{K}(Cf/S^n),$$

defined by j^*.

b) The homomorphism γ_1: $\tilde{K}(Cf) \approx K(Cf, S^n_+) \to K(B_1 \times B_2, S_1 \times B_2) \approx K(B_1, S_1) \approx \tilde{K}(S^n)$ sends u to $p\beta_n$. In fact, the map (f_0, f_1): $(B_1 \times B_2, S_1 \times B_2) \to$

(Cf, S^n_+) may be factorized up to homotopy into

$$(B_1 \times B_2, S_1 \times B_2) \longrightarrow (B_1 \times \{e\}, S_1 \times \{e\}) \longrightarrow (S^n, S^n_+) \longrightarrow (Cf, S^n_+).$$

The first morphism is a homotopy equivalence, and the third is induced by the inclusion of S^n in Cf. Finally, the second morphism is essentially the suspension of the map $x \mapsto m(x, e)$, hence is of degree p, when we identify B_1/S_1 and S^n/S^n_+ with S^n.

c) For the same reason, the homomorphism

$$\gamma_2 : \tilde{K}(Cf) \approx K(Cf, S^n_-) \longrightarrow K(B_1 \times B_2, B_1 \times S_2) \approx K(B_2, S_2) \approx \tilde{K}(S^n)$$

sends u to $q\beta_n$.

Let us denote the generators of the groups $K(B_1 \times B_2, S_1 \times B_2)$ and $K(B_1 \times B_2, B_1 \times S_2)$ by β'_n and β''_n, respectively. Then we have the following commutative diagram, where the horizontal arrows represent cup-products (II.5.8).

$$\begin{array}{ccc}
\tilde{K}(Cf) \times \tilde{K}(Cf) & \longrightarrow & \tilde{K}(Cf) \\
\wr\wr & & \uparrow \\
K(Cf, S^n_+) \times K(Cf, S^n_-) & \xrightarrow{\ \sigma\ } & K(Cf, S^n) \\
\Big\downarrow {\scriptstyle \gamma = (\gamma_1, \gamma_2)} & & \wr\wr \quad \nwarrow j^* \\
K(B_1 \times B_2, S_1 \times B_2) \times K(B_1 \times B_2, B_1 \times S_2) & \xrightarrow{\ \tau\ } & K(B_1 \times B_2, S_1 \times B_2 \cup B_1 \times S_2) \approx \tilde{K}(S^{2n})
\end{array}$$

The product of β'_n and β''_n, defined by τ, may be chosen as the generator β_{2n} of the group $K(B_1 \times B_2, S_1 \times B_2 \cup B_1 \times S_2)$ by III.1.3. The image of the pair (u, u) by $\tau\gamma$ (resp. σ) is $pq\beta_{2n}$ (resp. u^2) by b) and c) above. Since the image of β_{2n} by j^* is v, we have $h(f) = pq$.

Thus the assertions of 1.2 (for n even) are a consequence of the following theorem.

1.6. Theorem. *Let n be an even integer, and let $f: S^{2n-1} \to S^n$ be a continuous map with Hopf invariant an odd number. Then $n = 2, 4$, or 8. In particular, if S^{n-1} may be provided with an H-space structure, then n must be $2, 4$ or 8.*

Proof. From the general properties of the operations ψ^k (IV.7.19), we have $\psi^k(v) = k^{2r} v$ and $\psi^k(u) = k^r u + \sigma(k) v$, where $n = 2r$ and $\sigma(k) \in \mathbb{Z}$. On the other hand, since $\psi^2 = (\lambda^1)^2 - 2\lambda^2$ (IV.7.15), we have $\psi^2(u) = u^2 \bmod 2 = h(f) v \bmod 2$. Hence $\sigma(2)$ must have the same parity as $h(f)$, which is odd by hypothesis (put $k = 2$ in the relation above).

From the relation $\psi^k \psi^l = \psi^l \psi^k$ (IV.7.15), we write

$$k^r(k^r - 1)\sigma(l) = l^r(l^r - 1)\sigma(k).$$

In particular, if we choose $l=2$ and k odd, we see that 2^r must divide k^r-1 for each odd integer k. Therefore, in order to prove the theorem, it suffices to show that this special property of r implies $r=1, 2$, or 4.

If $r>1$, the group $(\mathbb{Z}/2^r\mathbb{Z})^*$ is of even order. Hence the identity $k^r \equiv 1 \bmod 2^r$ implies r even. If we choose $k=1+2^{r/2}$, we have $k^r \equiv 1+r2^{r/2} \bmod 2^r$ by binomial expansion. Hence r must be divisible by $2^{r/2}$, which is only possible if $r=2$ or 4. □

1.7. Remark. It is possible to prove (cf. Husemoller [1]) that for any *even* numbers n and λ, there exists a continuous map $f\colon S^{2n-1} \to S^n$ of Hopf invariant λ.

1.8. Theorem. *Let n be an odd integer, and let*

$$m\colon S^{n-1} \times S^{n-1} \longrightarrow S^{n-1}$$

be a continuous map of bidegree (p, q). Then p or q is equal to 0.

Proof. By III.1.3 and IV.3.24, we have $K_{\mathbb{C}}(S^{n-1} \times S^{n-1}) \approx K_{\mathbb{C}}(S^{n-1}) \otimes K_{\mathbb{C}}(S^{n-1}) \approx (\mathbb{Z} \oplus \mathbb{Z}u) \otimes (\mathbb{Z} \oplus \mathbb{Z}v)$, where u (resp. v) is the generator of the first factor $\tilde{K}_{\mathbb{C}}(S^{n-1})$ (resp. the second factor $\tilde{K}_{\mathbb{C}}(S^{n-1})$). If we write $K_{\mathbb{C}}(S^{n-1}) = \mathbb{Z} \oplus \mathbb{Z}w$, then the homomorphism

$$m^*\colon \mathbb{Z} \oplus \mathbb{Z}w \longrightarrow (\mathbb{Z} \oplus \mathbb{Z}u) \otimes (\mathbb{Z} \oplus \mathbb{Z}v)$$

sends w to an element of the form $pu \otimes 1 + 1 \otimes qv + su \otimes v$, for some integer s. Since m is a ring homomorphism, $0 = w^2$ must be sent to

$$(pu \otimes 1 + 1 \otimes qv + su \otimes v)^2 = 2pqu \otimes v.$$

Therefore $pq = 0$. □

2. The Solution of the Vector Field Problem on the Sphere

2.1. In this section, we wish to determine the maximum number of linearly independent vector fields on the sphere S^{t-1} (cf. I.5.5). The Gram-Schmidt orthonormalization procedure may be used to replace any field of $n-1$ linearly independent tangent vectors, by a field of $n-1$ tangent vectors of norm one, which are orthogonal to each other. Therefore, if $\mathrm{O}_{n,t}$ denotes the Stiefel manifold $\mathrm{O}(t)/\mathrm{O}(t-n)$, then the existence of a field of $n-1$ linearly independent tangent vectors on S^{t-1}, is equivalent to the existence of a continuous section $\sigma\colon \mathrm{O}_{1,t} = S^{t-1} \to \mathrm{O}_{n,t}$ of the natural projection $\mathrm{O}_{n,t} \to \mathrm{O}_{1,t}$, defined by $(a_1, \ldots, a_n) \mapsto a_n$ where (a_i) is an orthonormal system in $\mathbb{R}^t$.

2.2. In order to deal with this last problem, we use the following trick. Each element a of $\mathrm{O}_{n,t}$ defines a linear map $\varphi_a\colon \mathbb{R}^n \to \mathbb{R}^t$, which is injective, and depends continuously on a. Now let $\theta\colon S^{n-1} \times S^{t-1} \to S^{n-1} \times S^{t-1}$ be the continuous map defined by $\theta(v, b) = (v, \varphi_{\sigma(b)}(v))$. By identifying v with $-v$, we see that θ induces a continuous map $\overline{\theta}$ between $(S^{n-1}/\mathbb{Z}_2)\times S^{t-1}$ and $(S^{n-1}\times S^{t-1})/\mathbb{Z}_2$. These spaces may be identified with the sphere bundles $S(t\varepsilon)$ and $S(t\xi)$ where ε denotes the trivial bundle of rank one over $RP_{n-1}=S^{n-1}/\mathbb{Z}_2$, and ξ denotes the canonical line bundle over RP_{n-1} (I.2.4).

2.3. Proposition. *Let us assume that S^{t-1} admits $n-1$ linearly independent tangent vector fields. Then there is a continuous map $\overline{\theta}: S(t\varepsilon)\to S(t\xi)$, which makes the following diagram commutative:*

$$\begin{array}{ccc} S(t\varepsilon) & \xrightarrow{\overline{\theta}} & S(t\xi) \\ & \searrow \quad \swarrow & \\ & RP_{n-1} & \end{array}$$

Moreover, over each point x of RP_{n-1} the map

$$\overline{\theta}_x: S(t\varepsilon)_x \longrightarrow S(t\xi)_x$$

is a homotopy equivalence.

Proof. Only the last part of the proposition requires a proof. If $n=1$, then $\overline{\theta}$ is bijective and the proposition is obvious. If $n>1$, we must show that the maps $b\mapsto\varphi_{\sigma(b)}(v)$, for v in S^{n-1}, are homotopy equivalences from S^{t-1} to itself. Since S^{n-1} is arcwise connected, all these maps are homotopic. However, if we choose $v=e_n$ (the last vector of the canonical basis of $\mathbb{R}^n$), we see that $\varphi_{\sigma(b)}(v)=b$ since σ is a section of the map $O_{n,t}\to O_{1,t}$. □

2.4. The last proposition provides some important information about the Thom spaces of $t\varepsilon$ and $t\xi$. More generally, let V and W be vector bundles on a compact base X, and let $f: S(V)\to S(W)$ be a continuous map between sphere bundles, such that the diagram

$$\begin{array}{ccc} S(V) & \xrightarrow{f} & S(W) \\ & \searrow \quad \swarrow & \\ & X & \end{array}$$

is commutative. By radial extension, f induces a proper continuous map $\tilde{f}: V\to W$, hence a homomorphism $\tilde{f}^*: K_{\mathbb{R}}(W)\to K_{\mathbb{R}}(V)$. If we assume that $f_x: S(V_x)\to S(W_x)$ is a homotopy equivalence for each $x\in X$ [6], then f also induces a homotopy equivalence $\dot{f}_x: \dot{V}_x\to\dot{W}_x$ (note that $\dot{V}_x$ and $\dot{W}_x$ may be identified with the suspen-

[6] In this case we say that the sphere bundles $S(V)$ and $S(W)$ have the same *fiber homotopy type* (cf. Dold–Lashof [1]).

sions of $S(V_x)$ and $S(W_x)$, respectively). Therefore, the homomorphism $\tilde{f}^*$ has the property that for each point x, we have the commutative diagram

$$\begin{array}{ccc} K_{\mathbb{R}}(W) & \xrightarrow{\tilde{f}^*} & K_{\mathbb{R}}(V) \\ \downarrow & & \downarrow \\ K_{\mathbb{R}}(W_x) & \xrightarrow{\tilde{f}_x^*} & K_{\mathbb{R}}(V_x), \end{array}$$

where $\tilde{f}_x^*$ is an isomorphism.

Suppose now that V and W are of rank $8p$, and are provided with some spinorial structure. Let T_W be the Thom class of W (IV.5.13). Then $\tilde{f}^*(T_W)$ may be written as λT_V, where T_V is the Thom class of V, and $\lambda \in K_{\mathbb{R}}(X)$.

By restriction to a point x of X, we see that λ is an invertible element of $K_{\mathbb{R}}(X)$. Hence there exists a cover (X_i) of X with open and closed subsets such that $\lambda|_{X_i} = \varepsilon_i(1+y_i)$ where $\varepsilon_i = \pm 1$, $y \in K'_{\mathbb{R}}(X)$ and $y_i = y|_{X_i}$. Note that

$$\frac{\psi^k(1+y)}{1+y} = \frac{\psi^k(\lambda)}{\lambda}.$$

2.5. Proposition. *Let V and W be as above. Then there exists an element y of $K'_{\mathbb{R}}(X)$ such that for each k we have the relation*

$$\rho^k(V) = \rho^k(W)\frac{\psi^k(1+y)}{1+y},$$

where $\rho^k = \rho^k_{\mathbb{R}}$ is the operation defined in IV.7.26.

Proof. Let $\varphi_V : K_{\mathbb{R}}(X) \to K_{\mathbb{R}}(V)$ (resp. $\varphi_W : K_{\mathbb{R}}(X) \to K_{\mathbb{R}}(W)$) be the Thom isomorphism (cf. IV.5.14). Since $T_V = \varphi_V(1) = \tilde{f}^*[(1+y)T_W]$ where $y \in K'_{\mathbb{R}}(X)$, we have "formally"

$$\rho^k(V) = \varphi_V^{-1}(\psi^k(T_V)) = \frac{\psi^k(T_V)}{T_V} = \frac{\psi^k(T_W)\cdot\psi^k(\lambda)}{T_W\cdot\lambda} = \rho^k(W)\frac{\psi^k(1+y)}{1+y}. \quad \square$$

2.6. Corollary. *Let W be a spinorial bundle of rank $8p$ over RP_{n-1}, such that the sphere bundles $S(8p\varepsilon)$ and $S(W)$ have the same fiber homotopy type. Then $\rho^k(W) = k^{4p}$ if k is odd.*

Proof. Since k is odd, $\psi^k(\xi) = \xi$. Therefore, using the fact that $K'_{\mathbb{R}}(RP_{n-1}) \approx \tilde{K}_{\mathbb{R}}(RP_{n-1})$ is generated by $\xi - 1$ (IV.6.46), we see that the factor $\dfrac{\psi^k(1+y)}{1+y}$ is equal to one. Hence $\rho^k(V) = \rho^k(8p\varepsilon) = k^{4p}$. $\square$

2.7. Proposition. *Let a_n be the order of the group $\tilde{K}_{\mathbb{R}}(RP_{n-1})$,* that is, *$a_n = 2^f$ where f is the number of integers i such that $0 < i < n$ and $i = 0, 1, 2$ or $4 \bmod 8$ (IV.6.46). If S^{t-1} admits $n-1$ linearly independent tangent vector fields, then t is a multiple of a_n.*

Proof. We first prove that if the sphere bundles $S(t\varepsilon)$ and $S(t\xi)$ have the same fiber homotopy type, then $t\xi$ is stably trivial. For the case $n = 2$, the vector bundle

$\xi \oplus \xi$ is trivial. Hence, if t is even, there is nothing to prove. If t is odd, i.e. $t=2r+1$, then the Thom space $\widehat{t\xi}$ may be identified with $S^{2r}(RP_2)$, and $K_{\mathbb{R}}^{2r}(\dot{\widehat{t\xi}})=K_{\mathbb{R}}(RP_2)\neq K_{\mathbb{R}}^{2r}(\dot{\widehat{t\varepsilon}})$ (IV.6.45). Hence the Thom spaces $\dot{\widehat{t\xi}}$ and $\dot{\widehat{t\varepsilon}}$ do not have the same fiber homotopy type, which implies that $S(t\xi)$ and $S(t\varepsilon)$ do not have the same fiber homotopy type.

Let us now consider the case $n>2$. We first prove that $\dot{\widehat{t\varepsilon}}\sim\dot{\widehat{t\xi}}$ implies that t is a multiple of 4. The set of natural numbers u, such that $\dot{\widehat{u\xi+m\varepsilon}}$ has the same homotopy type as $\dot{\widehat{(u+m)\varepsilon}}$ for some m, has a smallest element α, which divides all such numbers (note that in general $\dot{\widehat{V\times W}}\approx\dot{V}\wedge\dot{W}$). Since $a_n=2^f$, the number α may be written as 2^g with $g\leqslant f$, so we must show that $g\neq 0, 1$. The Thom space $\dot{\widehat{u\xi}}$ can always be identified with RP_{n+u-1}/RP_{u-1} via the homeomorphism $(\mathbb{R}^n-\{0\})\times\mathbb{R}^u/\mathbb{R}^*\to RP_{n+u-1}-RP_{u-1}$, induced by the inclusion of $(\mathbb{R}^n-\{0\})\times\mathbb{R}^u$ in $\mathbb{R}^{n+u}-\{0\}$. Therefore, a homotopy equivalence between $\dot{\widehat{\alpha\xi+m\varepsilon}}$ and $\dot{\widehat{(\alpha+m)\varepsilon}}$ implies that $K_{\mathbb{R}}^{-\alpha}(RP_{n-1})\approx\tilde{K}_{\mathbb{R}}(S^{\alpha}(RP_{n-1}))\approx\tilde{K}_{\mathbb{R}}(RP_{n+\alpha-1}/RP_{\alpha-1})$. But, by IV.6.45, we have

$$K_{\mathbb{R}}^{-1}(RP_{n-1})=\mathbb{Z}/2;\qquad K_{\mathbb{R}}(RP_n/RP_0)\approx\mathbb{Z}/2^{\beta}\quad\text{with } \beta\geqslant 2;$$
$$\begin{aligned}K_{\mathbb{R}}^{-2}(RP_{n-1})&=\mathbb{Z}/2\quad\text{if } 3\leqslant n<7,\\&=\text{a group of order 2 or 4 if } n\geqslant 7;\end{aligned}$$

and by IV.6.42

$$\begin{aligned}K_{\mathbb{R}}(RP_{n+1}/RP_1)&=\operatorname{Coker}(K(C^{n+1,0})\to K(C^{1,0}))\\&=\mathbb{Z}/4\quad\text{if } 3\leqslant n<7,\\&=\mathbb{Z}/2^{\beta}\quad\text{with } \beta>2,\text{ if } n\geqslant 7.\end{aligned}$$

Therefore $\alpha\neq 1, 2$, i.e. t is a multiple of 4.

Assume now that $t=4l$. By 2.6, we must have $\rho^k(4l\xi+4l)=k^{4l}$ if k is odd. Therefore, applying proposition IV.7.30, we obtain the identity

$$1+\frac{k^{2l}-1}{2k^{2l}}\lambda=1$$

in the multiplicative group $1+\tilde{K}_{\mathbb{R}}(RP_{n-1})$, for every odd integer k. Now let $\theta\colon K_{\mathbb{R}}(RP_{n-1})\to\mathbb{Z}/2^{f+1}\mathbb{Z}$ be the ring homomorphism defined by $\theta(\lambda)=-2$. Then θ induces a group map

$$\theta'\colon 1+\tilde{K}_{\mathbb{R}}(RP_{n-1})\longrightarrow(\mathbb{Z}/2^{f+1}\mathbb{Z})^*,$$

such that $\theta'\left(1+\dfrac{k^{2l}-1}{2k^{2l}}\lambda\right)=(1/k)^{2l}$. As is well known (and easy to prove), the group $(\mathbb{Z}/2^{f+1}\mathbb{Z})^*$ may be identified with $\mathbb{Z}/2^{f-1}\mathbb{Z}\times\mathbb{Z}/2$, and as a generator of the factor $\mathbb{Z}/2^{f-1}\mathbb{Z}$ we may choose any element of $(\mathbb{Z}/2^{f+1}\mathbb{Z})^*$ of the form $4p+1$, with

p odd. If k is chosen so that $1/k$ is of this form, we see that $(1/k)^{2l}=1$ in $(\mathbb{Z}/2^{f+1}\mathbb{Z})^*$, if and only if $2l$ is a multiple of 2^{f-1}, thus $t=4l$ is a multiple of $a_n=2^f$. □

* **2.8.** *Remark.* To show that t is a multiple of 4 when $n>2$, we could also have used Stiefel-Whitney characteristic classes (IV.4.20 and Thom [1].*

2.9. Theorem. *The sphere S^{t-1} admits $n-1$ linearly independent tangent vector fields if and only if t is a multiple of a_n.*

Proof. By 2.7, it suffices to construct such linearly independent tangent vector fields, if t is a multiple of a_n. From the definition of a_n, we see that t is a multiple of a_n if and only if $\mathbb{R}^t$ is provided with a $C^{n-1,0}$-module structure, i.e. if there exist $n-1$ automorphisms $e_1, \ldots, e_{n-1}$ of $\mathbb{R}^t$, such that $(e_i)^2=-1$ and $e_i e_j + e_j e_i = 0$ for $i \neq j$. If G is the multiplicative finite group of order 2^n generated by $\pm e_i$, we may choose a metric on $\mathbb{R}^t$ so that G acts by orthogonal automorphisms. Hence, we may assume without loss of generality that $e_i^* = -e_i$. Thus for each vector v of S^{t-1}, the vectors $e_1 \cdot v, \ldots, e_{n-1} \cdot v$ are tangent vectors which are linearly independent. To see this last point, we set $e_0 = \mathrm{Id}$, and notice that the scalar products $\langle e_i \cdot v, e_j \cdot v\rangle = 0$, for $i \neq j$ and $i, j \in \{0, 1, \ldots, n-1\}$, since $e_j^* e_i = -e_i^* e_j$. □

2.10. Theorem (Adams [1]). *Let us write each integer t in the form $t=(2\alpha-1)\cdot 2^\beta$, where $\beta=\gamma+4\delta$ with $0 \leqslant \gamma \leqslant 3$, and let us define $\rho(t)=2^\gamma+8\delta$. Then the maximal number of linearly independent tangent vector fields on the sphere S^{t-1}, is $\rho(t)-1$.*

Proof. By 2.9, this number is $n-1$, where n is the greatest integer $\sigma(t)$ such that t is a multiple of a_n. Since a_n is a power of 2, this number depends only on β. On the other hand, $\sigma(16t)=\sigma(t)+8$ since $C^{p+8,0} \approx C^{p,0}(16)$ (III.3.21). Since $\rho(16t)=\rho(t)+8$, we need only check the cases $t=2$, 4, 8 and 16, where we observe that $\rho(2)=\sigma(2)=2$, $\rho(4)=\sigma(4)=4$, $\rho(8)=\sigma(8)=8$, and $\rho(16)=\sigma(16)=9$. □

3. Characteristic Classes and the Chern Character

3.1. In this section we assume that the reader has some basic knowledge of ordinary cohomology (cf. Greenberg [1], Eilenberg-Steenrod [1], Dold [1], Spanier [2]...). Recall that all cohomology theories satisfying the Eilenberg-Steenrod axioms, agree on finite CW-complexes, and hence on compact manifolds. In general, we will use Čech cohomology with $\mathbb{Z}$-coefficients, which is the theory best suited to K-theory. We define the Čech cohomology with compact support of a locally compact space X, as $\mathrm{Ker}[H^i(\dot{X}) \to H^i(\{\infty\})$, where $\dot{X}$ is the one point compactification of X. When there is no risk of confusion, we will simply write $H^i(X)$.

Our first objective in this section is to associate with each vector bundle V, some characteristic classes analogous to the classes constructed in Chapter IV.

3.2. Let V be a real vector bundle of rank n with base X. By definition, an orientation of V is given by an orientation ω_x of the real vector space V_x, for each point

x of X, which varies "continuously" with x. This means that for each point y in X, there exists a neighbourhood U of y and a trivialization $U \times \mathbb{R}^n \to V_U$ inducing an isomorphism of oriented vector spaces $\mathbb{R}^n \to V_x$ for every point x in U. It is easy to see that this definition agrees with the definition in IV.4.13 stated in terms of principal bundles.

An important example is given by a *complex* vector bundle. In fact, if E is a complex vector space of dimension p and if $e_1, \ldots, e_p$ is a basis of E, then the vectors $e_1, ie_1, \ldots, e_p, ie_p$ can be chosen as an oriented basis of the underlying real vector space. This orientation is independent of the choice of the basis, and invariant under complex automorphisms because $\mathrm{GL}_p(\mathbb{C})$ is connected. Moreover, if E' is another complex vector space, then the natural orientation of $E \oplus E'$ is the product of the orientations of E and E' in the obvious sense. Now, if V is a complex vector bundle, it may be given the orientation arising from the canonical orientations of its fibers.

For any oriented real vector bundle V of rank n, we have a canonical generator ω_x of the group $H^n(V_x)$, over each point x of X.

3.3. Theorem (Thom). *Let us assume that the base X is compact. Then there exists a unique cohomology class $U_V \in H^n(V)$, such that the restriction of U_V to each group $H^n(V_x)$ is the generator ω_x. Moreover, the cup-product with U_V defines an isomorphism*

$$\varphi_V : H^i(X) \longrightarrow H^{i+n}(V)$$

(U_V is called the Thom class in cohomology and φ_V the Thom isomorphism).

Proof. Using the Mayer-Vietoris exact sequence argument given in IV.1.3, we see that the existence of such a class U_V induces an isomorphism $H^i(X) \approx H^{i+n}(V)$. Now let (X_i), for $i=1, \ldots, p$, be a finite closed cover of X, such that $V|_{X_i}$ is a trivial vector bundle for each i. Our proof is by induction on p. If $Y = \bigcup_{i=1,\ldots,p-1} X_i$ and $Z = X_p$, we have the Mayer-Vietoris exact sequence

$$H^{n-1}(V|_{Y \cap Z}) \longrightarrow H^n(V) \longrightarrow H^n(V|_Y) \oplus H^n(V|_Z) \longrightarrow H^n(V|_{Y \cap Z}).$$

The first group of this sequence is 0 by the induction hypothesis. Thus we obtain the existence and uniqueness of U_V (hence the isomorphism φ_V) for a cover of p closed subsets X_i, such that $V|_{X_i}$ is trivial. □

3.4. Definition. The restriction $\chi(V)$ of the Thom class U_V to $H^n(V) \approx H^n(X)$, is called the *Euler class* of the oriented vector bundle V (more precisely, it is the *cohomological Euler class* as opposed to the K-theory Euler class defined in IV.1.13).

A justification of this terminology is given by the following proposition:

3.5. Proposition. *Let V be the tangent bundle of an oriented manifold X. Then $\chi(V)$, evaluated on the fundamental class of the manifold X, is equal to the Euler-Poincaré characteristic of the manifold.*

This proposition is proved in Husemoller [1], pp. 255–258. Since we do not require this result (except for an application in 3.28), we omit the proof.

3.6. Proposition. *Let V_1 and V_2 be vector bundles with bases X_1 and X_2, respectively. Then $U_{V_1 \times V_2} = U_{V_1} \cup U_{V_2}$. If $X_1 = X_2 = X$, then $\chi(V_1 \oplus V_2) = \chi(V_1) \cdot \chi(V_2)$.*

Proof. The first formula follows from the uniqueness of the Thom class as stated in 3.3, and from the elementary properties of the cup-product in cohomology. The second formula is obtained from the first via the diagonal map $X \to X \times X$. □

3.7. Proposition. *Using the notation of IV.2.3, we have the split exact sequence*

$$0 \longrightarrow H^r(P(V \oplus L), X) \xrightarrow{j^*} H^r(P(V \oplus L)) \longrightarrow H^r(X) \longrightarrow 0$$

$$\wr\wr$$

$$H^r(\xi_V^* \otimes \pi^* L)$$

$$\wr\wr$$

$$H^{r-2}(P(V)).$$

If U denotes the Thom class of the line bundle $\xi_V^* \otimes \pi^* L$, and if $r=2$, then $j^*(U)$ is the Euler class σ of the bundle $\xi_{V \oplus L}^* \otimes \pi_1^*(L)$, where $\pi_1 : P(V \oplus L) \to X$. Finally, if x is an element of Ker$(H^r(P(V \oplus L)) \to H^r(X)$, *and if $x' = x|_{H^r(P(V))}$, we have the formula $j^* \Phi(x') = x \cdot j^*(U)$.*

The proof of this proposition is analogous to the proof of IV.2.4.

3.8. Proposition. *Let X be a compact space, and let $P_n = P(\mathbf{C}^{n+1})$ be the complex projective space of dimension n. Then $H^*(X \times P_n)$ is a free $H^*(X)$-module with basis $1, t, \ldots, t^n$, where t is the Euler class of the line bundle $p^* \xi_n$ with $p : X \times P_n \to P_n$. Moreover, $t^{n+1} = 0$; hence $H^*(X \times P_n) \approx H^*(X)[t]/(t^{n+1})$.*

The proof of this proposition is analogous to the proof of IV.2.5.

3.9. Corollary. *Let P_n and P_m be complex projective spaces, and let $\eta_1 = \pi_1^* \xi_n^*$ and $\eta_2 = \pi_2^* \xi_m^*$, where $\pi_1 = P_n \times P_m \to P_n$ and $\pi_2 : P_n \times P_m \to P_m$. Let x and y be the Euler classes of the bundles η_1 and η_2. Then $H^*(P_n \times P_m) \approx \mathbf{Z}[x, y]/(x^{n+1})(y^{m+1})$.*

3.10. Proposition. *Let L_1 and L_2 be line bundles. Then $\chi(L_1 \otimes L_2) = \chi(L_1) + \chi(L_2)$, $\chi(L_1)$ is nilpotent, and $\chi(L_1^* \otimes L_2) = -\chi(L_1) + \chi(L_2)$.*

Proof. By I.7.10, it suffices to verify the proposition when the base space is $P_n \times P_m$, $L_1 = \pi_1^* \xi_n^*$, and $L_2 = \pi_2^* \xi_m^*$. By restriction to the factors P_n and P_m (choose base points in P_n and P_m), and by naturality, the corollary above shows that $\chi(L_1 \otimes L_2) = x + y = \chi(L_1) + \chi(L_2)$ in the group $H^2(P_n \times P_m)$. The nilpotence of $\chi(L_1)$ follows from the nilpotence of t in Proposition 3.8. Finally $0 = \chi(L_1^* \otimes L_1) = \chi(L_1^*) + \chi(L_1)$, and $\chi(L_1^* \otimes L_2) = \chi(L_1^*) + \chi(L_2) = -\chi(L_1) + \chi(L_2)$. □

3.11. *Remark.* If we let u denote the Euler class of $\pi^*\xi_n$, where $\pi: X\times P_n \to P_n$, then we have $u=-t$, and $H^*(X\times P_n)\approx H^*(X)[u]/(u^{n+1})$.

3.12. Proposition. *Let X be a compact space, and let V be a complex vector bundle of rank n over X. Let u be the Euler class of the line bundle $\xi=\xi_V$ over $P(V)$* (cf. IV.2.2). *Then $H^*(P(V))$ is a free $H^*(X)$-module with basis $1, u, \ldots, u^{n-1}$. In particular, the homomorphism $H^*(X)\to H^*(P(V))$ is injective.*

The proof of this proposition is analogous to the proof of Proposition IV.2.13.

3.13. Proposition. *Assume that V may be written in the form $L_1\oplus\cdots\oplus L_n$, where the L_i are line bundles. Then u^n is determined as a function of the u^i, for $i<n$, by the relation*

$$\prod_{i=1}^{n}(u-\chi(L_i))=0.$$

The proof of this proposition is analogous to the proof of IV.2.14 (where the unit element is simply 1).

The "splitting principle" in K-theory has the following analog in cohomology theory:

3.14. Proposition. *Let V be a complex vector bundle with compact base X, and let $\pi: F(V)\to X$ where $F(V)$ is the flag bundle of V* (cf. IV.3.3). *Then*

a) *the homomorphism $\pi^*: H^*(X)\to H^*(F(V))$ is injective; and* b) *the vector bundle $\pi^*(V)$ splits into the Whitney sum of line bundles.*

The proof of this proposition is analogous to the proof of IV.2.15, with the remarks made in IV.3.3.

3.15. Theorem. *To each complex vector bundle V of fixed rank with compact base X, we can uniquely associate cohomology classes $c_i(V)\in H^{2i}(X)$ called the Chern classes of V which satisfy the following axioms:*

1) *The $c_i(V)$ are "natural", i.e. $c_i(V)=f^*(c_i(V'))$ for any general morphism $V\to V'$ which is an isomorphism on the fibers, and induces $f: X\to X'$ on the bases* (cf. I.1.6).

2) *If V_1 and V_2 are vector bundles with the same base, then $c_k(V_1\oplus V_2)=\sum_{i+j=k} c_i(V_1)\cdot c_j(V_2)$.*

3) *If the rank of V is one, then $c_0(V)=1$, $c_1(V)=\chi(V)$, and $c_i(V)=0$ for $i\neq 0, 1$.*

The proof of this proposition is analogous to the proof of IV.2.17. Moreover, this proof gives us the following proposition:

3.16. Proposition. *Let X be a compact space, and let V be a complex vector bundle of rank n over X. Let u be the Euler class of the line bundle $\xi=\xi_V$, and let $c_i(V)$ be* *the Chern classes of* V (note that $c_i(V)=0$ for $i>n$). *Then we have the relation*

$$u^n - c_1(V)u^{n-1} + \cdots + (-1)^n c_n(V) = 0,$$

which completely determines the ring structure of $H^*(P(V))$ (cf. 3.12).

3.17. Proposition. *Let* V *be as above. Then* $c_n(V)$ *is equal to the Euler class* $\chi(V)$ *of* V.

Proof. By the splitting principle, we need only verify the proposition for $V = L_1 \oplus \cdots \oplus L_n$, where the L_i are line bundles. By 3.6, we then have

$$\chi(L_1 \oplus \cdots \oplus L_n) = \prod_{i=1}^{n} \chi(L_i) = \prod_{i=1}^{n} c_1(L_i) = c_n(V). \quad \square$$

3.18. *Remarks.* Let us write $c(V) = \sum_{i=0}^{n} c_i(V) \in H^{\text{even}}(X)$. This is called the "total Chern class" of V. Then relation 2) of 3.15 may be written more simply in the form $c(V_1 \oplus V_2) = c(V_1) \cdot c(V_2)$. We could also push the analogy with K-theory further by computing the cohomology of flag bundles, Grassmannians, etc. (cf. Dold [2]). This is left as an exercise for the reader.

3.19. By analogy with the characteristic classes $\psi^k(V)$ constructed in IV.7.15, we define classes $s_k(V) \in H^{2k}(X)$, such that $s_k(V_1 \oplus V_2) = s_k(V_1) + s_k(V_2)$, by putting $s_k(V) = Q_k(c_1, c_2, \ldots, c_n)$ where $c_i = c_i(V)$, and Q_k is the Newton polynomial. We make the convention that $s_0(V) = \text{rank}(V)$. The formal computations made in IV.7.15, may be repeated to prove the relation $s_k(V_1 \oplus V_2) = s_k(V_1) + s_k(V_2)$, which we will require.

3.20. Proposition. *Let* V_1 *and* V_2 *be complex vector bundles. Then we have the relation*

$$s_k(V_1 \otimes V_2) = \sum_{i+j=k} \frac{k!}{i!\,j!} s_i(V_1) s_j(V_2).$$

Proof. We provide the group $H^{\text{even}}(X)$ with a new multiplication law, denoted by $*$, which is defined on homogeneous elements x_i and x_j of degrees $2i$ and $2j$, respectively, by the formula

$$x_i * x_j = \frac{(i+j)!}{i!\,j!} x_i x_j.$$

Then $H^{\text{even}}(X)$, provided with this new multiplication, is still a commutative ring, and the formula required is

$$s_k(V_1 \otimes V_2) = \sum_{i+j=k} s_i(V_1) * s_j(V_2).$$

If we set $s(V)=\sum_{i=0}^{\infty} s_i(V)\in H^{\text{even}}(X)$, this is equivalent to proving that

$$s(V_1\otimes V_2)=s(V_1)*s(V_2).$$

Because of the splitting principle and the property stated in 3.19, it suffices to check this formula when V_1 and V_2 are of rank one. If we set $x_1=\chi(V_1)$, $x_2=\chi(V_2)$, then $x_1+x_2=\chi(V_1\otimes V_2)$ by 3.10. Hence, $s_k(V_1\otimes V_2)=(x_1+x_2)^k$, $s_i(V_1)=(x_1)^i$, and $s_j(V_2)=(x_2)^j$. Therefore, the identity required is simply

$$(x_1+x_2)^k=\sum_{i+j=k}(x_1)^i*(x_2)^j=\sum_{i+j=k}\frac{k!}{i!j!}(x_1)^i(x_2)^j,$$

which is the binomial identity. □

3.21. *Remark.* The definition of the classes $c_i(V)$ and $s_i(V)$ may be generalized to the case where the rank of V is not necessarily constant. If $X=\bigcup X_\alpha$ is a partition of X into open subsets such that the rank of $V|_{X_\alpha}$ is constant, then we define $c_i(V)$ (resp. $s_i(V)$) as the unique cohomology class such that $c_i(V)|_{X_\alpha}=c_i(V|_{X_\alpha})$ (resp. $s_i(V)|_{X_\alpha}=s_i(V|_{X_\alpha})$.

3.22. In order to avoid the above "twisting" of the multiplication in $H^{\text{even}}(X)$, we must work with rational cohomology, i.e. the cohomology theory with $\mathbb{Q}$ as the group of coefficients instead of $\mathbb{Z}$. As is well known (cf. Eilenberg-Steenrod [1]), $H^i(X;\mathbb{Q})\approx H^i(X)\otimes\mathbb{Q}$. Thus we may define

$$Ch_k(V)=\frac{1}{k!}s_k(V)=\frac{1}{k!}Q_k(c_1(V),c_2(V),\ldots,c_n(V))\in H^{2k}(X;\mathbb{Q}),$$

$$Ch(V)=\sum_{k=0}^{\infty}Ch_k(V)\in H^{\text{even}}(X;\mathbb{Q})$$

(we notice that $Ch_k(V)=0$ for k large enough, since the $c_i(V)$ are nilpotent by the splitting principle and Proposition 3.10).

3.23. Theorem. *Let V_1 and V_2 be vector bundles, with compact base X. Then we have the formulas*

$$Ch(V_1\oplus V_2)=Ch(V_1)+Ch(V_2),$$

and

$$Ch(V_1\otimes V_2)=Ch(V_1)\cdot Ch(V_2).$$

Moreover, if L is a line bundle, then $Ch(L)=\exp(\chi(L))$.

Proof. Since $s_k(V_1\oplus V_2)=s_k(V_1)+s_k(V_2)$, we clearly have $Ch_k(V_1\oplus V_2)=Ch_k(V_1)+Ch_k(V_2)$; hence, $Ch(V_1\oplus V_2)=Ch(V_1)+Ch(V_2)$. Similarly

$$Ch_k(V_1\otimes V_2)=\frac{1}{k!}s_k(V_1\otimes V_2)=\sum_{i+j=k}\frac{1}{i!}\frac{1}{j!}s_i(V_1)s_j(V_2)=\sum_{i+j=k}Ch_i(V_1)\cdot Ch_j(V_2).$$

Therefore $Ch(V_1 \otimes V_2) = Ch(V_1) \cdot Ch(V_2)$. Finally, if V is of rank one, then

$$Ch(V) = \sum_{k=0}^{\infty} \frac{1}{k!} s_k(V) = \sum_{k=0}^{\infty} \frac{1}{k!} \chi(V)^k = \exp(\chi(V)). \quad \square$$

3.24. Definition. We call the ring homomorphism

$$Ch: K_{\mathbb{C}}(X) \longrightarrow H^{\text{even}}(X; \mathbb{Q}) = \bigoplus_{i=0}^{\infty} H^{2i}(X; \mathbb{Q}),$$

defined by $Ch([E]-[F]) = Ch(E) - Ch(F)$, the *Chern character*.

3.25. Theorem. *The Chern character has the following properties:*

a) *If X is the sphere S^{2n}, then Ch is injective, and its image is $H^*(S^{2n}) \subset H^*(S^{2n}; \mathbb{Q})$.*

b) *For any compact space X, Ch induces an isomorphism*

$$K_{\mathbb{C}}(X) \otimes \mathbb{Q} \xrightarrow{\approx} H^{\text{even}}(X; \mathbb{Q})$$

(the cohomology used here is Čech cohomology).

A detailed and complete proof of this theorem is given in the Cartan-Schwartz seminar 1963/64, exposé 16 (Karoubi [1]).

3.26. Since $K_{\mathbb{C}}(X, Y) \approx \tilde{K}_{\mathbb{C}}(X/Y) = \operatorname{Ker}[K_{\mathbb{C}}(X/Y) \to K_{\mathbb{C}}(\mathit{Point})]$, and since $H^{\text{even}}(X, Y)\mathbb{Q}) \approx \tilde{H}^{\text{even}}(X/Y) = \operatorname{Ker}[H^{\text{even}}(X/Y) \to H^{\text{even}}(\mathit{Point})]$, we may extend the Chern character by a homomorphism, also denoted by Ch, from $K_{\mathbb{C}}(X, Y)$ to $H^{\text{even}}(X, Y; \mathbb{Q})$. Applying these observations to the pair $(X \times B^1, X \times S^0)$, we obtain a homomorphism from $K_{\mathbb{C}}^{-1}(X, Y)$ to $H^{\text{odd}}(X, Y; \mathbb{Q})$. If we define $K_{\mathbb{C}}^{\#}(X, Y) = K_{\mathbb{C}}^0(X, Y) \oplus K_{\mathbb{C}}^{-1}(X, Y)$, we finally obtain a homomorphism $Ch: K_{\mathbb{C}}^{\#}(X, Y) \to H^*(X, Y; \mathbb{Q})$, compatible with the multiplicative structures, and inducing an isomorphism $K_{\mathbb{C}}^{\#}(X, Y) \otimes \mathbb{Q} \approx H^*(X, Y; \mathbb{Q})$.

3.27. Theorem. *Let $\psi_H^k: H^{\text{even}}(X; \mathbb{Q}) \to H^{\text{even}}(X; \mathbb{Q})$ be the algebra homomorphism defined by $\psi_H^k(x) = k^r x$ for $x \in H^{2r}(X; \mathbb{Q})$. Then the following diagram is commutative*

$$\begin{array}{ccc} K_{\mathbb{C}}(X) & \xrightarrow{Ch} & H^{\text{even}}(X; \mathbb{Q}) \\ \downarrow{\scriptstyle \psi^k} & & \downarrow{\scriptstyle \psi_H^k} \\ K_{\mathbb{C}}(X) & \xrightarrow{Ch} & H^{\text{even}}(X; \mathbb{Q}), \end{array}$$

where ψ^k is the Adams operation (cf. IV.7.13).

Proof. It suffices to prove that for each vector bundle V of rank n, we have $Ch(\psi^k(V)) = \psi_H^k(Ch(V))$. By the splitting principle, we may further assume that

$n=1$. Then $\psi^k(V)=V^k$ and $c_1(V^k)=kc_1(V)$ (IV.7.13 and 3.10). Therefore

$$Ch(\psi^k(V))=\exp(kc_1(V))=1+\frac{k}{1!}c_1(V)+\frac{k^2}{2!}(c_1(V))^2+\cdots$$

$$=\psi_H^k\left(1+\frac{1}{1!}c_1(V)+\frac{1}{2!}(c_1(V))^2+\cdots\right)=\psi_H^k(Ch(V)).\quad \square$$

We now give an application of the theory of characteristic classes to a problem which was solved by Borel-Serre [1] using other methods.

3.28. Theorem. *Let TS^{2n} be the tangent bundle of the sphere. Then, if $n\neq 1$ or 3, TS^{2n} cannot be provided with a complex structure.*

Proof. If E is any complex vector bundle on S^{2n}, then $Ch(E)$ belongs to $H^*(S^{2n};\mathbb{Z})\subset H^*(S^{2n};\mathbb{Q})$ by 3.25. Since $c_i(E)=0$ for $i<n$, the Newton formulas imply that $Ch(E)=(-1)^{n-1}\dfrac{c_n(E)}{(n-1)!}$. By 3.5 and 3.17, $c_n(E)$ is twice the canonical generator of $H^{2n}(S^{2n};\mathbb{Z})$. Therefore $(n-1)!$ divides 2, which is only possible if $n=1$, 2, or 3.

If $n=2$, i.e. in the case of S^4, the computation above shows that if TS^4 is provided with a complex structure, then its class in $K_{\mathbb{C}}(X)$ is not trivial. But its class in $K_{\mathbb{R}}(S^4)$ is trivial, since $TS^4\oplus\theta_1$ is trivial (I.5.5). Since the homomorphism $K_{\mathbb{C}}(S^4)\to K_{\mathbb{R}}(S^4)$ is injective (IV.5.19), we have a contradiction. Hence $n\neq 2$, and the theorem is proved. $\square$

3.29. *Remark.* A purely "K-theoretical" proof of this last theorem is sketched in Exercise IV.8.15.

4. The Riemann–Roch Theorem and Integrality Theorems

4.1. Along the lines of Hirzebruch [2], we sketch a general method of constructing characteristic classes from formal power series. Let $f(x)$ be a formal power series with coefficients in a ring Λ, which may be written in the form $1+a_1x+a_2x^2+\cdots$. If $x_1,\ldots,x_n$ are n variables, then the product $f(x_1)\cdot f(x_2)\cdots f(x_n)$ is a formal power series which is symmetric in the variables $x_1,x_2,\ldots,x_n$. Therefore, we may write it in the form $\sum_{k=0}^{\infty} P_f^k(x_1,x_2,\ldots,x_n)$, where the P_f^k are well defined symmetric polynomials of degree k. If $\sigma_1,\ldots,\sigma_n$ denote the elementary symmetric functions of the x_i, then the polynomials $P_f^k(x_1,\ldots,x_n)$ may be uniquely expressed as $R_f^k(\sigma_1,\ldots,\sigma_n)$, where R_f^k is a polynomial of weight k in the variables $\sigma_1,\ldots,\sigma_n$.

Therefore, for each vector bundle V of rank n, the cohomology class $T_f^k(V)=R_f^k(c_1(V),c_2(V),\ldots,c_n(V))$ is a well-defined element of $H^{2k}(X)\otimes\Lambda$ (Note that $T_f^0(V)=1$).

4.2. Theorem. *Let* $T_f(V)=\sum_{k=0}^{\infty} T_f^k(V)\in H^{\text{even}}(X)$. *Then the characteristic class* T_f *is characterized by the following properties.*

a) T_f *is "natural"* (cf. 3.15).

b) $T_f(V_1\oplus V_2)=T_f(V_1)\cdot T_f(V_2)$.

c) $T_f(V)=f(\chi(V))$ *if* V *is of rank one, where* $\chi(V)=c_1(V)$ *denotes the Euler class of* V.

The proof of this theorem is analogous to the proof of Theorem 3.15.

4.3. Examples. When $f(x)=x$ and $\Lambda=\mathbb{Z}$, the class $T_f(V)$ is simply the total Chern class $c(V)=1+c_1(V)+\cdots+c_n(V)$ of the vector bundle V.

When $\Lambda=\mathbb{Q}$ and $f(x)=(1-e^{-x})/x$, the characteristic class T_f is called the "*Todd class*" of the bundle V, and is written as $\tau(V)$. A preliminary justification of the importance of this definition is the following theorem.

4.4. Theorem. *Let* V *be a complex vector bundle of rank* n, *and let* $\varphi_K: K_{\mathbb{C}}(X)\to K_{\mathbb{C}}(\overline{V})$ *and* $\varphi_H: H^*(X;\mathbb{Q})\to H^*(V;\mathbb{Q})$ *be the Thom isomorphisms in K-theory and cohomology, respectively (we let* H^* *denote Čech cohomology with compact support). Then* $\varphi_H^{-1}(Ch(\varphi_K(1))=\tau(V)$. *Furthermore, we have the formula*

$$Ch(\varphi_K(x))=\varphi_H(Ch(x)\cdot\tau(V))$$

for each element x *of* $K_{\mathbb{C}}(X)$ *(Note that* V *and* $\overline{V}$ *have the same underlying real vector bundle, hence the same underlying topological space).*

Proof. Let us temporarily denote the class $\varphi_H^{-1}(Ch(\varphi_K(1)))$ by $\tilde{\tau}(V)$. Then the method used in IV.7.18 shows that $\tilde{\tau}(V_1\oplus V_1)=\tilde{\tau}(V_1)\cdot\tilde{\tau}(V_2)$. By the splitting principle, it therefore suffices to check the first formula for the canonical line bundle L over $P_n=P_n(\mathbb{C})$. In this case, if we let s denote the zero section of L, and let $s_K^*: K_{\mathbb{C}}(\overline{L})\to K_{\mathbb{C}}(X)$ and $s_H^*: H^*(L;\mathbb{Q})\to H^*(X;\mathbb{Q})$ denote the induced homomorphism in K-theory and cohomology respectively, then we have the relations

$$\begin{aligned}\tilde{\tau}(L)\chi(L)&=s_H^*(Ch(\varphi_K(1)))=Ch(s_K^*(\varphi_K(1)))\\&=Ch(1-\overline{L})=1-\exp(-\chi(L))\end{aligned}$$

by 3.10. Since the relation $\tilde{\tau}(L)\chi(L)=1-\exp(-\chi(L))$ holds for each integer n, we see that the class $\tilde{\tau}(L)$ is obtained from the formal power series $(1-e^{-x})/x$ by the substitution $x\mapsto\chi(L)$. Therefore $\tilde{\tau}(L)=\tau(L)$, hence $\tilde{\tau}(V)=\tau(V)$ in general.

Furthermore,

$$\varphi_H^{-1}(Ch(\varphi_K(x)))=\varphi_H^{-1}(Ch(x)\cdot Ch(\varphi_K(1)))=Ch(x)\cdot\varphi_H^{-1}(Ch(\varphi_K(1)))=\tau(V)\cdot Ch(x).$$

Therefore $Ch(\varphi_K(x))=\varphi_H(\tau(V)\cdot Ch(x))$. □

4.5. *Remarks.* This is the first place where we must be careful about sign conventions in defining φ_K. The conventions we have adopted are the same as in Hirzebruch's book (for the definition of the Todd class), and are essentially motivated by algebraic geometry. In order to understand these conventions, it is convenient to think of a complex vector bundle as a real vector bundle provided with the ^{c}spinorial structure associated with the complex conjugate structure rather than the original complex structure.

Another aspect of these conventions that should be noted is that in the case of a trivial vector bundle V, we have $\tau(V)=1$. Hence Ch is compatible with the Thom isomorphisms in this case. In other words we have a commutative diagram

$$\begin{array}{ccc} K(X\times\overline{\mathbb{C}}^n) & \xrightarrow{Ch} & H^*(X\times\mathbb{C}^n) \\ \uparrow{\scriptstyle\varphi_K} & & \uparrow{\scriptstyle\varphi_H} \\ K(X) & \xrightarrow{Ch} & H^*(X) \end{array}$$

Finally, we must point out that the class which will turn out to be of importance for us is not $\tau(V)$, but rather its inverse $\tau'(V)=\tau(V)^{-1}$. In fact, the function $x/(1-e^{-x})$ may be expressed as

$$1+\frac{x}{2}+\sum_{s=1}^{\infty}(-1)^{s-1}\frac{B_s}{(2s)!}x^{2s},$$

where the B_s are the Bernoulli numbers (Hardy and Wright [1]). For example, $B_1=\frac{1}{6}$, $B_2=\frac{1}{30}$, $B_3=\frac{1}{42}$, $B_4=\frac{1}{30}$, etc. Therefore

$$\tau'_1(V)=\tfrac{1}{2}c_1,$$
$$\tau'_2(V)=\tfrac{1}{12}(c_2+c_1^2),$$
$$\tau'_3(V)=\tfrac{1}{24}c_2c_1,$$
$$\tau'_4(V)=\tfrac{1}{720}(-c_4+c_3c_1+3c_2^2+4c_2c_1^2-c_1^4),$$
$$\tau'_5(V)=\tfrac{1}{1440}(-c_4c_1+c_3c_1^2+3c_1c_2^2-c_2c_1^3),$$

and
$$\tau'_6(V)=\tfrac{1}{60480}(2c_6-2c_5c_1-9c_4c_2-5c_4c_1^2-c_3^2+11c_3c_2c_1+5c_3c_1^2 +10c_2^3+11c_2^2c_1^2-12c_2c_1^4+2c_1^6).$$

These computations are due to Hirzebruch [2].

4.6. Proposition. *Let* $\overline{V}$ *be the complex conjugate bundle of* V (I.4.8.e). *Then* $c_i(\overline{V})=(-1)^i c_i(V)$. *In particular, if* $V\approx\overline{V}$, *then the Chern classes* $c_i(V)$ *are elements of the 2-torsion of* $H^{2i}(X)$ *when* i *is odd.*

Proof. Let us define $c'_i(V)=(-1)^i c_i(\overline{V})$. In order to prove that $c'_i(V)=c_i(V)$, we must check that the $c'_i(V)$ satisfy the axioms of the Chern classes (3.15). Since the first two axioms are trivially satisfied, we need only verify that $c_1(\overline{V})=-c_1(V)$ if V is a complex line bundle. This follows immediately from 3.10.

Now if $V\approx\overline{V}$, we have $c_i(V)=(-1)^i c_i(V)$. Hence $2c_i(V)=0$ for i odd. □

4.7. Definition. Let W be a real vector bundle. We define its *Pontrjagin classes* as $(-1)^i c_{2i}(V)$, where $V = W \otimes \mathbb{C}$ is the complexification of W (I.4.8.e). They are denoted by $p_i(W)$.

The Pontrjagin classes satisfy some formal axioms, analogous to those of Chern classes. If $p(W) = 1 + p_1(W) + \cdots + p_n(W) + \cdots$, where $p_n(W) \in H^{4n}(X)$, denotes the "total Pontrjagin class", we have the relation $p(W_1 \oplus W_2) = p(W_1)p(W_2)$ mod 2 torsion. If W is the real vector bundle of rank 2 underlying a complex line bundle V, then $W \otimes \mathbb{C}$ may be identified with $V \oplus \bar{V}$ (I.4.8.e). Therefore, $p_1(W) = (c_1(V))^2$.

4.8. Assume now that W is a *real* vector bundle of rank $2n$, provided with a ^{c}spinorial structure (IV.4.25). By IV.5.14, we have a Thom isomorphism $\varphi_K: K_{\mathbb{C}}(X) \to K_{\mathbb{C}}(W)$, and hence the characteristic class $A(W) = \varphi_H^{-1}(Ch(\varphi_K(1)))$, which we call the "Atiyah-Hirzebruch class" of W. We would like to compute this class in terms of the Pontrjagin classes of W. Note that $A(W) = 1$ if W is trivial by our sign conventions (IV.5.8, IV.5.13, 5.4).

To do this, we consider the homomorphism $\mathrm{Spin}^c(2n) \to U(1)$ defined by $(\alpha, z) \to z^2$ (IV.4.30). This homomorphism induces a map $H^1(X; \mathrm{Spin}^c(2n)) \to H^1(X; U(1)) \approx \Phi_1^{\mathbb{C}}(X)$. Thus we may associate a complex line bundle L with the vector bundle W. We denote the Chern class $c_1(L)$ by $d(W)$. (∗ In fact it is not difficult to show that $d(W)$ determines the ^{c}spinorial structure of W.∗)

4.9. Proposition. *Let V be the complexified bundle of W. Then $A(W) = e^{d/2}\sqrt{\tau(V)}$, where $d = d(W)$, and $\tau(V)$ is the Todd class of V (4.3). More generally, $Ch(\varphi_K(x)) = \varphi_H(e^{d/2}\sqrt{\tau(V)} \cdot Ch(x))$.*

Proof. The class $A(W)$ has the following properties: $A(W_1 \oplus W_2) = A(W_1) \cdot A(W_2)$ (compare with 4.4), and $A(W) = 1$ if W is a trivial bundle.

Now the vector bundle $W \oplus W$ simultaneously has a complex structure V and a ^{c}spinorial structure. By IV.5.9, the corresponding Thom classes only differ by the factor L. Since $V \approx \bar{V}$, we have $A(W \oplus W) = e^d \tau(V)$. Therefore $A(W) = \sqrt{A(W)^2} = \sqrt{A(W \oplus W)} = e^{d/2}\sqrt{\tau(V)}$. The second part of the proposition may be proved in the same way as Proposition 4.4. □

4.10. In order to conveniently express the class $\sqrt{\tau(V)}$ in terms of the Pontrjagin classes of W, we formally write $c(V) = \prod_{i=1}^{n} (1 + x_i)(1 - x_i)$, so that the Pontrjagin classes of W, which was assumed to be of rank $2n$, appear as the elementary symmetric functions of the $(x_i)^2$. This is possible in rational cohomology, since $c_i(V) = 0$ mod 2 torsion for i odd (4.6). With this convention, formally we have

$$\sqrt{\tau(V)} = \prod \sqrt{\frac{1 - e^{-x_i}}{x_i} \cdot \frac{1 - e^{x_i}}{-x_i}} = \prod_{i=1}^{n} \frac{\sinh(x_i/2)}{(x_i/2)}$$

(which is a function of the $(x_i)^2$). From this computation, we obtain the following theorem which explicitly computes the class $A(W)$.

4.11. Theorem (Atiyah-Hirzebruch). *Let W be a ^{c}spinorial real vector bundle of rank $2n$, and let $d(W)$ be the cohomology class associated with W (4.8). For each element x of $K_{\mathbb{C}}(X)$, we have the relation*

$$Ch(\varphi_K(x)) = \varphi_H(A(W)\cdot Ch(x)),$$

where

$$A(W) = e^{d(W)/2}\prod_{i=1}^{n}\frac{\sinh(x_i/2)}{x_i/2},$$

and where we regard the Pontrjagin classes of W as the elementary symmetric functions of the $(x_i)^2$.

4.12. If W is the underlying real vector bundle of the complex vector bundle T, we provide it with the ^{c}spinorial structure associated with $\bar{T}$, according to our sign conventions. Then $d(W) = -c_1(T) = -\sum_{i=1}^{n} x_i$, and the Chern classes of T may be regarded as the elementary symmetric functions of the x_i. Therefore

$$A(W) = e^{d(W)/2}\prod_{i=1}^{n}\frac{\sinh(x_i/2)}{x_i/2} = \prod_{i=1}^{n}\frac{1-e^{-x_i}}{x_i} = \tau(T).$$

In general, the characteristic class $\prod_{i=1}^{n}\frac{\sinh(x_i/2)}{x_i/2}$ expressed in terms of the Pontrjagin classes, will be called the "reduced Atiyah-Hirzebruch class" of W and will be denoted by $\hat{A}(W)$. As in 4.5, we actually use its inverse more often. If we set $\hat{A}'(W) = 1/\hat{A}(W)$, we find that

$$\hat{A}'_1(W) = -\tfrac{1}{24}p_1$$
$$\hat{A}'_2(W) = \tfrac{7}{5760}p_1^2 - \tfrac{1}{1440}p_2$$
$$\cdots\cdots\cdots\cdots\cdots\cdots$$

4.13. The preceding observations may be extended to real K-theory by means of the Thom isomorphism in real K-theory (IV.5.14). More precisely, let W be a real vector bundle of rank $8n$, provided with a spinorial structure. Then we have Thom isomorphisms

$$\varphi_K\colon K_{\mathbb{R}}(X) \longrightarrow K_{\mathbb{R}}(V) \quad\text{and}\quad \varphi_H\colon H^*(X;\mathbb{Q}) \longrightarrow H^*(V;\mathbb{Q}).$$

For any locally compact space Y, we may consider the "Pontrjagin character" $P\colon K_{\mathbb{R}}(Y) \to \bigoplus_{i=0}^{\infty} H^{4i}(Y;\mathbb{Q})$, obtained by composing the complexification homomorphism $K_{\mathbb{R}}(Y) \to K_{\mathbb{C}}(Y)$ with the Chern character $K_{\mathbb{C}}(Y) \to H^*(Y;\mathbb{Q})$.

4.14. Theorem. *Let W be a real vector bundle of rank $8n$ provided with some spinorial structure. For each element x of $K_{\mathbb{R}}(X)$, we have the relation*

$$P(\varphi_K(x)) = \varphi_H(\hat{A}(W)\cdot P(x)),$$

where

$$\hat{A}(W) = \prod_{i=0}^{4n} \frac{\sinh(x_i/2)}{x_i/2}$$

and where again we regard the Pontrjagin classes of W as the elementary symmetric functions of the $(x_i)^2$.

4.15. *Remark.* Let x be an element of $K_{\mathbb{C}}(X)$ (resp. $K_{\mathbb{R}}(X)$), written in the form $[E]-[F]$, where E and F are vector bundles. Then we can define $\tau(x)=\tau(E)\tau(F)^{-1}$ (resp. $\hat{A}(x)=\hat{A}(E)\hat{A}(F)^{-1}$). Similarly, we set $d(x)=d(E)-d(F)$ if x is provided with a stable ^{c}spinorial structure (IV.5).

4.16. Now let X and Y be compact differentiable manifolds such that $\mathrm{Dim}(Y)-\mathrm{Dim}(X)=0 \bmod 2$, and let $f\colon X\to Y$ be a continuous map such that $v_f=f^*(TY)-TX$ is provided with a stable ^{c}spinorial structure (IV.5.23). In IV.5.27 we defined a Gysin homomorphism

$$f_* = f_*^K\colon K_{\mathbb{C}}(X) \longrightarrow K_{\mathbb{C}}(Y).$$

Along the same lines, we define a Gysin homomorphism in cohomology

$$f_*^H\colon H^*(X;\mathbb{Q}) \longrightarrow H^*(Y;\mathbb{Q})$$

which is actually dual (via Poincaré duality) to the homomorphism

$$f_*\colon H_*(X;\mathbb{Q}) \longrightarrow H_*(Y;\mathbb{Q}).$$

4.17. Theorem (the Atiyah-Hirzebruch version of the Riemann-Roch theorem). *Let $d=d(v_f)$ as in 4.15. Then, for each element x of $K_{\mathbb{C}}(X)$, we have the relation*

$$Ch(f_*^K(x)) = f_*^H(e^{d/2}\cdot\hat{A}(v_f)\cdot Ch(x)),$$

where $f\colon X\to Y$ satisfies the hypothesis of 4.16, and where $v_f=[f^*(TY)]-[TX]$.

Proof. Since both members of the formula above depend only on the homotopy class of f (cf. IV.5.24), we may assume without loss of generality that f is differentiable. First let us consider the case where f is an imbedding with N as normal bundle. Then the classes of v_f and N are equal in the group $K_{\mathbb{R}}(X)$. Hence

$\hat{A}(v_f)=\hat{A}(N)$. On the other hand, if we represent N as a tubular neighbourhood of X in Y, then the diagram

$$\begin{array}{ccc} K_{\mathbb{C}}(N) & \xrightarrow{u} & K_{\mathbb{C}}(Y) \\ \downarrow{\scriptstyle Ch} & & \downarrow{\scriptstyle Ch} \\ H^*(N;\mathbb{Q}) & \xrightarrow{v} & H^*(Y;\mathbb{Q}), \end{array}$$

where u and v are induced by the canonical map $Y \to \dot{N}$ (cf. IV.5.21), is commutative.

Let us now consider the diagram

$$\begin{array}{ccc} K_{\mathbb{C}}(X) & \xrightarrow{\varphi_K} & K_{\mathbb{C}}(N) \\ \downarrow{\scriptstyle Ch} & & \downarrow{\scriptstyle Ch} \\ H^*(X;\mathbb{Q}) & \xrightarrow{\varphi_H} & H^*(N;\mathbb{Q}). \end{array}$$

In general, this diagram is *not* commutative. In fact, if $x \in K_{\mathbb{C}}(X)$, then $Ch(\varphi_K(x))=\varphi_H(A(N)\cdot Ch(x))$ by 4.11. Therefore, $Ch(f_*^K(x))=Ch(u(\varphi_K(x))=v(Ch(\varphi_K(x)))=v(\varphi_H(e^{d/2}\cdot\hat{A}(v_f)\cdot Ch(x))=f_*^H(e^{d/2}\cdot\hat{A}(v_f)\cdot Ch(x))$, an identity which proves the theorem for this case.

In the general case, the differentiable map f may be factored into an imbedding followed by a projection, i.e. $f=p\cdot i$ where $i\colon x \rightarrowtail Y\times S^{2n}$ and $p\colon Y\times S^{2n} \twoheadrightarrow Y$ (projection onto the first factor). Hence $f_*^K=p_*^K\cdot i_*^K$ (IV.5.24), and similarly, $f_*^H=p_*^H\cdot i_*^H$. Also note that $v_f=v_i$ in the group $\widetilde{K}_{\mathbb{R}}(X)$. Since the diagram

$$\begin{array}{ccc} K_{\mathbb{C}}(Y\times S^{2n}) & \xrightarrow{p_*^K} & K_{\mathbb{C}}(Y) \\ \downarrow{\scriptstyle Ch} & & \downarrow{\scriptstyle Ch} \\ H^*(Y\times S^{2n};\mathbb{Q}) & \xrightarrow{p_*^H} & H^*(Y;\mathbb{Q}) \end{array}$$

is commutative by the multiplicative property of the Chern character, we have the relations

$$\begin{aligned} Ch(f_*^K(x)) &= Ch(p_*^K\cdot i_*^K(x))=p_*^H(Ch(i_*^K(x))) \\ &= p_*^H(i_*^H(e^{d/2}\cdot\hat{A}(v_f)\cdot Ch(x)))=f_*^H(e^{d/2}\cdot\hat{A}(v_f)\cdot Ch(x))). \quad \square \end{aligned}$$

4.18. Corollary. *Let us assume that v_f is provided with a stable complex structure (hence with a stable ^{c}spinorial structure; cf.* IV.5.27*). Then $e^{d/2}\cdot\hat{A}(v_f)$ is the Todd class $\tau(v_f)$ of v_f. Therefore, we have the relation*

$$Ch(f_*^K(x))=f_*^H(\tau(v_f)\cdot Ch(x)).$$

4.19. Theorem. *Let $f: X \to Y$ be a continuous map between compact differentiable manifolds, such that* $\mathrm{Dim}(Y)-\mathrm{Dim}(X)=0 \bmod 8$. *Suppose that $v_f = f^*(TY) - TX$ is provided with a stable spinorial structure. Then we have the relation*

$$P(f_*^K(x)) = f_*^H(\hat{A}(v_f) \cdot P(x)),$$

where f_*^K is the Gysin homomorphism in real K-theory, and P is the Pontrjagin character (4.13).

The proof of this theorem is analogous to the proof of Theorem 4.17.

4.20. We apply the above assertions to the case where Y is a point. Then the Gysin homomorphism $f_*^H: H^*(X; \mathbb{Q}) \to H^*(Y; \mathbb{Q})$ maps $H^i(X; \mathbb{Q})$ to 0 if $i \neq \mathrm{Dim}(X)$, and is an isomorphism $H^p(X; \mathbb{Q}) \approx H^0(Y; \mathbb{Q}) \approx \mathbb{Q}$ for $p = \mathrm{Dim}(X)$. In other words, $f_*^H(z)$ is the value of the cohomology class z on the fundamental class of X. Since up to isomorphism $Ch: K(Y) \to H^0(Y; \mathbb{Q})$ is the canonical inclusion of $\mathbb{Z}$ in $\mathbb{Q}$, we obtain the following integrality theorem as a consequence of 4.17.

4.21. Theorem (Atiyah-Hirzebruch [1]). *Let X be a compact differentiable manifold of even dimension, such that TX is provided with a stable ^{c}spinorial structure with associated cohomology class $d \in H^2(X; \mathbb{Z})$. Then for each element x of $K_{\mathbb{C}}(X)$, the value of $e^{-d/2}Ch(x)/\hat{A}(TX)$ on the fundamental class of X is an integer.*

4.22. Examples. If X is a 4 dimensional manifold, and if we choose $x=1$ then we find that $p_1 - 3d^2$ is divisible by 24. If X is a 6 dimensional manifold, we find that $d^3 - dp_1$ is divisible by 48. If X is an 8 dimensional manifold, we find that $15d^4 + 30p_1 d^2 + 7p_1^2 - 4p_2$ is divisible by 5760, etc.

4.23. Corollary. *Let X be a compact differentiable manifold of even dimension, such that TX is provided with a stable complex structure. Then for each element x of $K_{\mathbb{C}}(X)$, the value of $ch(x)/\tau(TX)$ on the fundamental class of X is an integer.*

Example. Again, if we choose $x=1$, the computations made in 4.5 show that some rational characteristic classes are integral. For example, if $\mathrm{Dim}(X)=8$, we find that the value of $c_4 - c_3c_1 - 3c_2^2 - 4c_2c_1^2 + c_1^4$ on the fundamental class of X is divisible by 720, where c_i are the Chern classes of TX (provided with its stable complex structure).

4.24. Theorem (cf. Hirzebruch [1]). *Let X be a compact manifold of dimension $= 4 \bmod 8$, such that TX is provided with a spinorial structure (i.e. $w_2(TX)=0$; cf. IV.4.20). Then for each element x of $K_{\mathbb{R}}(X)$, the value of $P(x)/\hat{A}(TX)$ on the fundamental class of X is an even integer.*

Proof. We apply Theorem 4.19 to the constant map from X to S^4. Then the Pontrjagin character $P: K_{\mathbb{R}}(S^4) \to H^*(S^4; \mathbb{Q})$ isomorphically maps $K_{\mathbb{R}}(S^4)$ onto $2H^4(S^4; \mathbb{Z}) = 2\mathbb{Z} \subset \mathbb{Q} = H^4(S^4; \mathbb{Q})$, since Ch isomorphically maps $K_{\mathbb{C}}(S^4)$ onto

$H^4(S^4;\mathbb{Z})$ (3.25), and since the complexification homomorphism $K_{\mathbb{R}}(S^4)\approx \mathbb{Z}\to K_{\mathbb{C}}(S^4)\approx\mathbb{Z}$ is multiplication by 2 (III.5.20). On the other hand, f_*^H may be factored into $\beta_*^H\alpha_*^H$, where $\alpha: X\to Point$ and $\beta: Point\to S^4$. Since β_* is an isomorphism from $H^0(Point)$ to $H^4(S^4)$, the expression $f_*^H(P(x)\hat{A}(v_f))$ is actually the value of $P(x)/\hat{A}(TX)$ on the fundamental class of X when we identify $H^4(S^4)$ with $\mathbb{Z}$. □

4.25. Example. Let X be of dimension 4. Then the value of p_1 is divisible by 48 if $w_2(TX)=0$ (note that $p_1(TX)$ is in general divisible by 24 if $w_2(TX)$ arises from an integral class; cf. IV.4.20 and IV.4.25). Similarly, if X is of dimension 8 and again $w_2(TX)=0$, then the value of $7p_1^2-4p_2$ is divisible by 11 520, etc. Here the p_i refer to the Pontrjagin classes of the tangent bundle.

4.26. *Remark.* The use of ^{c}spinoriality or spinoriality conditions may be avoided by considering the theory $K_{\mathbb{C}}(X)\otimes\mathbb{Z}[1/2]$ (cf. IV.8.13). We leave as an exercise for the reader, the fact that *up to a power of* 2, the reduced Atiyah-Hirzebruch class of *any* oriented manifold, evaluated on the fundamental class, is an integer. For example, if X is of dimension 4 (resp. 8), the value of p_1 (resp. $7p_1^2-4p_2$) on the fundamental class is divisible by 3 (resp. 45), etc.

4.27. In differential topology it is well known that the Pontrjagin classes of a manifold are not homotopy invariant. However, with the preceding method, we can prove that suitable classes are invariant mod m. More precisely, let p_i' denote the Pontrjagin classes considered in the quotient group $H^{4i}(X;\mathbb{Z})/\Gamma^{4i}(X)$, where $\Gamma^{4i}(X)$ is the torsion subgroup of $H^{4i}(X;\mathbb{Z})$. If $f: X\to Y$ is a homotopy equivalence, we have $p'(v_f)=f^*(p'(TY))\cdot p'(TX)^{-1})$. Since v_f is spinorial, we have $\hat{A}(v_f)\in P(K_{\mathbb{R}}(X))$ by Theorem 5.24. Here is an application:

4.28. Theorem (Hirzebruch [1]). *Let X be a compact manifold and let TX be its tangent bundle. Then the image of $p_1(TX)$ in the quotient group $H^4(X;\mathbb{Z})/\Gamma^4(X)$ is homotopy invariant mod* 24. *If $H^2(X;\mathbb{Z}/2)=0$, then the image is invariant mod* 48.

4.29. To conclude this section we give a final application of the Riemann-Roch theorem. Consider a differentiable fibration between compact manifolds

$$E\xrightarrow{\pi} B$$

with fiber F, a compact manifold of even dimension. We assume that TF and TB (hence v_π) are provided with ^{c}spinorial structures.

4.30. Theorem (Atiyah-Hirzebruch [3]). *Let us denote the class $e^{d(TF)}A(TF)$ evaluated on the fundamental class of F by $\mathscr{A}(F)$. Then the composition $\pi_*\pi^*$ is the $K_{\mathbb{C}}(B)$-module homomorphism defined by multiplication with an element of $K_{\mathbb{C}}(B)$. This element may be written as $\mathscr{A}(F)+x'$, where $x'\in K_{\mathbb{C}}'(B)$. In particular, if B is connected and if $\mathscr{A}(F)=\pm 1$, then the homomorphism $\pi^*: K_{\mathbb{C}}(B)\to K_{\mathbb{C}}(E)$ is injective.*

Proof. Since π^* and π_* are $K_{\mathbb{C}}(B)$-module homomorphisms, $\pi_*\pi^*$ is well-defined as multiplication by an element x of $K_{\mathbb{C}}(B)$. To find the desired expression of x, we use the commutative diagrams

$$\begin{array}{ccc} K_{\mathbb{C}}(B) & \xrightarrow{\pi^*} & K_{\mathbb{C}}(E) \\ \downarrow & & \downarrow \\ K_{\mathbb{C}}(P) & \xrightarrow{(\pi_F)^*} & K_{\mathbb{C}}(F) \end{array} \qquad \text{and} \qquad \begin{array}{ccc} K_{\mathbb{C}}(E) & \xrightarrow{\pi_*} & K_{\mathbb{C}}(B) \\ \downarrow & & \downarrow \\ K_{\mathbb{C}}(F) & \xrightarrow{(\pi_F)_*} & K_{\mathbb{C}}(P) \end{array}$$

where P denotes a point. Now we need only prove the theorem for the case where B is a point; but this follows directly from the definition of the Atiyah-Hirzebruch class. □

4.31. Examples. The reader can verify that the flag manifolds and complex projective spaces are examples of manifolds F such that $\mathscr{A}(F) = \pm 1$. This gives an a posteriori reason for the splitting principle in complex K-theory.

5. Applications of K-Theory to Stable Homotopy

5.1. Let E be a vector bundle with compact base X. Up to homotopy we may associate a metric with E (cf. I.8.5), hence a sphere bundle $S(E)$. If E' is another vector bundle, then the sphere bundles $S(E)$ and $S(E')$ are said to be fiber homotopy equivalent if there exists a continuous map $f: S(E) \to S(E')$, which makes the diagram

$$\begin{array}{ccc} S(E) & \xrightarrow{f} & S(E') \\ & \searrow \quad \swarrow & \\ & X & \end{array}$$

commutative, and such that for each point x of X, the map $f_x: S(E_x) \to S(E'_x)$ is a homotopy equivalence. According to a theorem of Dold-Lashof [1], the relation thus defined between E and E' is an equivalence relation; however, we do not use this result here. If we let $\Gamma(X)$ denote the quotient of $\Phi(X)$ by the equivalence relation generated by this relation, we see that $\Gamma(X)$ is an abelian monoid with respect to the Whitney sum of vector bundles. More precisely, $S(E \oplus F)$ may be identified with the quotient of $S(E) \times S(F) \times I$ by the relation $(x, y, 0) \sim (x', y, 0)$ and $(x, y, 1) \sim (x, y', 1)$. Hence, if $S(E) \sim S(E')$ and $S(F) \sim S(F')$, we have $S(E \oplus F) \sim S(E' \oplus F')$. The symmetrization of $\Gamma(X)$ is denoted by $J(X)$; it is a quotient of $K_{\mathbb{R}}(X)$ (cf. Atiyah [1]).

5.2. Certain observations about the group $K_{\mathbb{R}}(X)$ may be repeated for the group $J(X)$. For example, if X is connected we have $J(X) \approx \mathbb{Z} \oplus \tilde{J}(X)$, where $\tilde{J}(X)$

denotes the quotient of $\Gamma(X)$ by the equivalence relation generated by *stable* fiber homotopy type; i.e. $S(E)$ is stably equivalent to $S(E')$ if $S(E \oplus n)$ is equivalent to $S(E' \oplus n')$ for some n and n'. Then $J(X) \approx \text{inj lim}\, \Gamma_p(X)$ where $\Gamma_p(X)$ is the subset of $\Gamma(X)$ generated by bundles $S(E)$, where E has rank p.

5.3. Let us choose a base point y_0 in a compact space Y, and let us denote by $\mathrm{H}(p)$ the space of homotopy equivalences from S^{p-1} to S^{p-1}. Let $[Y, \mathrm{O}(p)]'$ (resp. $[Y, \mathrm{H}(p)]'$) be the set of homotopy classes of maps $f: Y \to \mathrm{O}(p)$ (resp. $f: Y \to \mathrm{H}(p)$) such that $f(y_0) = 1$. Let $X = SY$ be the suspension of Y. By computations made in I.3.9, the set $\Phi_p^{\mathbb{R}}(SY)$ may be identified with the quotient of $[Y, \mathrm{O}(p)]'$ by the action of $\mathbb{Z}/2 \approx \pi_0(\mathrm{O}(p))$. Let $[Y, \mathrm{O}(p)]''$ (resp. $[Y, \mathrm{H}(p)]''$) denote the quotient of $[Y, \mathrm{O}(p)]'$ (resp. $[Y, \mathrm{H}(p)]'$) by the action of $\pi_0(\mathrm{O}(p))$ (resp. $\pi_0(\mathrm{H}(p))$. This last action is defined by $(a, f) \mapsto afa^{-1}$, where a^{-1} is the homotopy inverse of a [7].

If $S(E)$ is the sphere bundle associated with a vector bundle E over SY, then $S(E)$ defines an element of $[Y, \mathrm{O}(p)]''$, hence an element of $[Y, \mathrm{H}(p)]''$.

5.4. Proposition. *Let E and E' be vector bundles of rank p over SY, such that the sphere bundles $S(E)$ and $S(E')$ are homotopy equivalent. Then they define the same class in $[Y, \mathrm{H}(p)]''$. Moreover, the map*

$$\Gamma_p(SY) \longrightarrow [Y, \mathrm{H}(p)]''$$

thus defined, is injective (but not surjective in general).

Proof. Let us denote the upper hemisphere (resp. lower hemisphere) of SY by S^+Y (resp. S^-Y). Then the vector bundle E (resp. E') is obtained by glueing together the trivial bundles $E_1 = S^+Y \times \mathbb{R}^p$ and $E_2 = S^-Y \times \mathbb{R}^p$, using a "transition function" $f: Y \to \mathrm{O}(p)$ (resp. $f': Y \to \mathrm{O}(p)$) such that $f(y_0) = 1$ (resp. $f'(y_0) = 1$). If $S(E)$ and $S(E')$ are homotopy equivalent, then we have a commutative diagram, analogous to the one considered in I.3.9,

$$\begin{array}{ccc} S(E) & \xrightarrow{\alpha} & S(E) \\ \big\downarrow{\scriptstyle \hat{f}} & & \big\downarrow{\scriptstyle \hat{f}'} \\ S(E') & \xrightarrow{\alpha'} & S(E') \end{array}$$

where the dotted arrows are only defined over Y, and where α and α' are fiber homotopy equivalences. From this diagram we see that f is homotopic to $\lambda f' \mu$ (where $\lambda, \mu \in \mathrm{H}(p)$) as maps from Y to $\mathrm{H}(p)$. Therefore the classes of f and f' in $[Y, \mathrm{H}(p)]''$ are equal, and the map $\Gamma_p(SY) \to [X, \mathrm{H}(p)]''$ is well-defined.

Now we prove that this map is injective. Assume that f and f' are maps from Y to $\mathrm{O}(p)$ which define the same class in $[Y, \mathrm{H}(p)]''$. If E and E' are the associated vector bundles, we define a fiber map from $S(E)$ to $S(E')$ by the same diagram as above, where $\alpha' = \mathrm{Id}$, and α is defined using the homotopy of $\hat{f}'^{-1}\hat{f}$ to the identity

[7] It can be proved that $\pi_0(\mathrm{H}(p)) \approx \pi_0(\mathrm{O}(p)) \approx \mathbb{Z}/2$, but we do not need this result here.

within the homotopy equivalences of S^{p-1} (cf. I.3.9). By its construction, this map is a fiber homotopy equivalence. □

5.5. Proposition. *Let* $\mathrm{H} = \operatorname{inj\,lim} \mathrm{H}(p)$, *and let* $[Y, \mathrm{H}]'$ *be the set of homotopy classes of maps* f *from* Y *to* H, *such that* $f(y_0) = 1$. *Then* $[Y, \mathrm{H}]' = \operatorname{inj\,lim}[Y, \mathrm{H}(p)]' = \operatorname{inj\,lim}[Y, \mathrm{H}(p)]''$. *In particular,* $\tilde{J}(SY)$ *is a subset of* $[Y, \mathrm{H}]'$.

Proof. It suffices to prove that the action of $\pi_0(\mathrm{H}(p))$ on $[X, \mathrm{H}(2p)]'$ is trivial. If a and b are homotopy equivalences from S^{p-1} to itself, then the pair (a, b) defines a homotopy equivalence from $S^{p-1} * S^{p-1} = S^{2p-1}$ to itself [8], which we denote by $\theta_{a,b}$. Then $\theta_{a,b} = \theta_{a,1} \cdot \theta_{1,b}$, and $\theta_{a,1}$ is homotopic to $\theta_{1,a}$ (use a rotation of S^{2p-1} which switches the two S^{p-1} factors). Therefore $\theta_{a,b}$ is homotopic to $\theta_{ab,1}$. Now if $f\colon Y \to \mathrm{H}(p)$ is a continuous map such that $f(y_0) = 1$, if $a \in \mathrm{H}(p)$ and if $a' \in \mathrm{H}(p)$ is a representative of the inverse homotopy equivalence to a, then $\theta_{af(y)a',1}$ is homotopic to $\theta_{af(y)a',a'a} = \theta_{a,a'} \cdot \theta_{f(y),1} \cdot \theta_{a',a} \sim \theta_{f(y),1}$ in a continuous fashion. □

5.6. Let $\mathrm{H}_1(p)$ (resp. $\mathrm{H}_{-1}(p)$) be the subset of $\mathrm{H}(p)$ consisting of homotopy equivalences which are homotopic to $\mathrm{Id}_{S^{p-1}}$ (resp. homotopic to the map $(x_1, \dots, x_p) \mapsto (x_1, \dots, -x_p)$). We let $\mathrm{H}_0(p)$ denote the set of continuous maps from S^{p-1} to S^{p-1}, which are homotopic to the constant map. Finally, we let $H_n^0(p)$ (for $n = 0, 1$, or -1) denote the subset of $\mathrm{H}_n(p)$, consisting of maps σ such that $\sigma(e) = e$, where e is the base point of S^{p-1}. In fact, $\mathrm{H}_n^0(p)$ is the fiber of a Hurewicz fibration

$$\mathrm{H}_n^0(p) \longrightarrow \mathrm{H}_n(p) \xrightarrow{\ \tau\ } S^{p-1},$$

where $\tau(\sigma) = \sigma(e)$. In particular, for any space Y with base point we have the following "exact sequence" of pointed sets

$$[Y, \Omega S^{p-1}]' \longrightarrow [Y, \mathrm{H}_n^0(p)]' \longrightarrow [Y, \mathrm{H}_n(p)]' \longrightarrow [Y, S^{p-1}]'.$$

If we define $\mathrm{H}_n^0 = \operatorname{inj\,lim} \mathrm{H}_n^0(p)$, we have $[Y, \mathrm{H}_n^0]' \approx [Y, \mathrm{H}_n]'$ since S^{p-1} is contractible in S^p. In particular, $[Y, \mathrm{H}_1^0] \approx [Y, \mathrm{H}_1] \approx \operatorname{inj\,lim}[Y, \mathrm{H}_1^0(p)]$.

5.7. Let $c\colon S^{p-1} \to S^{p-1} \vee S^{p-1}$ be the continuous map which sends S^{p-2} into the base point of $S^{p-1} \vee S^{p-1}$.

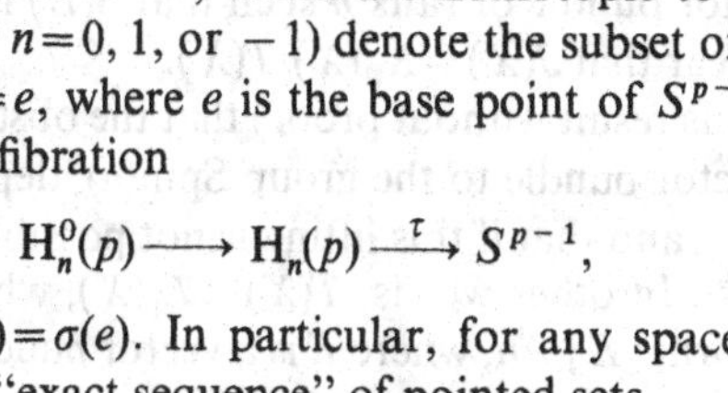

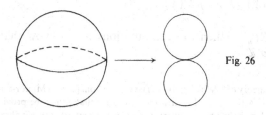

Fig. 26

[8] The join, $Z * T$, of two spaces Z and T is the quotient of $Z \times T \times I$ by the relations $(x, z, 0) \sim (x', z, 0)$ and $(x, z, 1) \sim (x, z', 1)$.

This map defines two continuous maps,

$$\alpha\colon \mathrm{H}_0^0(p)\times \mathrm{H}_1^0(p)\to \mathrm{H}_1^0(p) \quad\text{and}\quad \beta\colon \mathrm{H}_1^0(p)\times \mathrm{H}_{-1}^0(p)\longrightarrow \mathrm{H}_0^0(p).$$

If $a\in \mathrm{H}_1^0(p)$ and $b\in \mathrm{H}_{-1}^0(p)$, then elementary deformation arguments show that the maps $x\mapsto\alpha(x,a)$ and $y\mapsto\beta(y,b)$ define homotopy equivalences, inverse to each other, between $\mathrm{H}_0^0(p)$ and $\mathrm{H}_1^0(p)$. Since $[Y,\mathrm{H}_0^0(p)]'\approx[S^{p-1}\wedge Y, S^{p-1}]'$, we obtain the following theorem.

5.8. Theorem. *Let Y be an arcwise connected compact space. Then, via the maps defined above, $\tilde{J}(SY)$ may be identified with a subgroup of* $\operatorname{inj}\lim[S^{p-1}\wedge Y, S^{p-1}]'$. *In particular, $\tilde{J}(S^r)=\tilde{J}(SS^{r-1})$ is isomorphic to a subgroup of*

$$\pi_{r-1}^s=\operatorname{inj}\lim \pi_{p+r-1}(S^p).$$

5.9. The observations above provide the motivation for a systematic study of the groups $J(X)$ in general. The purpose of the next few paragraphs is to give a lower bound $J'(X)$ of the group $J(X)$. By "lower bound" we mean the existence of an epimorphism $J(X)\to J'(X)$. This group $J'(X)$ will be effectively computable in terms of K-theory.

More precisely, let $T(X)$ denote the subgroup of $K_{\mathbb{R}}(X)$ generated by elements of the form $[E]-n$, where E is a vector bundle of rank n such that $S(E)$ is of trivial stable fiber homotopy type. It is clear that $J(X)=K_{\mathbb{R}}(X)/T(X)$.

It is possible to prove (we use this result without proof) that the obstruction to lifting the structural group of a vector bundle to the group Spin(n), depends only on the fiber homotopy type of $S(E)$, and that if this lifting is not possible then the fiber homotopy type is not trivial[9)]. In other words, $T(X)\subset T_1(X)$, where $T_1(X)$ denotes the set of elements of the form $[E]-n$, where E is a vector bundle of rank $n=8k$, which is provided with a spinorial structure.

Now let $T'(X)$ be the subgroup of $T_1(X)$ consisting of elements of the form $x=[E]-n$, such that for each k, $\rho^k(x)=\psi^k(1+y)/1+y$, where $y\in K'_{\mathbb{R}}(X)$ independent of k. Of course, the definition of ρ^k implicitly assumes the choice of a Thom isomorphism $\varphi\colon K_{\mathbb{R}}(X)\to K_{\mathbb{R}}(E)$. But, as was pointed out in IV.7.29, if φ and φ' are two Thom isomorphisms, then the classes $(\varphi^{-1}\psi^k\varphi)(1)$ and $(\varphi'^{-1}\psi^k\varphi')(1)$ differ only by a multiplicative factor of the form $\psi^k(1+y)/1+y$, where $y\in \tilde{K}_{\mathbb{R}}'(X)$. The inclusion $T(X)\subset T'(X)$, which was proved in 2.5, induces an epimorphism $J(X)\twoheadrightarrow J'(X)$, where $J'(X)=K_{\mathbb{R}}(X)/T'(X)$.

5.10. Example. If $X=RP_{n-1}$, then the computations in 2.7 show that $J(X)=J'(X)=K_{\mathbb{R}}(X)=\mathbb{Z}\oplus\mathbb{Z}/a_n\mathbb{Z}$.

---

9) This result follows from the theory of Stiefel–Whitney classes (cf. Thom [1] and Milnor-Stasheff [1]). However, if X is a sphere S^r with $r\geqslant 2$, then it is easy to prove that E may be provided with a spinorial structure by considering the homotopy exact sequence associated with the fibration

$$\mathbb{Z}/2\longrightarrow \mathrm{Spin}(n)\longrightarrow \mathrm{SO}(n)$$

and by describing the oriented bundles over S^r with $\pi_{r-1}(\mathrm{SO}(n))$.

5.11. In order to make explicit computations, we use cohomological methods. The following proposition is a step in this direction.

5.12. Proposition. *Let E be a complex vector bundle of rank n. Then $Ch(\rho_{\mathbb{C}}^k(\bar{E}))= k^n\psi_H^k(\tau(E))/\tau(E)$ where $\tau(E)$ is the Todd class of E* (cf. 4.3).

Proof. By the splitting principle, it suffices to verify the proposition for the case where E is a line bundle. In this case $\rho_{\mathbb{C}}^k(\bar{E})=1+\bar{E}+\cdots+\bar{E}^{k-1}$, and $Ch(\rho_{\mathbb{C}}^k(\bar{E}))= 1+e^{-x}+\cdots+e^{-(k-1)x}$, where $x=c_1(E)$. On the other hand,

$$k\psi_H^k(\tau(E))/\tau(E)=k\psi_H^k\left(\frac{1-e^{-x}}{x}\right)\cdot\frac{x}{1-e^{-x}}=k\frac{1-e^{-kx}}{kx}\cdot\frac{x}{1-e^{-x}}$$

$$=1+e^{-x}+\cdots+e^{-(k-1)x}. \quad \square$$

5.13. Recall that the Bernoulli numbers B_s are defined from the series

$$x/1-e^{-x}=\sum_{s=0}^{\infty}\beta_s x^s/s!,$$

by the formula $B_s=(-1)^{s-1}\beta_{2s}$ (note that $\beta_{2s+1}=0$ for $s>0$).

5.14. Lemma. *Let $\sum_{t=1}^{\infty}\alpha_t x^t/t!$ be the series expansion of the function* $\operatorname{Log}\left(\frac{1-e^{-x}}{x}\right)$. *Then $\alpha_t=\beta_t/t$* for $t>1$.

Proof. Derive the series expansion of the function $\operatorname{Log}\left(\frac{1-e^{-x}}{x}\right)$. $\square$

5.15. Proposition. *For each element x of $K_{\mathbb{C}}(X)$, we have the formula*

$$\operatorname{Log}(\tau(x))=\sum_{t=1}^{\infty}(-1)^t\alpha_t Ch_t(x)$$

where $\tau(x)$ is the Todd class of x, where $Ch_t(x)\in H^{2t}(X;\mathbb{Q})$ is the component of degree $2t$ of $Ch(x)$, and where $\alpha_1=\frac{1}{2}$, $\alpha_{2s+1}=0$ for $s>0$, and $\alpha_{2s}=(-1)^{s-1}B_s/2s$.

Proof. This is an immediate consequence of the splitting principle and the observations above. $\square$

5.16. Theorem. *Let $x=E-n$ be an element of $K_{\mathbb{R}}(X)$ where* $\operatorname{rank}(E)=n$. *If the class of x is 0 in the group $J(X)$, then there exists an element $y\in K'_{\mathbb{R}}(X)$, such that*

$$\sum_{t=1}^{\infty}\frac{\alpha_t}{2}Ch_t(cx)=\operatorname{Log}(Ch(1+cy)). \quad (1)$$

If $X=SY$, then this condition may be simply written as

$$\frac{\alpha_t}{2}\,Ch_t(cx)=Ch_t(cy).$$

Finally, if $K_{\mathbb{R}}(X)$ has no torsion, then condition (1) *supplies a necessary and sufficient condition for $x\in T'(X)$.*

Proof. Let us write the equation $\rho^k_{\mathbb{R}}(x)=\psi^k(1+y)/1+y$, where $y\in K'_{\mathbb{R}}(X)$. By IV.7.28, we have the formula $c(\rho^k_{\mathbb{R}}(x))^2=\rho^k_{\mathbb{C}}(cx)=\psi^k(1+cy)^2/(1+cy)^2$. Therefore $Ch(\rho^k_{\mathbb{C}}(cx))=\psi^k_H(\tau(cx))/\tau(cx)=\psi^k_H(Ch(1+cy)^2)/Ch(1+cy)^2$, or

$$\psi^k_H\left(\frac{\tau(cx)}{Ch(1+cy)^2}\right)=\frac{\tau(cx)}{Ch(1+cy)^2}$$

by 4.12. Since the only invariant elements of $H^{\text{even}}(X;\mathbb{Q})$ under ψ^k_H are the constants, we see that

$$\tau(cx)=Ch(1+cy)^2.$$

If we apply the logarithm function to both sides of this equation, we obtain

$$\sum_{t=1}^{\infty}\frac{\alpha_t}{2}\,Ch_t(cx)=\mathrm{Log}(Ch(1+cy))$$

as desired.

If $X=SY$, all cup-products in cohomology are 0. Hence $\mathrm{Log}(Ch(1+cy))=\mathrm{Log}(1+Ch(cy))=Ch(cy)$, and $\frac{\alpha_t}{2}\,Ch_t(cx)=Ch_t(cy)$. □

5.17. Theorem. *Let d_n be the denominator of $B_n/4n$, where B_n is the nth Bernoulli number* (cf. 5.13). *Then $\tilde{J}'(S^{4n})$ is cyclic of order d_n, and thus we have an epimorphism $\tilde{J}(S^{4n})\to\mathbb{Z}/d_n\mathbb{Z}$. In particular, π^s_{4n-1} has a subquotient, i.e. a quotient of a subgroup, isomorphic to $\mathbb{Z}/d_n\mathbb{Z}$.*

Proof. We apply Theorem 5.16 to the case $X=S^{4n}$. Then the condition $\frac{\alpha_t}{2}\,Ch_t(cx)=Ch_t(cy)$ is simply $\left(\frac{\alpha_n}{2}\right)\cdot x=y$, or equivalently $\left(\frac{B_n}{4n}\right)\cdot x=y$ which is equivalent to x being divisible by d_n. □

5.18. Examples. Here is a short table of Bernoulli numbers. From this table we obtain, for instance, that $|\pi^s_{31}|\geqslant 16\,230$.

n	B_n	d_n	π^s_{4n-1}
1	1/6	24	π^s_3
2	1/30	240	π^s_7
3	1/42	504	π^s_{11}
4	1/30	480	π^s_{15}
5	5/66	264	π^s_{19}
6	691/2730	65 520	π^s_{23}
7	7/6	24	π^s_{27}
8	3617/510	16 230	π^s_{31}

5.19. To conclude this section, we briefly show how K-theory may be used in homotopy theory from another point of view. Let $f: X \to S^{2p}$ be a continuous map such that the induced homomorphism $\tilde{K}^*(S^{2p}) \to \tilde{K}^*(X)$ is zero (here K denotes complex K-theory $K_{\mathbb{C}}$). Then we consider the Puppe sequence of f (II.3.29)

$$X \xrightarrow{f} S^{2p} \xrightarrow{i} Cf \xrightarrow{j} SX \xrightarrow{Sf} S^{2p+1},$$

and see that $\tilde{K}(Cf) \approx \mathbb{Z} \oplus \tilde{K}(SX)$ (compare with 1.3). If x is an element of $\tilde{K}(Cf)$ such that $i^*(x) = \beta_{2p}$, generator of $\tilde{K}(S^{2p})$, then $\psi^k(x)$ may be written in the form $k^p x + j^*(a_k)$, where a_k is an element of $\tilde{K}(SX)$. If we replace x by $x + j^*(y)$ then we see that a_k is transformed into $a_k + (\psi^k - k^p)(y)$. Hence the class of a_k in $\tilde{K}(SX)$ is well-defined modulo the subgroup $(\psi^k - k^p)(\tilde{K}(SX))$, and depends only on the homotopy class of f. This provides an invariant of the homotopy class of f.

In fact, the classes a_k are not independent of each other, since the relation $\psi^k \psi^l = \psi^l \psi^k$ (cf. IV.7.15) implies $(\psi^l - l^p)(a_k) = (\psi^k - k^p)(a_l)$. If $H^{2p-1}(X; \mathbb{Q}) = 0$, then $\psi^k - k^p$ and $\psi^l - l^p$ are isomorphisms modulo torsion in $\tilde{K}(SX)$ (cf. 3.25 and 3.27). Hence $(\psi^k - k^p)^{-1}(a_k) = (\psi^l - l^p)^{-1}(a_l)$ is a well-defined element of $\tilde{K}(SX) \otimes \mathbb{Q}/\mathbb{Z}$, independent of k. When X is an odd dimensional sphere, this invariant has been studied, notably by J. F. Adams. It has been used to prove that $J(S^{4n}) = J'(S^{4n})$, which implies that $\mathbb{Z}/d_n\mathbb{Z}$ is a direct factor of π^s_{4n-1}.

6. Historical Note

The problem of finding maps between spheres of odd Hopf invariant was a long outstanding problem in topology (cf. appendix in Steenrod [1]). The complete solution was first given by Adams using secondary cohomology operations. Indeed he showed that there is no map of odd Hopf invariant between S^{2n-1} and S^n unless

$n=2, 4$ or 8. This implies that the only spheres which may be provided with an H-space structure are the classical examples S^1, S^3 and S^7. Another proof of these results was found by Adams and Atiyah [1]. This proof relies on K-theory, and is much simpler than the first. It is reproduced here with some minor modifications due to Husemoller [1].

The vector field problem on the sphere was also solved by Adams [1] using secondary operations, and later on, using K-theory. His proof relied on the earlier work of James [1] and Atiyah [1]. The proof presented here is based on the same ideas, but with a notable simplification due to Woodward [1].

Characteristic classes have some history, beginning with the work of Chern, Pontrjagin, Stiefel, and Whitney (see the preface of the Milnor-Stasheff book [1]). Sections 3 and 4 present those aspects of this theory which are connected with the subject of this book. These aspects are essential in the application of K-theory to integrality theorems (Hirzebruch [2], Borel-Hirzebruch [1], Atiyah-Hirzebruch [1]). These integrality theorems are intimately connected with the Atiyah-Singer index theorem [2].

Section 5 only provides a sketch of some potential applications of K-theory to stable homotopy. More complete results are found in Adams [2].

Bibliography

Adams, J. F.: [1] Vector fields on the sphere. Ann. Math. **75**, 603–632 (1962).

Adams, J. F.: [2] On the groups J(X) I, II, III, IV. Topology **2**, 181–195 (1963); **3**, 137–171, 193–222 (1965), **5**, 21–71 (1966).

Adams, J. F.: [3] Compact Lie groups. New York: Benjamin 1969.

Adams, J. F.: [4] Algebraic Topology. A student's guide. New York and London: Cambridge University Press 1972.

Adams, J. F., Atiyah, M. F.: [1] *K*-Theory and the Hopf invariant. Quart. J. Math. Oxford (2), **17**, 31–38 (1966).

Adams, J. F., Walker, G.: [1] On complex Stiefel manifolds. Proc. Cambridge Phil. Soc. **61**, 81–103 (1965).

Anderson, D. W.: [1] The real *K*-theory of classifying spaces. Proc. Nat. Acad. Sci. **51**, 634–636 (1964).

Atiyah, M. F.: [1] Thom complexes. Proc. London Math. Soc. **11**, 291–310 (1961).

Atiyah, M. F.: [2] Vector bundles and the Kunneth formula. Topology **1**, 245–248 (1962).

Atiyah, M. F.: [3] *K*-theory. New York: Benjamin 1964.

Atiyah, M. F.: [4] On the *K*-theory of compact Lie groups. Topology **4**, 95–99 (1965).

Atiyah, M. F.: [5] Power operations in *K*-theory. Quart. J. Math. Oxford (2) **17**, 165–193 (1966).

Atiyah, M. F.: [6] *K*-theory and Reality. Quart. J. Math. Oxford (2) **17**, 367–386 (1966).

Atiyah, M. F.: [7] Bott periodicity and the index of elliptic operators. Quart. J. Math. Oxford **74**, 113–140 (1968).

Atiyah, M. F., Bott, R.: [1] On the periodicity theorem for complex vector bundles. Acta Math. **112**, 229–247 (1964).

Atiyah, M. F., Bott, R., Shapiro, A.: [1] Clifford modules. Topology **3**, 3–38 (1964).

Atiyah, M. F., Hirzebruch, F.: [1] Riemann–Roch theorems for differentiable manifolds. Bull. Amer. Math. Soc. **65**, 276–281 (1959).

Atiyah, M. F., Hirzebruch, F.: [2] Quelques théorèmes de non-plongement pour les variétés différentiables. Bull. Soc. Math. France **87**, 383–396 (1959).

Atiyah, M. F., Hirzebruch, F.: [3] Vector bundles and homogeneous spaces. Proc. Symposium in Pure Maths, Amer. Math. Soc. **3**, 7–38 (1961).

Atiyah, M. F., Singer, I. M.: [1] The index of elliptic operators on compact manifolds. Bull. Amer. Math. Soc. **69**, 422–433 (1963).

Atiyah, M. F., Singer, I. M.: [2] The index of elliptic operators I, II, III. Ann. of Math. **87**, 484–530, 531–545 (with G. Segal), 546–604 (1968).

Atiyah, M. F., Singer, I. M.: [3] Index theory for skew-adjoint Fredholm operators. Publ. Math. I.H.E.S. **37**, 5–26 (1969).

Bass, H.: [1] Algebraic *K*-theory. New York: Benjamin 1968.

Bass, H., Heller, A., Swan, R. G.: [1] The Whitehead group of a polynomial extension. Publ. Math. I.H.E.S. **22**, 61–79 (1964).

Borel, A., Hirzebruch, F.: [1] Characteristic classes and homogeneous spaces I, II, III. Amer. J. Math. **80**, 458–538 (1958); **81**, 315–382 (1959); **82**, 491–504 (1960).

Borel, A., Serre, J. P.: [1] Groupes de Lie et puissances réduites de Steenrod. Amer. J. Math. **75**, 409–448 (1953).

Borel, A., Serre, J. P.: [2] Le théorème de Riemann–Roch (d'après Grothendieck). Bull. Soc. Math. France **86**, 97–136 (1958).

Bott, R.: [1] An application of the Morse theory to the topology of Lie groups. Bull. Soc. Math. France **84**, 251–281 (1956).

Bott, R.: [2] The stable homotopy of the classical groups. Ann. of Math. **70**, 313–337 (1959).

Bott, R.: [3] Lectures on $K(X)$ (mimeographed notes). Cambridge, Mass.: Harvard University 1962.

Bott, R.: [4] A note on the KO-theory of sphere bundles. Bull. Amer. Math. Soc. **68**, 395–400 (1962).

Bourbaki, N. [1] Topologie Générale. Chapitres 1, 2 et 9. Paris: Hermann (1958) (1965).

Bourbaki, N.: [2] Algèbre. Chapitre 9. Paris: Hermann (1959).

Cartan, H.: [1] Espaces fibrés et homotopie. Séminaire 1949/1950. New York: Benjamin 1967.

Cartan, H., Eilenberg, S.: [1] Homological algebra. Princeton, N.J.: Princeton University Press, 1956.

Cartan, H., Schwartz, L.: [1] Le théorème d'Atiyah–Singer. Séminaire 1963/1964. New York: Benjamin 1967.

Chevalley, C.: [1] Theory of Lie groups. Princeton, N.J.: Princeton University Press 1946.

Chevalley, C.: [2] The construction and study of certain important algebras. Tokyo: Publ. of the Mathematical Society of Japan 1955.

Conner, P. E., Floyd, E. E.: [1] The relation of cobordism to K-theories. Lecture Notes in Math. **28**. Berlin–Heidelberg–New York: Springer 1966.

Dold, A.: [1] Lectures on Algebraic Topology. Berlin–Heidelberg–New York: Springer 1972.

Dold, A.: [2] Chern. classes in general cohomology—Symposia Mathematica—385—410 (1970).

Dold, A., Lashof, R.: [1] Principal quasifibrations and fibre homotopy equivalences of bundles. Illinois J. Math. **3**, 285–305 (1959).

Dupont, J.: [1] K-theory. Aarhus, Danemark: Publ. of the Matematisk Institut 1968.

Dyer, E.: [1] Cohomology theories. New York: Benjamin 1969.

Eilenberg, S., Steenrod, N.: [1] Foundations of algebraic topology. Princeton, N.J.: Princeton University Press 1952.

Frenkel, J.: [1] Cohomologie non abélienne et espaces fibrés. Bull. Soc. Math, France **85**, 135–220 (1957).

Godbillon, C.: [1] Géométrie différentielle et mécanique analytique. Paris: Hermann 1969.

Godbillon, C.: [2] Eléments de Topologie Algébrique. Paris: Hermann 1971.

Green, P. S.: [1] A cohomology theory based upon self-conjugacies of complex vector bundles. Bull. Amer. Math. Soc. **70**, 522–524 (1964).

Greenberg, M. J.: [1] Lectures on Algebraic Topology. New York: Benjamin 1967.

Grothendieck, A.: [1] Théorie des classes de Chern. Bull. Soc. Math. France **86**, 137–154 (1958).

Grothendieck, A.: [2] Séminaire Chevelley, exposé 4. Paris: Société Mathématique de France, 11 rue P. et M. Curie 1958.

Hardy, G. H., Wright, E. M.: [1] An introduction to the theory of Numbers. London: Oxford University Press 1960.

Hilton, P. J.: [1] An introduction to homotopy theory. Cambridge Tracts in Mathematics and Mathematical Physics **43**. New York: Cambridge University Press 1953.

Hilton, P. J.: [2] General cohomology theory and K-theory. New York: Cambridge University Press 1971.

Hirzebruch, F.: [1] A Riemann–Roch theorem for differentiable manifolds. Séminaire Bourbaki **177**, (1959).

Hirzebruch, F.: [2] Topological methods in Algebraic Geometry. Berlin–Heidelberg–New York: Springer 1965.

Hodgkin, L.: [1] On the K-theory of Lie groups. Topology **6**, 1–36 (1967).

Hu, S. T.: [1] Homotopy theory. New York: Academic Press, Inc. 1959.

Hurewicz, W., Wallman, H.: Dimension theory. Princeton, N.J.: Princeton University Press 1941.

Husemoller, D.: [1] Fibre bundles. New York: McGraw-Hill Book Company 1966. 2nd edition: GTM, vol. 20. Berlin–Heidelberg–New York: Springer 1975.

James, I. M.: [1] Spaces associated with Stiefel manifolds. Proc. London Math. Soc. (3) **9**, 115–140 (1959).

Janich, K.: [1] Vektorraumbündel und der Raum der Fredholm Operatoren. Math. Ann. **161**, 129–142 (1965).

Karoubi, M.: [1] Séminaire Cartan–Schwartz 1963/64, exposé 16. New York: Benjamin 1967.

Karoubi, M.: [2] Algèbres de Clifford et K-théorie. Ann. Sci. Ec. Norm. Sup. 4ᵉ sér. **1**, 161–270 (1968).

Karoubi, M.: [3] Espaces classifiants en *K*-théorie. Trans. Amer. Math. Soc. **14**, 74–115 (1970).
Karoubi, M.: [4] Périodicité de la *K*-théorie hermitienne. Lecture Notes in Math. **343**, (Springer-Verlag), 301–411 (1973).
Karoubi, M.: [5] La périodicité de Bott en *K*-théorie générale. Ann. Sci. Ec. Norm. Sup. 4ᵉ série **1**, 63–95 (1971).
Karoubi, M.: [6] *K*-théorie équivariants des fibrés en sphères. Topology **12**, 275–281 (1973).
Karoubi, M., Villamayor, O.: [1] *K*-théorie algébrique et *K*-théorie topologique I. Math. Scand. **28**, 265–307 (1971).
Kelley, J.: [1] General Topology. Princeton, N.J.: Van Nostrand 1955.
Kuiper, N. H.: [1] The homotopy type of the unitary group of Hilbert space. Topology **3**, 19–30 (1965).
Lang, S.: [1] Algebra. Reading, Mass.: Addison-Wesley 1964.
Lang, S.: [2] Introduction to differentiable manifolds. New York: Interscience 1962.
Lorch, E.: [1] Spectral theory. University Tests in Mathematical Sciences. New York: Oxford University Press 1962.
Lundell, A. T., Weingram, S.: [1] The topology of CW-complexes. Princeton, N.J.: Van Nostrand 1969.
Michael, E.: [1] Convex structures and continuous selections. Canad. J. Math. **11**, 556–575 (1959).
Milnor, J.: [1] On spaces having the homotopy type of a CW-complex. Trans. Amer. Math. Soc. **90**, 272–280 (1959).
Milnor, J.: [2] Construction of universal bundles I, II. Ann. of Math. **63**, 272–284 and 430–436 (1963).
Milnor, J.: [3] Introduction to Algebraic *K*-theory. Ann. of Math. Studies **72**. Princeton, N.J.: Princeton University Press 1971.
Milnor, J., Kervaire, M. A.: [1] Bernoulli numbers, homotopy groups and a theorem of Rohlin. Proc. Intern. Congr. Math. (1958).
Milnor, J., Stasheff, J.: [1] Lectures on characteristic classes. Ann. of Math. Studies **197**. Princeton, N.J.: Princeton University Press 1974.
Milnor, J.: [4] Morse theory. Ann. of Math. Studies **51**. Princeton, N.J.: Princeton University Press 1963.
Mitchell, B.: [1] Theory of categories. New York: Academic Press 1965.
Northcott, D. G.: [1] An introduction to homological algebra. Cambridge and London: Cambridge University Press 1962.
Palais, R.: [1] Seminar on the Atiyah–Singer index theorem (with contributions by A. Borel, E. E. Floyd, R. T. Seeley, W. Shih and R. Solovay). Ann. of Math Studies **57**, Princeton 1965.
Quillen, D.: [1] Higher algebraic *K*-theory. Lecture Notes in Math. **341**, 85–147. Berlin–Heidelberg–New York: Springer 1973.
Segal, G.: [1] Equivariant *K*-theory. Publ. Math. I.H.E.S. **34**, 129–151 (1968).
Serre, J. P.: [1] Représentations linéaires des groupes finis. Paris: Hermann 1967.
Serre, J. P.: [2] Modules projectifs et espaces fibrés à fibre vectorielle, exposé 23. Paris: Séminaire Dubreil-Pisot, 11 Rue P. et M. Curie 1958.
Shih, W.: [1] Une remarque sur les classes de Thom. C.R. Acad. Sc. Paris **260**, 6259–6262 (1965).
Spanier, E.: [1] A formula of Atiyah and Hirzebruch. Math. Z. **80**, 154–162 (1962).
Spanier, E.: [2] Algebraic Topology. New York: McGraw-Hill Book Company 1966.
Spivak, M.: [1] Calculus on Manifolds. New York: Benjamin 1965.
Steenrod, N.: [1] Topology of Fibre Bundles (5th printing). Princeton Mathematical Series. Princeton, N.J.: Princeton University Press 1965.
Swan, R. G.: [1] Vector bundles and projective modules. Trans. Amer. Math. Soc. **105**, 264–277 (1962).
Thom, R.: [1] Espaces fibrés en sphères et carrés de Steenrod. Ann. Sc. Ecole Norm. Sup. **69**, 109–182 (1952).
Thom, R.: [2] Quelques propriétés globales des variétés différentiables. Comment. Math. Helv. **28**, 17–86 (1954).
Toda, H.: [1] Composition methods in homotopy groups of spheres. Ann. of Math. Studies **49**, Princeton 1962.
Whitehead, G. W.: [1] Generalized homology theories. Trans. Amer. Math. Soc. **102**, 227–283 (1962).
Wood, R.: [1] Banach algebras and Bott periodicity. Topology **4**, 371–389 (1966).
Woodward, L. M.: [1] Vector fields on spheres and a generalization. Quart. J. Math. Oxford (2) **24**, 357–366 (1973).

List of Notation

Index

K-Theory. An Introduction
Postface

Since the publication of this book thirty years ago, K-theory has considerably enlarged its scope. It is now essentially divided in two main areas of research which are sketched very briefly in this postface.

1. K-theory of Banach algebras and more specifically C^*-algebras (also called noncommutative spaces) has also been generalized as a bifunctor $KK(A, B)$ which is the KK-theory of Kasparov [Kas]. This generalization is an important tool in the development of the so called noncommutative geometry by Connes [Co]. The usual Chern character has been generalized in this framework first by Connes [Co] and the author [Kar] and later on by J. Cuntz [Cu]. Many of applications may be found in [Co] which countains hundreds of references on the subject.
2. On the algebraic side, after the main contributions of Quillen [Q], many important results have been obtained by A. Suslin and V. Voevoedsky among others. A nice overview of the subject may be found in the Handbook of K-theory [HK] (for instance in the papers of E. Friedlander and B. Kahn).

[Co] CONNES A.: Noncommutative Geometry. Academic Press (1994).

[Cu] CUNTZ J.: Cyclic theory and the bivariant Chern-Connes character. Noncommutative geometry, 73-135, Lecture Notes in Math., 1831, Springer (2004).

[HK] GRAYSON D. and FRIEDLANDER E. (editors): Handbook of K-theory. Springer Verlag (2005).

[Kar] KAROUBI M.: Homologie cyclique et K-théorie. Astérisque 149. Société Mathématique de France (1987).

[Kas] KASPAROV G.: The operator K-functor and extension of C^*-algebras. Izv. Akad. Nauk. SSSR, Ser. Math. 44 (1980) 571–636.

[Q] QUILLEN D.: Higher Algebraic K-theory. Lectures Notes in Maths Nr 341. Springer (1973).

Addenda and Errata

There were few misprints in the original version which are listed here. There is also an important simplification made later on for the calculation of the K-theory of real projective spaces or bundles (IV.6), which is important in the solution of the vector field problem on the sphere by Adams. This simplification (which applies also to the equivariant case) is included in the following paper of the author

KAROUBI M. : Equivariant K-theory of real vector spaces and real projective spaces. Topology and its applications, 122 (2002) 531–546.

In $(XIV; -4)$ says: $[X, 0]$,
it should say: $[X, O]$

In $(10; 4)$ says: $X \in U_i \cap U_j \cap U_r \cap U_s$,
it should say: $x \in U_i \cap U_j \cap U_r \cap U_s$

In $(10; -1)$ says: $\alpha : E \to E'$,
it should say: $\alpha : E \to F$

In $(11; 9)$ says: $(h_s(x))^{-1} h^r(x) g_i^r(x)$,
it should say: $(h_s(x))^{-1} h_r(x) g_i^r(x)$

In $(11; -6)$ says: *The associated*,
it should say: *the associated*

In $(11; -1)$ says: $E_j|_{U_i \cap U_j} \xrightarrow{\lambda_i|_{U_i \cap U_j}} F_j|_{U_i \cap U_j}$,

it should say: $E_j|_{U_i \cap U_j} \xrightarrow{\lambda_j|_{U_i \cap U_j}} F_j|_{U_i \cap U_j}$

In $(12; 4)$ says: $g_{hi}(x) = \lambda_j(x)^{-1} \lambda_i(x)$,
it should say: $g_{ji}(x) = \lambda_j(x)^{-1} \lambda_i(x)$

In $(23; -14)$ says: $s(x) = \sum_{\alpha \in I} \alpha_i(x) s_i'(x)$,
it should say: $s(x) = \sum_{\alpha \in I} \alpha_i(x) s_i(x)$

In (25; 14) says: *for* $y \in U_x$,
it should say: *for* $y \in V_x$

In (26; −12) says: $g_\alpha : U \to \mathscr{E}(M, N)$,
it should say: $g_\alpha : U \to \mathscr{E}(M_\alpha, N)$

In (27; 3) says: $p_{x_0} f(x) = f(x) p_x$,
it should say: $p_x f(x) = f(x) p_{x_0}$

In (27; 5) says:

$$\begin{CD}
0 @>>> \operatorname{Ker} p @>>> X \times M @>p_0>> X \times M\,, \\
@. @VVV @V\widehat{f}VV @V\widehat{f}VV \\
0 @>>> X \times \operatorname{Ker} p_{x_0} @>>> X \times M @>p>> X \times M
\end{CD}$$

it should say:

$$\begin{CD}
0 @>>> \operatorname{Ker} p_{x_0} @>>> X \times M @>p>> X \times M \\
@. @VVV @V\widehat{f}VV @V\widehat{f}VV \\
0 @>>> X \times \operatorname{Ker} p_{x_0} @>>> X \times M @>p_0>> X \times M
\end{CD}$$

In (30; −2) says: *in 6.9 is* ,
it should say: *in 6.10 is*

In (31; 5) says: *given in 6.10* ,
it should say: *given in 6.9*

In (37; −5) says: $\xi_f = \operatorname{Im} p$,
it should say: $\xi_g = \operatorname{Im} p$

In (40; 13) says: $= (1 - p - p + 2qp)$,
it should say: $= (1 - p - q + 2qp)$

In (53; −20) says: $\widehat{E \oplus F}$,
it should say: $\dot{\widehat{E \oplus F}}$

In (57; 16) says: $\operatorname{Ker}\big[K(X) \to H^0(X; \mathbb{Z})$,
it should say: $\operatorname{Ker}\big[K(X) \to H^0(X; \mathbb{Z})\big]$

In (59; 17) says: $\approx [X, 0]'$,
it should say: $\approx [X, O]'$

In (62; 9) says: $d(E, F, \alpha^{-1})$,
it should say: $d(F, E, \alpha^{-1})$

In (65; 7) says: $f(\sigma(t)) = \sigma'(t)$,
it should say: $\widetilde{f}(\sigma(t)) = \sigma'(t)$

In (70; 13) says: $\widetilde{K}(X/Y) \longrightarrow K(X) \longrightarrow$,
it should say: $K(X/Y) \longrightarrow K(X) \longrightarrow$

In (72; 14) says: *and* (E_1, φ_1) ,
it should say: *and* (E_1, α_1)

In (74; 20) says: *category (I.6.9)*,
it should say: *category (I.6.10)*

In (78; −4) says: $K(X/Y) \xrightarrow{\approx} K(X/Y)$,
it should say: $K(X'/Y') \xrightarrow{\approx} K(X/Y)$

In (80; −5) says: $\begin{array}{ccc} K(C'f) & \longrightarrow & \widetilde{K}(X) \\ \Vert & & \end{array}$

it should say: $\begin{array}{ccc} \widetilde{K}(C'f) & \longrightarrow & \widetilde{K}(X) \\ \Vert & & \end{array}$

In (81; −4) says: $\longrightarrow \widetilde{K}(S'(Y)) \longrightarrow \widetilde{K}(Cf)$,
it should say: $\longrightarrow \widetilde{K}(S'(Y)) \longrightarrow \widetilde{K}(C'f)$

In (83; −14) says: $\widehat{(Z-T)} \approx$,
it should say: $\overset{\cdot}{\widehat{(Z-T)}} \approx$

In (83; −2) says: $\widehat{(X-Y)} \approx$,
it should say: $\overset{\cdot}{\widehat{(X-Y)}} \approx$

In (84; −3) says: $\dot{Z} \approx Y \times [0,1]/Y \vee [0,1]$,
it should say: $\dot{Z} \approx \dot{Y} \times [0,1]/\dot{Y} \vee [0,1]$

In (84; −2) says: *class of* $(y, (1 + (1-t)u)$,
it should say: *class of* $(y, 1 + (1-t)u)$

In (85; 10) says: $Z - Y \times \overline{[0,1)} \simeq$,
it should say: $Z - Y \times [0,1) \simeq$

In (85; −1) says: $\begin{array}{cc} \downarrow & \\ K(\dot{Y} \times \mathbb{R}) & \longrightarrow \end{array}$,

it should say: $\begin{array}{cc} \downarrow & \\ K(Y \times \mathbb{R}) & \longrightarrow \end{array}$

In (93; 15) says: *Let* $\dot{X}$ *and* Y,
it should say: *Let* $\dot{X}$ *and* $\dot{Y}$

In (95; 17) says: *in I.1.29*,
it should say: *in 1.29*

In (97; −9) says: *$D|_Y$ is an automorphism of $E|_Y$*,
it should say: *$D|_{X'}$ is an automorphism of $E|_{X'}$*

In (102; 7) says: $= (v, \partial_v(w) \cdot d_v(\lambda))$,
it should say: $= (v, \partial_v(w), d_v(\lambda))$

In (102; 13) says: $K_{\mathbb{C}}(B^2, S^2)$,
it should say: *of* $K_{\mathbb{C}}(B^2, S^1)$

In (112; 10) says: $K^{-n}(X \times B^2, X \times S^{-1} \cup Y \times B^2)$,
it should say: $K^{-n}(X \times B^2, X \times S^1 \cup Y \times B^2)$

In (137; −9) says: $f \cdot \rho(\lambda) = \rho(\lambda) \cdot f$,
it should say: $f \cdot \rho(\lambda) = \rho'(\lambda) \cdot f$

In (155; −1) says: $ie_1 = \begin{pmatrix} 0 & -i \\ i & 0 \end{pmatrix}$,
it should say: $ie_1 = \begin{pmatrix} 0 & i \\ -i & 0 \end{pmatrix}$

In (175; 14) says: $\widetilde{K}_{\mathbb{R}}(P_2(\mathbb{C}) \approx \mathbb{Z}$,
it should say: $\widetilde{K}_{\mathbb{R}}(P_2(\mathbb{C})) \approx \mathbb{Z}$

In (179; −16) says: *Atiyah ([6]; cf. also 7.14*,
it should say: *Atiyah [6]; (cf. also 7.14)*

In (182; −9) says:
$$K^{q-1}(P_S \times \mathbb{R})^n \oplus K^{q-1}(P_T \times \mathbb{R})^n \longrightarrow$$
$$K^{q-1}(P_{S\cap Y} \times \mathbb{R})^n \xrightarrow{\Delta} K^2(P_{S\cup Y})^n \ ,$$
it should say:
$$K^{q-1}(P_S \times \mathbb{R}) \oplus K^{q-1}(P_T \times \mathbb{R}) \longrightarrow$$
$$K^{q-1}(P_{S\cap Y} \times \mathbb{R}) \xrightarrow{\Delta} K^2(P_{S\cup Y})$$

In (184; 3) says: *Since* $(D_{x,v})^2 = Q_x(v)$,
it should say: *Since* $(\Delta_{x,v})^2 = Q_x(v)$

In (209; −13) says: ${}^t(\widetilde{\rho}_x(v)) = \rho_x(v)$,
it should say: ${}^t(\widetilde{\rho}_x(v)) = \widetilde{\rho}_x(v)$

In (210; 12) says: $-v(\lambda v + w')v^{-1} = \lambda v + w$ *since* v *and* w',
it should say: $-v(\lambda v + v')v^{-1} = \lambda v + v'$ *since* v *and* v'

In (211; −6) says: $(C(V) \times_X C(V) \to C(V)$,
it should say: $C(V) \times_X C(V) \to C(V)$

In (211; −2) says: *map* $V \times_X E$,
it should say: *map* $V \times_X E \to E$

In (212; −1) says: *resp.* $(\beta \in H^1(X; \mathrm{Spin}(n))$,
it should say: *resp.* $(\beta \in H^1(X; \mathrm{Spin}(n)))$

In $(214;5)$ says: $Z^2(X;\mathbb{Z}/2)$,
it should say: $H^2(X;\mathbb{Z}/2)$

In $(214;5)$ says: *of* $H^2(X,\mathbb{Z}/2)$,
it should say: *of* $H^2(X;\mathbb{Z}/2)$

In $(215;-1)$ says: *a principle bundle*,
it should say: *a principal bundle*

In $(216;-6)$ says: $(\widetilde{\tau}'(1),\overline{\gamma}(1),$ *or*,
it should say: $(\widetilde{\tau}'(1),\overline{\gamma}(1))$, *or*

In $(221;10)$ says: $+(\lambda_{p+1})^2+\cdots+(\lambda_{p+2})^2$,
it should say: $+(\lambda_{p+1})^2+\cdots+(\lambda_{p+q})^2$

In $(225;6)$ says: *where* V,
it should say: *where* $\dot{V}$

In $(235;-2)$ says: *such that* $Y_{Y\subset}\bigcup U_i$,
it should say: *such that* $Y\subset\bigcup U_i$

In $(243;10)$ says: *in the homermorphism*,
it should say: *in the homeomorphism*

In $(243;-9)$ says: $+\eta'\sin\theta_1\sin\theta_1$,
it should say: $+\eta'\sin\theta_2\sin\theta_1$

In $(247;7)$ says: *and let* u,
it should say: *and let* $\widetilde{u}$

In $(247;-9)$ says: $\pi_1(S^+(W\oplus 1),S^+(W\oplus 1)|_Y\cup S(W)\to$,
it should say: $\pi_1(S^+(W\oplus 1),S^+(W\oplus 1)|_Y\cup S(W))\to$

In $(247;-6)$ says: $K_r^{\xi\oplus n,n}(P(W\oplus 1),P(W\oplus 1)|_Y\cup P(W)$,
it should say: $K_r^{\xi\oplus n,n}(P(W\oplus 1),P(W\oplus 1)|_Y\cup P(W))$,

In $(306;$ *column*$2;-16)$ says: $Z_\wedge T$, $Z_\wedge T$
it should say: $Z_\wedge T$, $Z_\vee T$

In $(306;$ *column*$1;20)$ says: $\frac{S(Z)}{S'(Z)}$,
it should say: $\frac{S'(Z)}{S(Z)}$

CIM

M. Aigner Combinatorial Theory ISBN 978-3-540-61787-7
A. L. Besse Einstein Manifolds ISBN 978-3-540-74120-6
N. P. Bhatia, G. P. Szegő Stability Theory of Dynamical Systems ISBN 978-3-540-42748-3
J. W. S. Cassels An Introduction to the Geometry of Numbers ISBN 978-3-540-61788-4
R. Courant, F. John Introduction to Calculus and Analysis I ISBN 978-3-540-65058-4
R. Courant, F. John Introduction to Calculus and Analysis II/1 ISBN 978-3-540-66569-4
R. Courant, F. John Introduction to Calculus and Analysis II/2 ISBN 978-3-540-66570-0
P. Dembowski Finite Geometries ISBN 978-3-540-61786-0
A. Dold Lectures on Algebraic Topology ISBN 978-3-540-58660-9
J. L. Doob Classical Potential Theory and Its Probabilistic Counterpart ISBN 978-3-540-41206-9
R. S. Ellis Entropy, Large Deviations, and Statistical Mechanics ISBN 978-3-540-29059-9
H. Federer Geometric Measure Theory ISBN 978-3-540-60656-7
S. Flügge Practical Quantum Mechanics ISBN 978-3-540-65035-5
L. D. Faddeev, L. A. Takhtajan Hamiltonian Methods in the Theory of Solitons ISBN 978-3-540-69843-2
I. I. Gikhman, A. V. Skorokhod The Theory of Stochastic Processes I ISBN 978-3-540-20284-4
I. I. Gikhman, A. V. Skorokhod The Theory of Stochastic Processes II ISBN 978-3-540-20285-1
I. I. Gikhman, A. V. Skorokhod The Theory of Stochastic Processes III ISBN 978-3-540-49940-4
D. Gilbarg, N. S. Trudinger Elliptic Partial Differential Equations of Second Order ISBN 978-3-540-41160-4
H. Grauert, R. Remmert Theory of Stein Spaces ISBN 978-3-540-00373-1
H. Hasse Number Theory ISBN 978-3-540-42749-0
F. Hirzebruch Topological Methods in Algebraic Geometry ISBN 978-3-540-58663-0
L. Hörmander The Analysis of Linear Partial Differential Operators I – Distribution Theory and Fourier Analysis ISBN 978-3-540-00662-6
L. Hörmander The Analysis of Linear Partial Differential Operators II – Differential Operators with Constant Coefficients ISBN 978-3-540-22516-4
L. Hörmander The Analysis of Linear Partial Differential Operators III – Pseudo-Differential Operators ISBN 978-3-540-49937-4
L. Hörmander The Analysis of Linear Partial Differential Operators IV – Fourier Integral Operators ISBN 978-3-642-00117-8
K. Itô, H. P. McKean, Jr. Diffusion Processes and Their Sample Paths ISBN 978-3-540-60629-1
T. Kato Perturbation Theory for Linear Operators ISBN 978-3-540-58661-6
S. Kobayashi Transformation Groups in Differential Geometry ISBN 978-3-540-58659-3
K. Kodaira Complex Manifolds and Deformation of Complex Structures ISBN 978-3-540-22614-7
Th. M. Liggett Interacting Particle Systems ISBN 978-3-540-22617-8
J. Lindenstrauss, L. Tzafriri Classical Banach Spaces I and II ISBN 978-3-540-60628-4
R. C. Lyndon, P. E Schupp Combinatorial Group Theory ISBN 978-3-540-41158-1
S. Mac Lane Homology ISBN 978-3-540-58662-3
C. B. Morrey Jr. Multiple Integrals in the Calculus of Variations ISBN 978-3-540-69915-6
D. Mumford Algebraic Geometry I – Complex Projective Varieties ISBN 978-3-540-58657-9
O. T. O'Meara Introduction to Quadratic Forms ISBN 978-3-540-66564-9
G. Pólya, G. Szegő Problems and Theorems in Analysis I – Series. Integral Calculus. Theory of Functions ISBN 978-3-540-63640-3
G. Pólya, G. Szegő Problems and Theorems in Analysis II – Theory of Functions. Zeros. Polynomials. Determinants. Number Theory. Geometry ISBN 978-3-540-63686-1
W. Rudin Function Theory in the Unit Ball of $\mathbb{C}^n$ ISBN 978-3-540-68272-1
S. Sakai C*-Algebras and W*-Algebras ISBN 978-3-540-63633-5
C. L. Siegel, J. K. Moser Lectures on Celestial Mechanics ISBN 978-3-540-58656-2
T. A. Springer Jordan Algebras and Algebraic Groups ISBN 978-3-540-63632-8
D. W. Stroock, S. R. S. Varadhan Multidimensional Diffusion Processes ISBN 978-3-540-28998-2
R. R. Switzer Algebraic Topology: Homology and Homotopy ISBN 978-3-540-42750-6
A. Weil Basic Number Theory ISBN 978-3-540-58655-5
A. Weil Elliptic Functions According to Eisenstein and Kronecker ISBN 978-3-540-65036-2
K. Yosida Functional Analysis ISBN 978-3-540-58654-8
O. Zariski Algebraic Surfaces ISBN 978-3-540-58658-6